Workbook

Introduction to Hydraulics for Industry Professionals

Dr. Medhat Kamel Bahr Khalil, Ph.D, CFPHS, CFPAI.
Director of Professional Education and Research Development,
Applied Technology Center, Milwaukee School of Engineering,
Milwaukee, WI, USA.

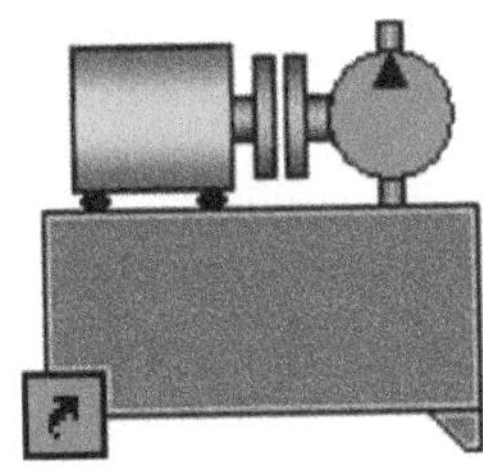

CompuDraulic LLC
www.CompuDraulic.com

CompuDraulic LLC

Workbook

Introduction to Hydraulics for Industry Professionals

ISBN: 978-0-692-65544-3

Printed in the United States of America
First Published by 2016
Revised by July 2018

All rights reserved for CompuDraulic LLC.
3850 Scenic Way, Franksville, WI, 53126 USA.
www.compudraulic.com

No part of this book may be reproduced or utilized in any form or by any means, electronic or physical, including photocopying and microfilming, without written permission from CompuDraulic LLC at the address above.

Disclaimer

It is always advisable to review the relevant standards and the recommendations from the system manufacturer. However, the content of this book provides guidelines based on the author's experience.

Any portion of information presented in this book could be not applicable for some applications due to various reasons. Since errors can occur in circuits, tables, and text, the publisher assumes no liability for the safe and/or satisfactory operation of any system designed based on the information in this book.

The publisher does not endorse or recommend any brand name product by including such brand name products in this book. Conversely the publisher does not disapprove any brand name product by not including such brand name in this book. The publisher obtained data from catalogs, literatures, and material from hydraulic components and systems manufacturers based on their permissions. The publisher welcomes additional data from other sources for future editions.

Workbook
Introduction to Hydraulics for Industry Professionals

PREFACE

This Workbook is a complementary part to the textbook of the same title. This book is used as a workbook for students to take notes during the course delivery. It contains colored printout of the PowerPoint slides that are designed to present the course. Each chapter is followed by a number of review questions and assignments for homework.

Dr. Medhat Kamel Bahr Khalil

Chapter 1
Hydraulic System Overview

Objectives:

This chapter presents a comparative study of different power transmission and control systems; namely: Mechanical, Electrical, Hydraulic and Pneumatic. This chapter explains the main features of hydraulic systems and concludes by exploring the structure of a generic hydraulic system.

0

Brief Contents:

1

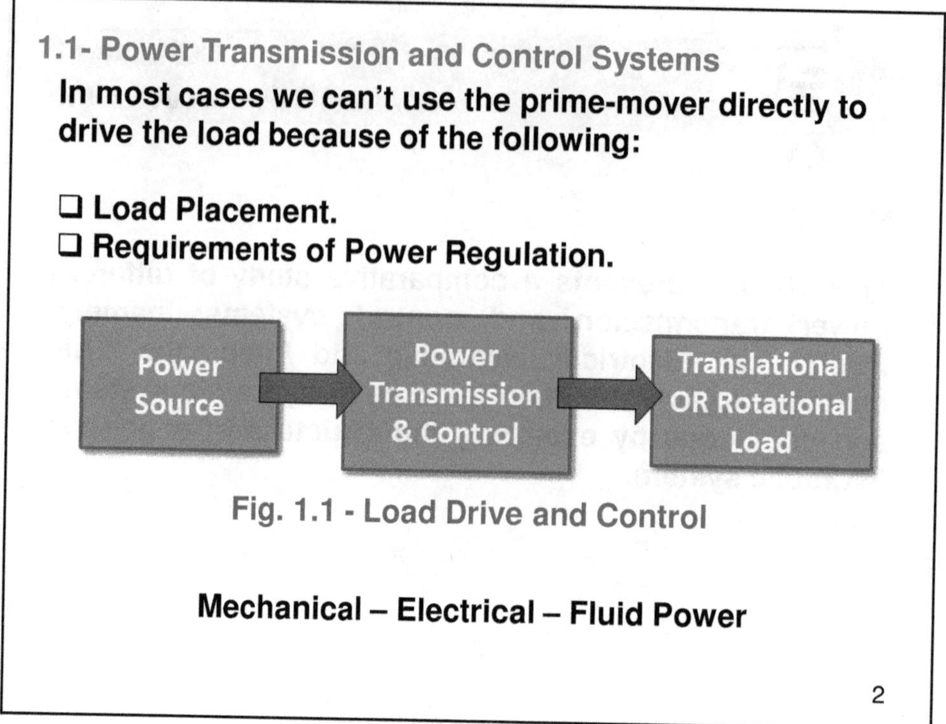

Fig. 1.2 - Elements of Mechanical Power Transmission and Control System

Fig. 1.3 - Automotive Gear-Shifting Power Transmission

Mechanical Power Transmission & Control

3

Fig. 1.4 - CNC Machine Tool

Electrical Power Transmission & Control

4

Fig. 1.5 – Hydraulic-Driven Earth Moving Machines

Fluid Power Transmission & Control

5

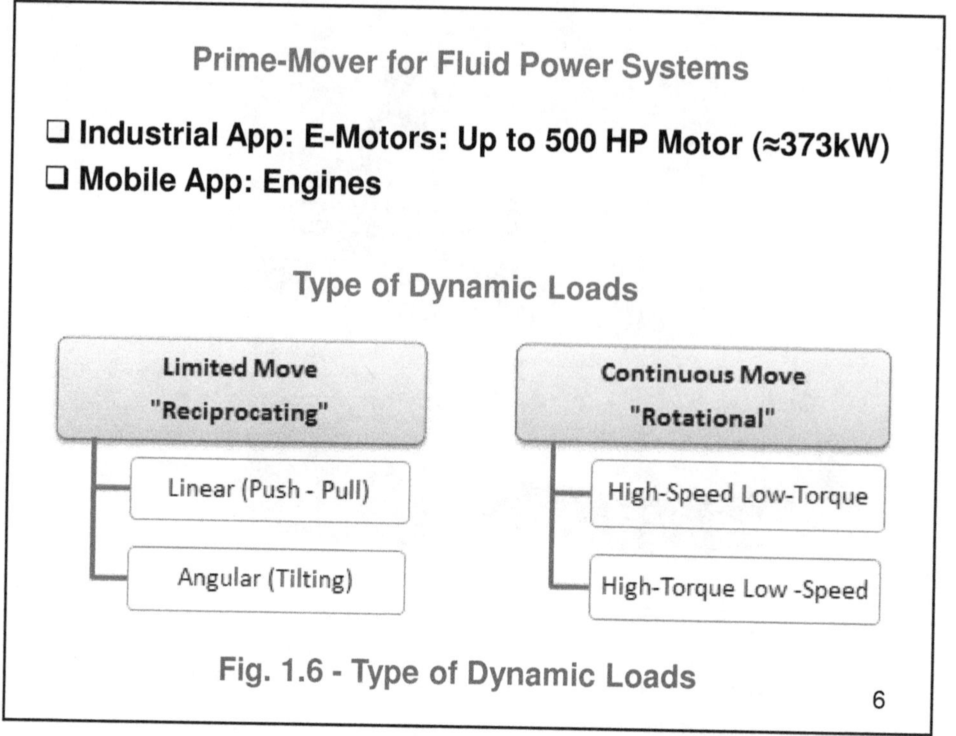

Fig. 1.6 - Type of Dynamic Loads

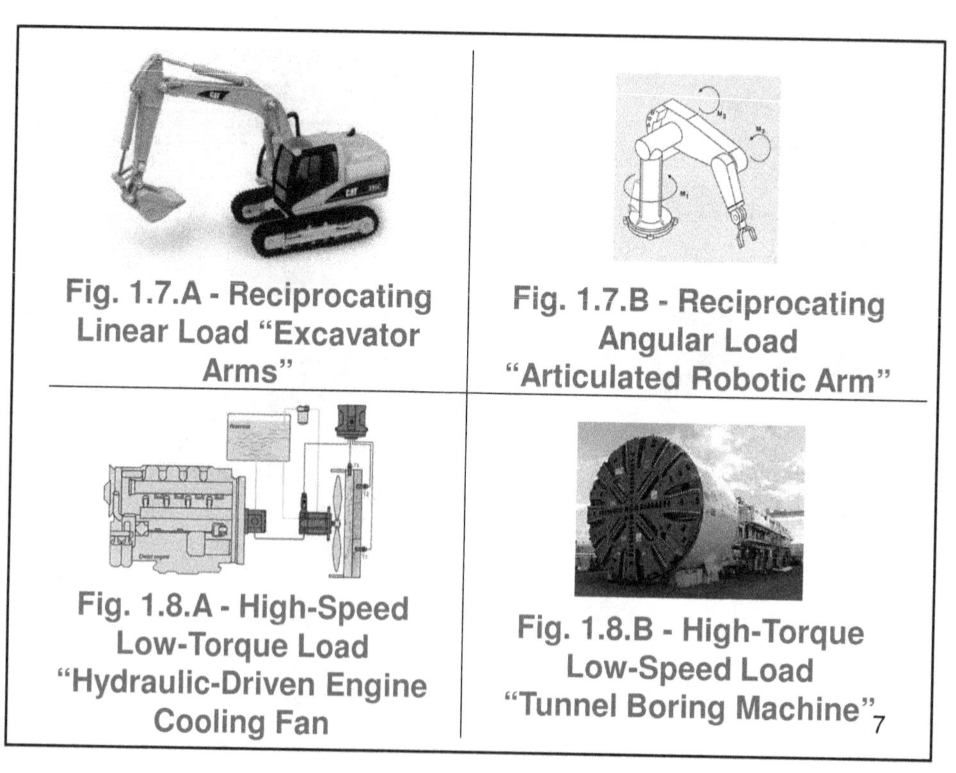

Fig. 1.7.A - Reciprocating Linear Load "Excavator Arms"

Fig. 1.7.B - Reciprocating Angular Load "Articulated Robotic Arm"

Fig. 1.8.A - High-Speed Low-Torque Load "Hydraulic-Driven Engine Cooling Fan

Fig. 1.8.B - High-Torque Low-Speed Load "Tunnel Boring Machine"

1.2- Power Transmission and Control Systems Selection

Which system is the best for power transmission and control?

Absolute "Quantitative" Measures

e.g. Speed, Load and Power Capacity - Dynamic Performance - Control Strategies

Relative "Qualitative" Measures

e.g. Compactness - Maintainability - Reliability - Initial & Running cost

External Influences

e.g. Know-How - Available Resources - Economical Reasons

Fig. 1.9 - Decision Making for Power Trans. System Selection

8

1.3- Power Trans. Systems General Features Comparison
1.3.1- Application Characteristics

Mechanical – Electrical – Hydraulic – Pneumatic

- ❑ **Transmission Distance.**
- ❑ **Speed ratio between prime-mover and actuator.**
- ❑ **Actuator Speed.**
- ❑ **Controllability with large Power Transmission.**
- ❑ **Adequacy for Industrial versus Mobile Applications.**
- ❑ **Clean Work Environmental.**

9

1.3.2- Initial and Running Cost

	Fluid Power	*Electrical*
Logic Functions are done by:	Valves	Relays
Conductors:	Pipes/Tubes/Hoses	Wires
Energy Carriers:	Oil/Air	Electrons
Maintenance:	Mandatory	Maintenance Free Elements

Table 1.1 – Fluid Power versus Electric Drive and Control System

Running Cost in Order of Magnitude:

Cost: Electric = 1unit, Hydraulic = 3-5 Pneumatic = 5-7

10

1.3.3- Precision of Motion

Hydraulic Systems:
Motion precision affected by oil large bulk modulus.
It offers good precision even with large transmitted power.

Pneumatic Systems:
Motion precision affected by air compressibility.
It offers good precision only with small transmitted power.

Electrical Systems:
Motion precision affected by the physical properties of the elements that consist the system.

Mechanical Systems:
Motion precision affected by the quality of the machined elements such as gears, indexing pins, ball and screw, etc..

11

1.3.4- Power Density and Energy Density

	Hydraulic Motor		Electric Motor
Same Power Capacity Motors		1 : 10 By Weight and Size	
Same Size Motors		HM + 300 psi (17 bar) = EM	

Fig. 1.10 - Power Density

- ❑ **Power Density - Power to Weight Ratio - Specific Power.**
- ❑ **Energy Density.**

12

High Power/weight ratio
(≈ 10 times higher than electric motors).

H. Motor (250 psi ≈ 17 bar) = E. Motor of the same weight

H. Motor (3000 psi ≈ 200 bar) = 10 x E. Motor of the same weight

1.3.5- Providing Linear Motion

Electrical Linear Actuators - Hydraulic & Pneumatic Cylinders

- ❑ **Stroke Length.**
- ❑ **Load Carrying Capacity.**
- ❑ **Price of Linear Actuators.**

13

1.3.6- Efficiency of Energy Transmission

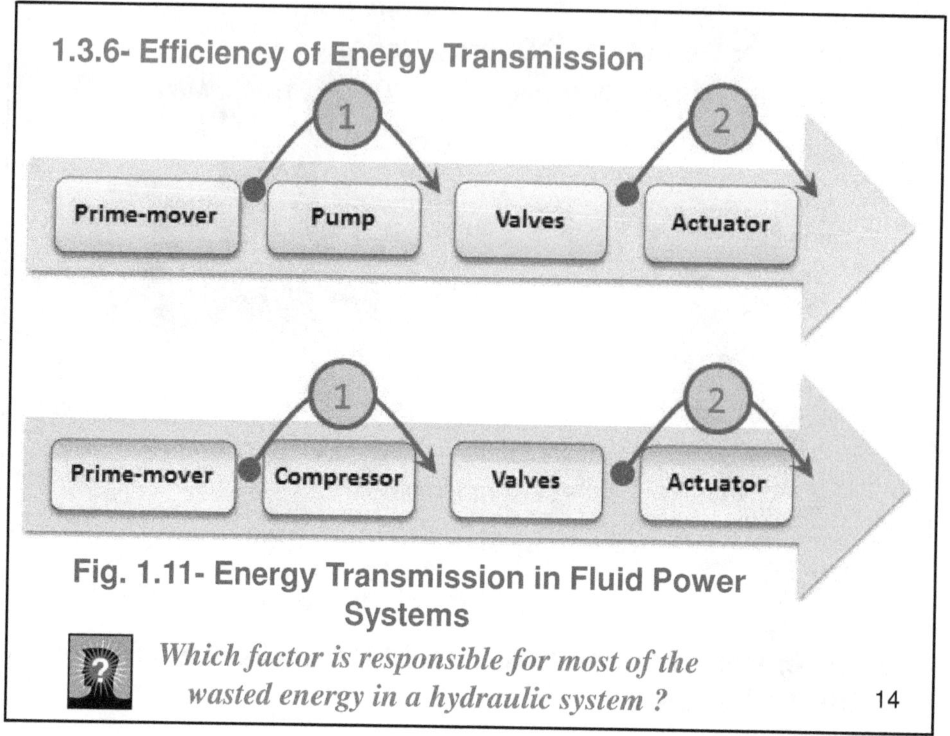

Fig. 1.11- Energy Transmission in Fluid Power Systems

Which factor is responsible for most of the wasted energy in a hydraulic system ?

14

1.3.7- Conclusion Table

Point of Comparison	*Power Transmission System*			
Energy Transmission Distance	E	P	H	M
Actuator Speed	E	M	P	H
Power Density	H	E	P	M
Energy Density	E	H	P	M
Energy Transmission Efficiency	E	M	H	P
Magnitude of Energy Transmitted to an Actuator	H	E	M	P
Motion Controllability with Large Power Transmission	H	E	M	P
Easiness of Linear Motion	H	P	E	M
Cleanliness of Operating Conditions	E	P	M	H
Maintenance Requirements	H	M	P	E

E (Electrical), H (Hydraulic), P (Pneumatic), and M (Mechanical)

Table 1.2 – Power Transmission System Comparison

15

1.4- Fluid Power versus Fluid Mechanics

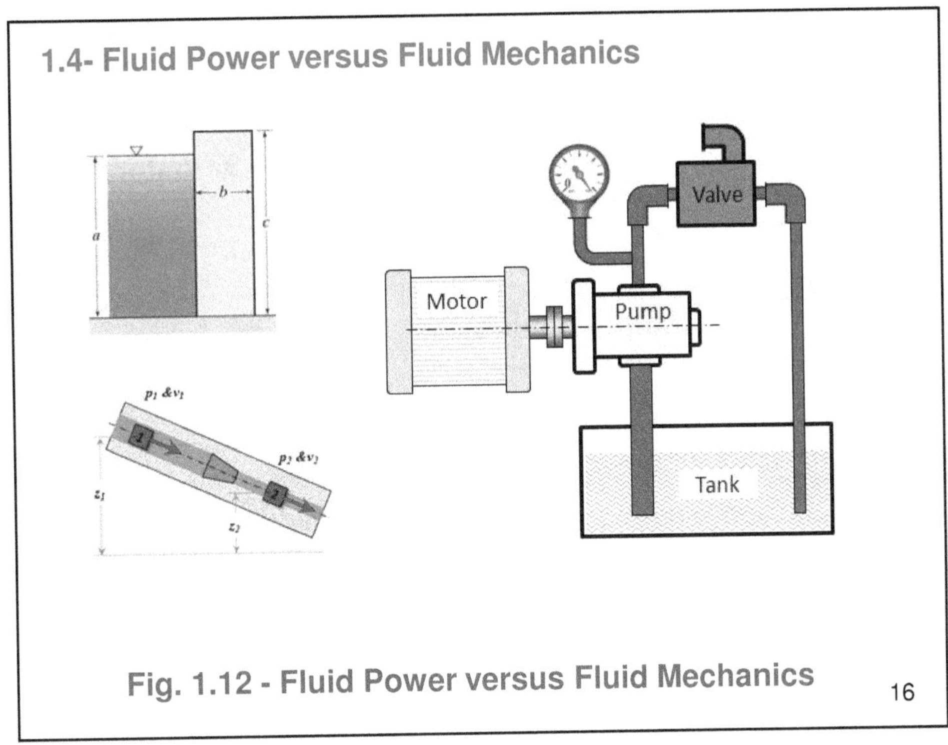

Fig. 1.12 - Fluid Power versus Fluid Mechanics

16

1.5- Fluid Power Systems Classification

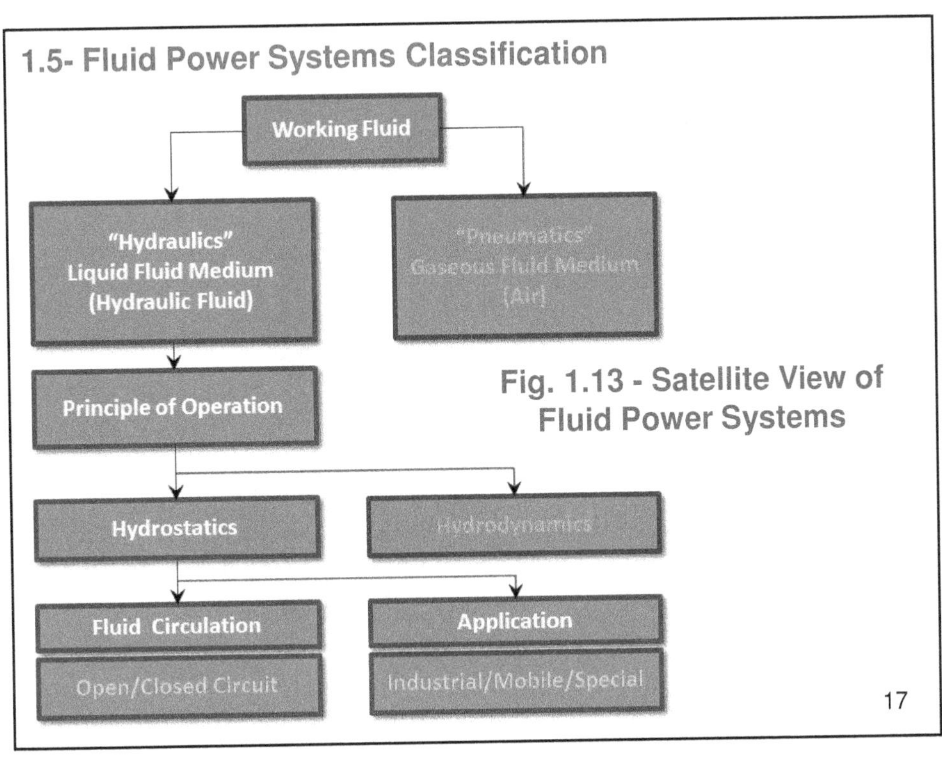

Fig. 1.13 - Satellite View of Fluid Power Systems

17

1.5.1- Hydraulic versus Pneumatic

Similarities:

Designation - Circuit Structure - Valves - Actuators

Differences:

- ❑ **Fluid Compressibility**
- ❑ **Working Pressure**
- ❑ **Magnitude of Transmitted Power**
- ❑ **Precision of Motion Control**
- ❑ **Actuators Speed**
- ❑ **Cleanliness**

Table 1.3 - Basic Differences between Hydraulic and Pneumatic Systems

18

1.5.2- Hydrostatic versus Hydrodynamic

A- Hydraulic Jack.

B- Centrifugal Pump

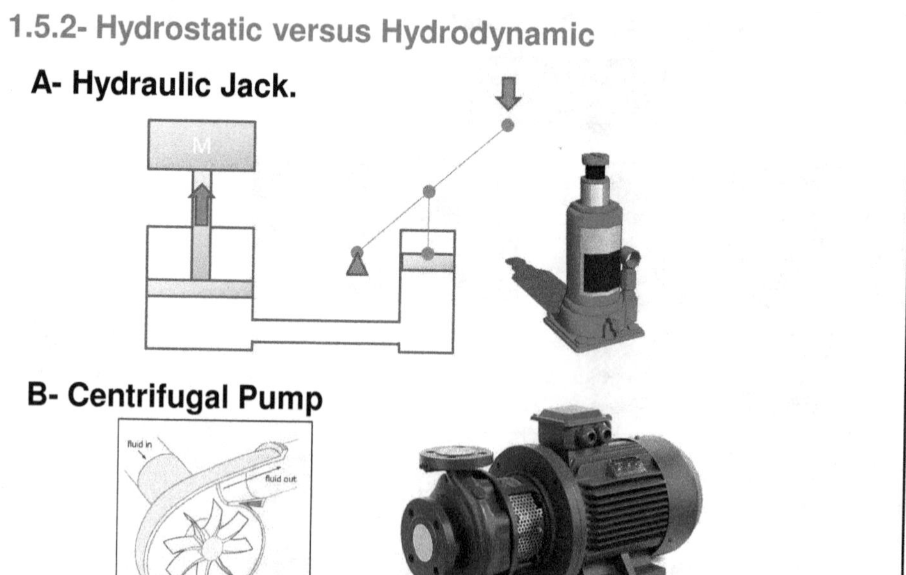

Fig. 1.14 - Hydrostatic versus Hydrodynamic Devices

19

	Hydrostatic and Aerostatic Devices		*Hydrodynamic and Aerodynamic Device*	
	Generator	***Consumers***	***Generator***	***Consumers***
Hydraulic	Positive Displacement Pump	Hydraulic Motors	Centrifugal Pumps	Hydraulic Turbines
Pneumatic	Positive Displacement Compressors	Air Motors	Blower, fans & non-positive Displacement Compressors	Gas Turbines

Table 1.4 – Static and Dynamic Devices in Hydraulic and Pneumatic Systems

20

1.5.3- Open Circuit versus Closed Circuit

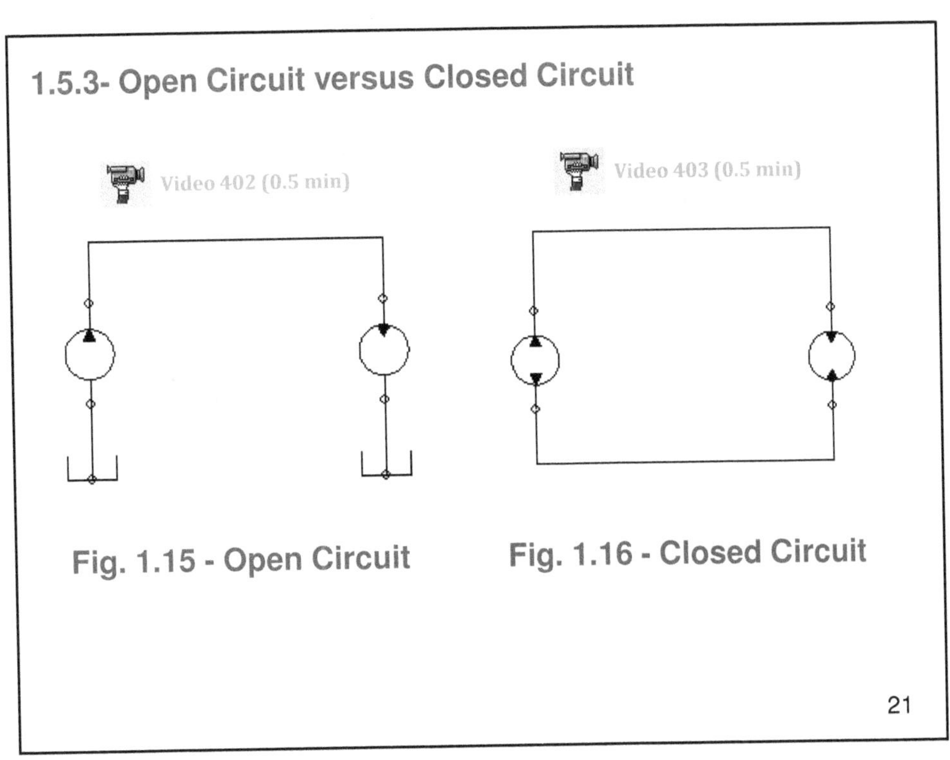

Fig. 1.15 - Open Circuit **Fig. 1.16 - Closed Circuit**

21

1.5.4- Industrial versus Mobile Application

OIL

OIL

Video 001 (0.5 min)

Fig. 1.17 - Hydraulic Industrial Applications

22

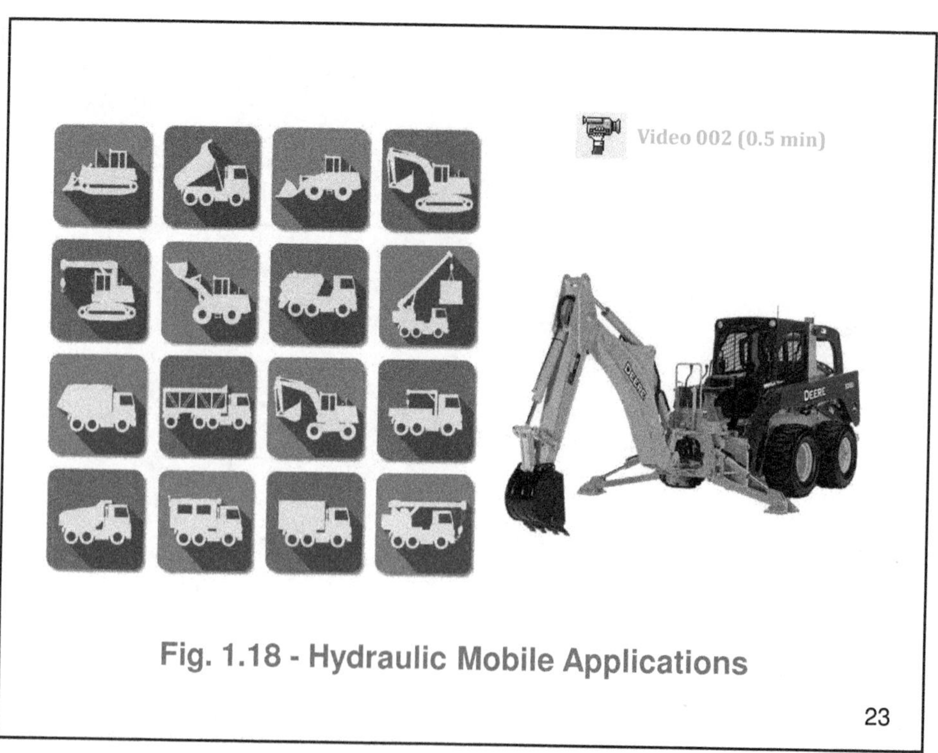

Fig. 1.18 - Hydraulic Mobile Applications

23

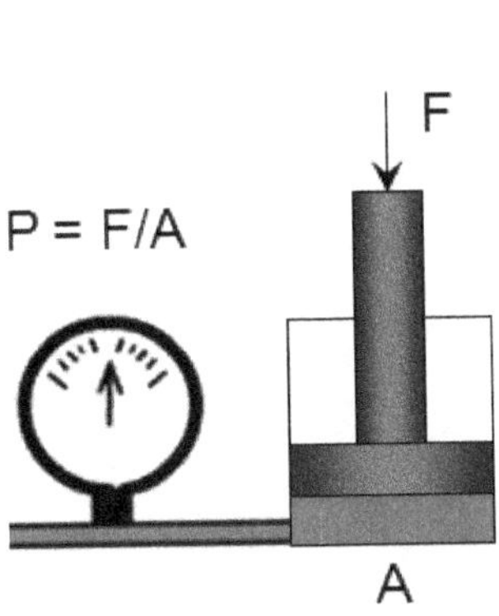

Fig. 1.19 – Concept of Sizing a Hydraulic System

24

	Industrial Application	*Mobile Application*
Working Pressure	Low to Medium (1000-3000) psi	Medium to High (2000-4000) psi
Pump Driving Speed	Standard electric motor RPM e.g. 1200, 1800 RPM	Engine RPM Up to 2500 RPM
Working Temperature	Well controlled indoor temperature around 160 °F	Outdoor temperature could varies very widely
Working Environment	Indoor relatively clean environment	Harsh outdoor environment e.g. mud, clay, snow, etc.
Valve System	Discrete valves	Combined control blocks
Components Used	Standard and interchangeable	Mostly custom and compact
Fail safe Design	Required	Has high priority because machine operator is inside

Table 1.5 - Industrial versus Mobile Hydraulic Applications

25

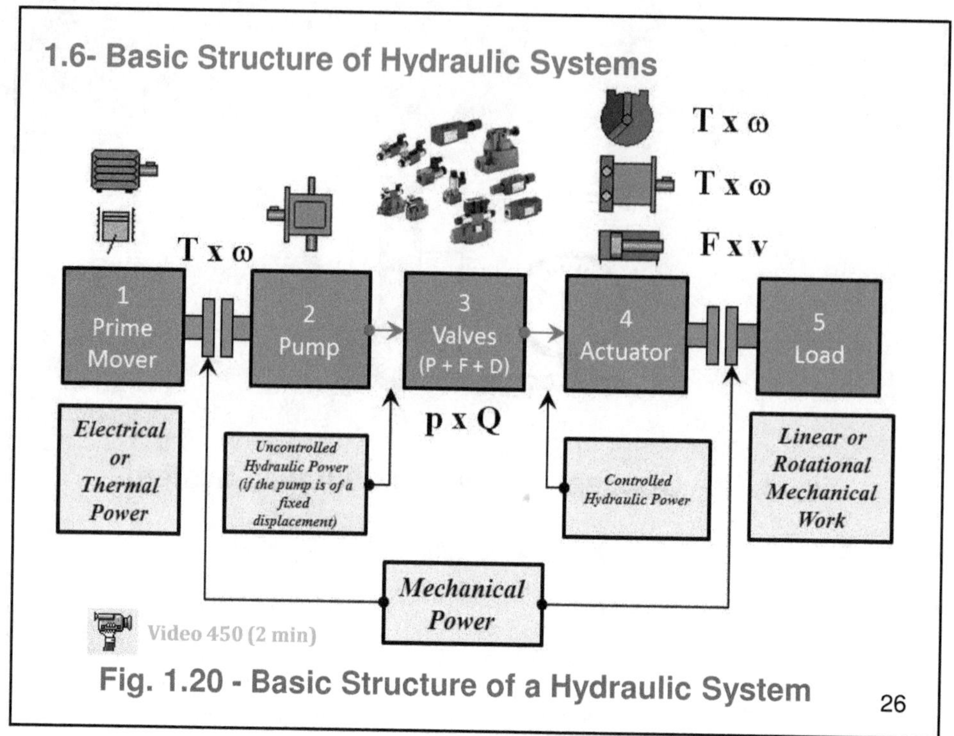

Fig. 1.20 - Basic Structure of a Hydraulic System

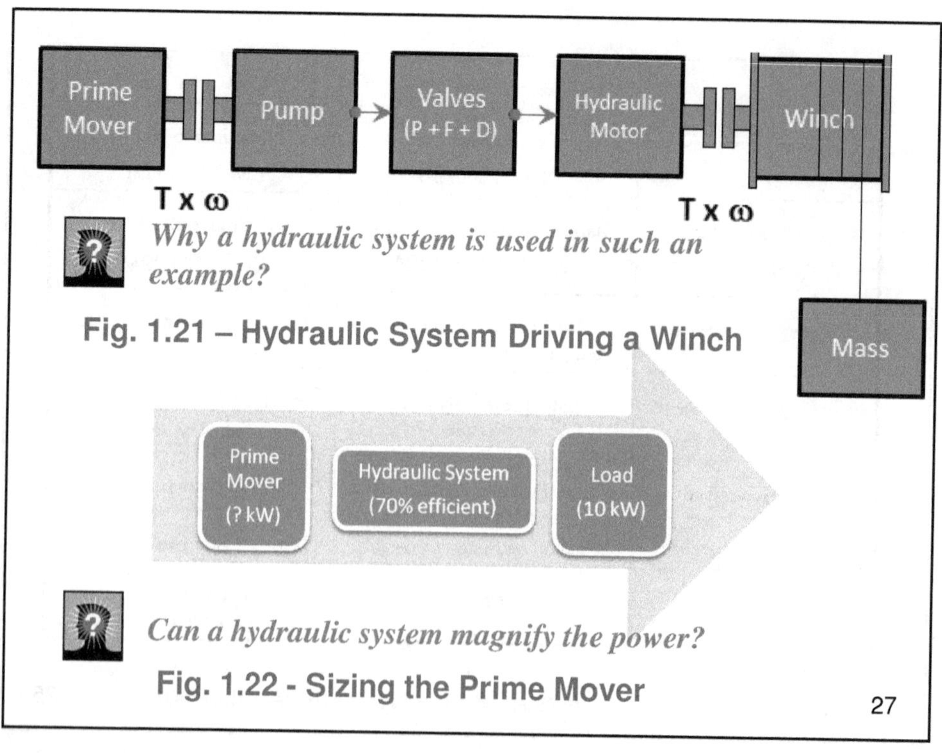

Fig. 1.21 – Hydraulic System Driving a Winch

Fig. 1.22 - Sizing the Prime Mover

1.7- Advantages and Limitations of Hydraulic Systems

1.7.1- Use of Hydraulic Fluids

Advantages	*Limitations*
▪ Different Types. ▪ Fill in spaces. ▪ Compressibility (0.5% volume reduction for 1000 psi pressure increase). ▪ Cooling. ▪ Self lubrication.	▪ Viscosity - Friction - losses. ▪ Mineral oil is flammable. ▪ Ecologically hard to recycle. ▪ Working temperature. ▪ Possible internal & external leak. ▪ Possible cavitation. ▪ Good carrier of contaminants.

Table 1.6 - Features of Hydraulic Systems
(Use of Hydraulic Fluids)

28

1.7.2- Power Transmission

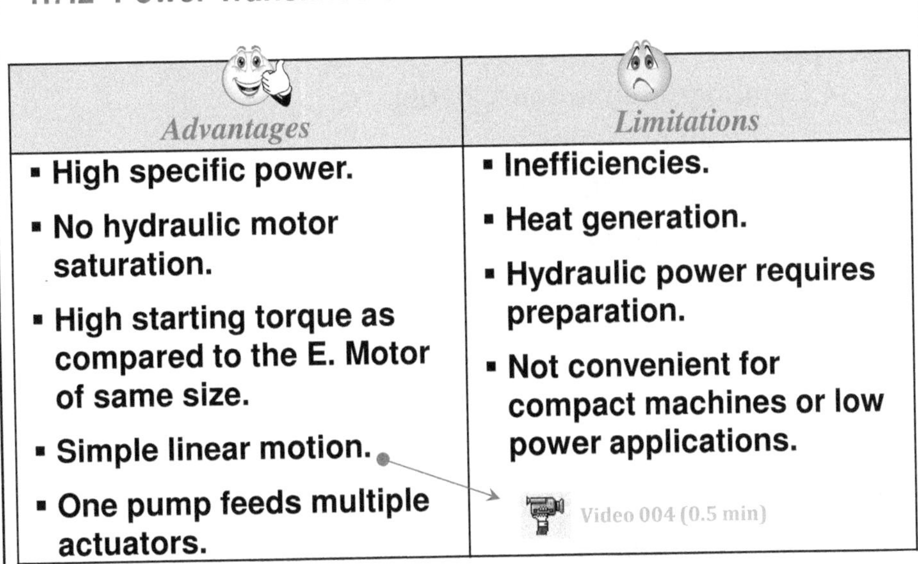

Advantages	*Limitations*
▪ High specific power. ▪ No hydraulic motor saturation. ▪ High starting torque as compared to the E. Motor of same size. ▪ Simple linear motion. ▪ One pump feeds multiple actuators.	▪ Inefficiencies. ▪ Heat generation. ▪ Hydraulic power requires preparation. ▪ Not convenient for compact machines or low power applications. Video 004 (0.5 min)

Table 1.7 - Features of Hydraulic Systems
(Power Transmission)

29

1.7.3- Controllability and System Design

Advantages	*Limitations*
▪ Wide variety of components. ▪ Pressure (load) control. ▪ Flow (speed) control. ▪ Fail-Safe design. ▪ Standardization. ▪ EH sys. & automatic control. ▪ Rugged systems that work in harsh conditions.	▪ Know-How. ▪ Multiple designs to do the same function. ▪ Pressure spikes. ▪ Nonlinearity of valves. ▪ Modeling and simulation is a challenge. ▪ Energy saving is a challenge.

Table 1.8 - Features of Hydraulic Systems
(Controllability and System Design)

30

1.8- Industry Standards for Hydraulic Systems

Benefits:

- Component interchangeably.
- Component testing.
- Component selection.
- Symbols and terminology.
- Improve safety and performance in industry.
- Improve communication.
- Guide people recently introduced to the technology.

Types:

- Communication Standards.
- Design Standards.
- Performance Standards.

31

Source of Standards	Year Founded	Website
Society of Automotive Engineers (SAE).	1905	www.sae.org
Joint Industry Conference (JIC).	1950	-
National Fluid Power Association (NFPA).	1953	www.nfpa.com
International Fluid Power Society (FPS).	1960	www.ifps.org
American National Standards Institute (ANSI) B93	1961	www.ansi.org
International Standards Organization (ISO) TC131.	1970	http://www.iso.org

Table 1.9 - standards that contain information about fluid power systems

32

1219 ISO standard

▪ Rotating Device		▪ Shifting Device	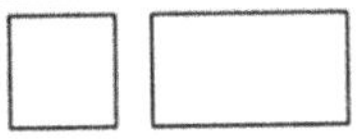
▪ Treatment Device		▪ Adjustable Device	
▪ Main line		▪ Outline Enclosure	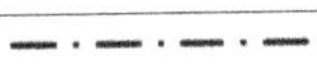
▪ Pilot/Drain line	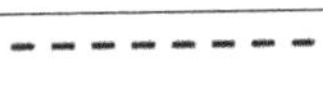	▪ Flexible Line	
▪ Crossing Lines	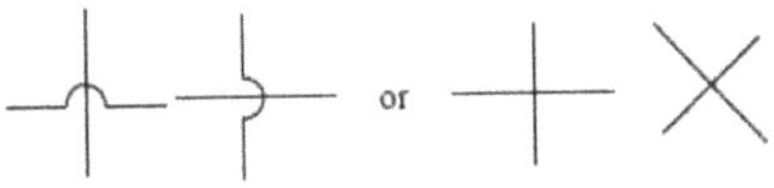 or		
▪ Joining Lines	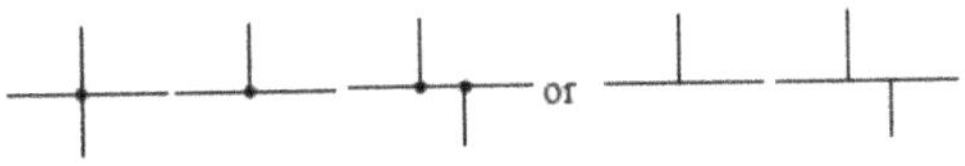 or		

Table 1.10 - Rules for Symbols

33

Chapter 1 Reviews

1. Which statement is **True**?
    A. Pneumatic systems are more suitable than hydraulic systems to drive powerful machines.
    B. Mechanical systems transfer energy for longer distances than hydraulic systems.
    C. Hydraulic systems are limited to use only one type of hydraulic fluid.
    D. None of the above statements is true.

2. Which statement is **False?**
    A. Pneumatic systems are most usable for logic control rather than power drive systems.
    B. Hydraulic systems are the most energy-efficient systems among the other energy transmission and control systems.
    C. Hydraulic systems have the highest power/weight ratio among the other energy transmission and control systems.
    D. Mechanical transmission systems are generally used when the speed ratio between the prime mover and the actuator is not large.

3. Which power transmission system has the highest **Power Density?**
    A. Pneumatic System.
    B. Mechanical System.
    C. Hydraulic System.
    D. Electrical System.

4. Which component has the highest **Energy Density?**
    A. Accumulator.
    B. Battery.
    C. Capacitor.
    D. Spring.

5. Which energy transmission system is best suited to produce **powerful linear motion?**
    A. Pneumatic System.
    B. Mechanical System.
    C. Hydraulic System.
    D. Electrical System.

6. If your application is to drive a rotational load at 10,000 rpm continuously, **which actuator should be used?**
    A. Hydraulic Motor.
    B. Rotary Actuator.
    C. Linear Motor.
    D. Electrical Motor.

7. If your application is to drive a translational load of 2 tons at low velocity, **which actuator should be used?**
    A. Hydraulic Motor.
    B. Rotary Actuator.
    C. Hydraulic Cylinder.
    D. Electrical Motor.

8. What are **the main constraints of a mechanical system that is used to transmit and control power?**
    A. Working temperature and transmission distance.
    B. Transmission distance and the operating pressure.
    C. Working temperature and speed ratio between the prime-mover and the load.
    D. Transmission distance and speed ratio between the prime-mover and the load.

9. If in your application, a pump is receiving fluid from the reservoir and the return oil gets back to the reservoir, **the hydraulic circuit is called a (an)?**
    A. Open-center system.
    B. Closed-center system.
    C. Open Circuit.
    D. Closed Circuit.

10. If you are asked to build a hydraulic system for a mobile machine, **which design parameter would you first consider to start with?**
    A. Open-center system.
    B. Hydraulic fluid viscosity.
    C. High working pressure for the sake of system compactness.
    D. Type of the conductors.

Chapter 1 Assignment

Student Name: -- Student ID: ------------------

Date: --- Score: -----------------------

A. Draw a block diagram that shows the basic structure of a hydraulic system.

B. List two reasons why a hydraulic system is used instead of a mechanical system to transmit and control the power between a prime-mover and a load.

Chapter 2
Basic Concepts Review

Objectives:

This chapter presents reviews of fundamental concepts of multidisciplinary systems; namely: Mechanics, Physics, Fluid Mechanics and Hydraulics. The fundamental concepts reviewed are the ones that are useful in understanding some properties of hydraulic systems. In reviewing the concepts of fluid mechanics, this chapter explains how to quantify the frictional and local pressure losses in a hydraulic line based on laminar or turbulent flow patterns in the line.

0

Brief Contents:

1

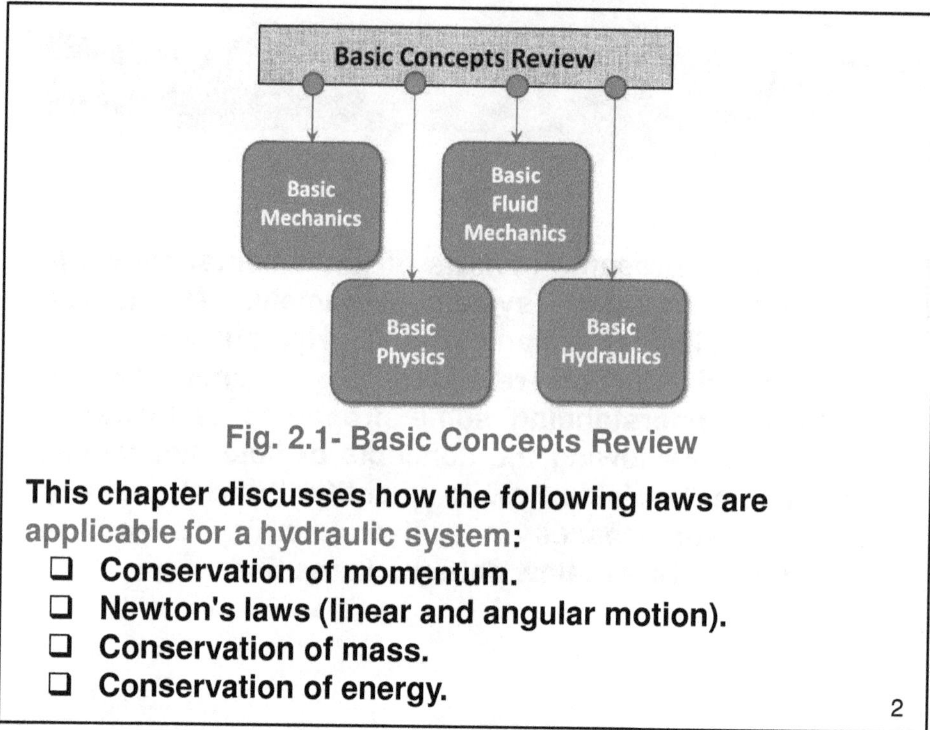

Fig. 2.1- Basic Concepts Review

This chapter discusses how the following laws are applicable for a hydraulic system:

- ❑ **Conservation of momentum.**
- ❑ **Newton's laws (linear and angular motion).**
- ❑ **Conservation of mass.**
- ❑ **Conservation of energy.**

2

2.1- Basic Mechanics.

2.1.1- Effort and Flow Variables for Physical Systems.

	Effort Variable	*Flow Variable*	
Mechanical "Translational"	Force *F*	Linear Velocity *v*	+*F* Mass *m* -*F*

Table 2.1 - Effort Variable and Flow Variable for Physical Systems - A

- ❑ **Unbalanced** forces **generate** translational **motion.**
- ❑ **If the** forces balance each other**, this means that the mass is** stopped or moving at constant speed**.**

3

	Effort Variable	*Flow Variable*	
Mechanical "Rotational"	Torque T	Angular Velocity ω	+T, Mass Moment Of Inertia J, -T

Table 2.1 - Effort Variable and Flow Variable for Physical Systems - B

- **Unbalanced** torques **generate** rotational **motion.**
- **If the** torques balance each other, **this means that the mass moment of inertia is** stopped or rotating at constant speed.

4

	Effort Variable	*Flow Variable*	
Hydraulic	Pressure p = Force/area	Fluid Flow Q	p_1, p_2, Q, $\Delta p = p_1 - p_2$

Table 2.1 - Effort Variable and Flow Variable for Physical Systems - C

- **Differential** pressures **generates** fluid flow.
- **If the** differential pressure equals zero, **this means that the** fluid flow equals zero.

5

atmospheric psi = 14.7

	Effort Variable	*Flow Variable*	
Electrical	Voltage v	Electrical Current i	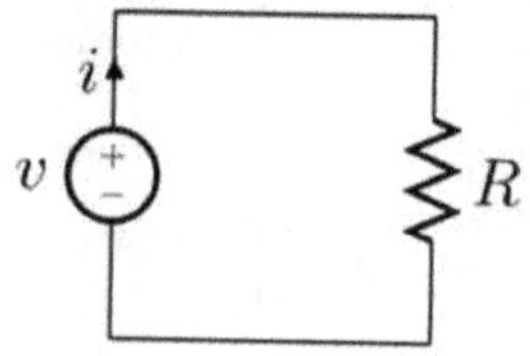 i v R

Table 2.1 - Effort Variable and Flow Variable for Physical Systems - D

- ❑ **Voltage difference generates electrical current.**
- ❑ **If the voltage difference equals zero, this means that the electrical current equals zero.**

6

	Effort Variable	*Flow Variable*	
Thermal	Temperature T	Heat Flow Q	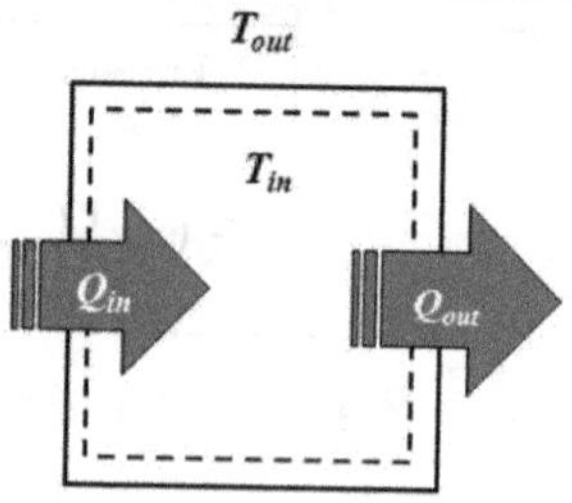 T_{out} T_{in} Q_{in} Q_{out}

Table 2.1 - Effort Variable and Flow Variable for Physical Systems - E

- ❑ **Temperature difference generates flow of heat.**
- ❑ **If the temperature difference equals zero, this means that the heat flow equals zero.**

7

2.1.2- Force

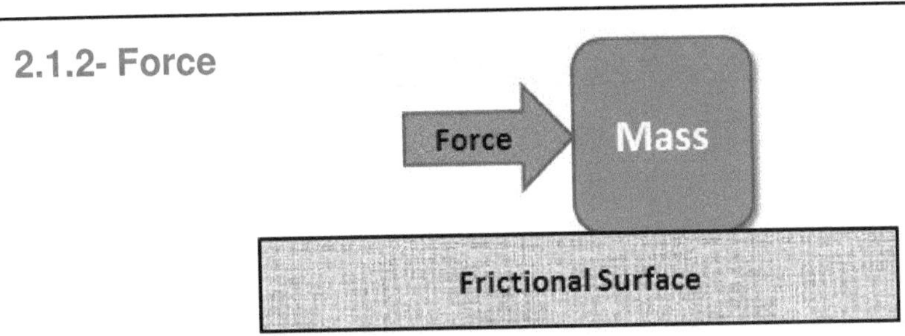

Fig. 2.2- Force Definition

Force can be defined as the effort required to change the status of a translational mass.

- **English System of Units:**
- **Poundal:** is the force required to accelerate one pound mass with an acceleration of 1ft/sec^2
- **Pound force (lb_f):** is the force required to accelerate one pound mass with acceleration equal to the gravitational acceleration (32.2 ft/s^2).

8

- **Metric System of Units:**
- **Newton (N):** is the force required to accelerate one kg mass with an acceleration of 1m/sec^2 .
- **kg force:** is the force required to accelerate one kg mass with an acceleration equal to the gravitational acceleration (9.81 m/s^2).

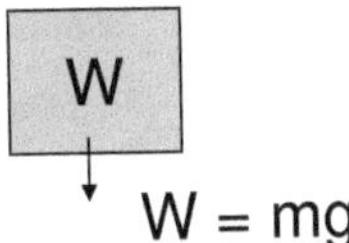

W = mg

- **Weight:**
- **Weight of an object equals the product of the object's mass times the gravitational acceleration.**

- **Then:**
  - Weight of 1 lb_m = 1 lb_f = 32.2 Poundal.
  - Weight of 1 kg_m = 1 kg_f = 9.81 N.

9

Newton's Second Law as applied to a translational mass:

$$\sum F = ma \qquad 2.1$$

Where:

F: **Force acting positively or negatively against the motion of a translational mass.**

m: **Translational mass.**

a: **Linear acceleration.**

10

2.1.3- Torque

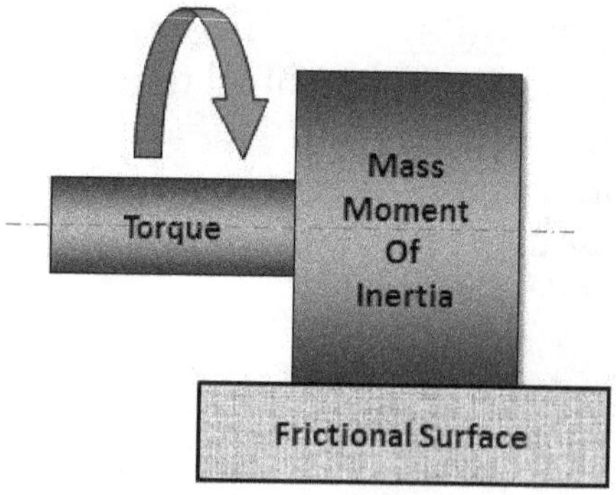

Fig. 2.3- Torque Definition

Torque can be defined as the effort required to change the status of a rotational mass.

Torque is quantified as follows:

Torque = force x torque arm

English System of units: $lb_f.ft.$

Metric System of units: N.m.

11

Newton's Second Law as applied to a rotational mass:

$$\sum T = J\alpha \qquad 2.2$$

Where:

T: **Any torque acting positively or negatively against the rotation of the rotational mass.**

J: **Mass moment of inertia.**

α: **Angular acceleration.**

12

2.1.4- Energy

Matter of facts:

- **Energy can't be generated from nothing.**
- **Energy can't be vanished.**
- **Energy can take different forms.**
- **Energy can transfer from one form to another form.**

Physical Elements:

- **Inductive Elements: are the elements that have an inertial effect and can gain kinetic energy due to the increase of the flow variable (not applicable for thermal systems).**
- **Capacitive Elements: are the elements that store energy inside its limited capacitance by increasing the effort variable.**
- **Resistive Elements: are the elements that dissipate (waste) the energy due to resisting the flow variable.**

13

Physical System ------ *Element*	*Mechanical "Translational"*	*Mechanical "Rotational"*	*Hydraulic*	*Electrical*	*Thermal*
Inductive (Inertial) Elements	Mass	Mass Moment of Inertia	Hydraulic Inductance	Electrical Inductance	NA
Capacitive Elements	Linear Spring	Torsion Spring	Accumulator	Capacitor	Heat Sink
Resistive (Dissipative) Elements	Linear Damper	Rotational Damper	Hydraulic Resistance	Electrical Resistor	Heat Shields

Table 2.2 – Types of Elements within a Physical System

14

For a hydraulic system:

Kinetic energy of moving Fluid in a conductor =

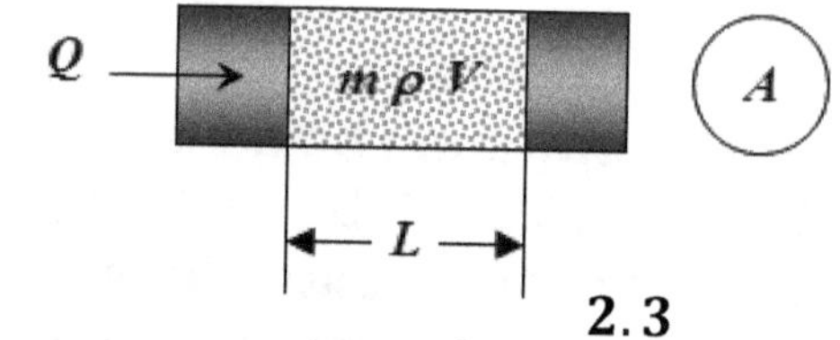

$$E_K = \frac{1}{2} I Q^2 = \frac{1}{2}\left(\frac{\rho L}{A}\right) Q^2 \qquad 2.3$$

Stored energy in an accumulator =

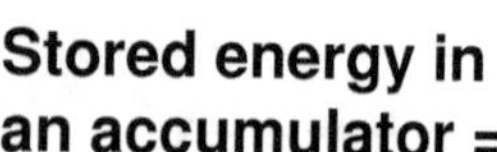

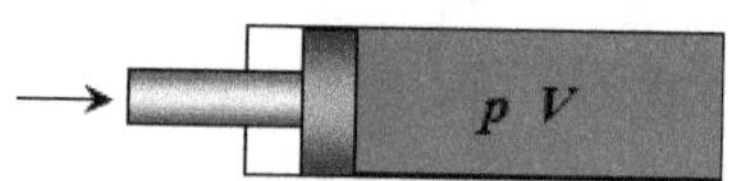

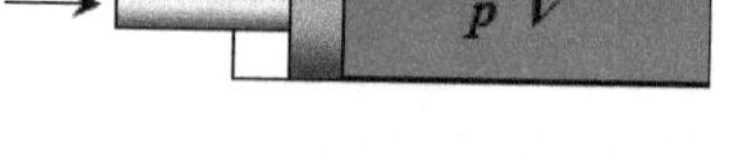

$$E_S = pV \qquad 2.4$$

Wasted energy in a hydraulic resistance =

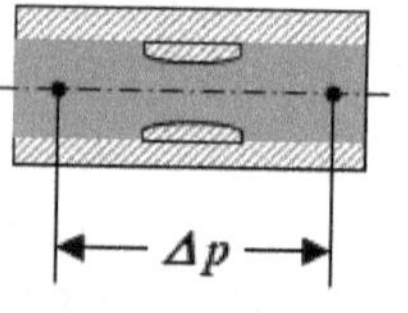

$$E_W = \Delta p V \qquad 2.5$$

15

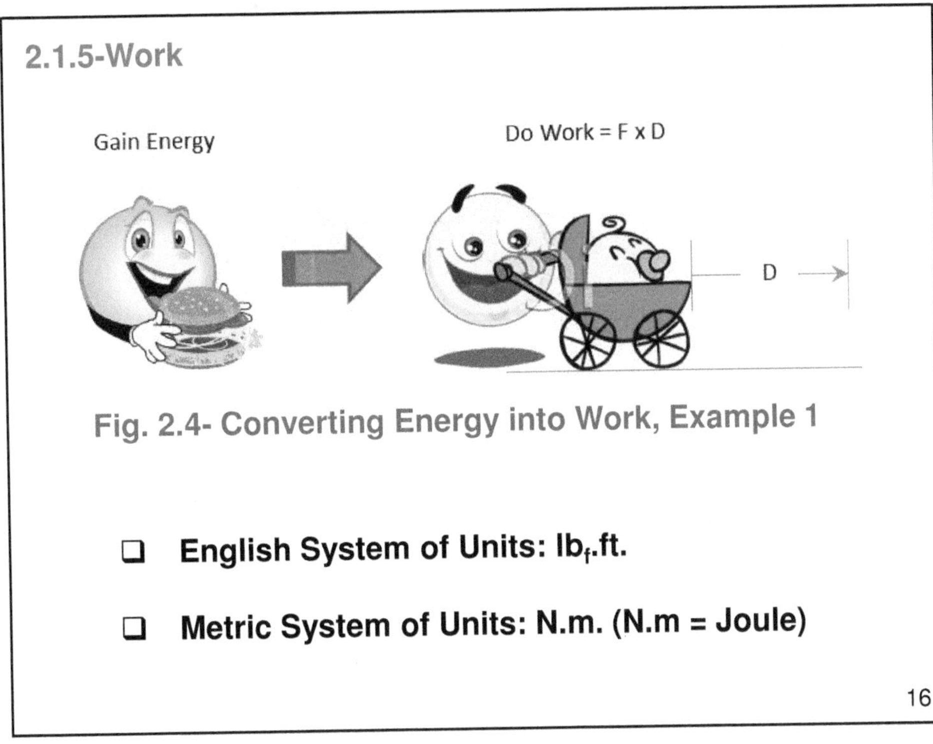

Fig. 2.4- Converting Energy into Work, Example 1

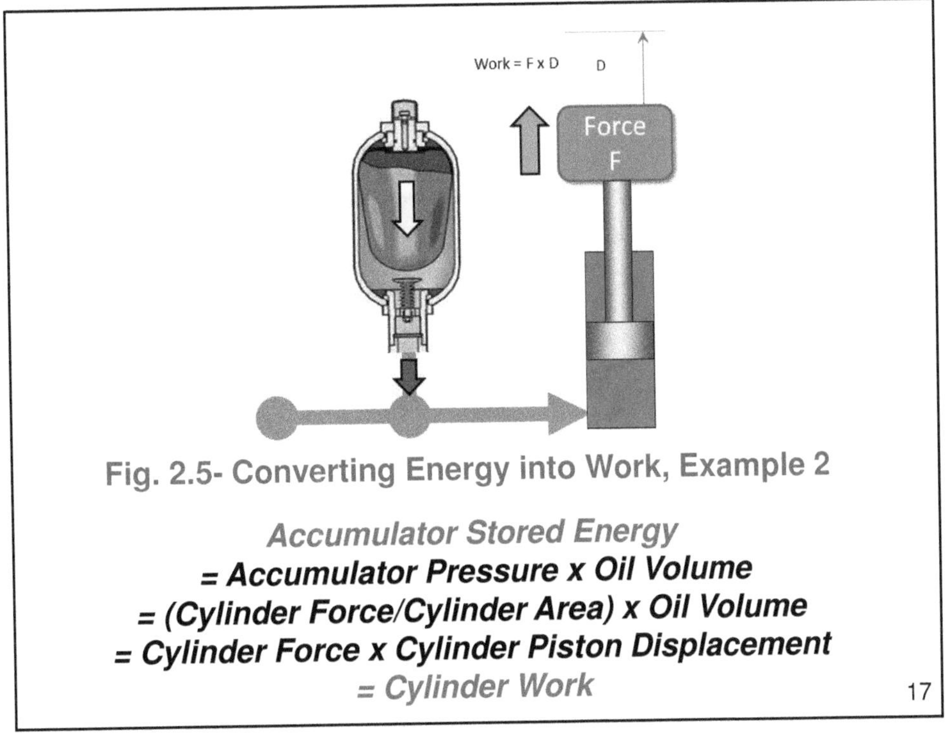

Fig. 2.5- Converting Energy into Work, Example 2

2.1.6- Power

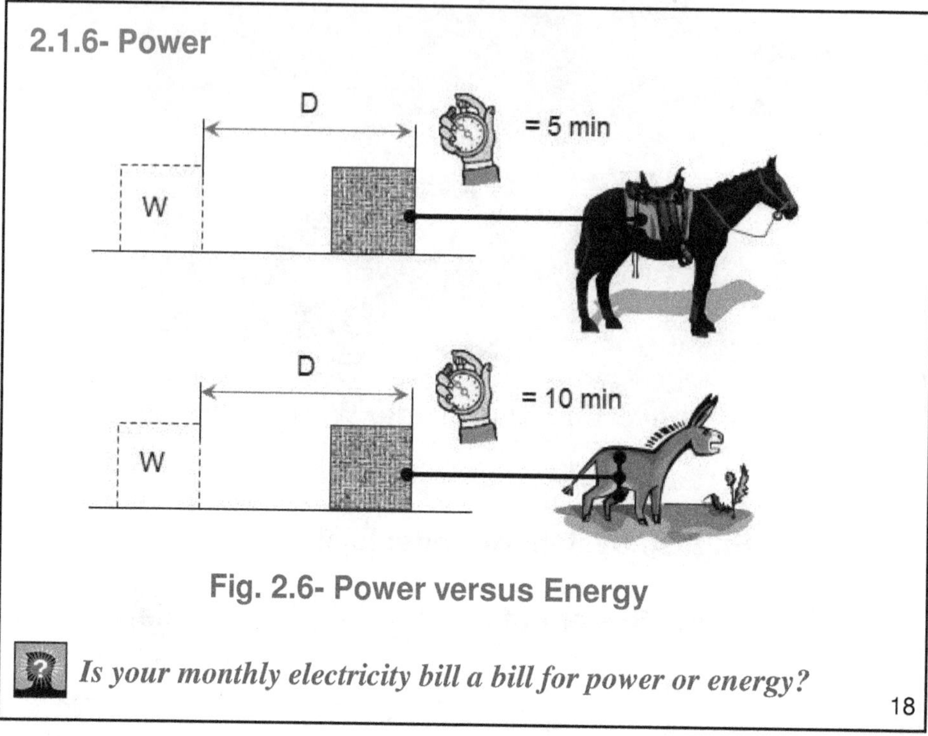

Fig. 2.6- Power versus Energy

Is your monthly electricity bill a bill for power or energy?

18

2.1.6.1- Power of a Linear Actuator (Hydraulic Cylinder)

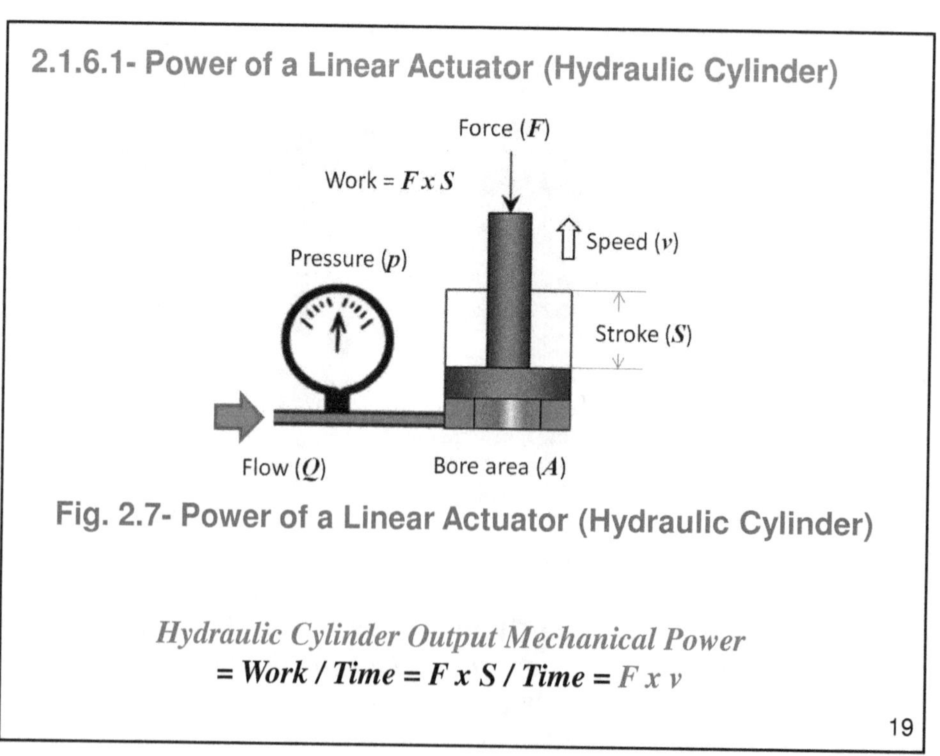

Fig. 2.7- Power of a Linear Actuator (Hydraulic Cylinder)

Hydraulic Cylinder Output Mechanical Power
= Work / Time = F x S / Time = F x v

19

In Metric System of Units:

$$\text{Cylinder Output Mechanical Power } (kW) = P_C$$

$$= \frac{Force\ [N] x\ Speed\ [m/s]}{1000} \qquad 2.6$$

In English System of Units:

$$\text{Cylinder Output Mechanical Power } (\mathrm{HP}) = P_C$$

$$= \frac{Force\ [lb_f] x\ Speed\ [ft/s]}{550} \qquad 2.7$$

20

For a 100% efficient cylinder & no back pressure:

Cylinder Output Mechanical Power

$$= F\ x\ v = (F / A)\ x\ (v\ x\ A)$$

$$= p\ x\ Q$$

= Cylinder Input Fluid Power

- **Force : Pressure**
- **Speed : Flow**

If the available power is less than required, can a cylinder do the required work?

21

2.1.6.2- Power of a Rotational Actuator (Hydraulic Motor)

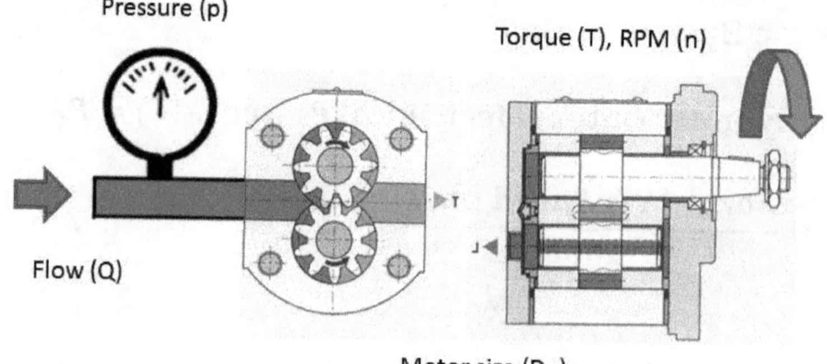

Fig. 2.8- Power of a Rotational Actuator (Hydraulic Motor)

Hydraulic Motor Output Mechanical Power
= Work / Time = Torque x angular Displacement / Time
= Torque x Angular Speed

22

In Metric System of Units:

$$\textbf{Motor Output Mechanical Power } (kW) = P_M$$

$$= \frac{T\,[N.m]\; x\; 2\pi\; x\; n[rpm]}{60\; x\; 1000} \qquad 2.8$$

In English System of Units:

$$\textbf{Motor Output Mechanical Power } (HP) = P_M$$

$$= \frac{T\,[lb_f.ft]\; x\; n[rpm]}{5252} \qquad 2.9$$

23

For a 100% efficient motor:

Hydraulic Motor Output Mechanical Power

$$= 2\pi x\ T\ x\ n = (2\pi T / D_m)\ x\ (n\ x\ D_m)$$

$$= \Delta p\ x\ Q$$

= Hydraulic Motor Input Fluid Power

- **Torque : Pressure**
- **RPM : Flow**

24

2.1.6.3- Hydraulic Power Calculations

Hydraulic power developed, consumed, or wasted by a hydraulic component is calculated as follows:

In Metric System of Units:

$$\textbf{Hydraulic Power } (kW)\ \frac{\Delta p\ [bar]\ x\ Q\ [l/min]}{600} \qquad 2.10$$

In English System of Units:

$$\textbf{Hydraulic Power } (HP) = \frac{\Delta p\ [psi]\ x\ Q[gpm]}{1714} \qquad 2.11$$

$$\text{Proof: Pressure} \times \text{Flow} = \frac{\text{Force}}{\text{Area}} \times \frac{\text{Volume}}{\text{Time}}$$

$$= \frac{\text{Force} \times \text{Length}}{\text{Time}} = \frac{\text{Energy}}{\text{Time}} = \text{Power}$$

25

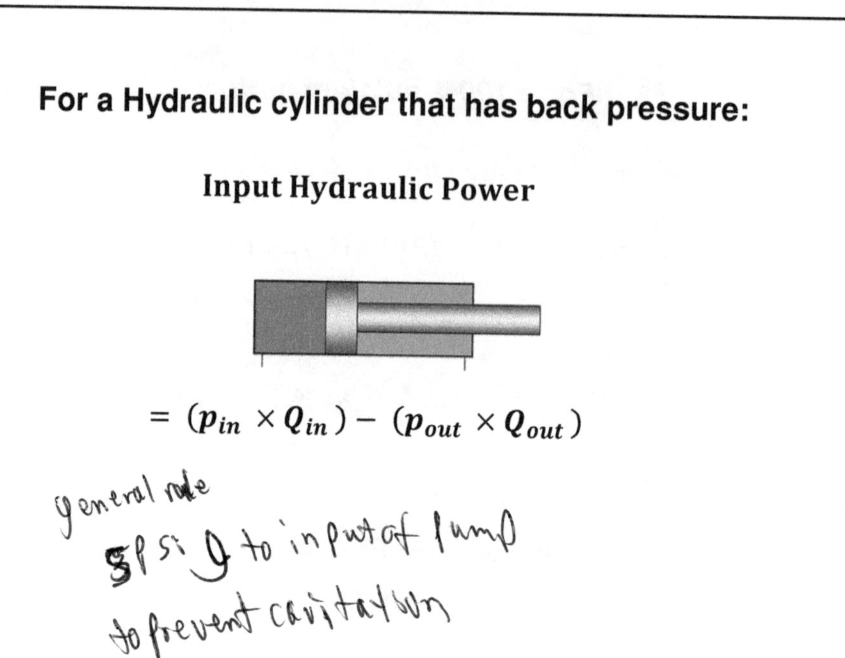

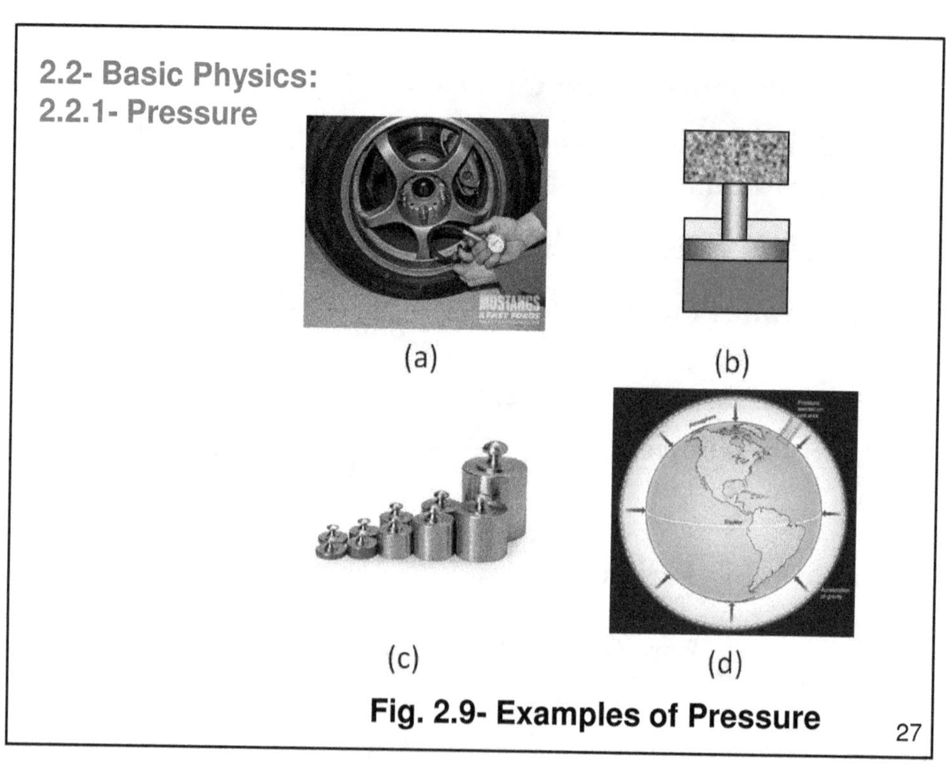

Fig. 2.9- Examples of Pressure

The ISO standard unit of pressure is the Pascal.
Pascal is a very small value = $1 N/m^2$.

Industrial Unit:

- **1 [bar] = 10^5 [N/m^2]= 10^5 [Pa].**
- **1 [bar] = 10^5 / (9.81 x 10000) = 1.0193 [kg/cm^2]**
- **1 [bar] = 14.5 [psi].**

28

2.2.2- Atmospheric Pressure

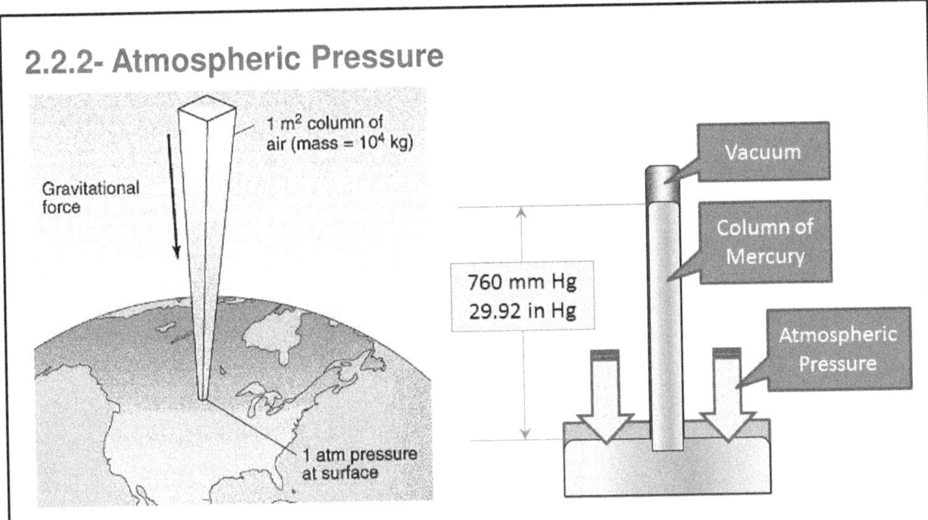

Fig. 2.10- Atmospheric Pressure   Fig. 2.11- Barometer

Atmospheric pressure = 1.01325 bar (14.7 psi).

29

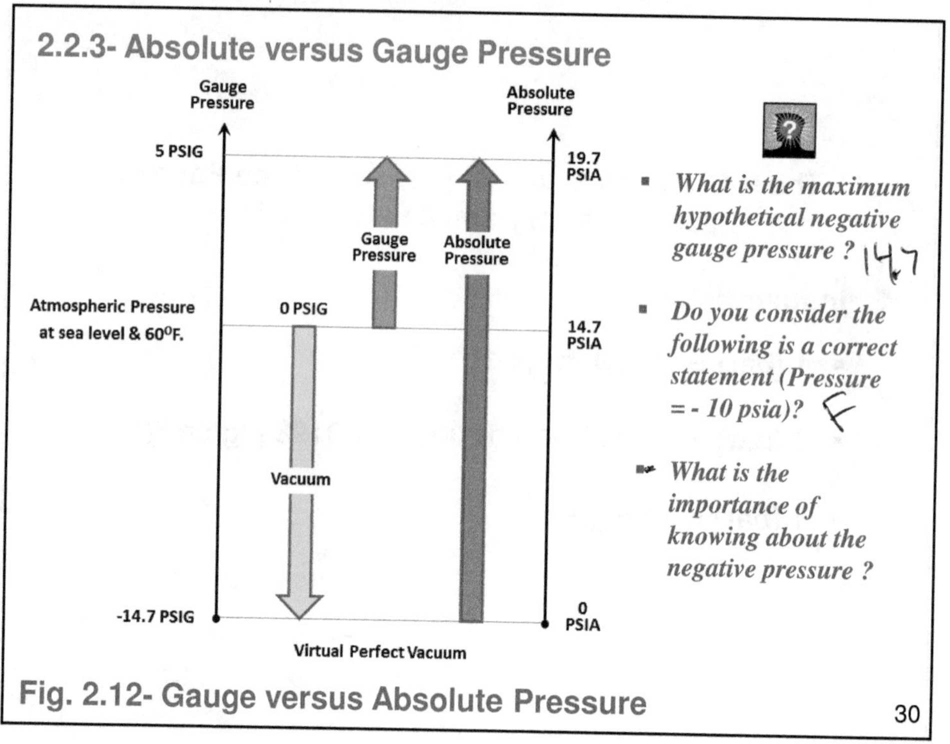

Fig. 2.12- Gauge versus Absolute Pressure

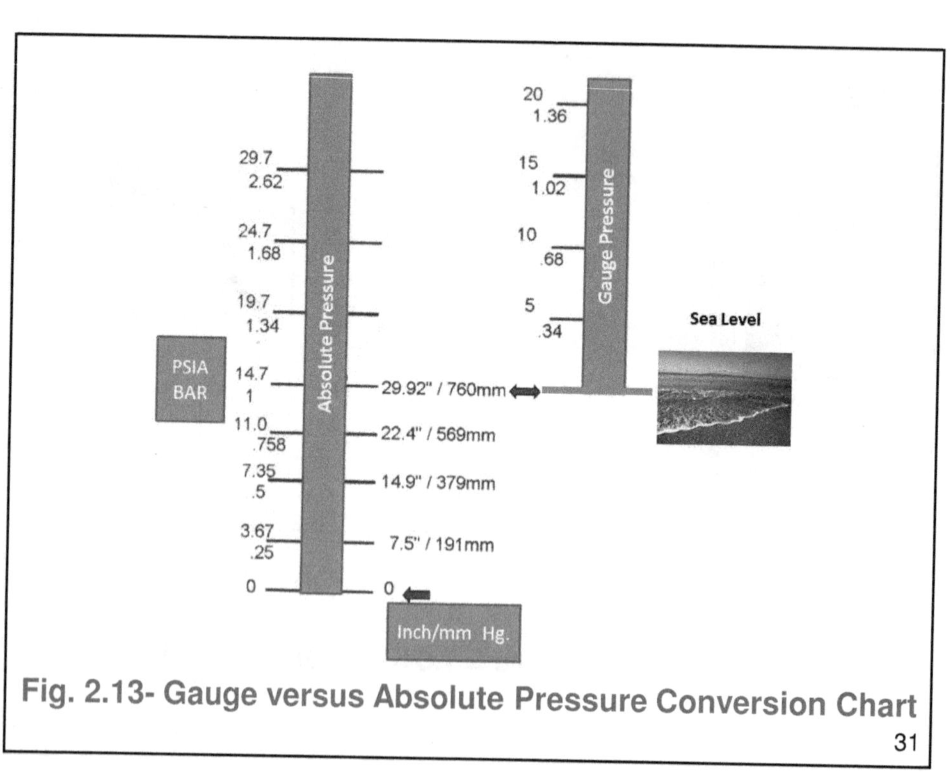

Fig. 2.13- Gauge versus Absolute Pressure Conversion Chart

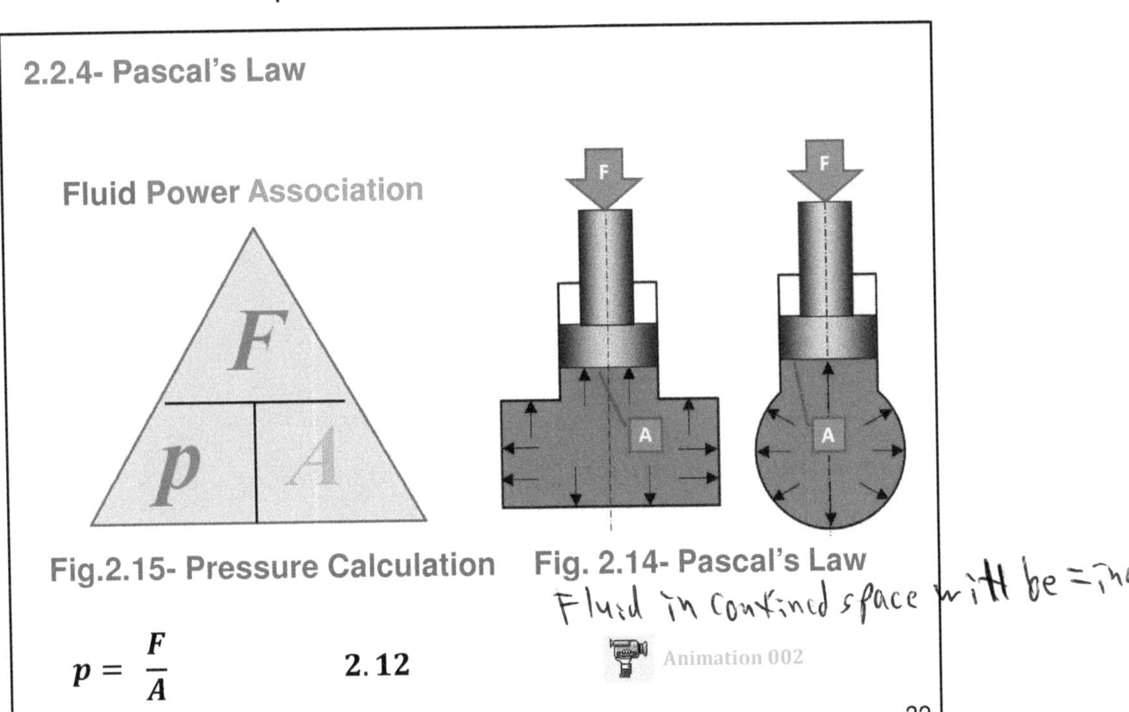
2.2.4- Pascal's Law
Fluid Power Association
F
p
A
F
F
A
A
Fig.2.15- Pressure Calculation
Fig. 2.14- Pascal's Law
Fluid in confined space will be = in all surfaces
$p = \frac{F}{A}$ 2.12
Animation 002
32

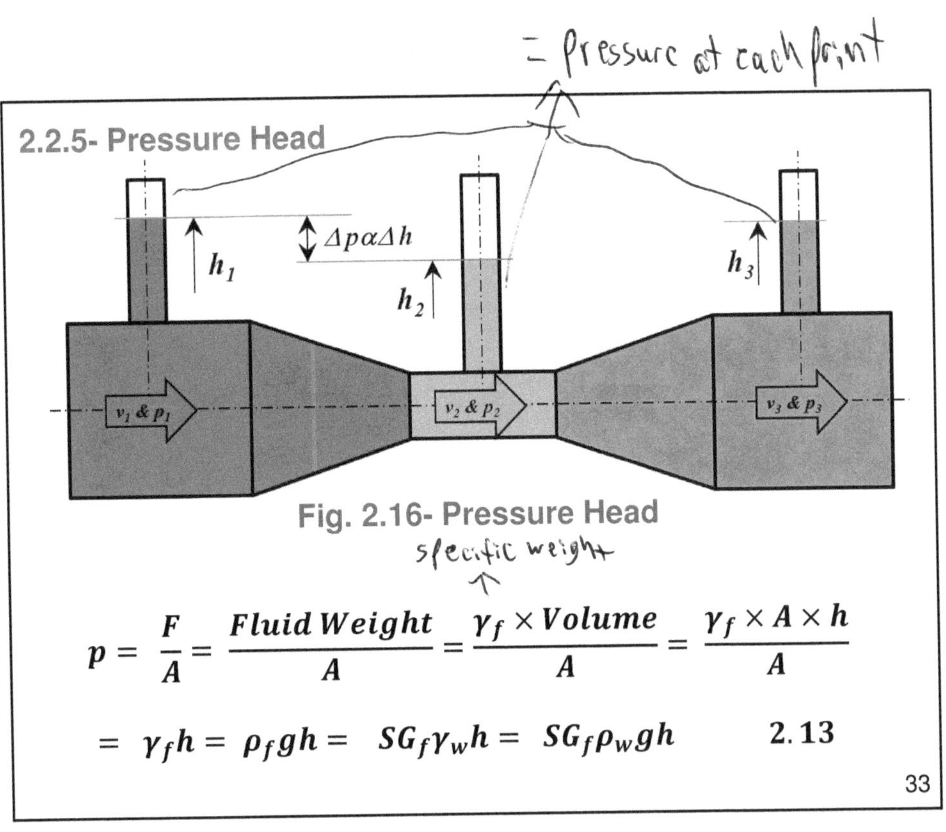
2.2.5- Pressure Head
= Pressure at each point
h_1
$\Delta p \alpha \Delta h$
h_2
h_3
$v_1 \& p_1$
$v_2 \& p_2$
$v_3 \& p_3$
Fig. 2.16- Pressure Head
specific weight
$p = \frac{F}{A} = \frac{Fluid\,Weight}{A} = \frac{\gamma_f \times Volume}{A} = \frac{\gamma_f \times A \times h}{A}$
$= \gamma_f h = \rho_f g h = SG_f \gamma_w h = SG_f \rho_w g h$ 2.13
33

SG

ρ_f = ***Fluid's Desnsity.***

γ_f = ***Fluid's Specific Weight.***

SG_f = ***Fluid's Specific Gravity*** $= \frac{\gamma_f}{\gamma_w} = \frac{\rho_f}{\rho_w}$

ρ_w = ***Water's Desnsity = = 62.4*** $[lb_m/ft^3] = 1000\ [kg_m/m^3]$.

γ_w = ***Water's Specific Weight = 62.4*** $[lb_f/ft^3] = 1000\ [kg_f/m^3]$.

 ***What is the importance of studying the pressure HEAD?***

34

Example:

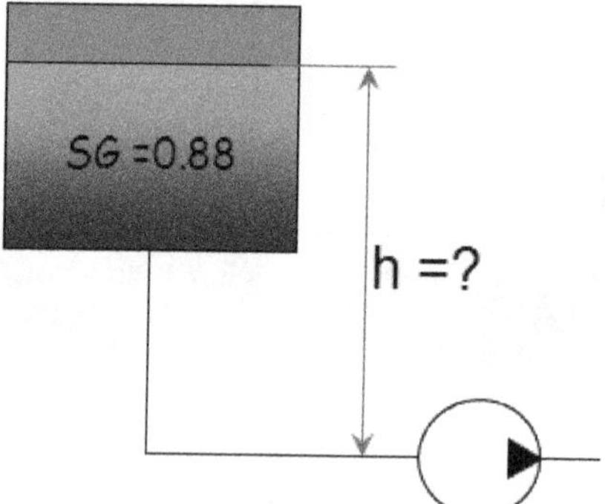

$$h\ (ft) = \frac{p}{\gamma_f} = \frac{p}{SG_f \times \gamma_w} = \frac{2\ [lb/in^2]}{0.88 \times 62.4\ [lb/ft^3]}\left(\frac{144in^2}{ft^2}\right) = 5.245$$

35

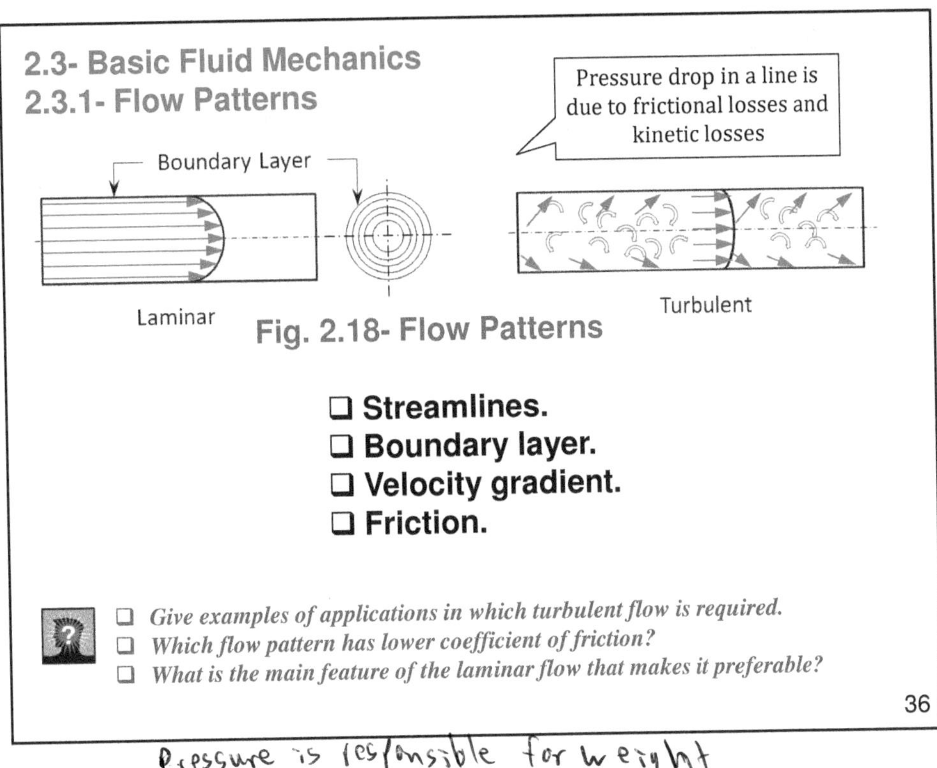

Fig. 2.18- Flow Patterns

Pressure is responsible for weight
Flow rate controls speed

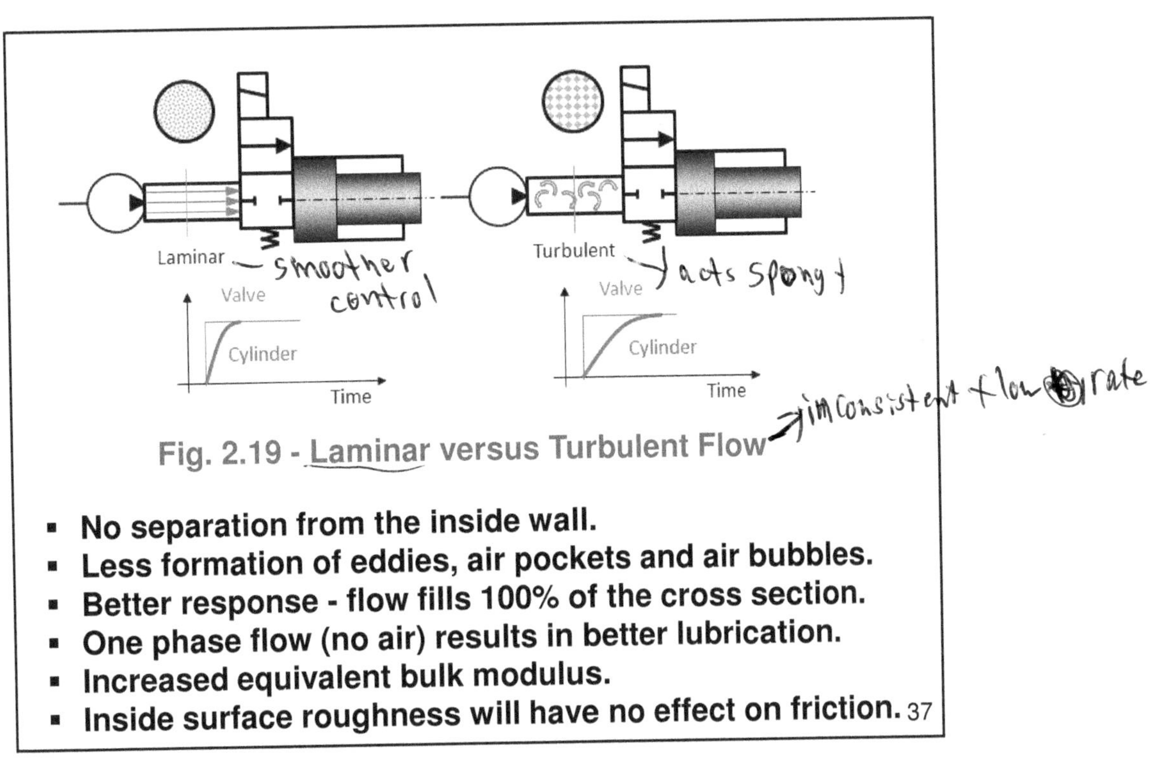

Fig. 2.19 - Laminar versus Turbulent Flow

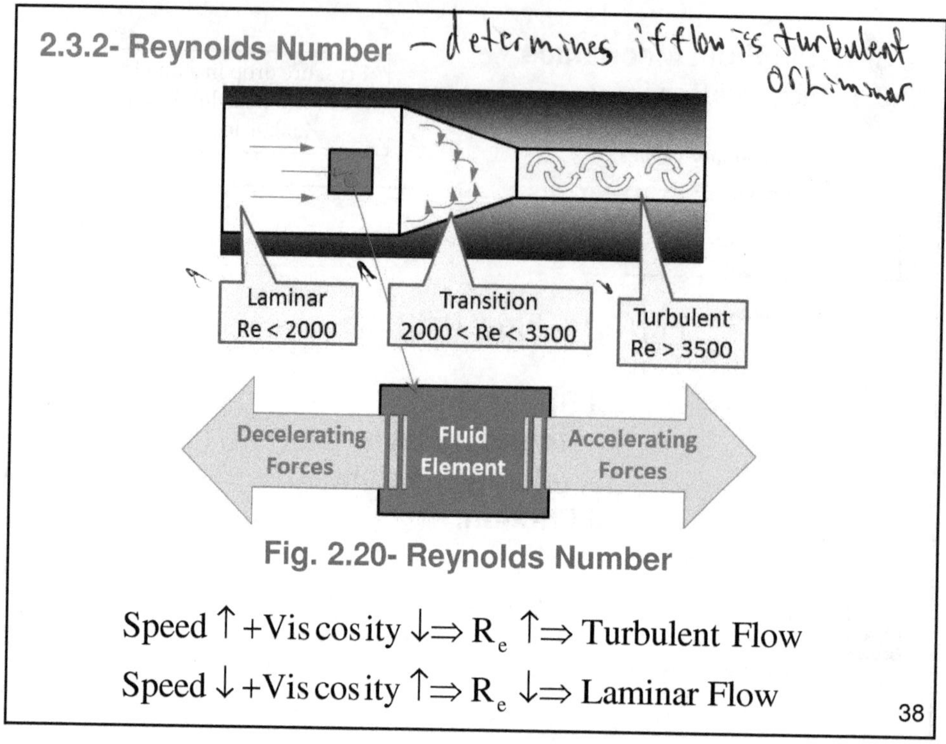
2.3.2- Reynolds Number
Laminar
Re < 2000
Transition
2000 < Re < 3500
Turbulent
Re > 3500
Decelerating Forces
Fluid Element
Accelerating Forces
Fig. 2.20- Reynolds Number
$\text{Speed} \uparrow + \text{Viscosity} \downarrow \Rightarrow R_e \uparrow \Rightarrow \text{Turbulent Flow}$
$\text{Speed} \downarrow + \text{Viscosity} \uparrow \Rightarrow R_e \downarrow \Rightarrow \text{Laminar Flow}$
38

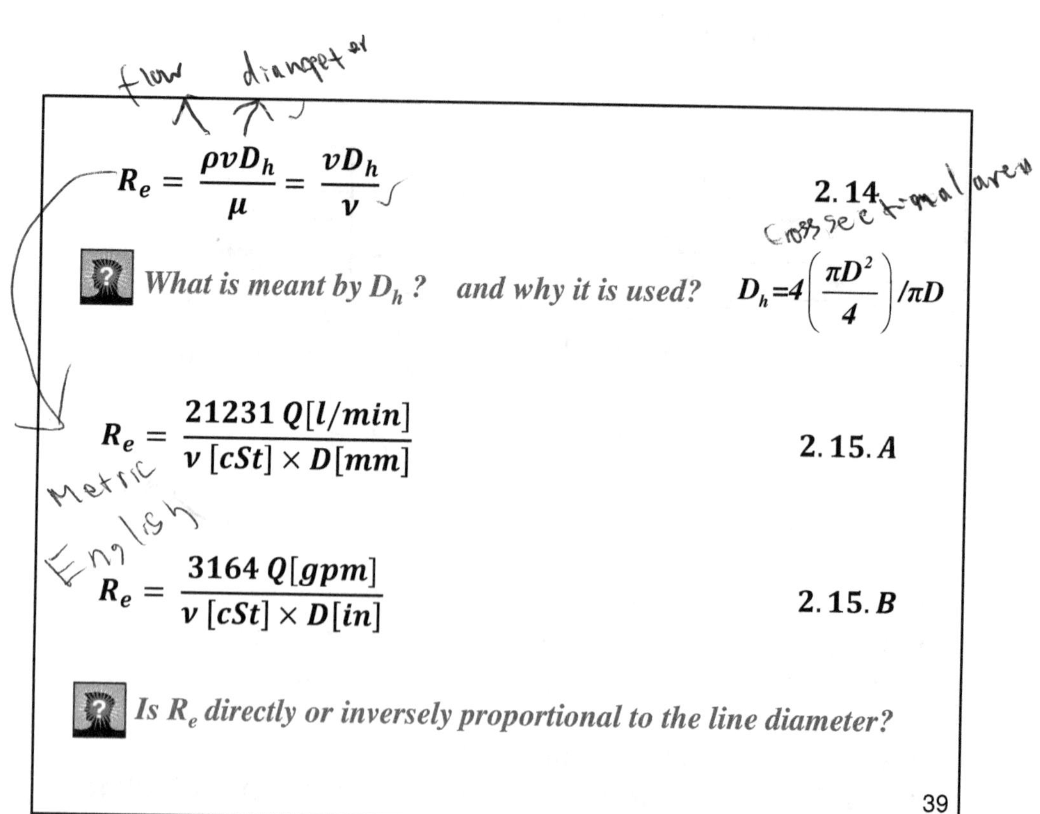
$R_e = \frac{\rho v D_h}{\mu} = \frac{v D_h}{\nu}$ 2.14
What is meant by D_h ? and why it is used? $D_h = 4\left(\frac{\pi D^2}{4}\right)/\pi D$
$R_e = \frac{21231\, Q[l/min]}{\nu\,[cSt] \times D[mm]}$ 2.15.A
$R_e = \frac{3164\, Q[gpm]}{\nu\,[cSt] \times D[in]}$ 2.15.B
Is R_e directly or inversely proportional to the line diameter?
39

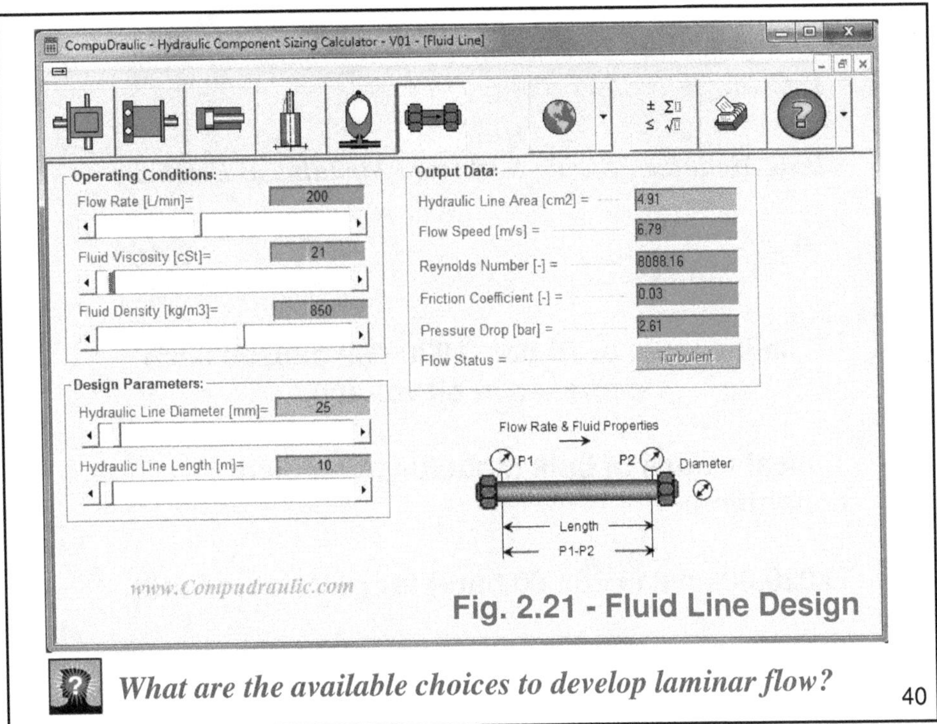
CompuDraulic - Hydraulic Component Sizing Calculator - V01 - [Fluid Line]
Operating Conditions:
Flow Rate [L/min]= 200
Fluid Viscosity [cSt]= 21
Fluid Density [kg/m3]= 850
Design Parameters:
Hydraulic Line Diameter [mm]= 25
Hydraulic Line Length [m]= 10
Output Data:
Hydraulic Line Area [cm2] = 4.91
Flow Speed [m/s] = 6.79
Reynolds Number [-] = 8088.16
Friction Coefficient [-] = 0.03
Pressure Drop [bar] = 2.61
Flow Status = Turbulent
Flow Rate & Fluid Properties
P1
P2
Diameter
Length
P1-P2
www.Compudraulic.com
Fig. 2.21 - Fluid Line Design
What are the available choices to develop laminar flow?
40

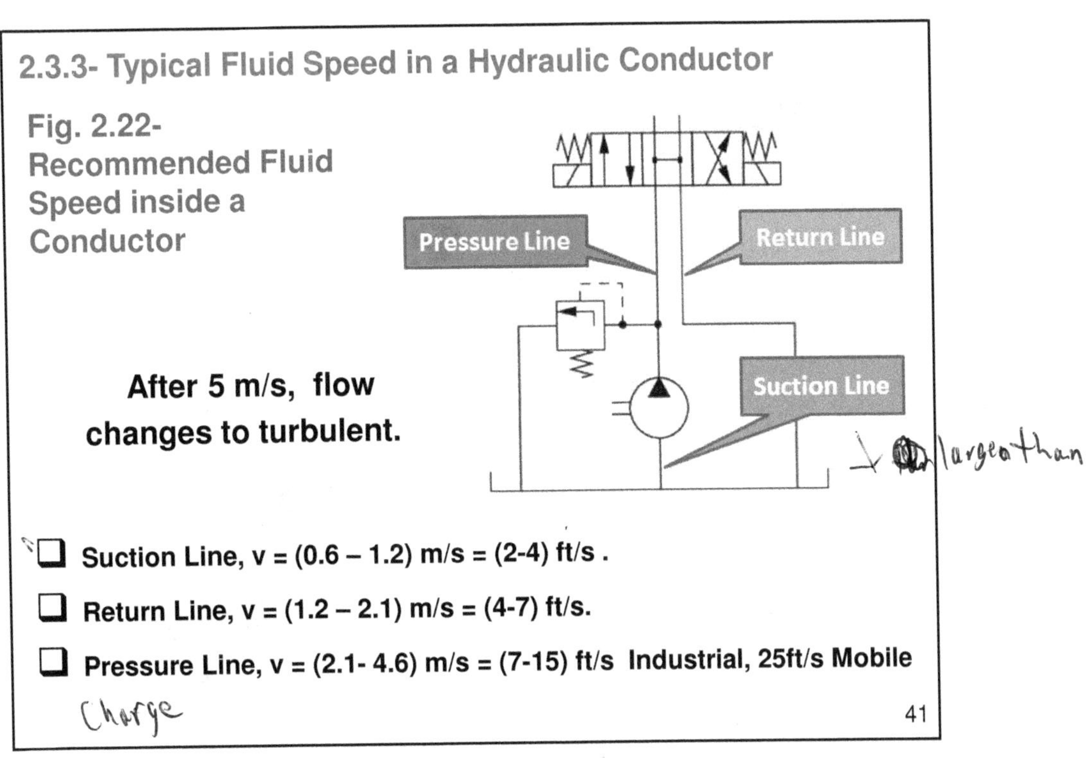
2.3.3- Typical Fluid Speed in a Hydraulic Conductor
Fig. 2.22- Recommended Fluid Speed inside a Conductor
Pressure Line
Return Line
Suction Line
After 5 m/s, flow changes to turbulent.
Suction Line, v = (0.6 – 1.2) m/s = (2-4) ft/s .
Return Line, v = (1.2 – 2.1) m/s = (4-7) ft/s.
Pressure Line, v = (2.1- 4.6) m/s = (7-15) ft/s Industrial, 25ft/s Mobile
41

2.3.4- Conservation of Mass in Fluid Flow
2.3.4.1- Compressible Flow

Bulk Modulus - Spring Stiffness - Modulus of Elasticity

$$B = -\frac{\Delta p}{\Delta V / V} \qquad 2.16$$

An increase of 70 bar (1000 psi) causes 0.5% decrease in oil volume.

Typical values of bulk modulus at ambient conditions:

- **220,000 psi (≈ 15,000 bars) for petroleum fluid.**
- **300,000 psi (≈ 20,000 bars) for water.**

42

1. High amounts of fluid (press)
2_amount of entrapped air
1 psi could be ?

higher # means less compressible

What are the applications in which the fluid compressibility effect will be more noticeable ?

Large oil volume

Precise dynamic motion Control

High Operating Pressure

High Operating Temperature

Fig. 2.23- Examples where Fluid Compressibility are Considered

43

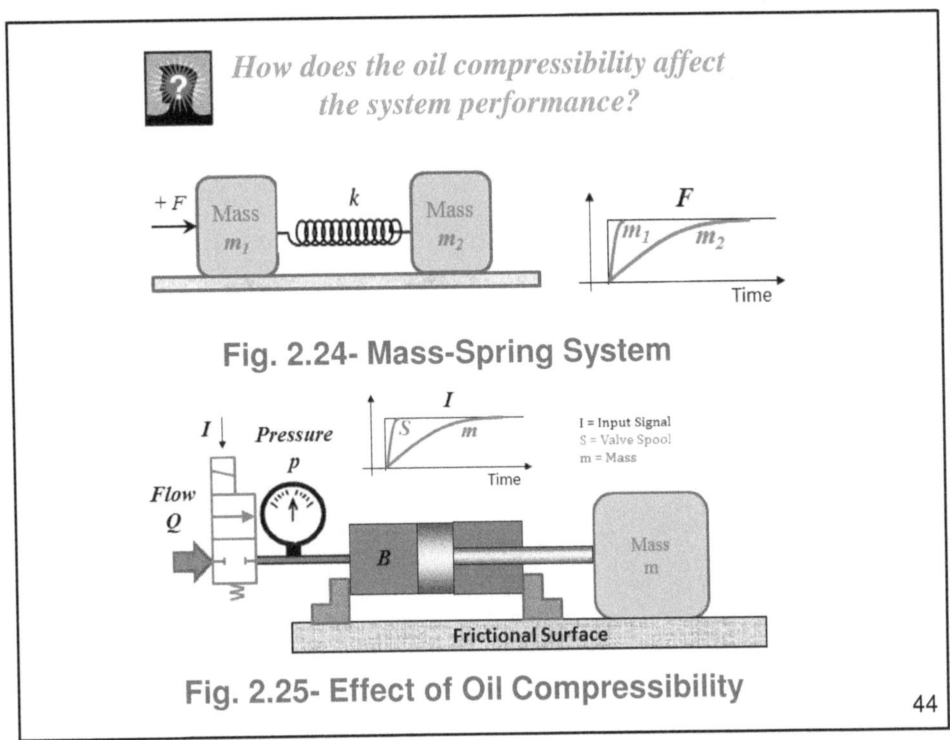

Fig. 2.24- Mass-Spring System

Fig. 2.25- Effect of Oil Compressibility

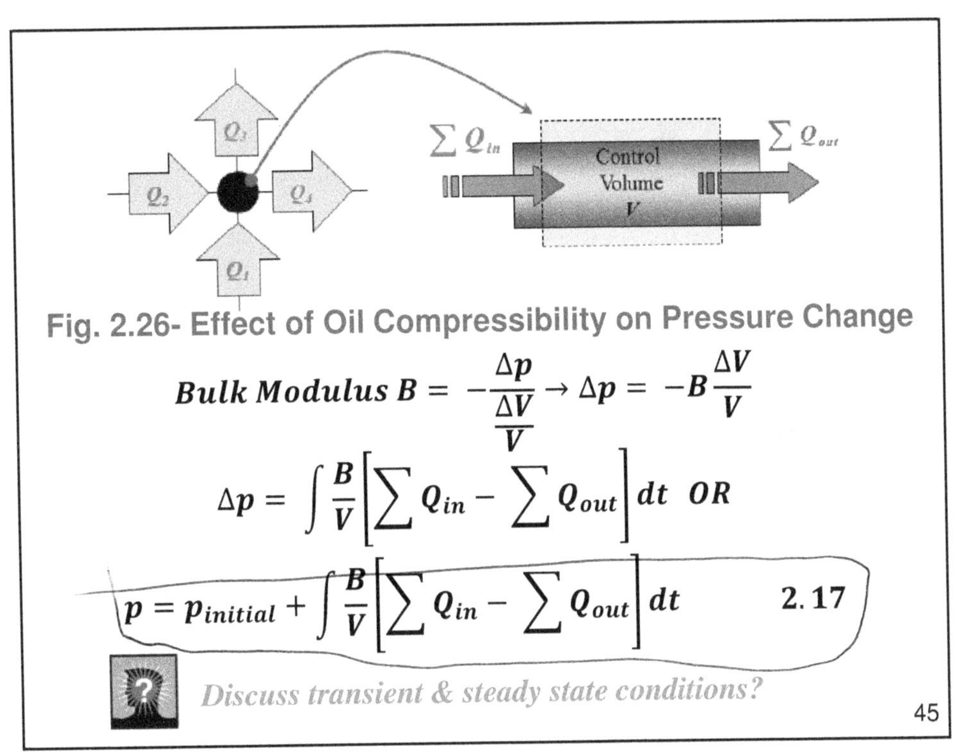

Fig. 2.26- Effect of Oil Compressibility on Pressure Change

$$\text{Bulk Modulus } B = -\frac{\Delta p}{\frac{\Delta V}{V}} \rightarrow \Delta p = -B\frac{\Delta V}{V}$$

$$\Delta p = \int \frac{B}{V}\left[\sum Q_{in} - \sum Q_{out}\right] dt \quad OR$$

$$p = p_{initial} + \int \frac{B}{V}\left[\sum Q_{in} - \sum Q_{out}\right] dt \qquad 2.17$$

Discuss transient & steady state conditions?

45

2.3.4.2- Incompressible Flow

$$Eq. 2.17 \rightarrow \quad \sum Q_{in} - \sum Q_{out} = \frac{V}{B} \times \frac{dp}{dt} \qquad 2.18$$

**Incompressible Flow → Infinite Bulk Modulus
→ Continuity Equation**

$$\sum Q_{in} - \sum Q_{out} = 0$$

$$OR \quad \sum Q_{in} = \sum Q_{out} \qquad 2.19$$

46

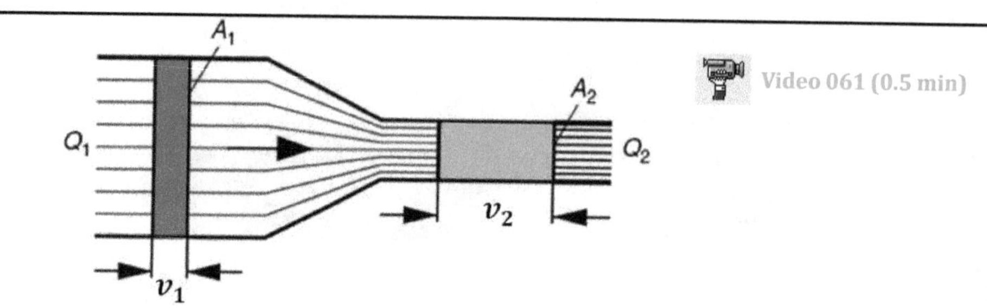

Video 061 (0.5 min)

$$Q_1 = Q_2 \rightarrow A_1 v_1 = A_2 v_2 \;\; \& \quad A \uparrow \rightarrow v \downarrow, \qquad A \downarrow \rightarrow v \uparrow$$

Fig. 2.27- Continuity Equation

$$v\,[cm/s] = \frac{1000\, Q[l/min]}{60 \times A[cm^2]} \qquad 2.20.A$$

$$v\,[fps] = \frac{0.321\, Q[gpm]}{A[in^2]} \qquad 2.20.B$$

47

Flow $Q = V \cdot A$

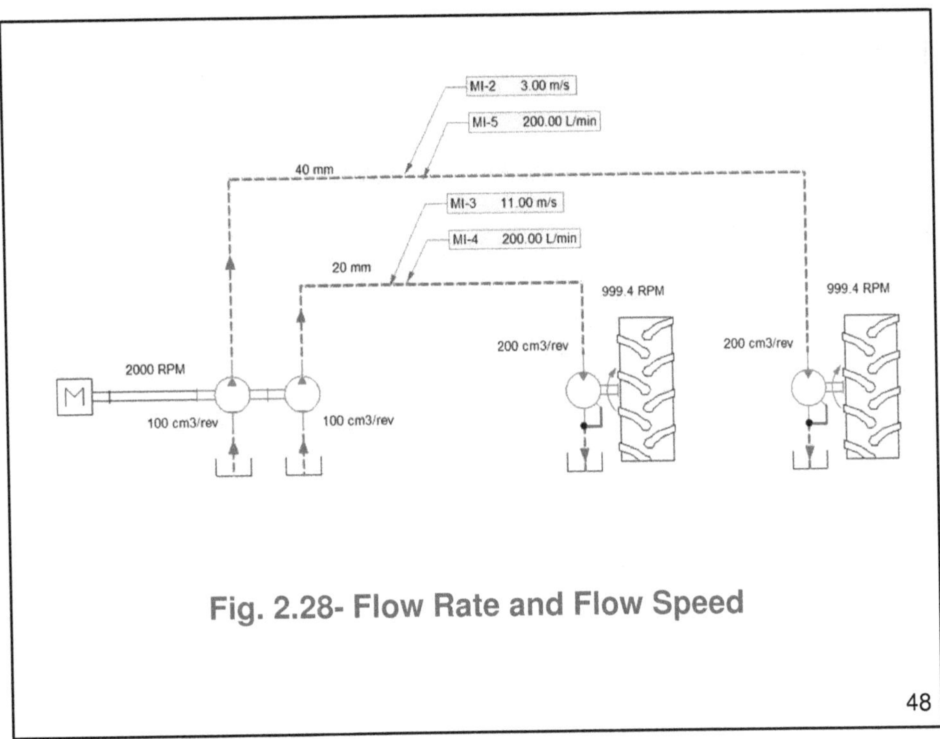

Fig. 2.28- Flow Rate and Flow Speed

48

2.3.5- Conservation of Energy in Fluid Flow
2.3.5.1- Bernoulli Equation for Ideal Fluid Flow

Pressure Energy = pV
Kinetic Energy = $mv^2/2$
Potential Energy = mgz

Divide by unit weight:

Pressure Head = $p / \rho g$
Kinetic Head = $v^2 / 2g$
Potential Head = z

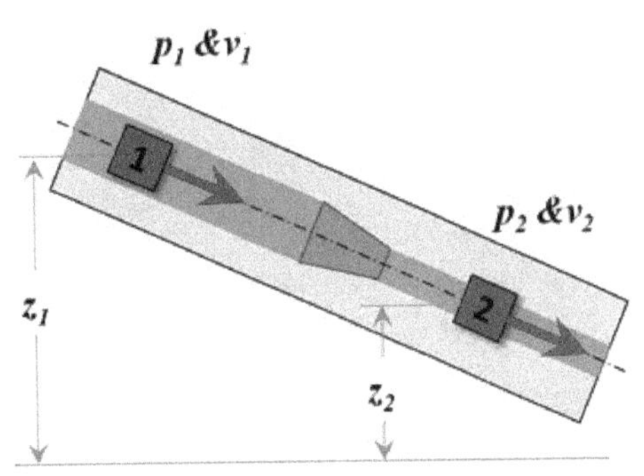

Fig. 2.29- Ideal Fluid Flow

$$\left(\frac{p}{\rho g} + \frac{v^2}{2g} + z\right)_1 = \left(\frac{p}{\rho g} + \frac{v^2}{2g} + z\right)_2 = Constant \qquad 2.21$$

49

First special case:

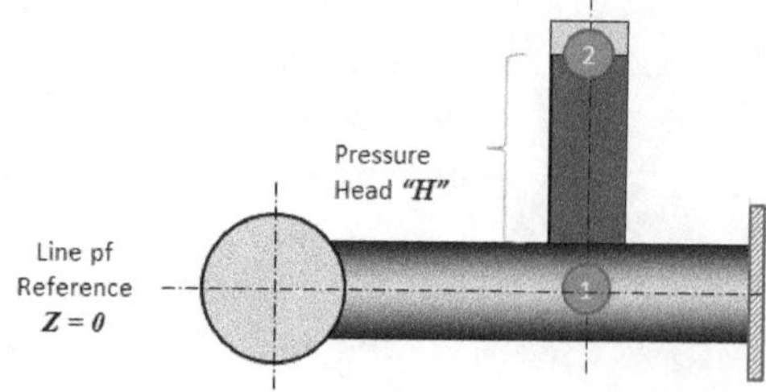

Fig. 2.30- Pressure Head

$$\Delta p = p_1 - p_2 = \rho g\,(z_2 - z_1) = \rho g H \qquad 2.22.A$$

Since the kinetic head is the same at points 1 and 2, the pressure head is equivalent to the potential head.

50

Second special case:

For which line in the system is this observation important?

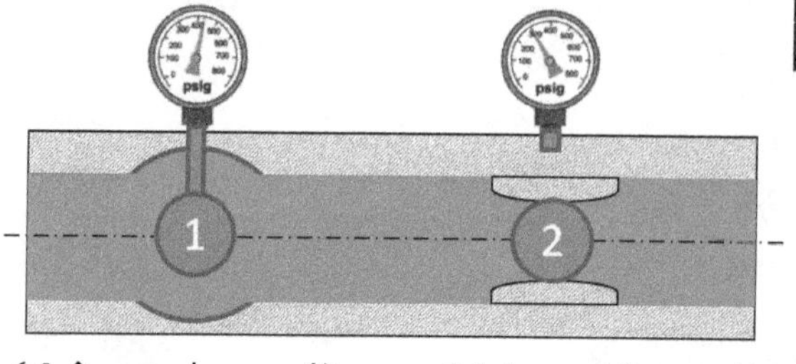

$(A \uparrow \rightarrow v \downarrow \rightarrow p\ \uparrow)_1 \qquad (A \downarrow \rightarrow v \uparrow \rightarrow p \downarrow)_2$

Fig. 2.31- Kinetic Head

$$\Delta p = p_1 - p_2 = \frac{\rho}{2}\,(v_2^2 - v_1^2) \qquad 2.22.B$$

Since the potential head is the same at point 1 and 2, the reduction in the pressure head is equivalent to the increase in the kinetic head.

51

Third special case:

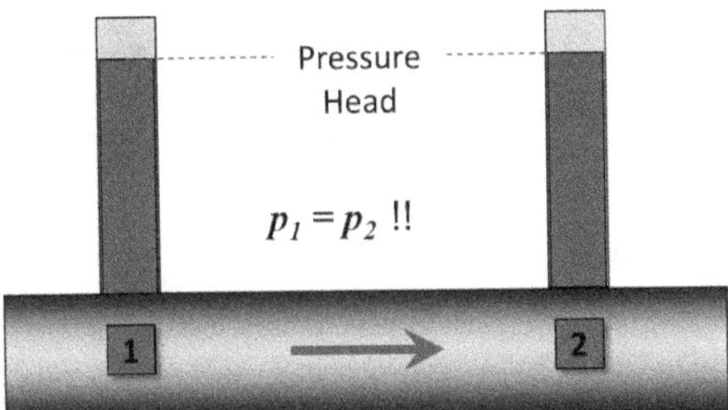

Fig. 2.32- Practical Observation of Continuity Equation and Bernoulli's Equation

$$\textit{Continuity Equation} \rightarrow v_1 = v_2$$
$$\textit{Bernoulli's Equation} \rightarrow p_1 = p_2 \text{ !!}$$

Is this situation practically correct?, If Not, what is missing ?

52

2.3.5.2- Bernoulli Equation for Real Fluid Flow

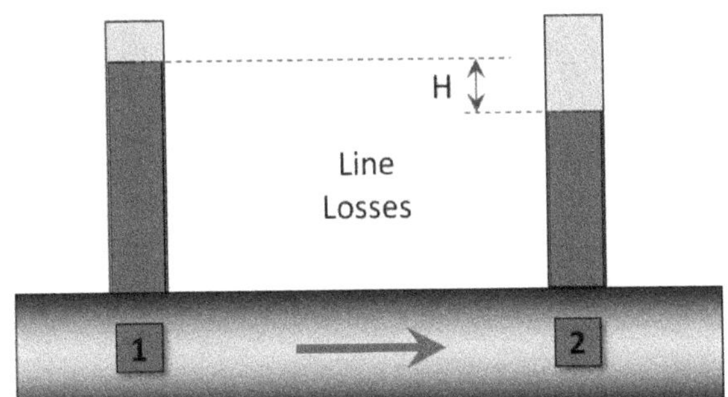

Fig. 2.33- Real Fluid Flow

$$\left(\frac{p}{\rho g} + \frac{v^2}{2g} + z\right)_1 - \left(\frac{p}{\rho g} + \frac{v^2}{2g} + z\right)_2 = H_{1\rightarrow 2} \qquad 2.23.A$$

$$\Delta p = p_1 - p_2 = \rho g H_{1\rightarrow 2} \qquad 2.23.B$$

53

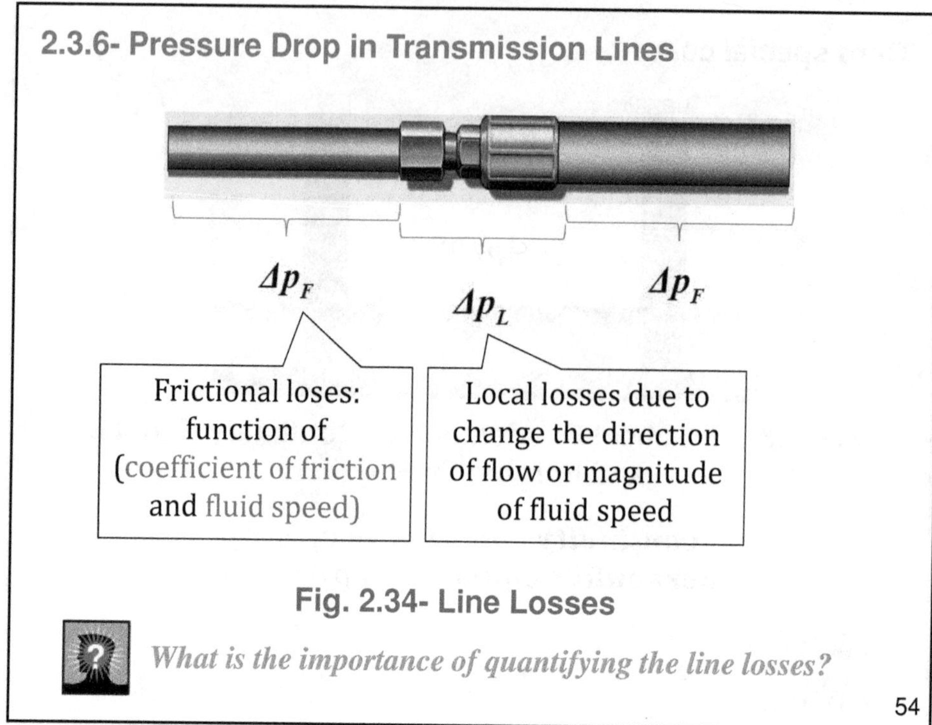

Fig. 2.34- Line Losses

2.3.6.1 – Frictional Pressure Losses in Hydraulic Lines

$$\Delta p_F = p_1 - p_2 = \rho g h_{L1\rightarrow 2} = \rho g \lambda \frac{L}{D}\frac{v^2}{2g} = \lambda \frac{L}{D}\frac{\rho v^2}{2g} \quad 2.24$$

- **Dimensional parameters *(D & L)***
- **Fluid property *(ρ)***
- **Operational conditions *(λ & v)***

Pressure drop is a combination of frictional losses and kinetic losses

In other words, pressure drop in a transmission line depends on the friction (fluid-to-fluid and fluid-to-wall) and how fast the fluid moves in the line.

$$\Delta p_F(bar) = \frac{2254 \times \lambda \times L(m) \times SG \times [Q(\frac{lit}{min})]^2}{[D(mm)]^5} \quad 2.25.A$$

$$\Delta p_F(psi) = \frac{\lambda \times L(ft) \times SG \times [Q(gpm)]^2}{74.3 \times [D(in)]^5} \quad 2.25.b$$

55

In case of Laminar flow:

$$\lambda = \frac{64}{R_e} \qquad 2.25$$

$$\Rightarrow \Delta p_F = \frac{128\,\mu\,L}{\pi\,D^4}\,Q = RQ \qquad 2.26$$

"Linear Relationship"

In case of Turbulent flow & Smooth Pipe:

$$\lambda = \frac{0.3164}{R_e^{0.25}} \qquad 2.27$$

In all cases Moody Diagram can be used.

56

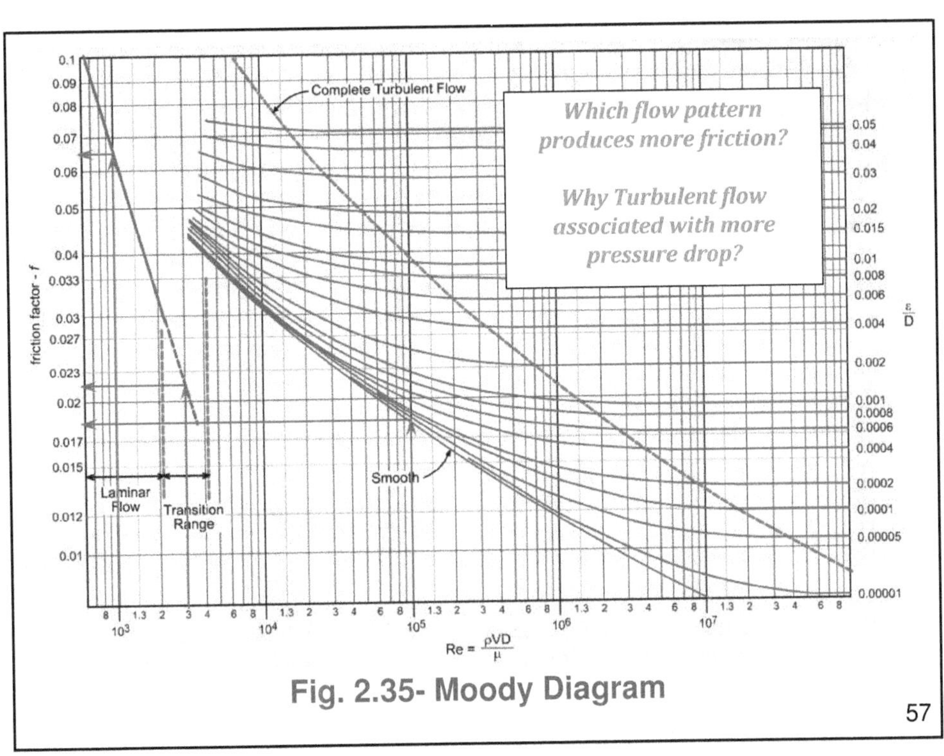

Fig. 2.35- Moody Diagram

57

Example 1 (Frictional Losses Metric System of Units):

Given: A flow in a smooth pipe has the following variables.

$$Q\left(\frac{lit}{min}\right) = 76, \quad \nu\,(cSt) = 64, \quad D(mm) = 25$$

$$, \quad SG = 1.055 \quad and\ L\,(m) = 3.$$

Solution:

$$Eq.2.15.A \rightarrow R_e = \frac{21231 \times 76}{64 \times 25} = 1009$$

$$Laminar\ Flow \rightarrow Moody\ Diagram \rightarrow \lambda \approx 0.06$$

$$Eq.2.25.A \rightarrow \Delta p_F(bar) = \frac{2245 \times 0.06 \times 3 \times 1.055 \times 76^2}{25^5}$$
$$= 0.27$$

58

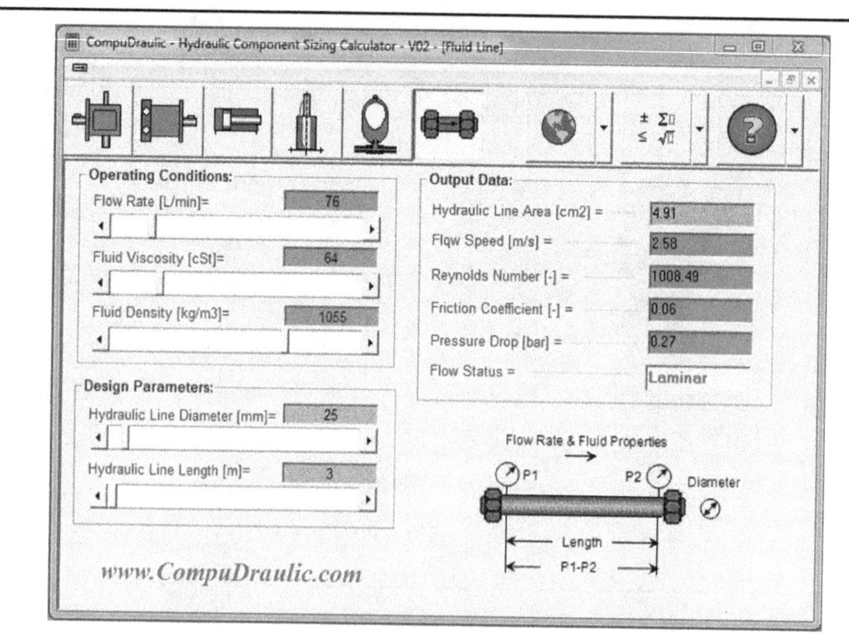

Fig. 2.36- Frictional Pressure Losses Calculation in Metric System of Units

59

Example 2 (Frictional Losses in English System of Units):

Given: A flow in a smooth pipe has the following variables.

$$Q\,(gpm) = 20.08, \quad \nu\,(cSt) = 64, \quad D(in) = 0.98, \quad SG = 1.055 \quad and\; L(ft) = 9.84.$$

Solution:

$$Eq. 2.15. A \rightarrow R_e = \frac{3164 \times 20.08}{64 \times 0.98} = 1013$$

$$Laminar\ Flow \rightarrow Moody\ Diagram \rightarrow \lambda \approx 0.06$$

$$Eq. 2.24. B \rightarrow \Delta p_F(psi) = \frac{0.06 \times 9.84 \times 1.055 \times 20.08^2}{74.3 \times 0.98^5} = 3.82$$

60

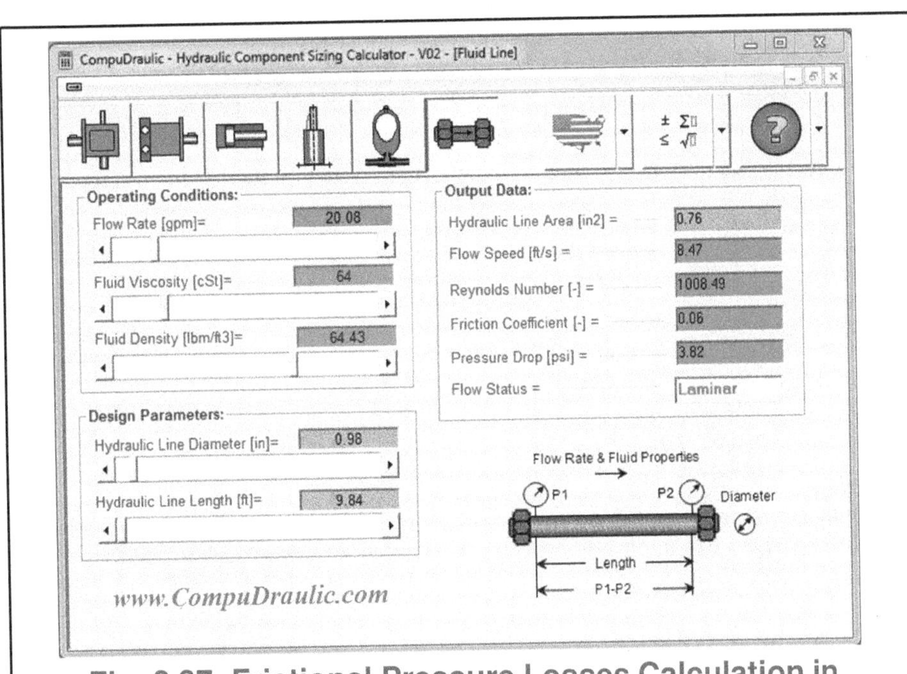

Fig. 2.37- Frictional Pressure Losses Calculation in English System of Units

61

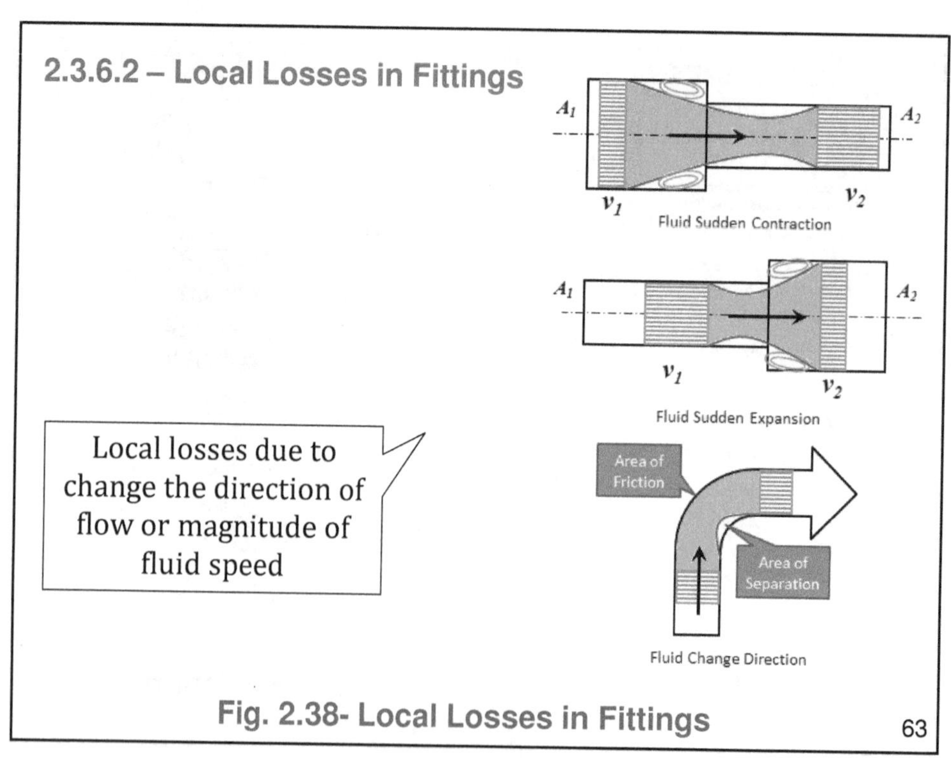

Fig. 2.38- Local Losses in Fittings

$$\Delta p_L = p_{in} - p_{out} = K\,\frac{\rho v^2}{2} \qquad 2.28$$

$$\Delta p_L(bar) = \frac{2.254 \times K \times SG \times [Q(\frac{Lit}{min})]^2}{[D(mm)]^4} \qquad 2.29.A$$

$$\Delta p_L(psi) = \frac{K \times SG \times [Q(gpm)]^2}{891.3 \times [D(in)]^4} \qquad 2.29.B$$

64

Fitting	90° bend	90° angle	T-Piece	Double angle	Valve
K	0.5 - 1	1.2	1.3	2	5 - 15
Sudden Contraction	Well-Rounded	Slightly Rounded	Sharp Edged	Shoulder	
K	0.04	0.2	0.5	0.8	
Sudden Expansion	Well-Rounded	Slightly Rounded	Sharp Edged	Shoulder	
K	1	1	1	1	

Table 2.3 – Losses factor in Fittings

65

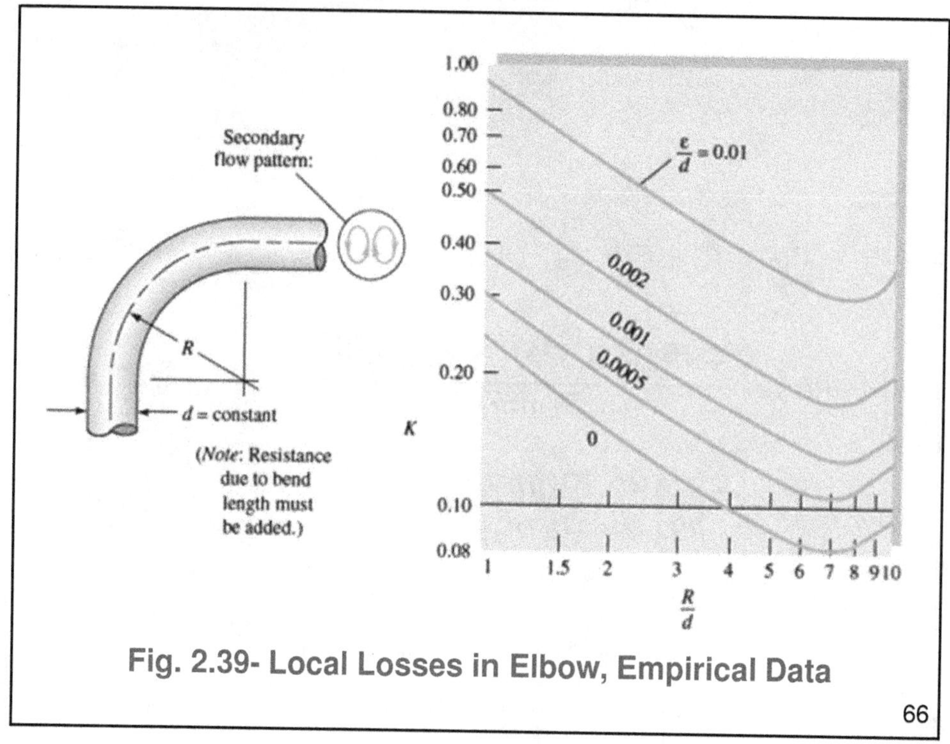

Fig. 2.39- Local Losses in Elbow, Empirical Data

66

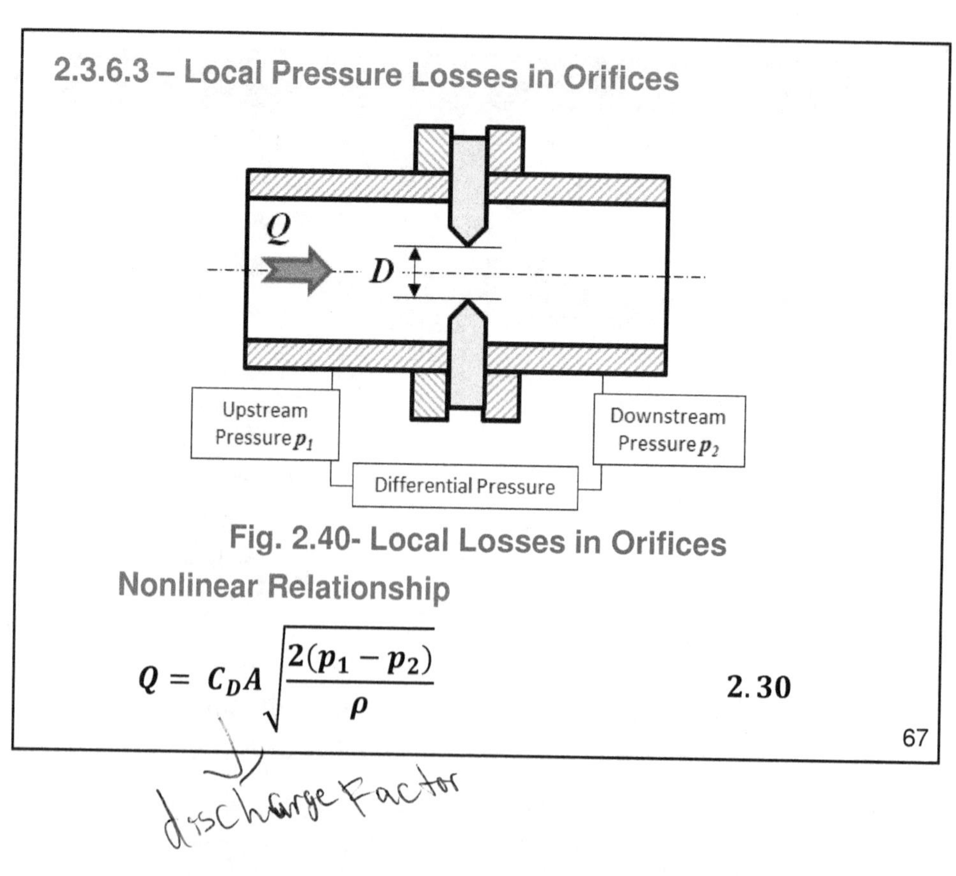

2.3.6.3 – Local Pressure Losses in Orifices

Fig. 2.40- Local Losses in Orifices

Nonlinear Relationship

$$Q = C_D A \sqrt{\frac{2(p_1 - p_2)}{\rho}} \qquad 2.30$$

67

discharge Factor

Throttling Coefficient

$$k_{th} = \frac{\Delta p}{Q^2} = \frac{\rho}{2C_D^2 A^2} \qquad 2.31$$

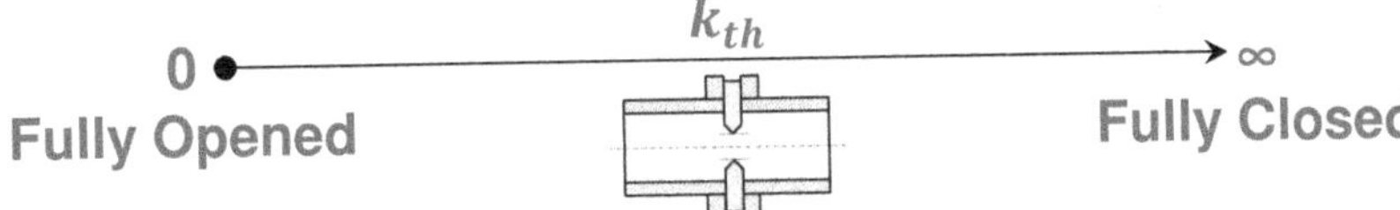

$$k_{th}\left[\frac{bar}{(Lit/min)^2}\right] = \frac{2.254 \times SG}{C_D^2[D(mm)]^4} \qquad 2.32.A$$

$$k_{th}\left[\frac{psi}{(gpm)^2}\right] = \frac{5.34 \times 10^3 \times SG}{C_D^2[D(in)]^4} \qquad 2.32.b$$

68

2.3.6.4 – Local Pressure Losses in Valves

Valve Coefficient

$$Q = C_v\sqrt{\frac{\Delta p}{SG}} \rightarrow C_v = Q\sqrt{\frac{SG}{\Delta p}} \qquad 2.33$$

69

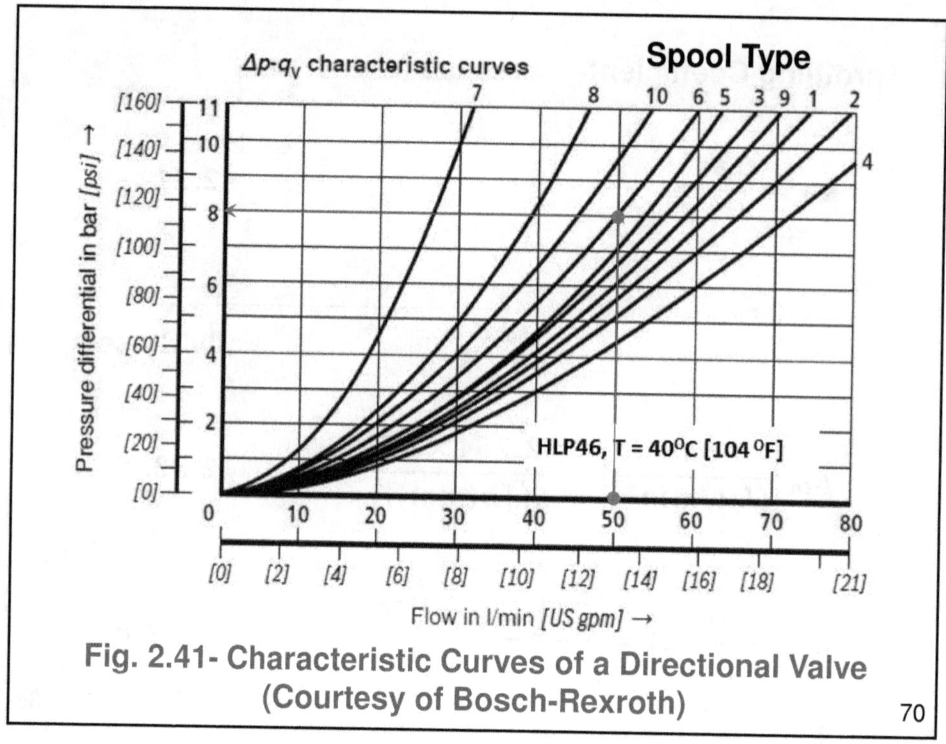

Fig. 2.41- Characteristic Curves of a Directional Valve (Courtesy of Bosch-Rexroth)

70

Viscosity %	75	100	150	200	300	400
Correction Factor	0.93	1	1.11	1.19	1.32	1.41

Table 2.4- Correction Factor

71

Q=V~A

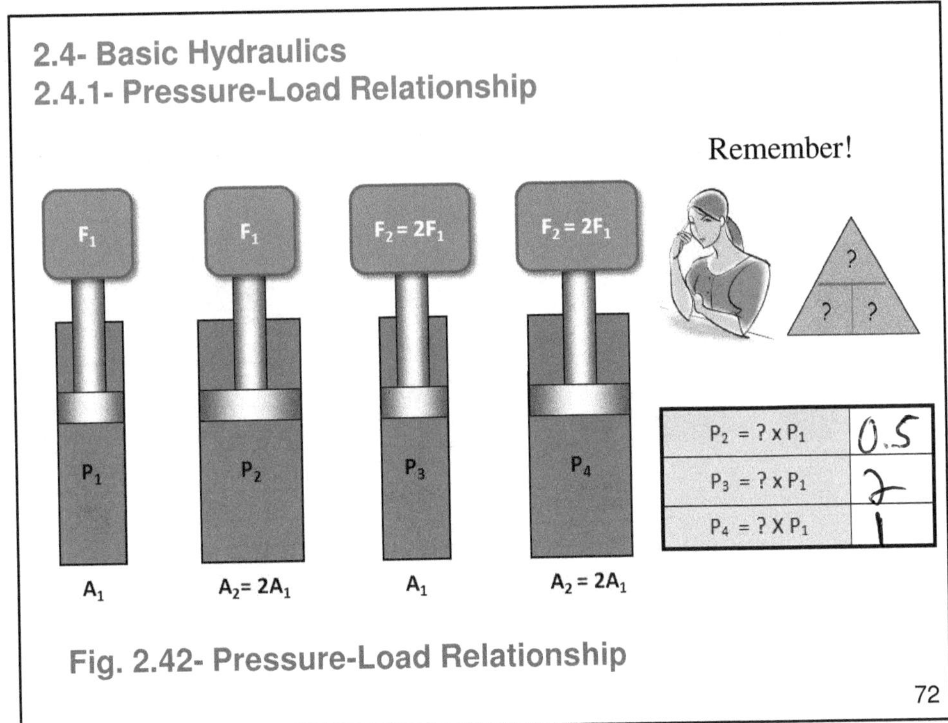

Fig. 2.42- Pressure-Load Relationship

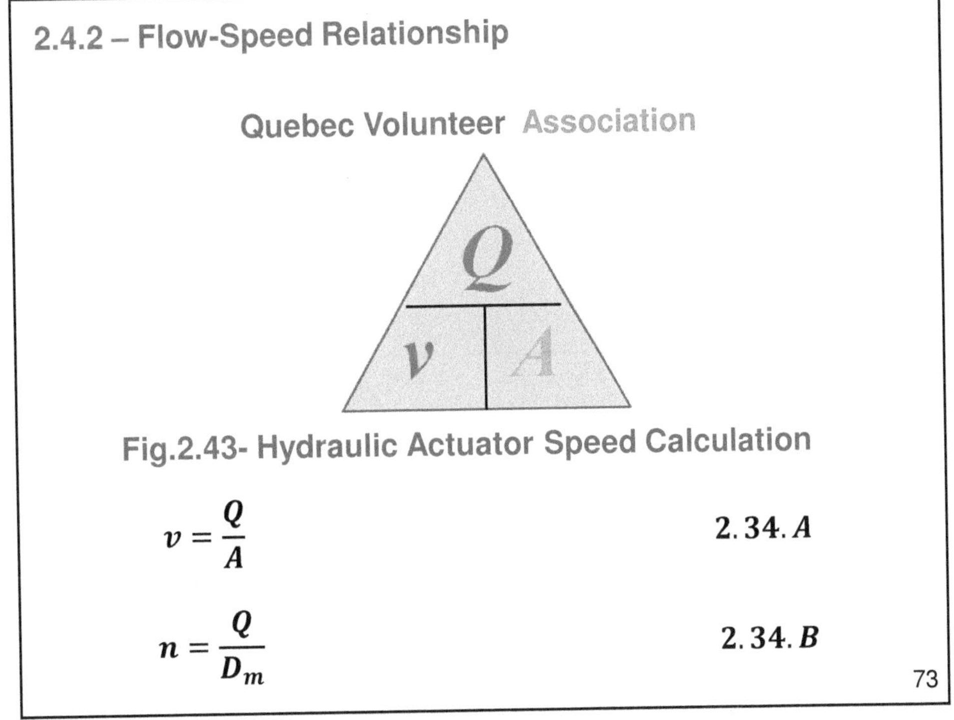

Fig.2.43- Hydraulic Actuator Speed Calculation

$$v = \frac{Q}{A} \qquad 2.34.A$$

$$n = \frac{Q}{D_m} \qquad 2.34.B$$

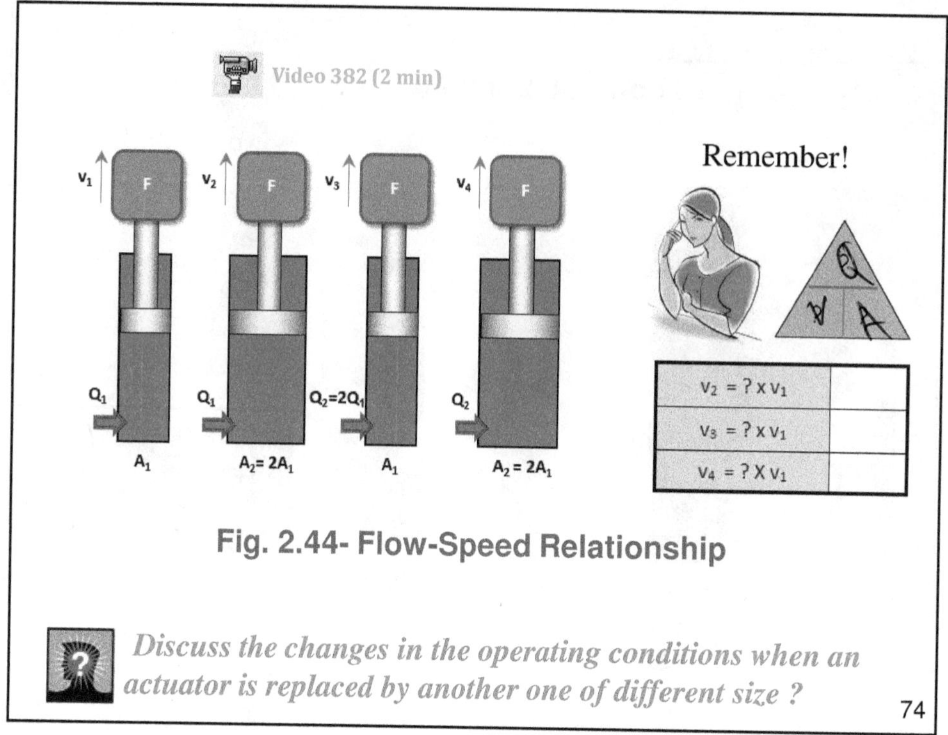

Fig. 2.44- Flow-Speed Relationship

Discuss the changes in the operating conditions when an actuator is replaced by another one of different size ?

74

2.4.3 – Force Intensifiers

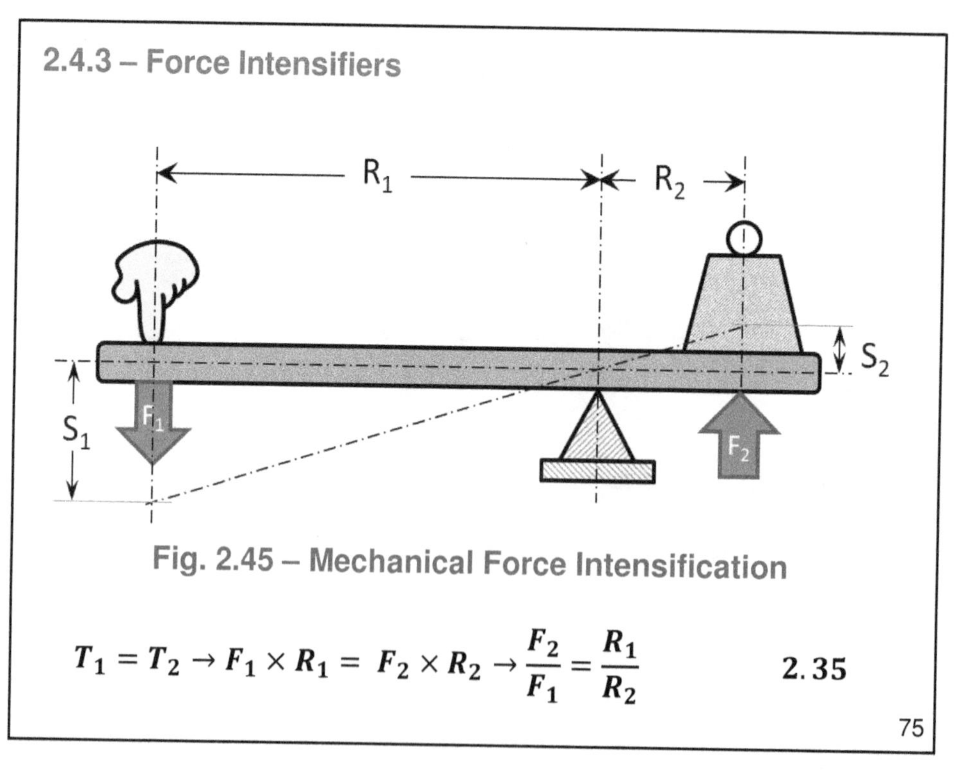

Fig. 2.45 – Mechanical Force Intensification

$$T_1 = T_2 \rightarrow F_1 \times R_1 = F_2 \times R_2 \rightarrow \frac{F_2}{F_1} = \frac{R_1}{R_2} \quad 2.35$$

75

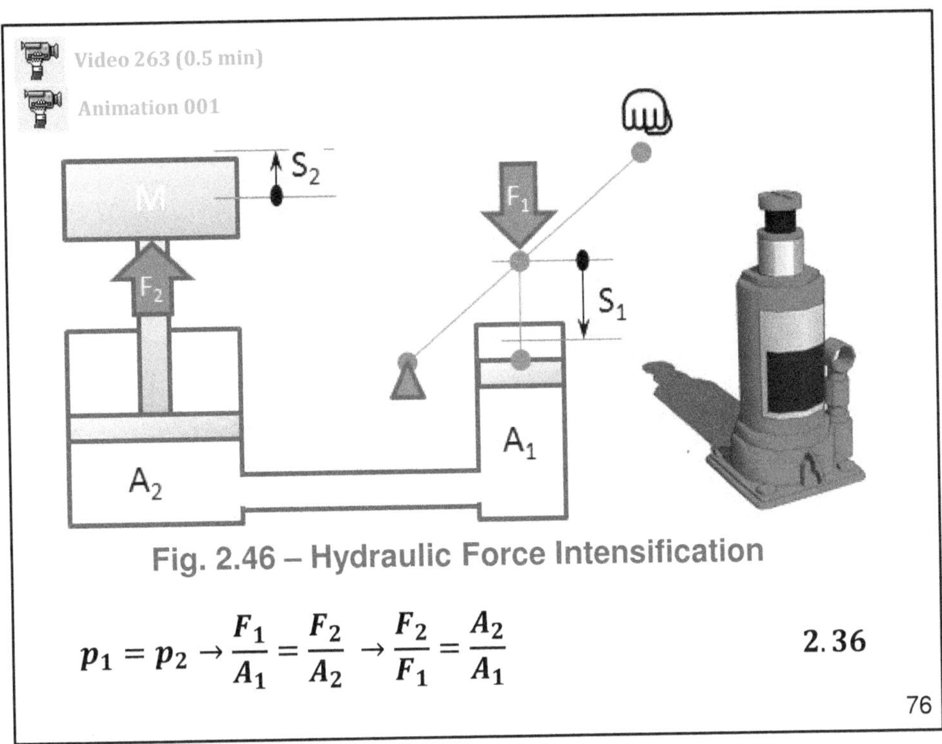

Fig. 2.46 – Hydraulic Force Intensification

$$p_1 = p_2 \rightarrow \frac{F_1}{A_1} = \frac{F_2}{A_2} \rightarrow \frac{F_2}{F_1} = \frac{A_2}{A_1} \qquad 2.36$$

76

2.4.4 – Pressure Intensifiers

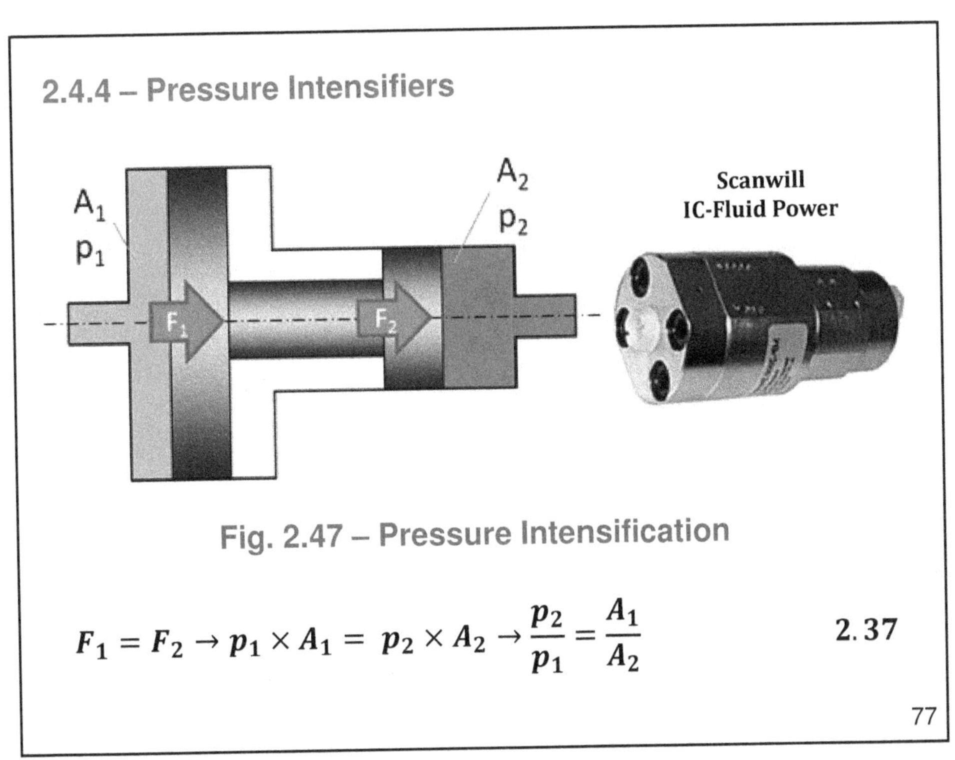

Fig. 2.47 – Pressure Intensification

$$F_1 = F_2 \rightarrow p_1 \times A_1 = p_2 \times A_2 \rightarrow \frac{p_2}{p_1} = \frac{A_1}{A_2} \qquad 2.37$$

77

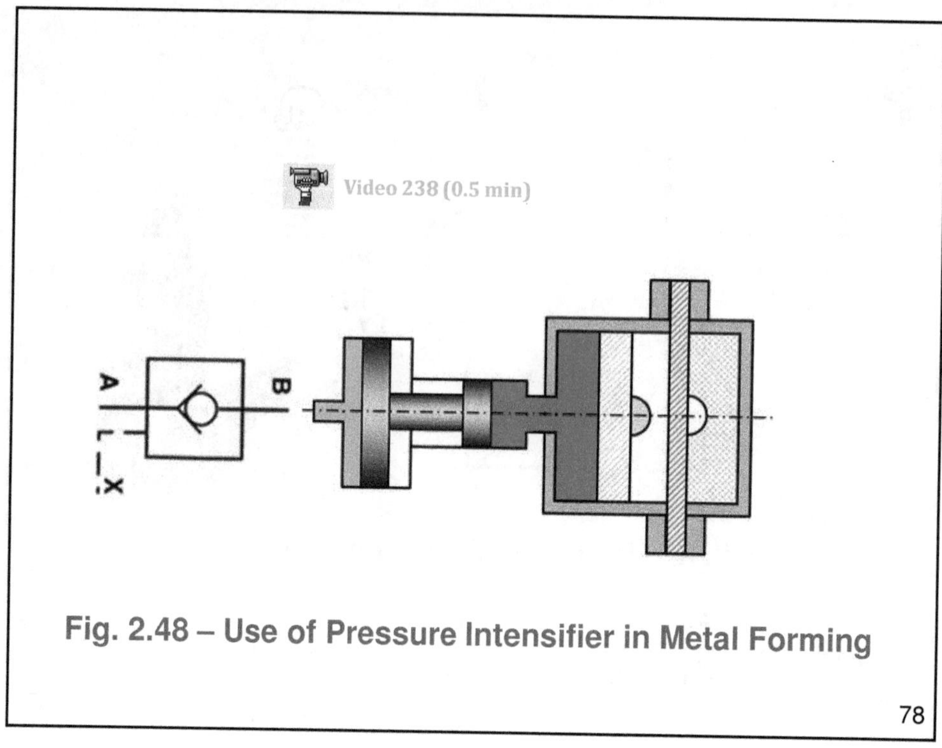

Fig. 2.48 – Use of Pressure Intensifier in Metal Forming

78

2.4.5 – Series versus Parallel Resistance

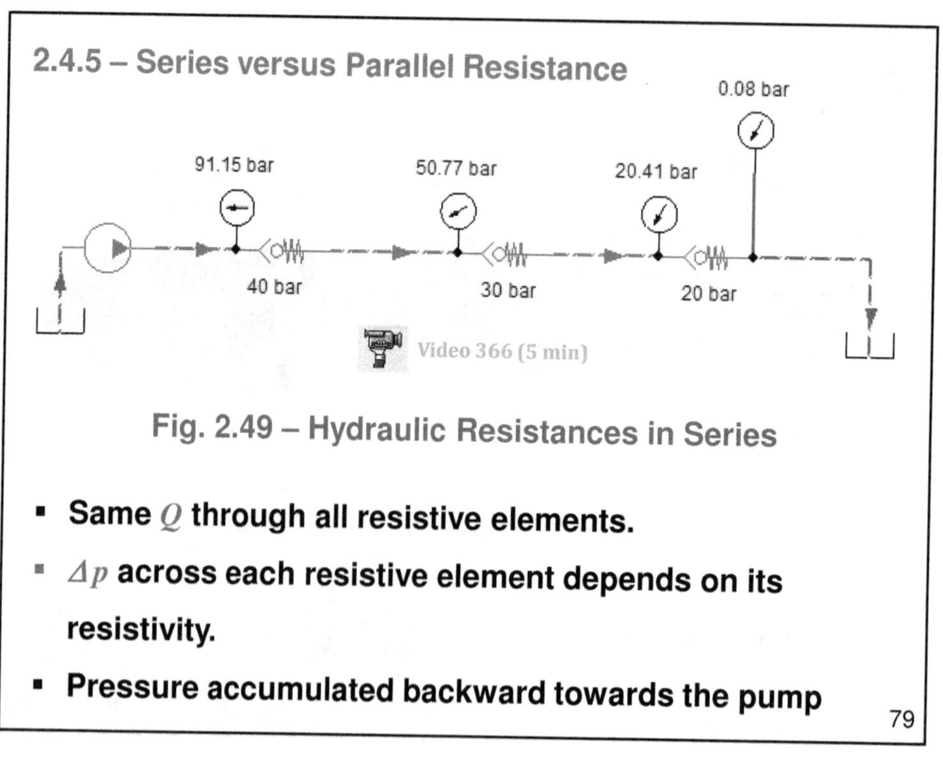

Fig. 2.49 – Hydraulic Resistances in Series

- **Same Q through all resistive elements.**
- **Δp across each resistive element depends on its resistivity.**
- **Pressure accumulated backward towards the pump**

79

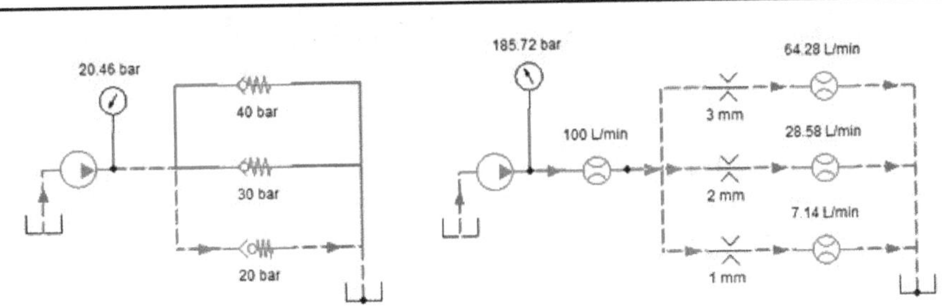

Fig. 2.50 – Hydraulic Resistances in Parallel

- Δp **across all resistive elements will be the same.**
- Q **is distributed naturally based on the resistivity.**
- **Highest flow passes through the least resistance**
- **If the least resistant passage is able to swallow the full flow at this differential pressure, there will be no flow left for the other passages.**

80

Chapter 2 Reviews

1. A certain cylinder is being extended at a speed of 12 in/s against a specific external resistive load using a prime mover that has a power of 20 HP. If the external load is doubled and the cylinder speed is reduced to 6 in/s, what **prime mover power** is required to perform the new job?
   A. 20 HP.
   B. 10 HP.
   C. 30 HP.
   D. 40 HP.

2. Which statement represents the correct match between the shown below component's names and the component's **basic functions**?

   A. (I=1, II=2, III=3, IV=4)
   B. (I=2, II=3, III=4, IV=1)
   C. (I=1, II=3, III=2, IV=4)
   D. (I=1, II=4, III=3, IV=2)

   1- Pump.
   2- Accumulator.
   3- Pressure Relief Valve.
   4- Cylinder.

   I- Hydraulic power generation.
   II-System overload protection.
   III- Hydraulic power storage.
   IV- Converting hydraulic power to mechanical power.
   ----

3. If pressure is measured at the pump inlet port and found to be 10 psia, which of the following **pressure value and units** can be used to express the pump suction pressure?
   A. 4.7 psig.
   B. - 4.7 psig.
   C. 4.7 psid.
   D. 4.7 psia.

4. Which mathematical expression explains **Pascal's law?**
   A. Fluid pressure = Force x Speed.
   B. Fluid pressure = Force x Area on which the force is applied.
   C. Fluid pressure = Power ÷ Energy.
   D. Fluid pressure = Force ÷ Area on which the force is applied.

5. In the figure below, which point within the flow stream has the lowest **pressure?**

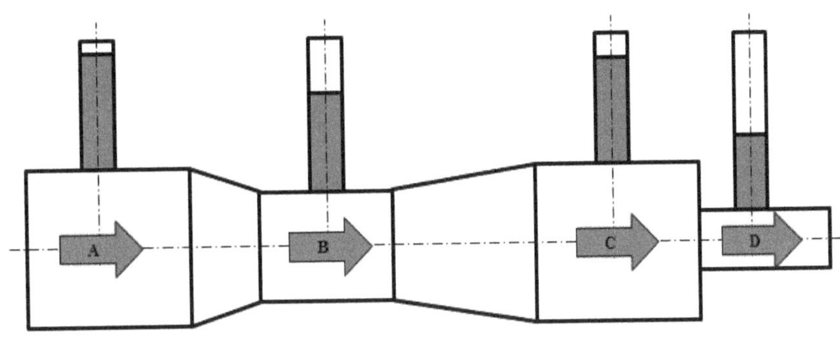

6. In the above figure, which point within the flow stream has the highest **fluid velocity?**

7. In the above shown figure, assume the cross-sectional area at point A equals twice the area at point B. If the flow at point A is 10 gpm, then flow at point B is?
   A. 5 gpm.
   B. 10 gpm.
   C. 15 gpm.
   D. 20 gpm.

8. A flow stream has a Reynolds Number of 6000, what is the **flow type?**
   A. Laminar.
   B. Transition.
   C. Turbulent.
   D. Vortex.

9. In a turbulent flow stream, what should be done to change it to **laminar flow?**
   A. Increase the diameter of the conductor.
   B. Reduce the flow rate.
   C. Increase the oil viscosity.
   D. All the above.

10. From your understanding of the **relation between an actuator's speed and its flowrate**, if a motor size is doubled and the supplied flow rate to the motor is also doubled, what would be the new motor's speed as compared to the original one?
   A. Doubled.
   B. Reduced to half.
   C. Reduced to quarter.
   D. None of the above.

Chapter 2 Assignment

Student Name: -- Student ID: ------------------

Date: -- Score: ------------------------

A hydraulic system has the following:

- Line pressure losses of 12 bar.
- Pump discharge is 15 liters/min.
- Kinematic viscosity 32 cSt.
- Smooth pipe is used.

Find:

- Power losses in the line (kW)?
- Minimum line diameter for laminar flow?
- Line length?

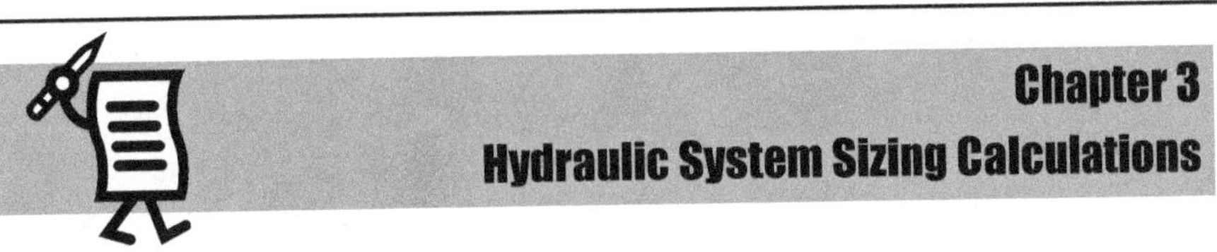

Objectives:
This chapter presents a case study to demonstrate the method to size a hydraulic system to achieve desired operating conditions. The chapter presents the step-by-step procedure to perform the calculations required to size the components in the system.

Brief Contents:
3.1 – Forward Path Power Losses in Hydraulic Systems
3.2 – Backward Path Power Accumulation in Hydraulic Systems.

0

3.1- Forward Path Power Losses in Hydraulic Systems

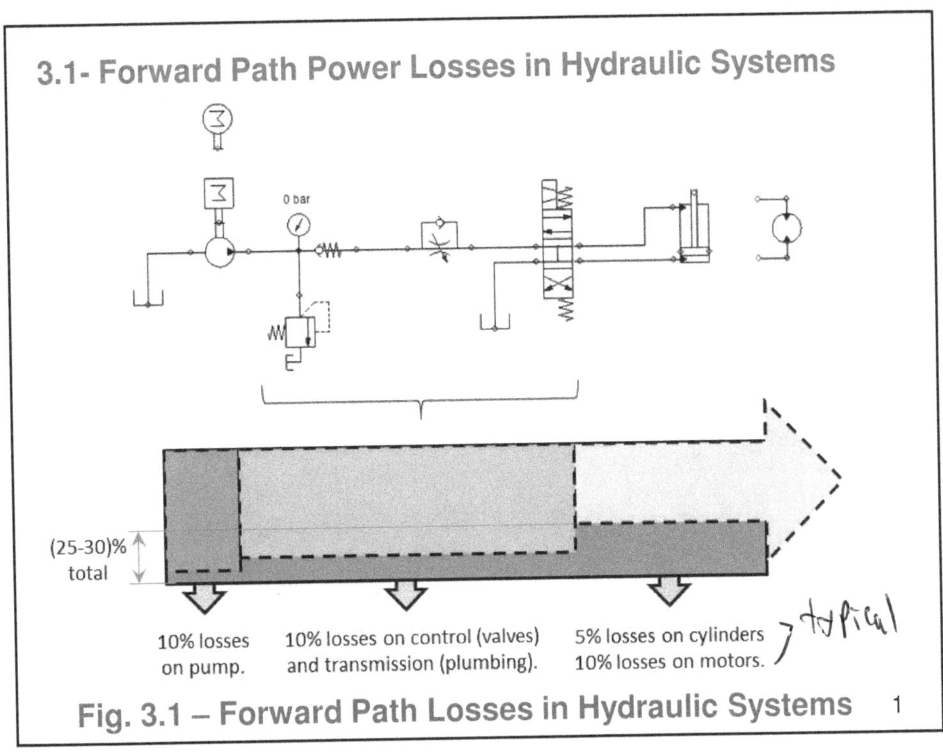

Fig. 3.1 – Forward Path Losses in Hydraulic Systems

1

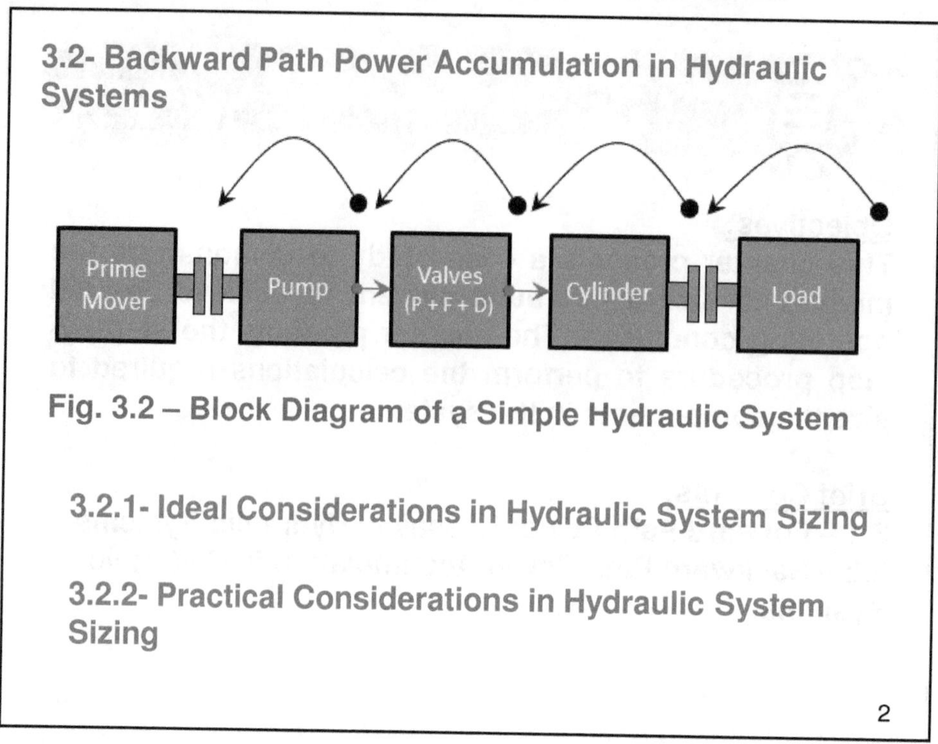

Fig. 3.2 – Block Diagram of a Simple Hydraulic System

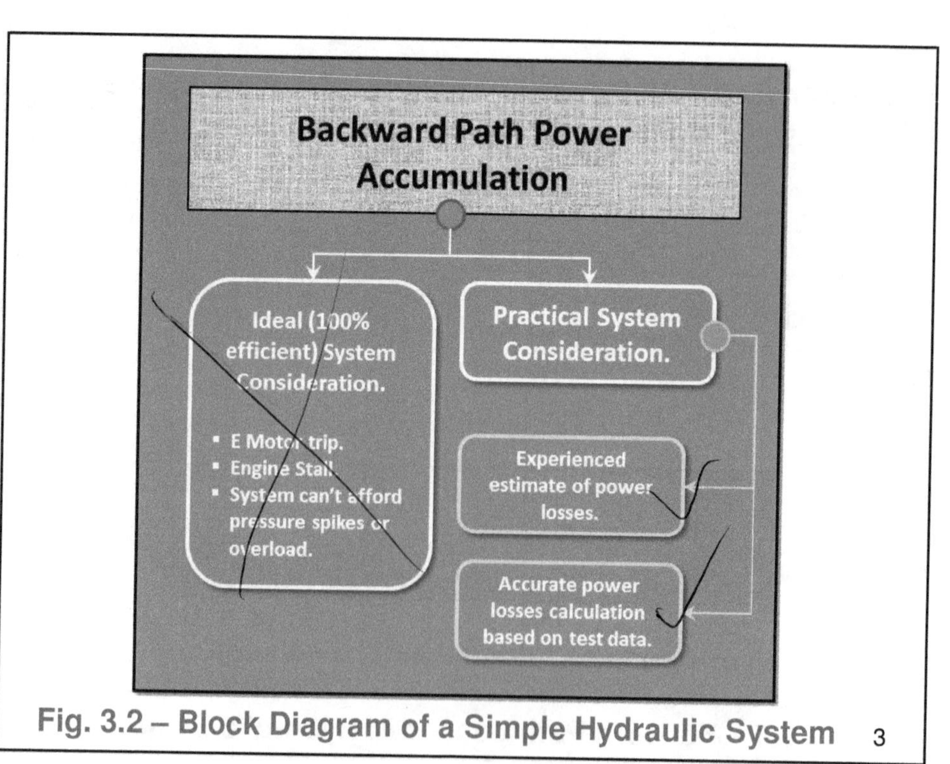

Fig. 3.2 – Block Diagram of a Simple Hydraulic System

Fig. 3.4 – Dump Truck

Space constraints #1 step

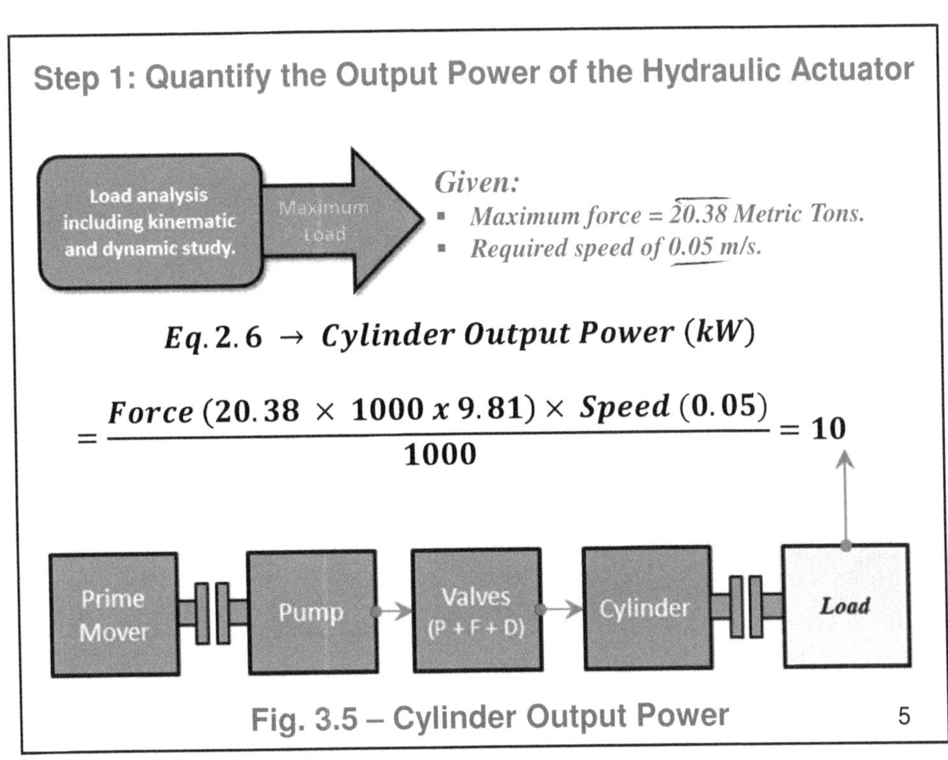

Fig. 3.5 – Cylinder Output Power

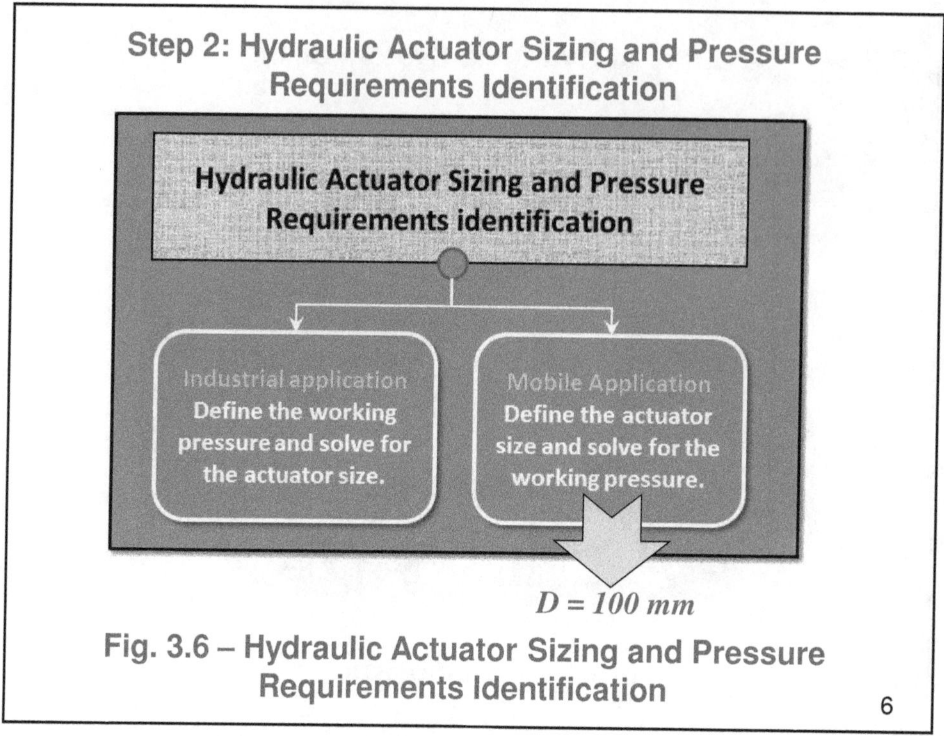

Fig. 3.6 – Hydraulic Actuator Sizing and Pressure Requirements Identification

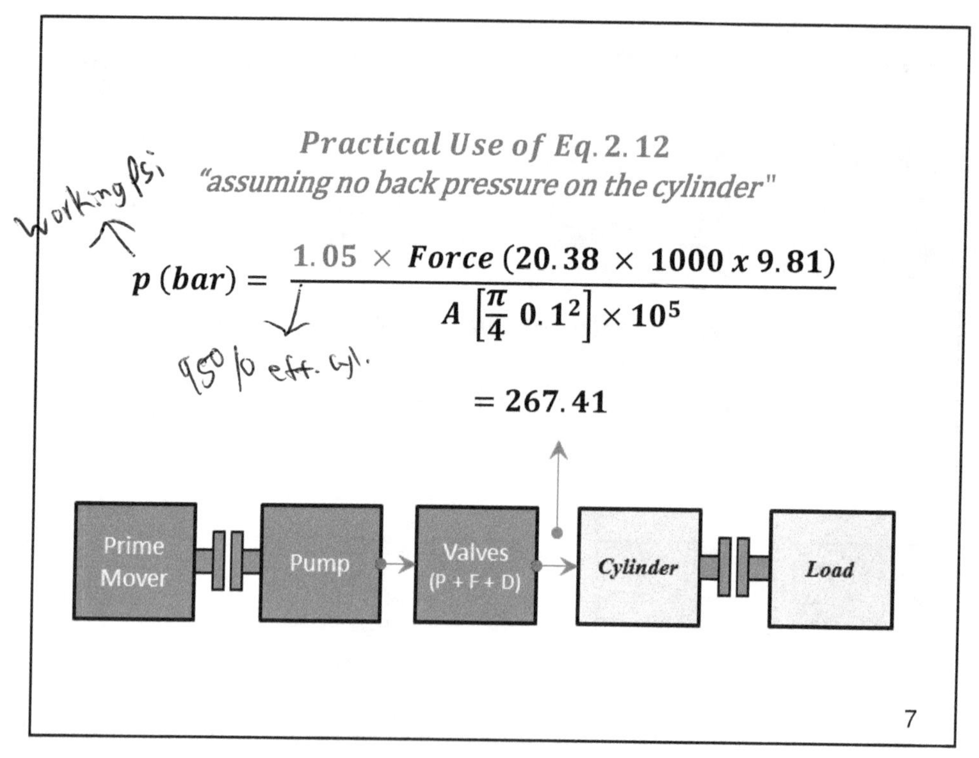

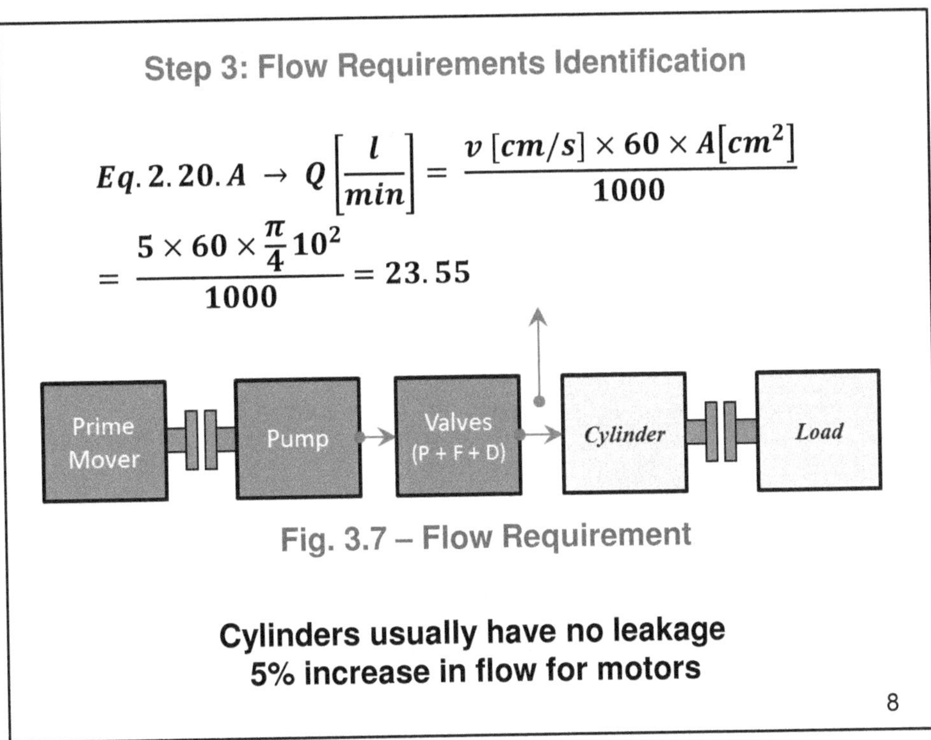

Fig. 3.7 – Flow Requirement

Step 4: Hydraulic Pump Sizing

Pump Size D_p = f (flow rate Q_p + driving speed n_p)
Equation will be posted in chapter 4

Assume a standard pump size was found to give the required flow (23.55 l/min) under reasonable <u>maximum</u> working pressure

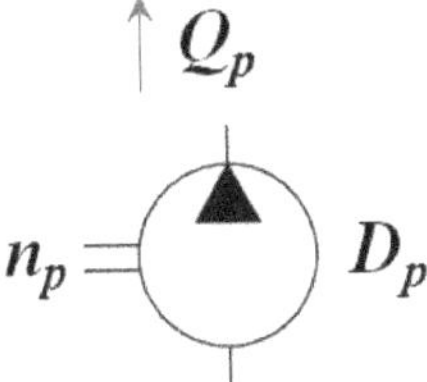

9

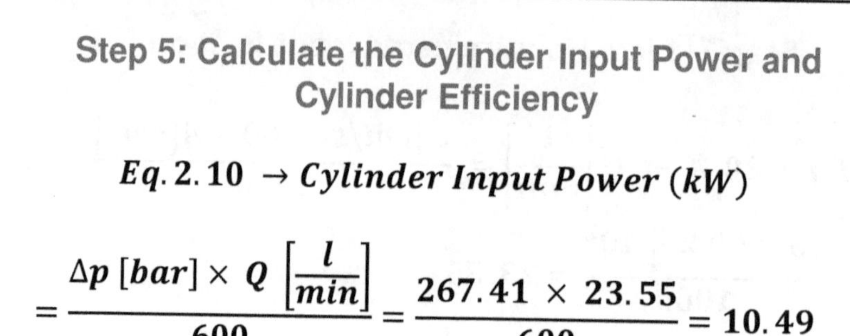

$$Cylinder\ Efficiency\ \frac{10}{10.49} = 0.95$$

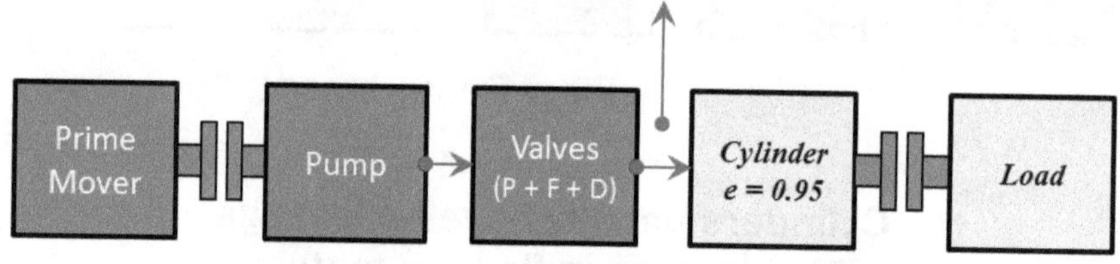

Fig. 3.8 – Flow Requirement

10

Step 6: Flow Distribution Analysis

flow distribution analysis <u>during all phases</u> of the machine operation.

Out of this analysis, maximum flow passes through each valve, hydraulic conductor and other components that will be defined.

Consider <u>differential</u> cylinders.

11

Step 7: Hydraulic Conductors Sizing and Line Losses Calculations

Assume fluid viscosity = 32 *Cst*
Assume Reynold's number = 2000

Minimum diameter of the conductor

$$Eq.2.15.A \rightarrow D[mm] = \frac{21231\, Q\left[\frac{l}{min}\right]}{\nu\,[Cst] \times R_e} = \frac{21231 \times 23.55}{32 \times 2000}$$
$$= 7.8$$

Select closest upper standard size
Assume 10 mm "inner diameter"

Each line in the circuit is individually sized based on the flow distribution

12

Given: smooth pipe has the following variables.

$$Q\left(\frac{lit}{min}\right) = 23.55, v\,(cSt) = 32, D(mm) = 10,$$
$$SG = 0.9, and\ L\,(m) = 3.$$

$$Eq.2.15.A \rightarrow R_e = \frac{21231 \times 23.55}{32 \times 10} = 1562$$

$$Laminar\ Flow \rightarrow Moody\ Diagram \rightarrow \lambda \approx 0.04$$

$$Eq.2.25.A \rightarrow \Delta p_F(bar) = \frac{2245 \times 0.04 \times 3 \times 0.9 \times 23.55^2}{10^5}$$
$$= 1.35$$

Pressure losses in each line are independently calculated based on flow distribution.
Assume 5 bar total pressure loss in all lines

13

Step 8: Valve Sizing and Valve Losses Calculations

flow distribution analysis during all phases of the machine operation.

Published data by valve manufacturer.

Pressure drop in the valve.

Assume a total of 15 bar total pressure drop have been found including the two paths inside the directional valve.

14

Step 9: Quantification of the Pump Output Power

New pressure after considering the losses

$$Eq. 2.10 \rightarrow Pump\ Ouput\ Power\ (kW) = \frac{\Delta p\ [bar] \times Q\ \left[\frac{l}{min}\right]}{600} = \frac{287.41 \times 23.55}{600} = 11.28$$

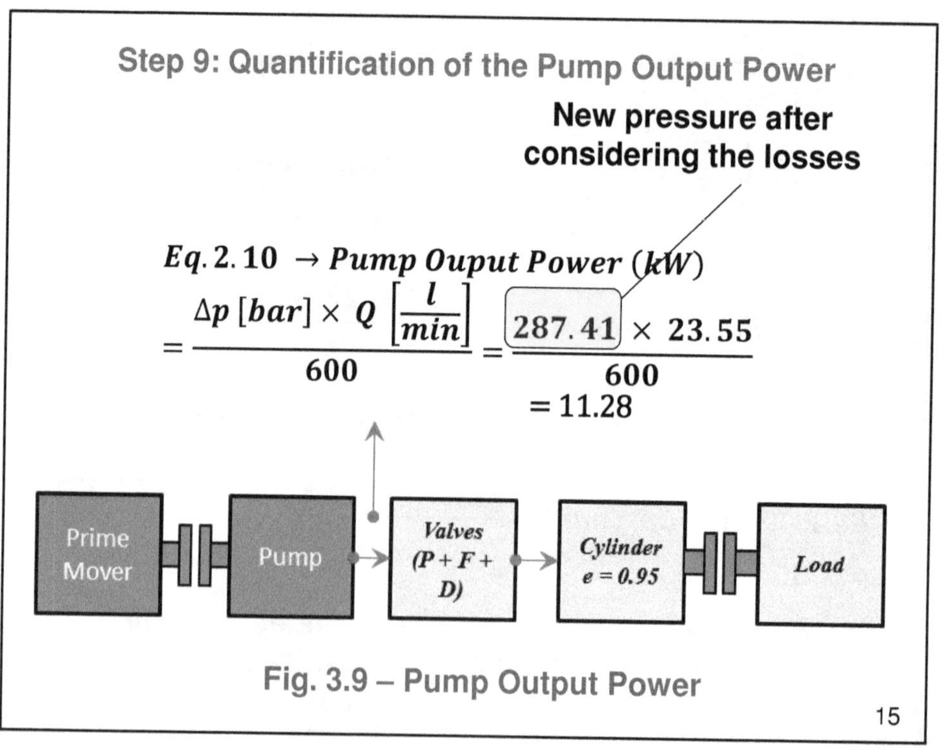

Fig. 3.9 – Pump Output Power

15

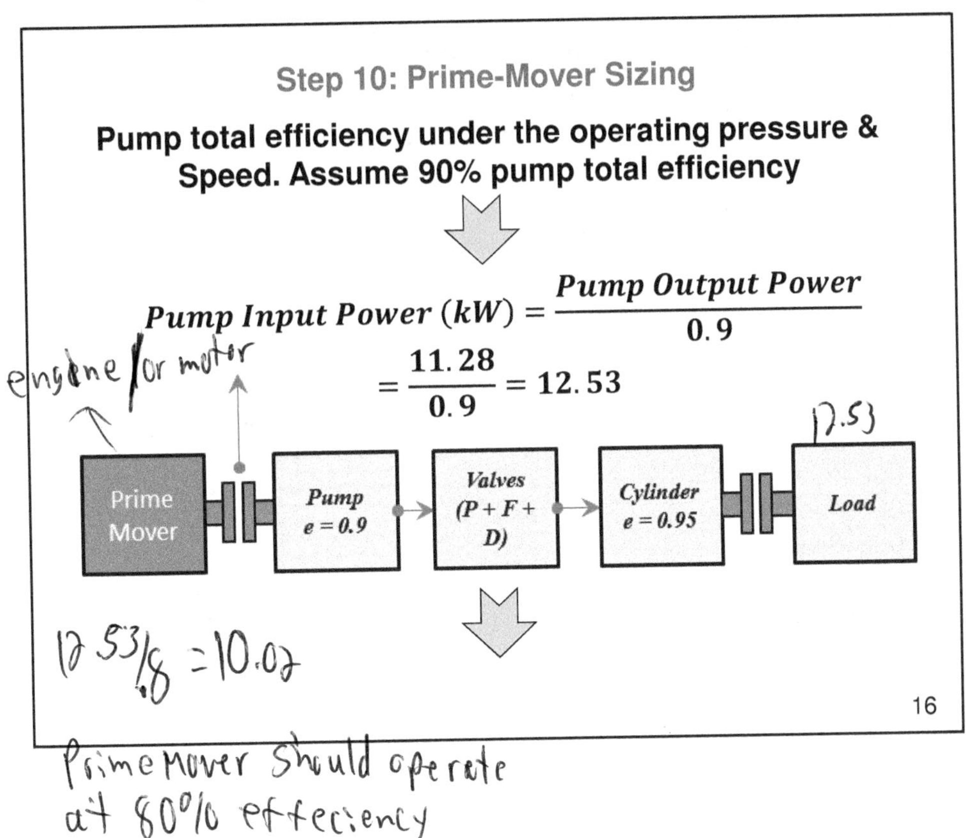

The calculation shows that the total power loss in the hydraulic system is approximately 25%.

For proper selection of the prime-mover, pump input power should be (80-90)% of the prime-mover rated power in order to afford intermittent overload conditions due to pressure spikes OR due to adjusting the PRV 10% higher than the load requirements.

17

Chapter 3 Reviews

1. In order to quantify the mechanical output power of a cylinder, which of the following parameters must be known?
   A. Cylinder flowrate and speed.
   B. Cylinder flowrate and pressure.
   C. Cylinder force and stroke.
   D. Cylinder force and speed.

2. For a 100% efficient cylinder that has an effective area of 10 square inches, what pressure is required to support a 10,000 pound vertical load?
   A. 1000 psi.
   B. 1000 bar.
   C. 1000 Pascal.
   D. 1000 kPa.

3. The flowrate required to extend a cylinder at a certain speed is function of?
   A. The cylinder piston area and stroke.
   B. The cylinder rod diameter and flowrate.
   C. The cylinder piston diameter and extension velocity.
   D. The cylinder piston diameter and flowrate.

4. In order to quantify the hydraulic input power to a cylinder, which of the following parameters must be known?
   A. Cylinder flowrate and speed.
   B. Cylinder flowrate and pressure.
   C. Cylinder force and stroke.
   D. Cylinder force and speed.

5. The cylinder efficiency is affected by?
   A. The cylinder speed.
   B. The cylinder internal leakage and seal friction.
   C. The cylinder pressure.
   D. The cylinder size.

Chapter 3 Assignment

Student Name: -- Student ID: ------------------

Date: -- Score: ------------------------

Given:

- A cylinder supports a resistive force of 10 Metric Tons.
- The cylinder should be extended at a speed of 5 cm/s.
- The piston diameter is selected as 6 cm.
- Oil kinematic viscosity is 48 cSt.
- Total line and valve losses are 15 bar.

Required:

Follow the 10 step procedure of sizing the system to find the proper size of the prime-mover.

Objectives:

This chapter has three main parts. The first part presents an overview of: positive displacement machines versus hydrodynamic machines, pumps versus motor mechanisms, fixed versus variable displacement machines, and symbols. The second part focuses on investigating the power losses in pumps and motors. This investigation is used to evaluate the efficiency of a pump and a motor under certain operating conditions using data published by the manufacturers. The third part of the chapter presents the construction and operating principles of gear, vane, and piston type positive displacement machines. The chapter concludes by the selection criteria for pumps and motors.

0

Brief Contents:

1

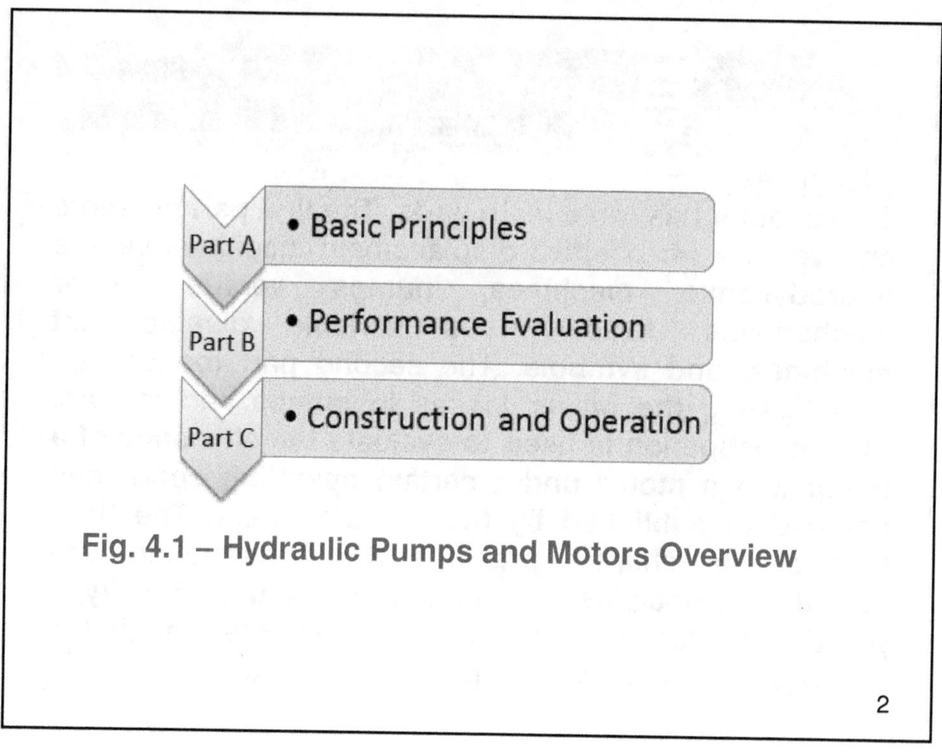

Fig. 4.1 – Hydraulic Pumps and Motors Overview

4.1- Basic Function of Hydraulic Pumps and Motors

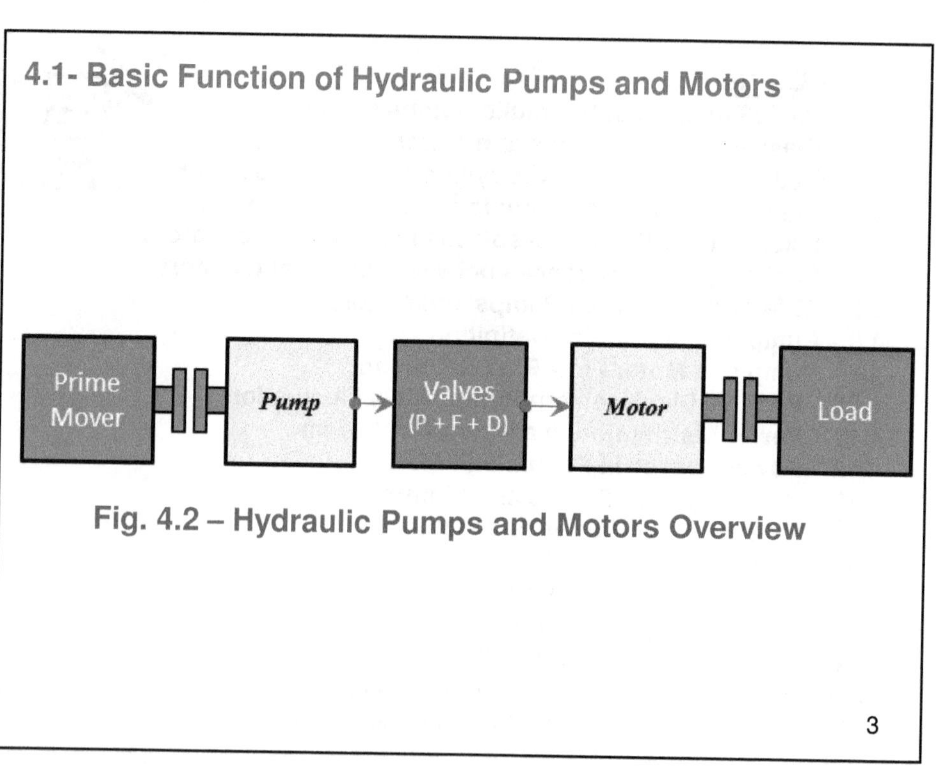

Fig. 4.2 – Hydraulic Pumps and Motors Overview

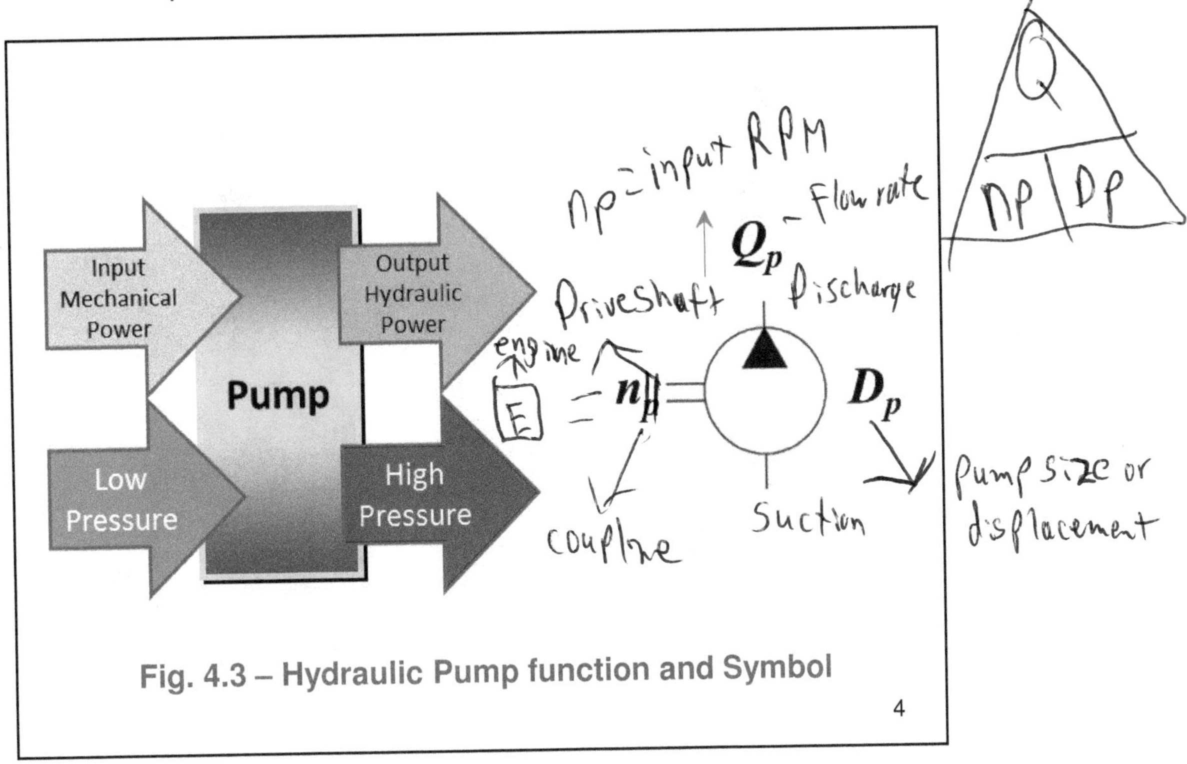

Fig. 4.3 – Hydraulic Pump function and Symbol

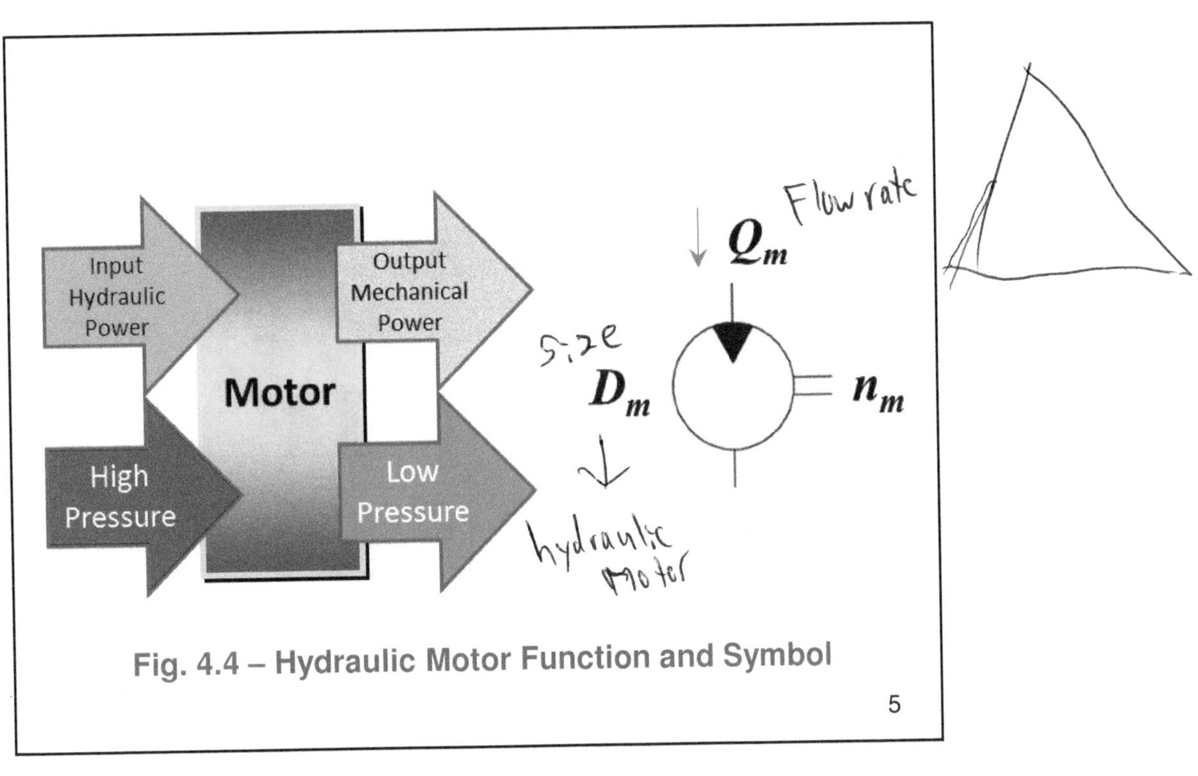

Fig. 4.4 – Hydraulic Motor Function and Symbol

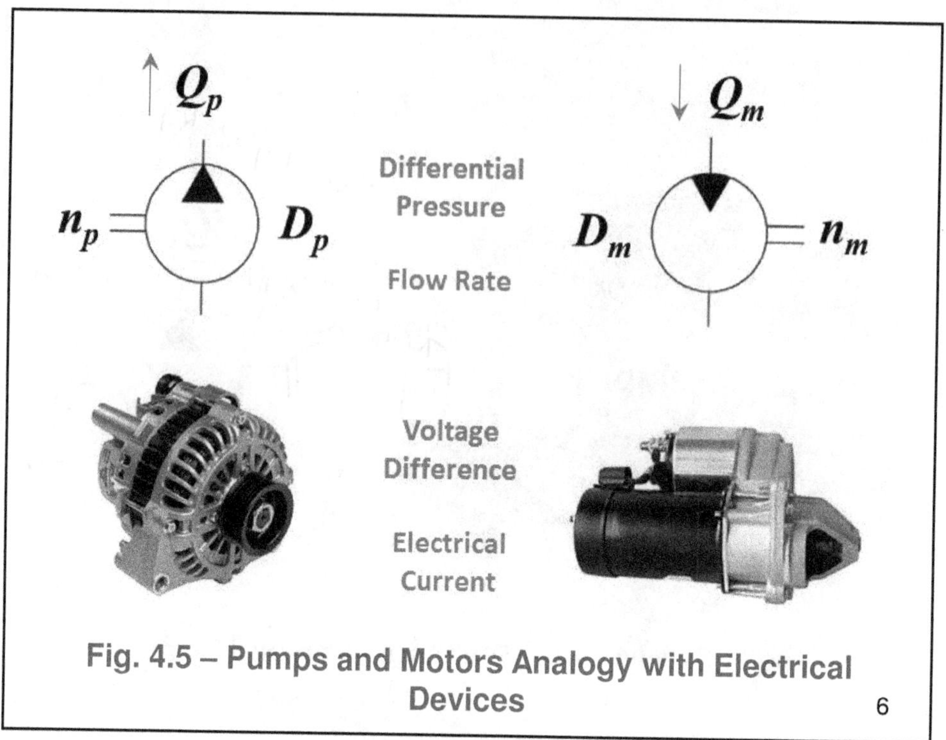

Fig. 4.5 – Pumps and Motors Analogy with Electrical Devices

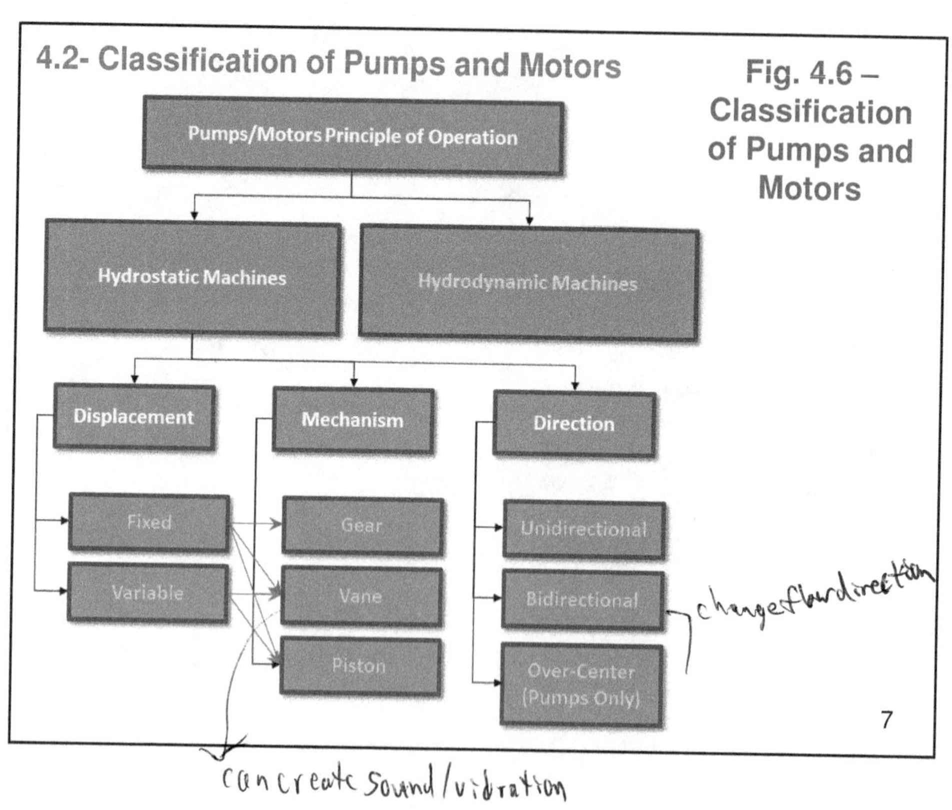

Fig. 4.6 – Classification of Pumps and Motors

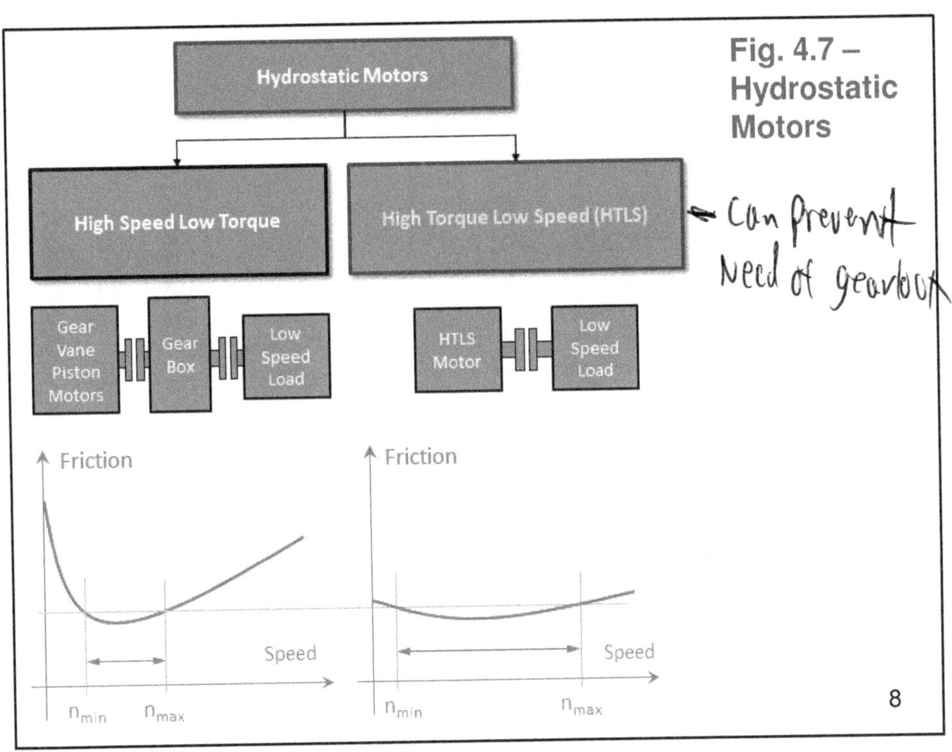

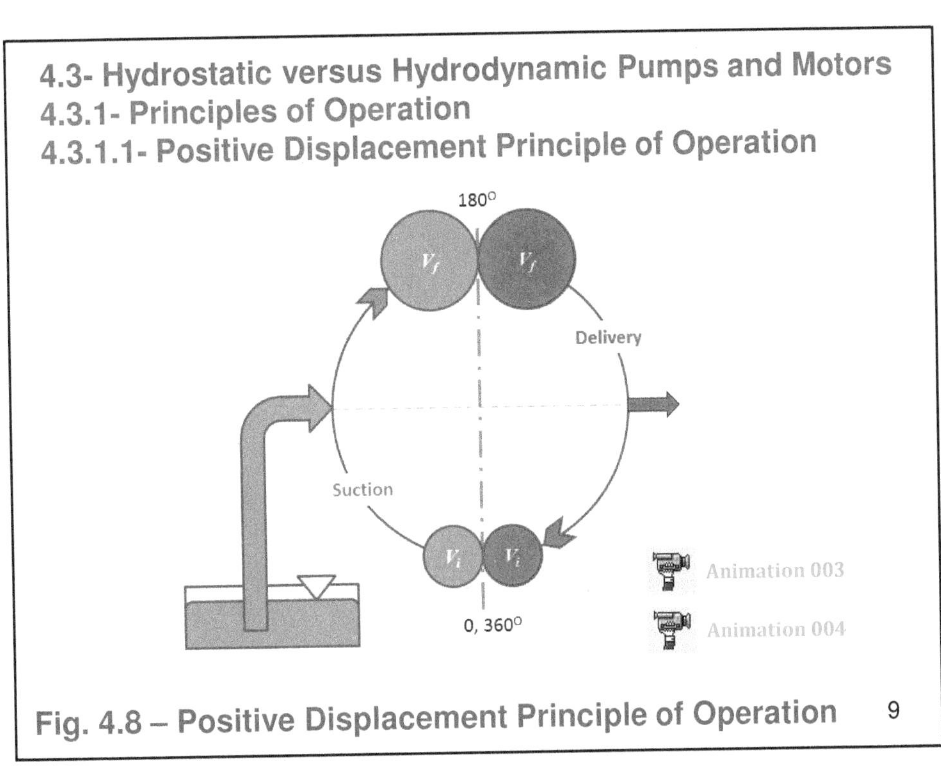

Fig. 4.8 – Positive Displacement Principle of Operation

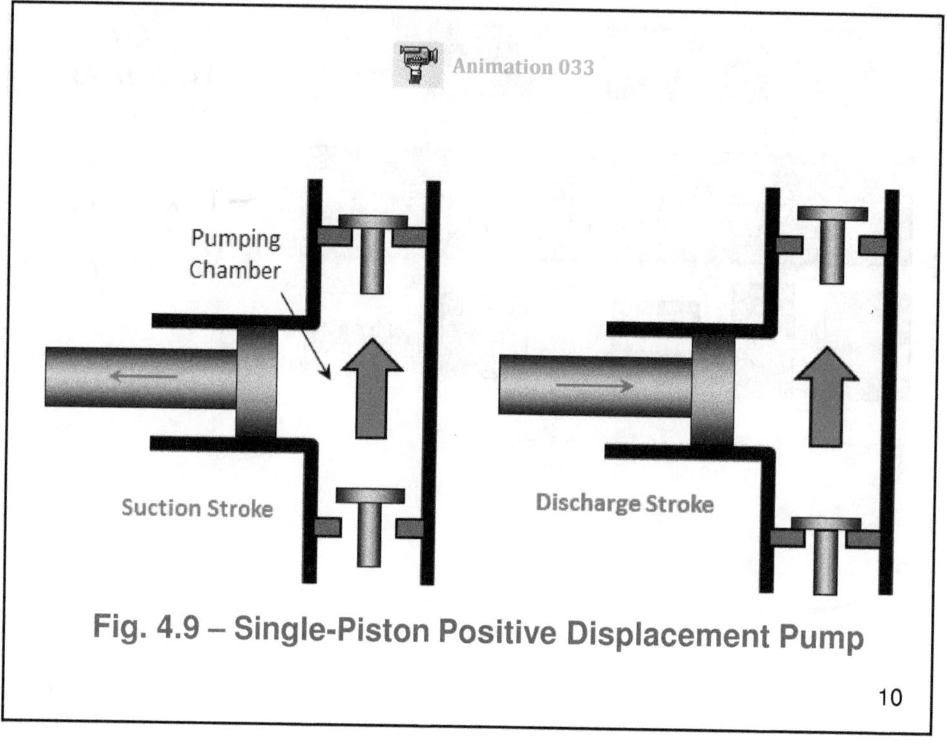

Fig. 4.9 – Single-Piston Positive Displacement Pump

10

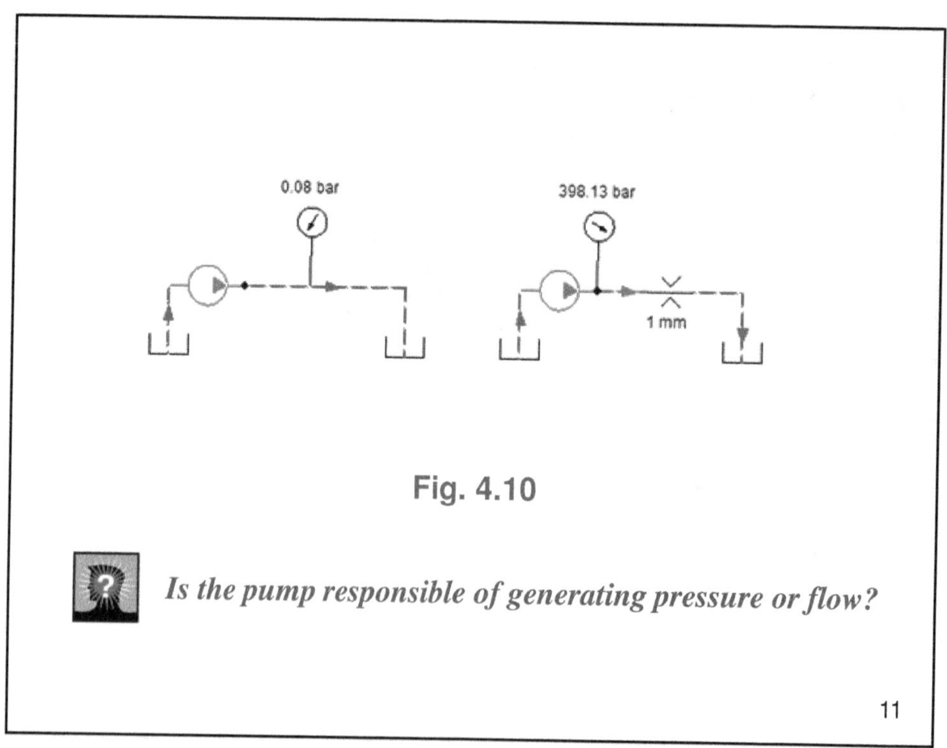

Fig. 4.10

Is the pump responsible of generating pressure or flow?

11

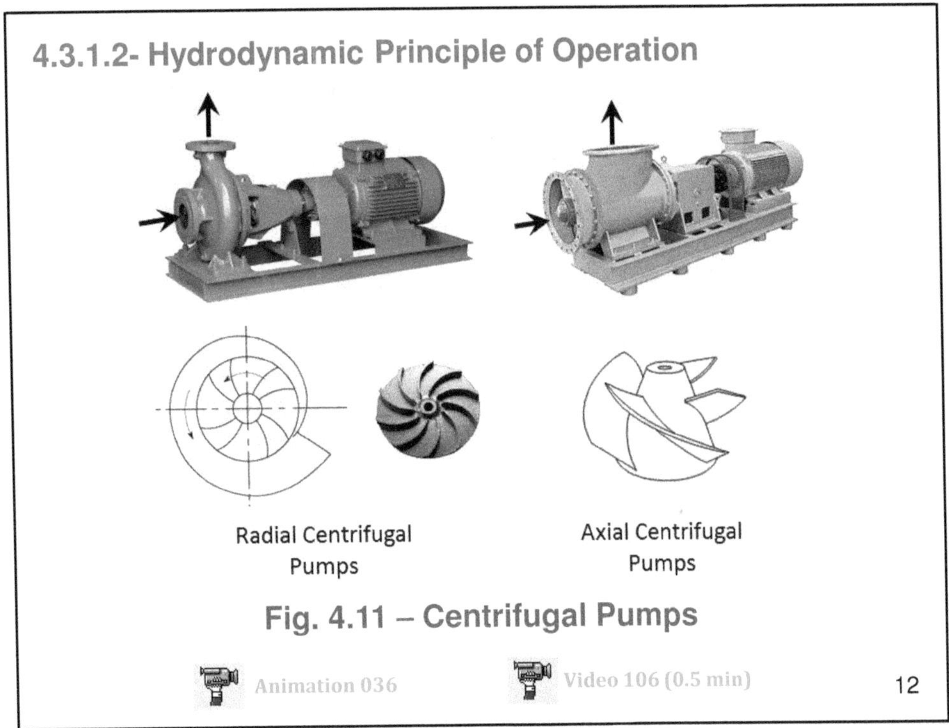

Fig. 4.11 – Centrifugal Pumps

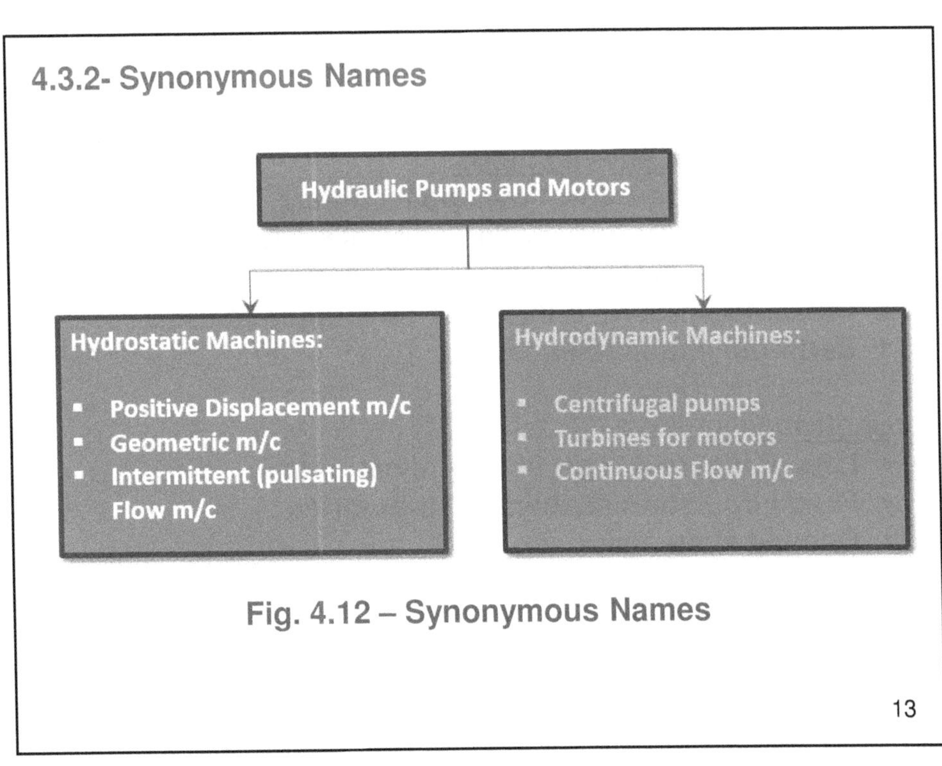

Fig. 4.12 – Synonymous Names

4.3.3- Characteristic Curves

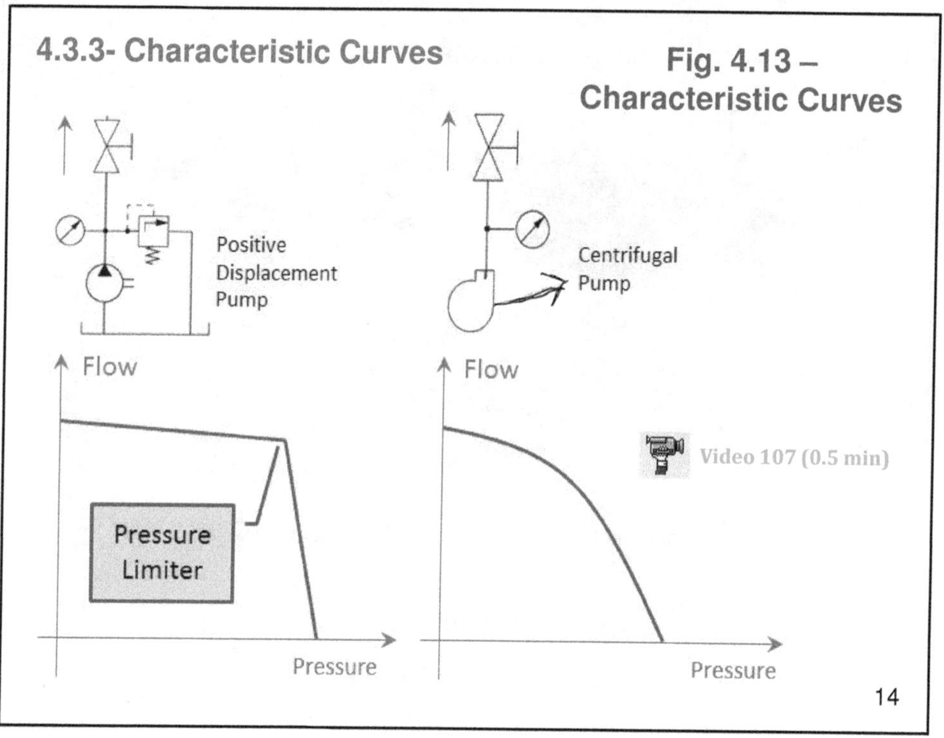

Fig. 4.13 – Characteristic Curves

14

4.3.4- Applications

Positive displacement pumps:

- **High pressure & low flow applications.**
- **Hydraulic drive and motion control systems.**
- **Lubrication systems.**
- **Pressing systems & injection molding machines.**
- **Earth-moving machines, etc.**

Centrifugal pumps:

- **Low pressure & high flow applications.**
- **Fluid circulation in chemical industries.**
- **Cooling systems.**
- **Water supply utilities.**
- **Irrigation systems, fire-fighting, etc.**

15

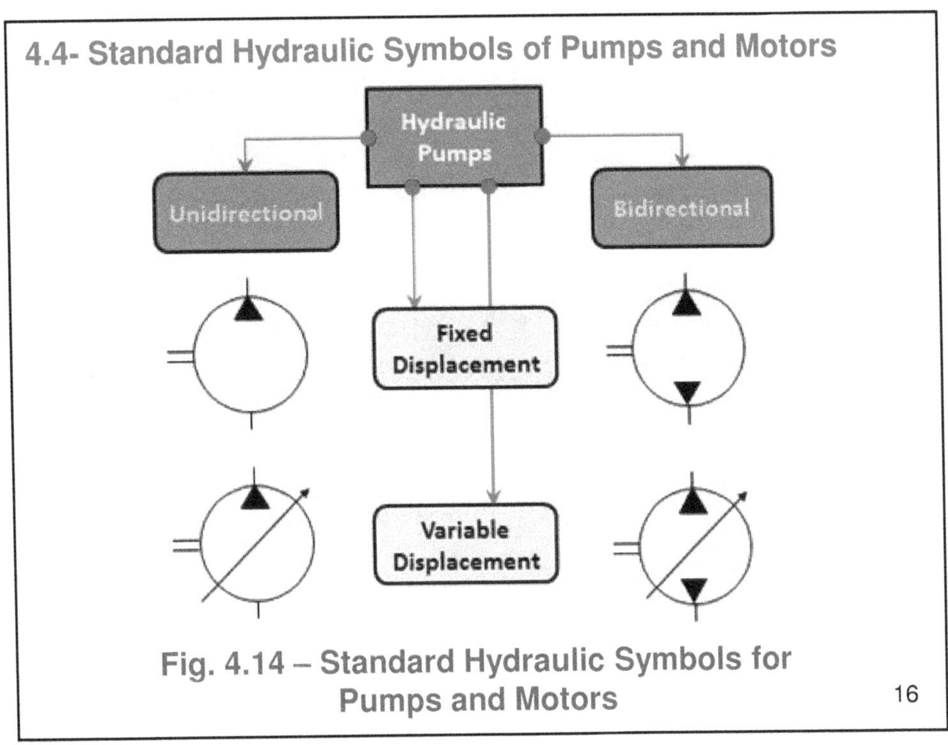

Fig. 4.14 – Standard Hydraulic Symbols for Pumps and Motors

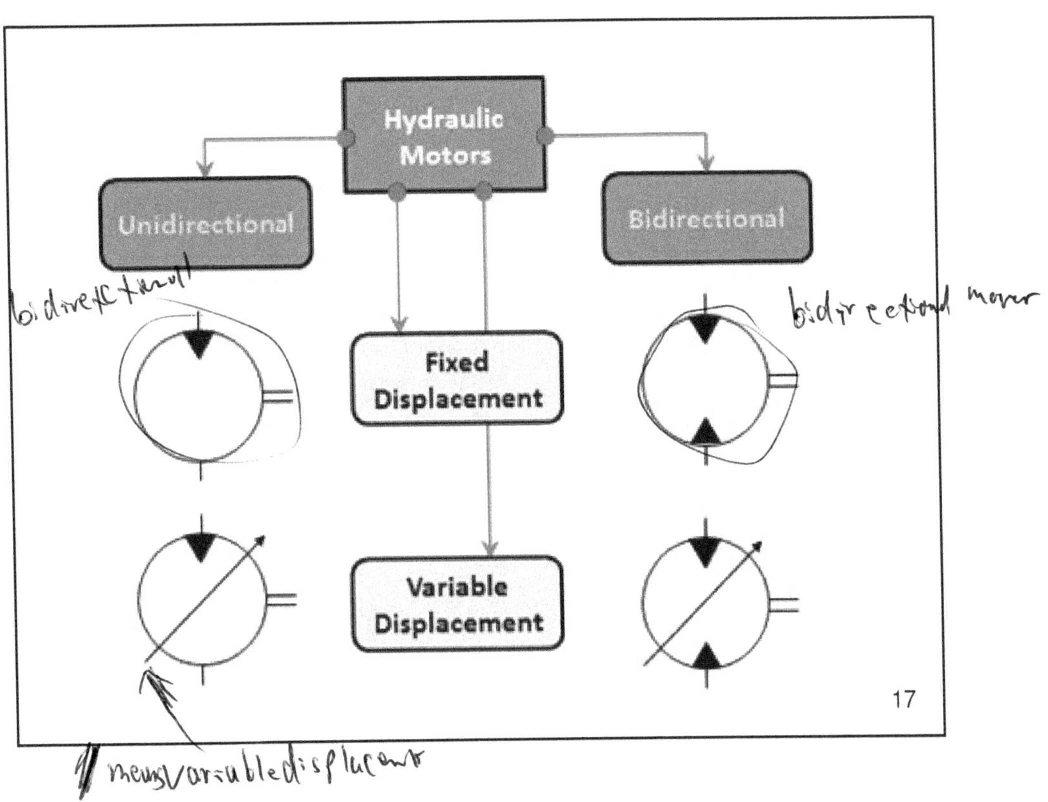

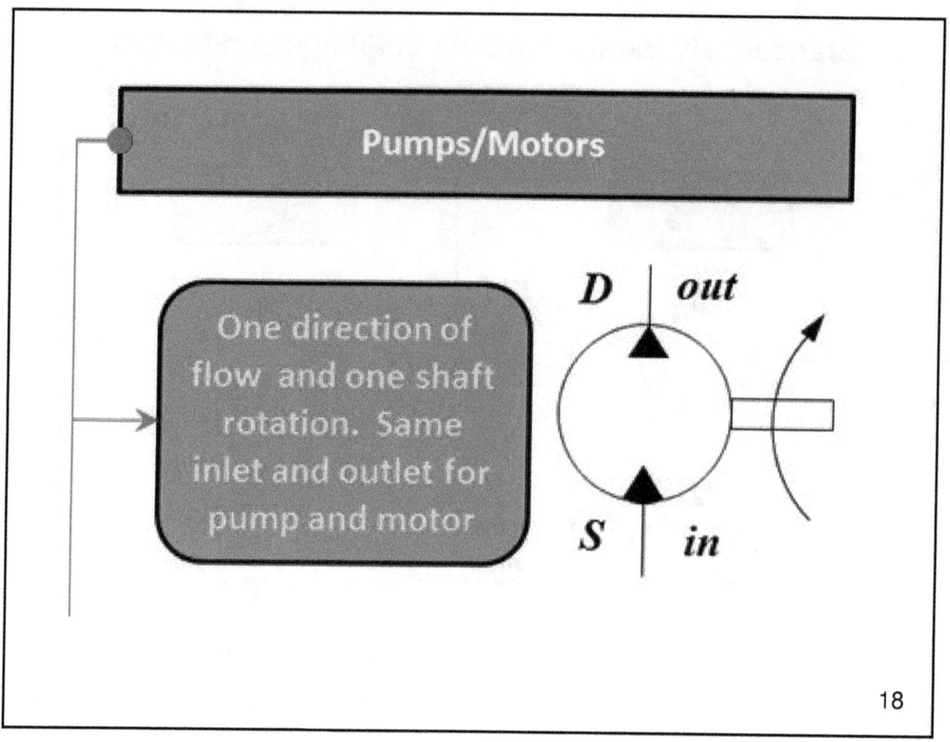
Pumps/Motors
One direction of flow and one shaft rotation. Same inlet and outlet for pump and motor
D
out
S
in
18

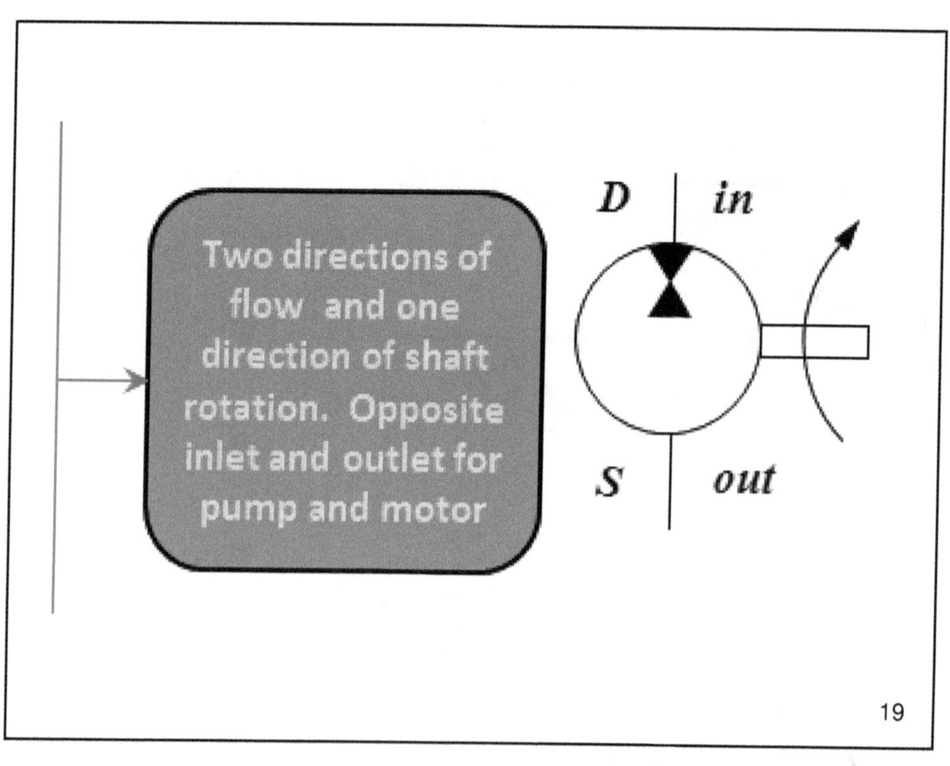
Two directions of flow and one direction of shaft rotation. Opposite inlet and outlet for pump and motor
D
in
S
out
19

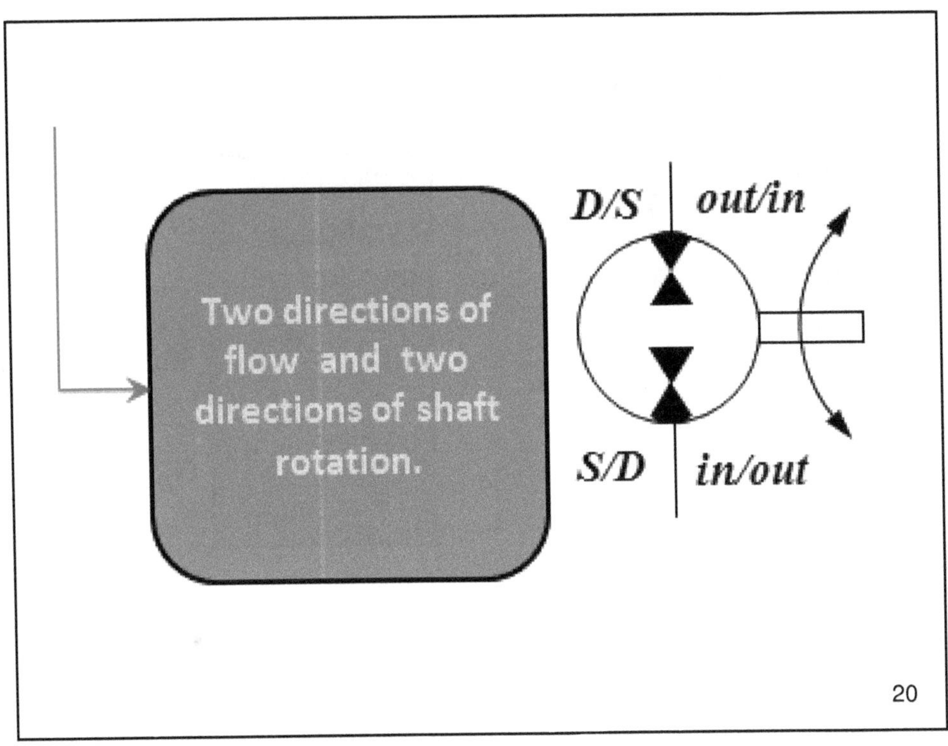

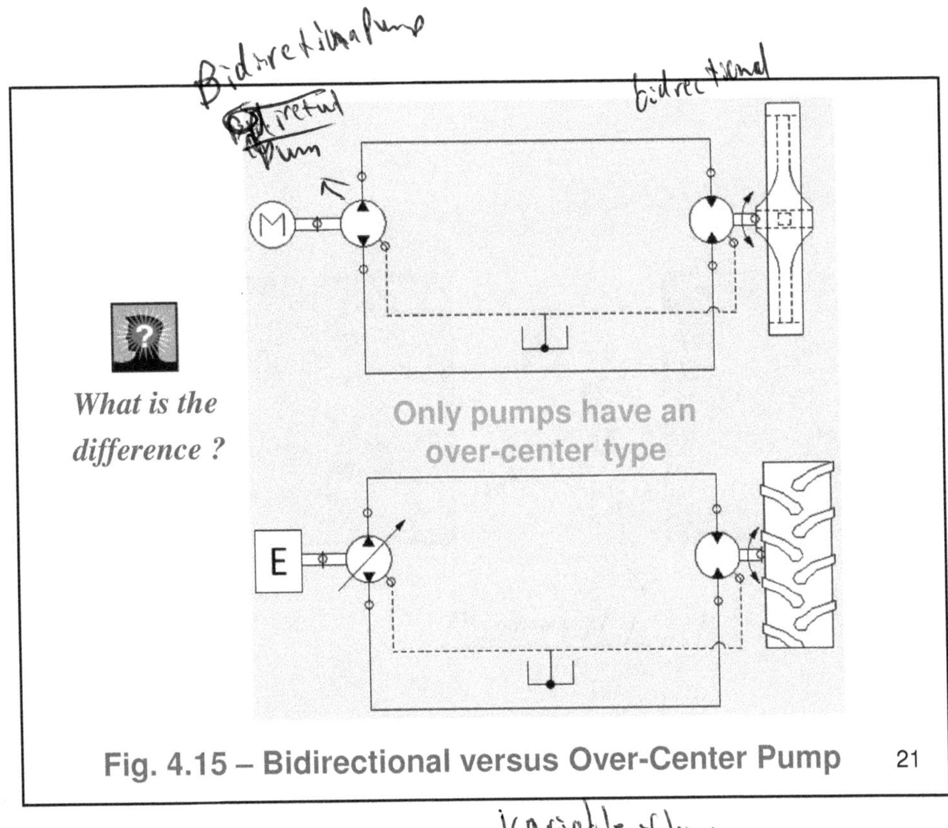

Fig. 4.15 – Bidirectional versus Over-Center Pump

$Q = D \times n$

4.5- Fixed versus Variable Displacement Pumps and Motors
4.5.1- Functional Comparison

Fixed

Pump:

Discharges constant flow rate whenever the input speed is constant.

Q_p D_p n_p

Variable

Pump:

Satisfies the system variable flow demand even with constant input speed.

Q_p D_p n_p

Fig. 4.16 – Standard Hydraulic Symbols for Pumps and Motors

22

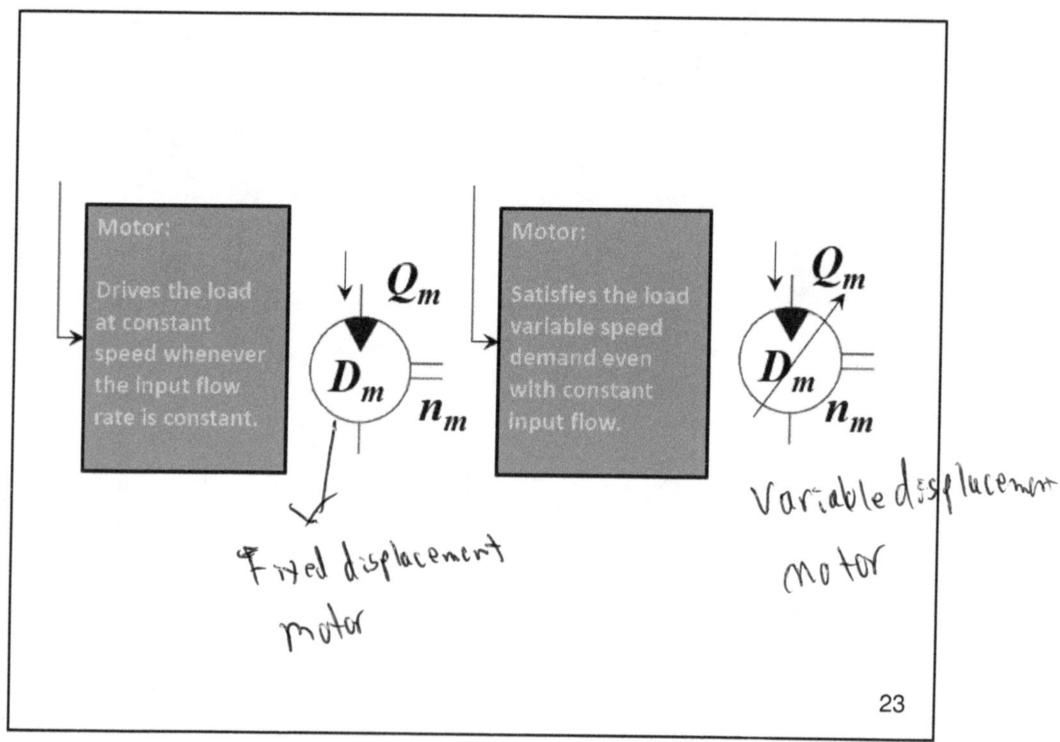

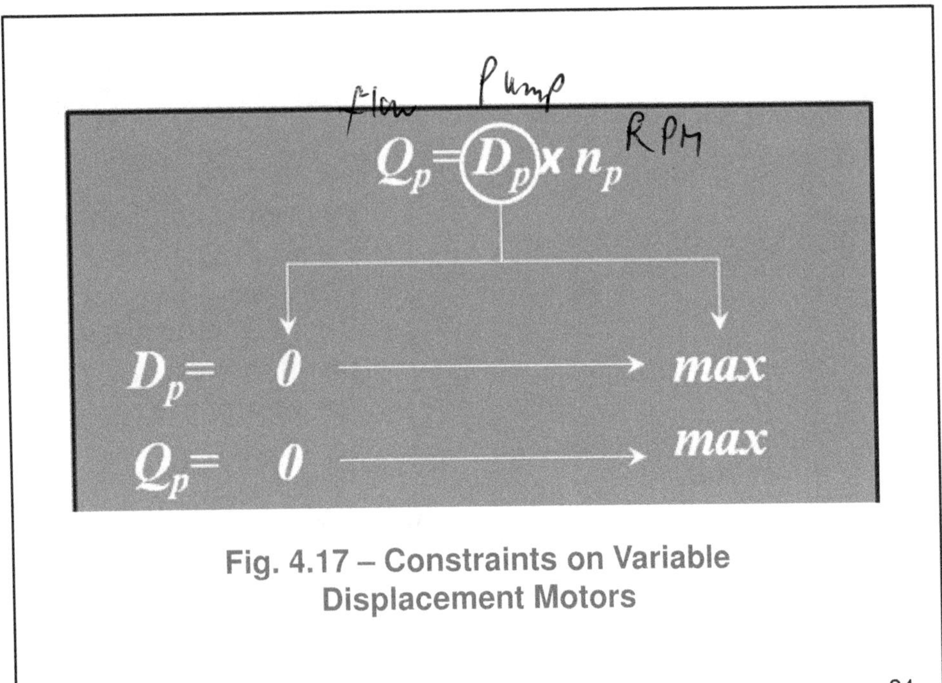

Fig. 4.17 – Constraints on Variable Displacement Motors

24

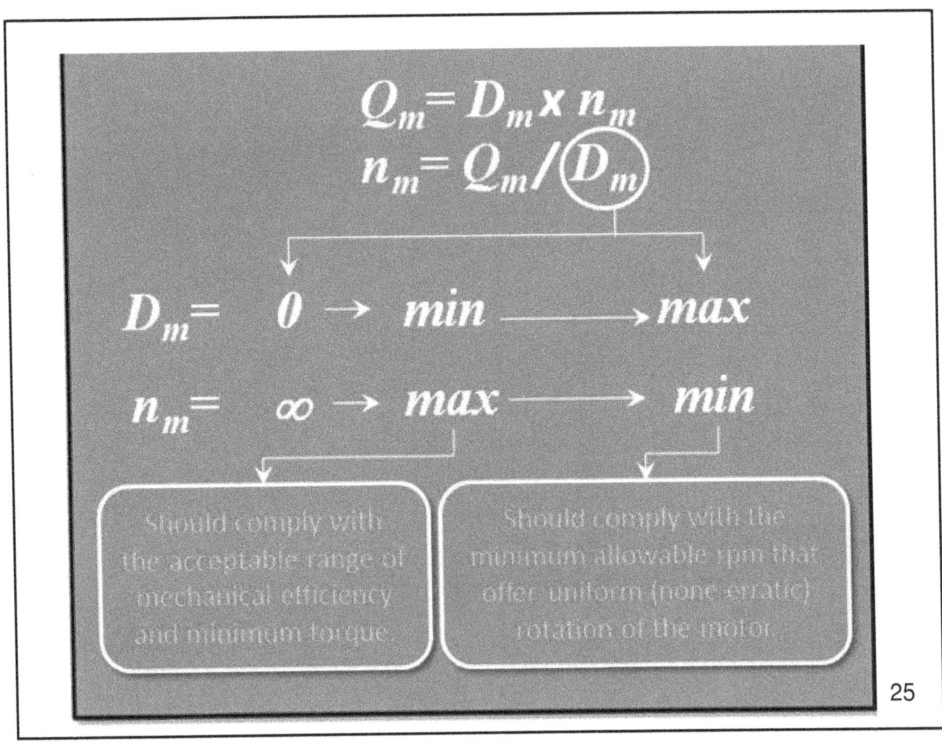

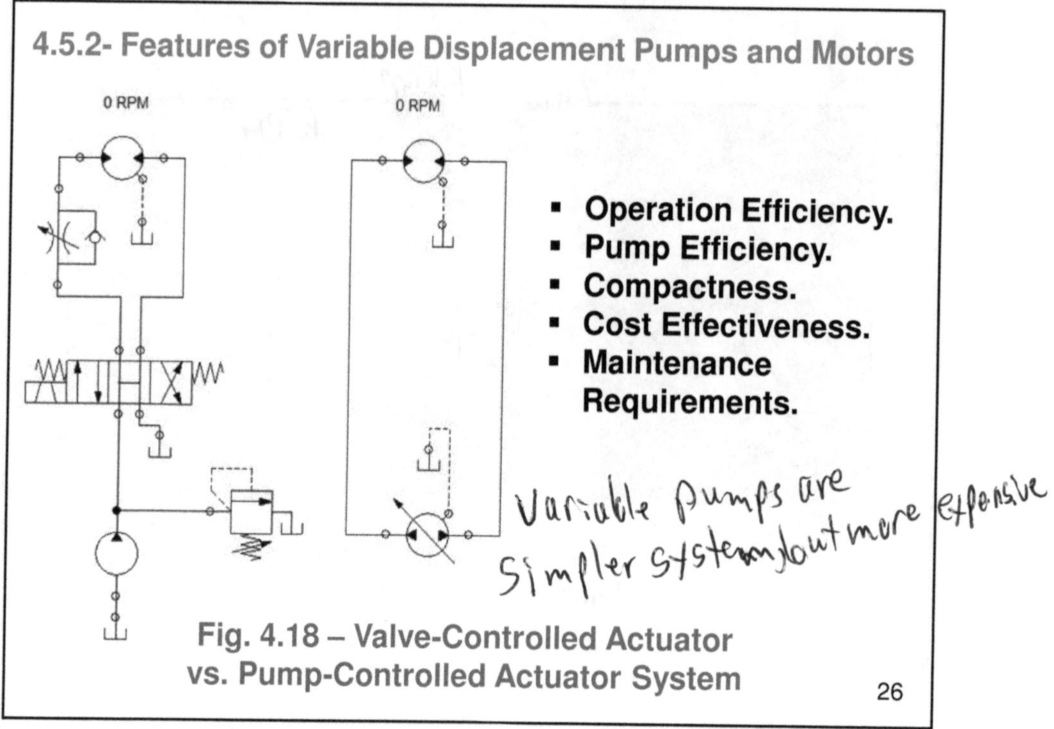

Fig. 4.18 – Valve-Controlled Actuator vs. Pump-Controlled Actuator System

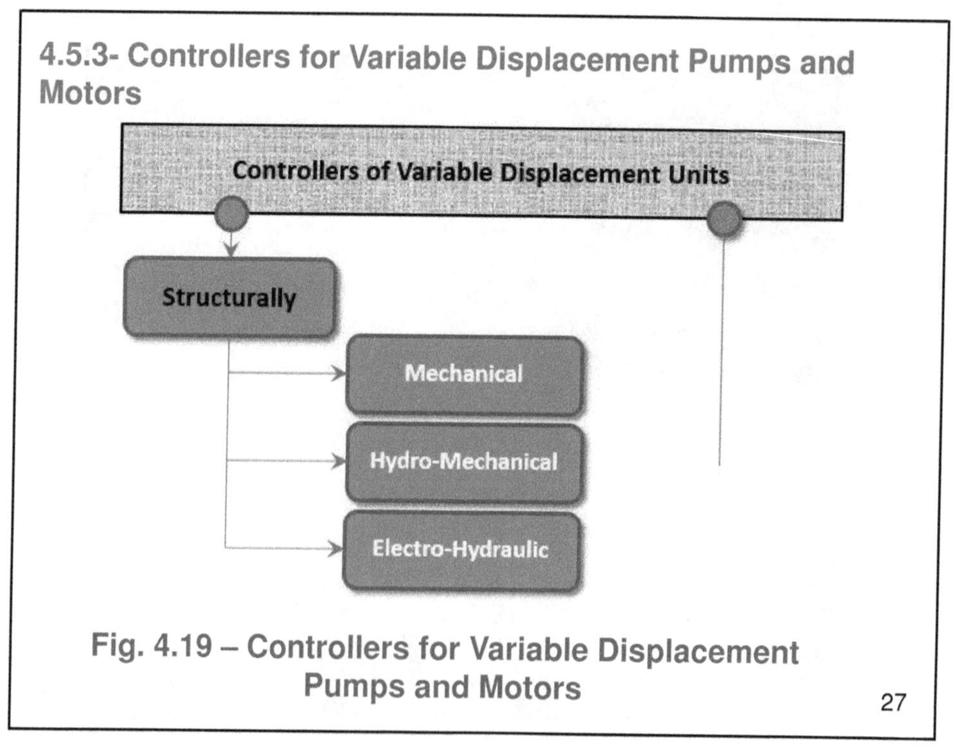

Fig. 4.19 – Controllers for Variable Displacement Pumps and Motors

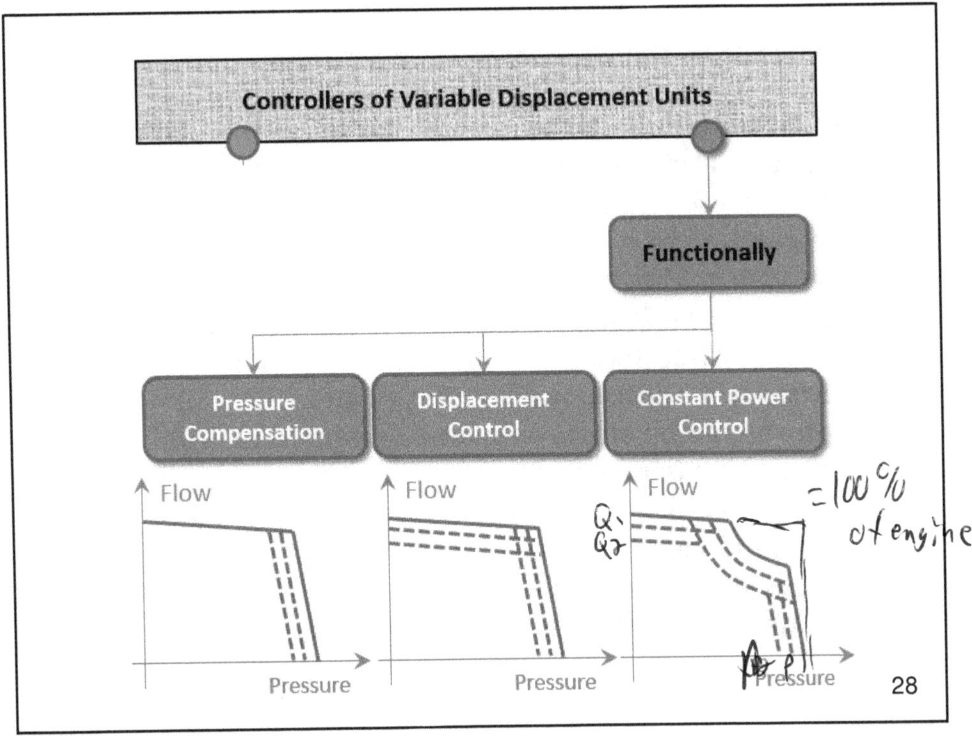

Power = P x Q

4.6- Similarities Between Pumps and Motors

The following are the similarities between pumps and motors:

- **Both are positive displacement rotating machines.**
- **Both could be of gear or vane or pistons type.**
- **Both could be of a fixed or variable displacement.**
- **Both could be subjected to cavitation conditions.**
- **Both could be of unidirectional or bi-directional type.**
- **In both pumps and motors, pressure is proportional to the torque on the shaft and size.**
- **In both pumps and motors, flow is proportional to the speed and size.**

29

4.7- Differences Between Pumps and Motors

The following bullets summarize the differences between pumps and motors:

- **Inverse function.**
- **Volumetric & mechanical efficiency.**
- **Pump (+Δp) – Motors (- Δp).**
- **Leakage direction.**
- **Pumps run continuously at high speed.**
- **Motors may be slowed down or stopped on idle.**
- **Pump shaft is guarded.**
- **Motor shaft is subjected to side forces from the attached load.**
- **Motors only have HTLS versions.**
- **Pumps only have Over-center version.**

30

4.8- Pumps and Motors Size Definition

10cc Pump = 10 cubic centim per revolution

10 gpm - volume per unit of time

$$\uparrow D_p = V_{oil} \qquad \downarrow D_m = V_{oil}$$

$$n_p = 1 \qquad D_m \qquad n_m = 1$$

Fig. 4.20 – Pump and Motor Size Definition

Units:

Range from fraction of in³/rev to 61 in³/rev (1000 cc/rev).

How to Evaluate:

- **Theoretical calculation from the geometry.**
- **Direct measurement in the test lab.**

31

4.9- Pumps and Motors Flow Rate Definition

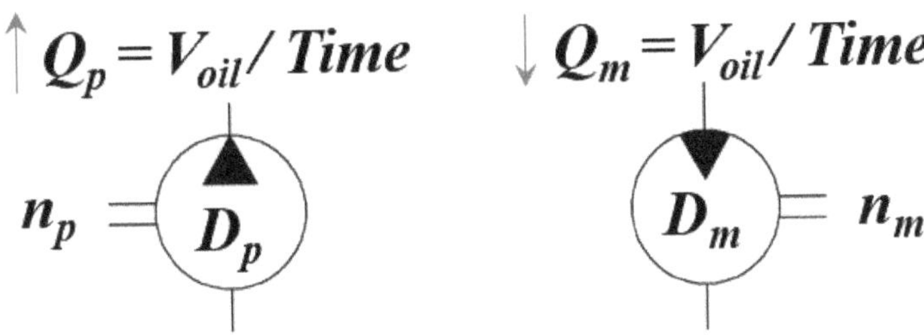

Fig. 4.21 – Pump and Motor Size Definition

Units:

Gallons per minute (gpm) "English" – Liters/min "SI"

How to Evaluate:

- **Theoretical Calculation :** $D \times n$.
- **Direct measurement in the test lab.**

32

In Metric System of Units:

displacement RPM

$$Q_{thp}(Lit/min) = \frac{D_p\left(\frac{cc}{rev}\right) \times n_p\left(\frac{rev}{min}\right)}{1000} \quad 4.1$$

$$Q_{thm}(Lit/min) = \frac{D_m\left(\frac{cc}{rev}\right) \times n_m\left(\frac{rev}{min}\right)}{1000} \quad 4.2$$

Do you consider knowing that a pump discharges 10 lit/min is complete information?

33

In English System of Units:

$$Q_{thp}(\text{gpm}) = \frac{D_p\left(\frac{in^3}{rev}\right) \times n_p\left(\frac{rev}{min}\right)}{12^3} \times 7.48 = \frac{D_p\left(\frac{in^3}{rev}\right) \times n_p\left(\frac{rev}{min}\right)}{231} \qquad 4.3$$

$$Q_{thm}(\text{gpm}) = \frac{D_m\left(\frac{in^3}{rev}\right) \times n_m\left(\frac{rev}{min}\right)}{12^3} \times 7.48 = \frac{D_m\left(\frac{in^3}{rev}\right) \times n_m\left(\frac{rev}{min}\right)}{231} \qquad 4.4$$

So far "Theoretical"

34

4.10- Positive Displacement Pumps Flow Fluctuation

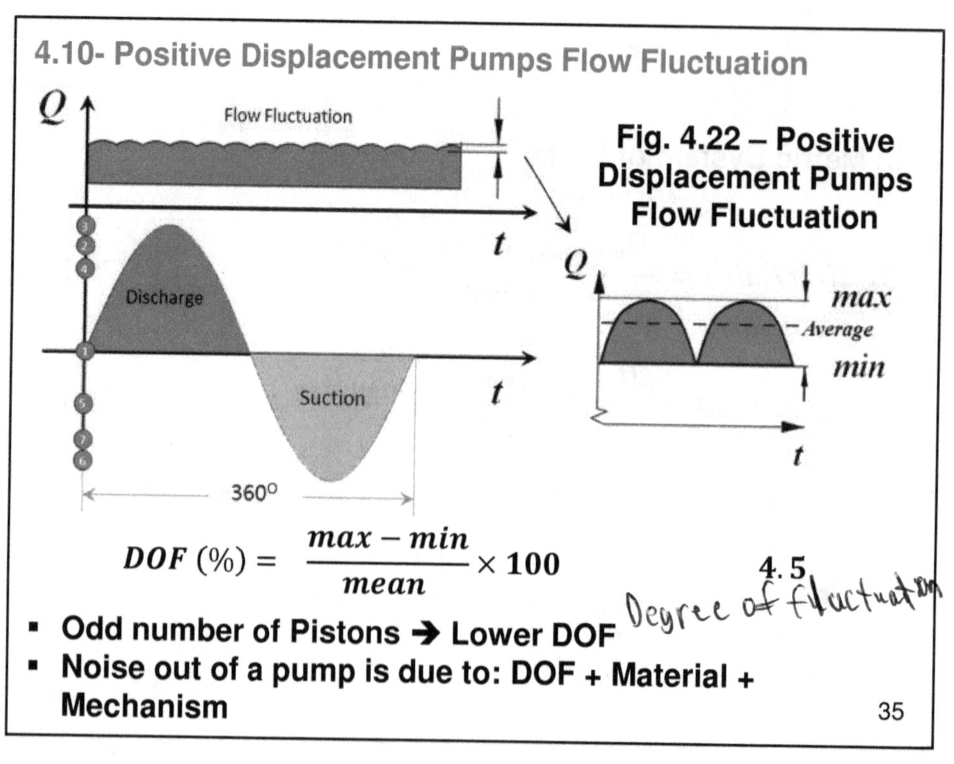

Fig. 4.22 – Positive Displacement Pumps Flow Fluctuation

$$DOF\ (\%) = \frac{max - min}{mean} \times 100 \qquad 4.5$$

Degree of fluctuation

- Odd number of Pistons → Lower DOF
- Noise out of a pump is due to: DOF + Material + Mechanism

35

4.11- Power Distribution in a Hydraulic System

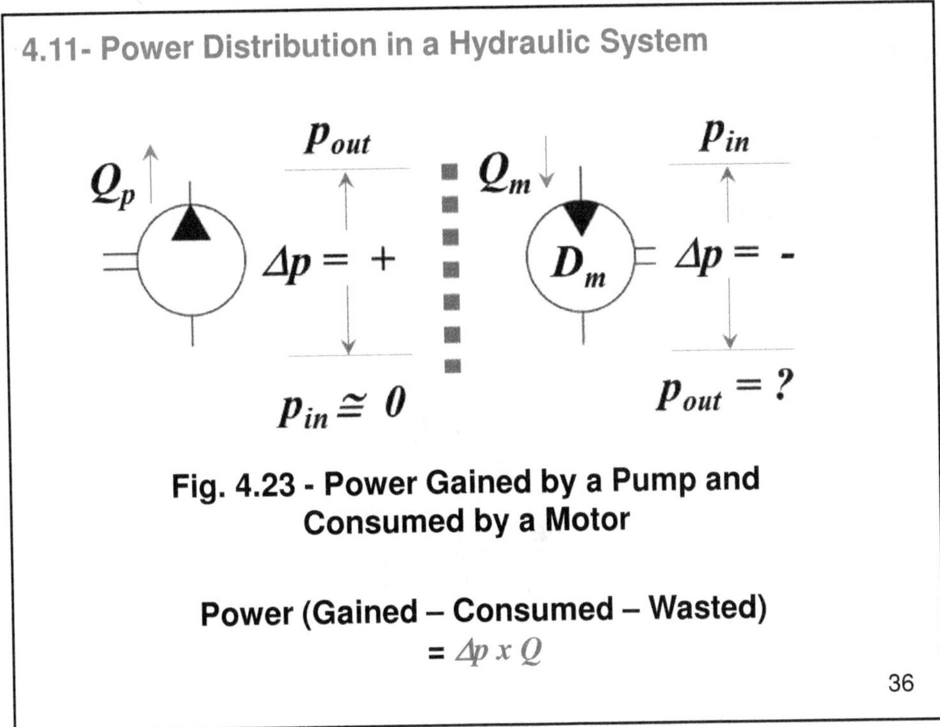

Fig. 4.23 - Power Gained by a Pump and Consumed by a Motor

Power (Gained – Consumed – Wasted)
$= \Delta p \times Q$

36

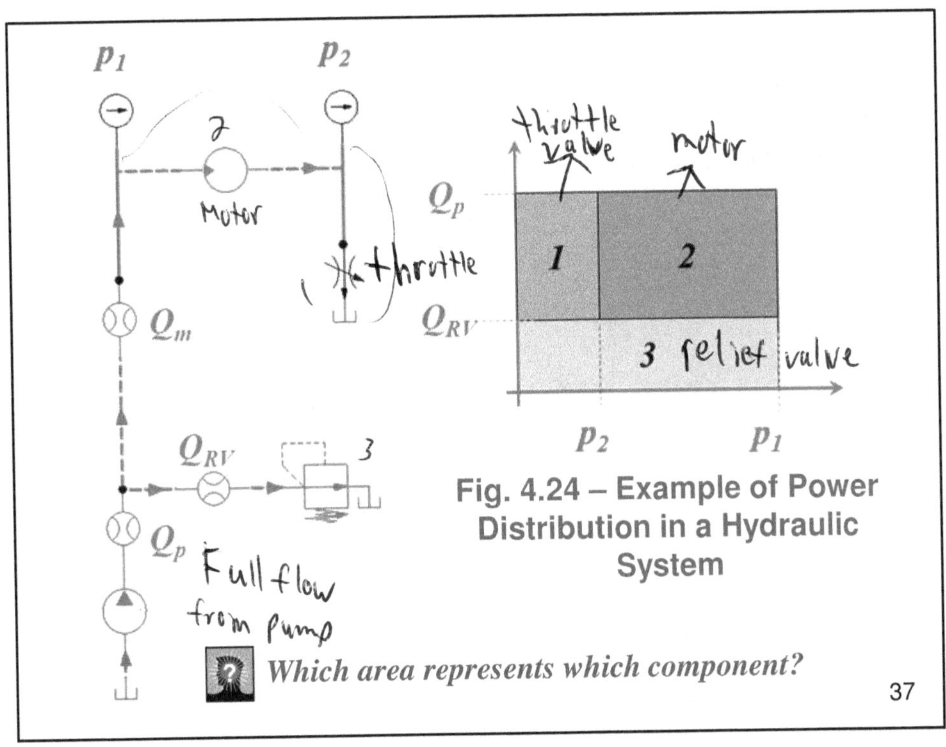

Fig. 4.24 – Example of Power Distribution in a Hydraulic System

Which area represents which component?

37

$$P_{RV} = p_1 \times Q_{RV} \qquad 4.6$$

$$P_m = (p_1 - p_2) \times (Q_p - Q_{RV}) = (p_1 - p_2) \times Q_m \qquad 4.7$$

$$P_{thv} = p_2 \times Q_m \qquad 4.8$$

$$P_p = p_1 \times Q_p \qquad 4.9$$

Ex.2

38

4.12- Power Losses in Pumps and Motors

Studying this topic is important for:

- **System Design (size of prime mover & heat exchanger)**
- **System Operation (productivity or energy consumption)**

4.12.1- Matter of Facts

Is it better to have a pump or a motor with zero leakage?

NO, wont cool

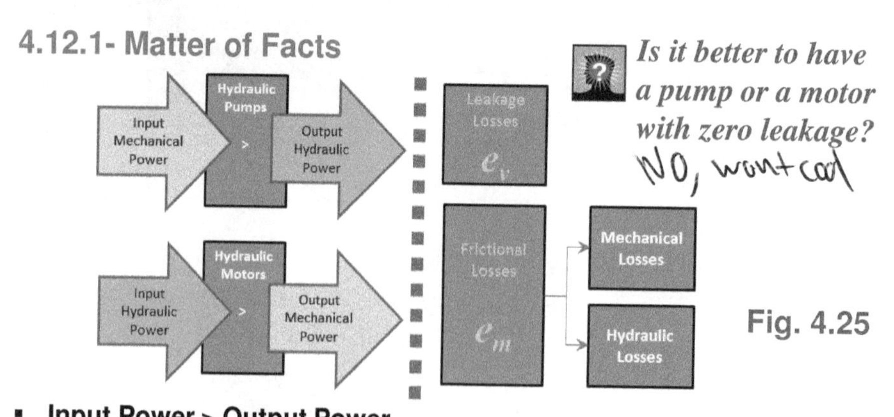

Fig. 4.25

- **Input Power > Output Power.**
- **Input Power = Output Power + Total Losses.**
- **Total Efficiency = Output Power/ Input Power.**
- **Power wasted is converted into heat, not vanished.**

39

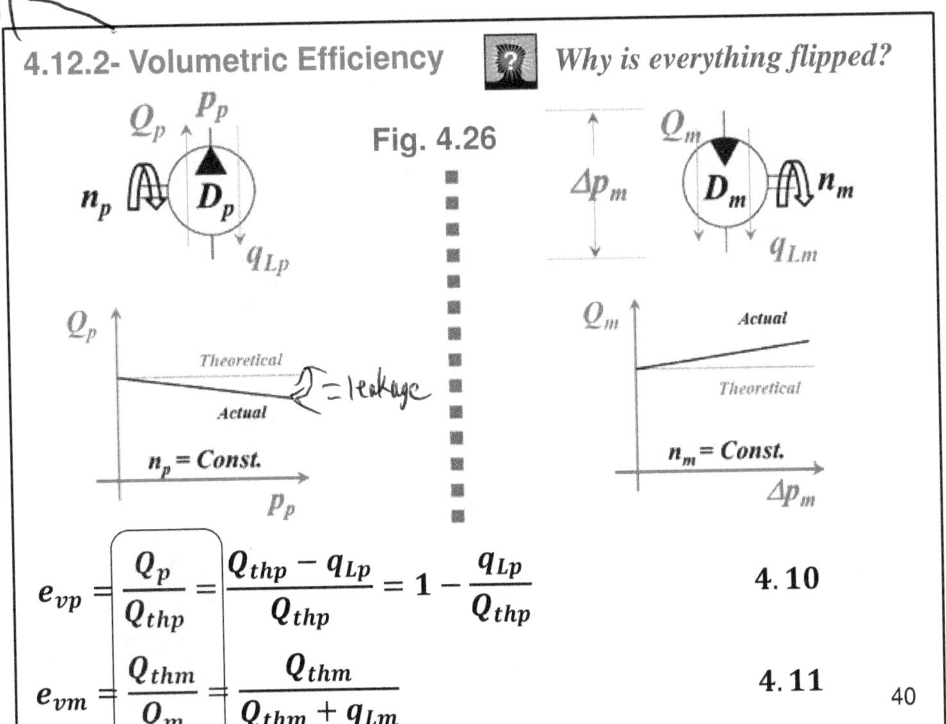
4.12.2- Volumetric Efficiency
Why is everything flipped?
Fig. 4.26
Q_p p_p n_p D_p q_{Lp}
Q_m Δp_m D_m n_m q_{Lm}
Theoretical
Actual
$n_p = Const.$
Actual
Theoretical
$n_m = Const.$
$e_{vp} = \frac{Q_p}{Q_{thp}} = \frac{Q_{thp} - q_{Lp}}{Q_{thp}} = 1 - \frac{q_{Lp}}{Q_{thp}}$ 4.10
$e_{vm} = \frac{Q_{thm}}{Q_m} = \frac{Q_{thm}}{Q_{thm} + q_{Lm}}$ 4.11
40

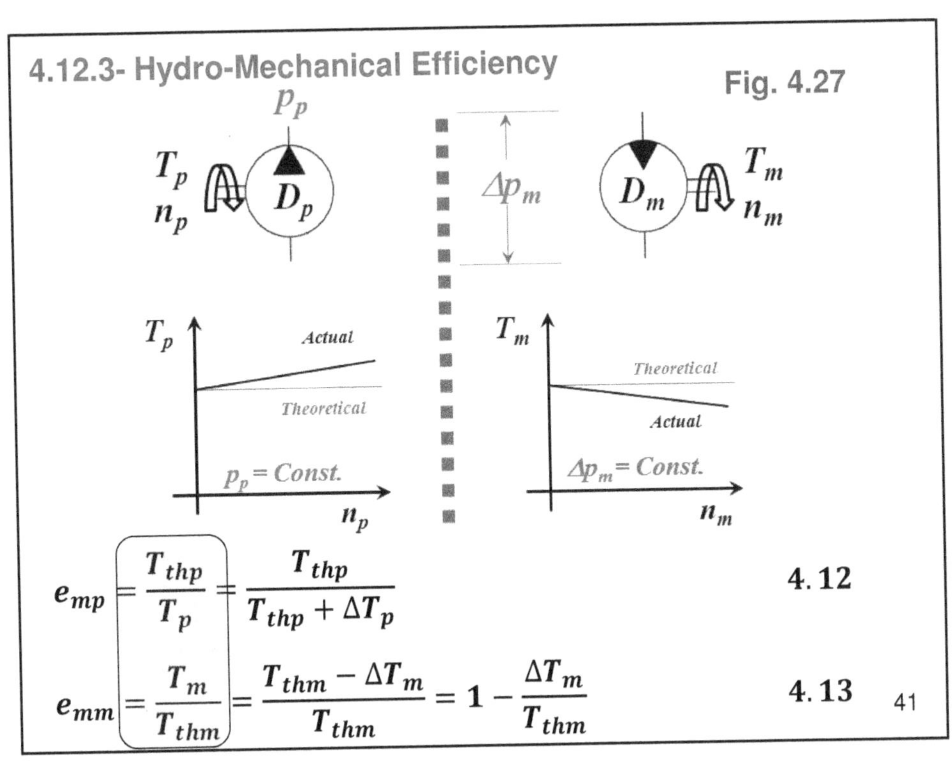
4.12.3- Hydro-Mechanical Efficiency
Fig. 4.27
p_p T_p n_p D_p
Δp_m D_m T_m n_m
Actual
Theoretical
$p_p = Const.$
Theoretical
Actual
$\Delta p_m = Const.$
$e_{mp} = \frac{T_{thp}}{T_p} = \frac{T_{thp}}{T_{thp} + \Delta T_p}$ 4.12
$e_{mm} = \frac{T_m}{T_{thm}} = \frac{T_{thm} - \Delta T_m}{T_{thm}} = 1 - \frac{\Delta T_m}{T_{thm}}$ 4.13
41

In Metric System of Units:

Pump Theoretical Input Torque T_{thp} (N.m)

$$= \frac{p_p\,[bar]\;x\;D_p[cc/rev]}{20\pi} \qquad 4.14$$

Motor Theoretical Output Torque T_{thm} (N.m)

$$= \frac{\Delta p_m\,[bar]\;x\;D_m[cc/rev]}{20\pi} \qquad 4.15$$

In English System of Units:

Pump Theoretical Input Torque T_{thp} (lb.in)

$$= \frac{p_p\,[psi]\;x\;D_p[in^3/rev]}{2\pi} \qquad 4.16$$

Motor Theoretical Output Torque T_{thm} (lb.in)

$$= \frac{\Delta p_m\,[psi]\;x\;D_m[in^3/rev]}{2\pi} \qquad 4.17$$

42

4.12.4- Total Efficiency

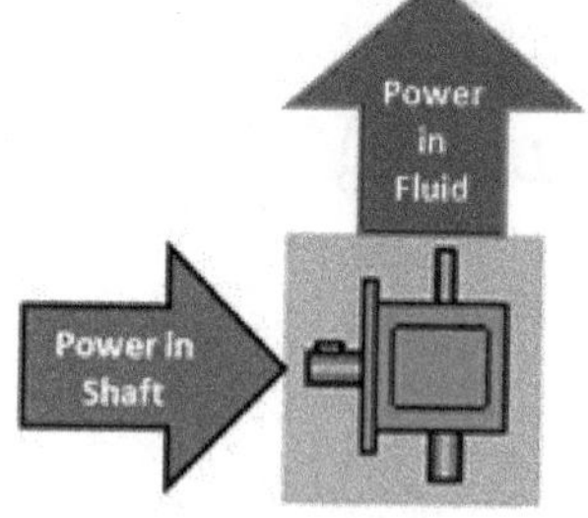

Fig. 4.28 –Total Efficiency

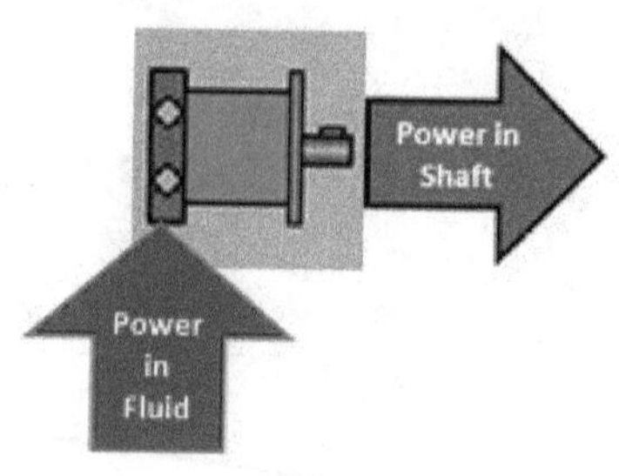

$$e_{tp} = \frac{Power\ in\ the\ Fluid}{Power\ in\ the\ Shaft}$$

$$= \frac{p_p \times Q_p}{T_p \times 2\pi n_p} = \frac{p_p \times Q_{thp} \times e_{vp}}{T_p \times 2\pi n_p}$$

torque

$$= \frac{p_p \times D_p \times e_{vp}}{T_p \times 2\pi} = e_{mp} e_{vp} \quad 4.18$$

$$e_{tm} = \frac{Power\ in\ the\ Shaft}{Power\ in\ the\ Fluid}$$

$$= \frac{T_m \times 2\pi n_m}{p_m \times Q_m} = \frac{T_m \times 2\pi n_m \times e_{vm}}{p_m \times Q_{thm}}$$

$$= \frac{T_m \times 2\pi \times e_{vm}}{p_m \times D_m} = e_{mm} \times e_{vm} \quad 4.19$$

43

Mechanical eff   Volumetric eff.

4.12.5- Effect of Operating Conditions on Efficiencies

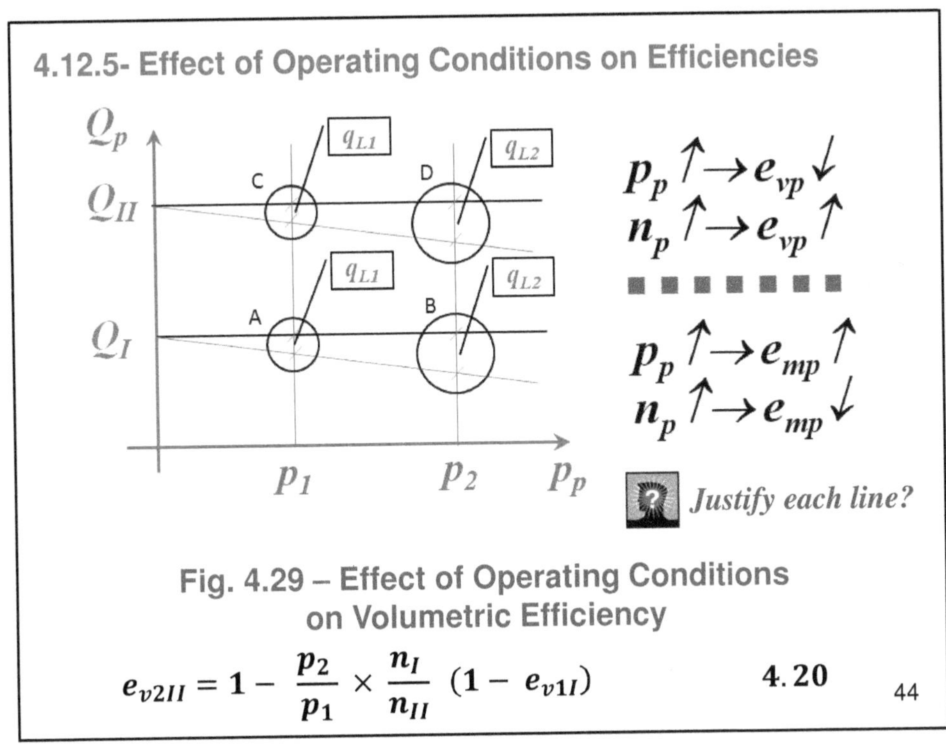

Fig. 4.29 – Effect of Operating Conditions on Volumetric Efficiency

$$e_{v2II} = 1 - \frac{p_2}{p_1} \times \frac{n_I}{n_{II}} (1 - e_{v1I}) \qquad 4.20$$

44

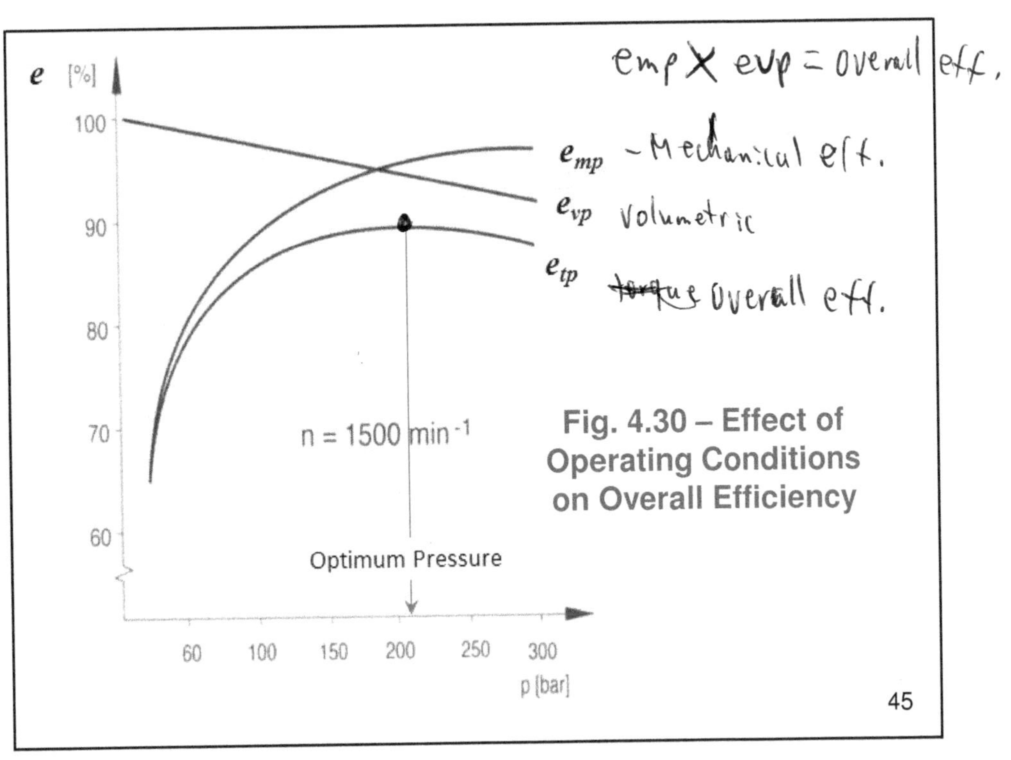

Fig. 4.30 – Effect of Operating Conditions on Overall Efficiency

45

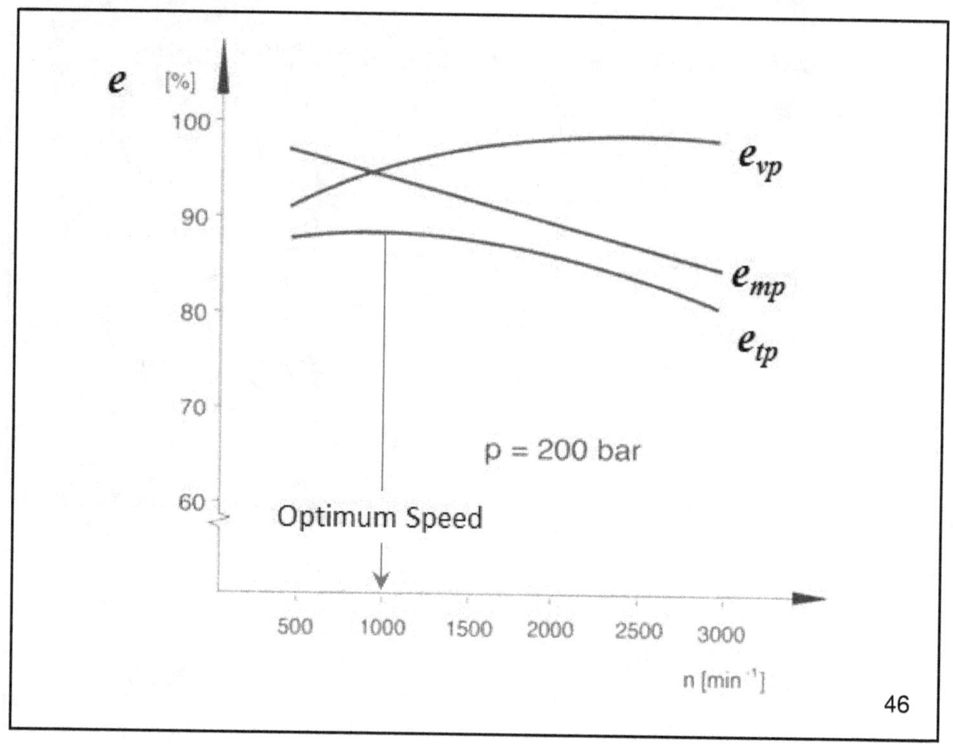

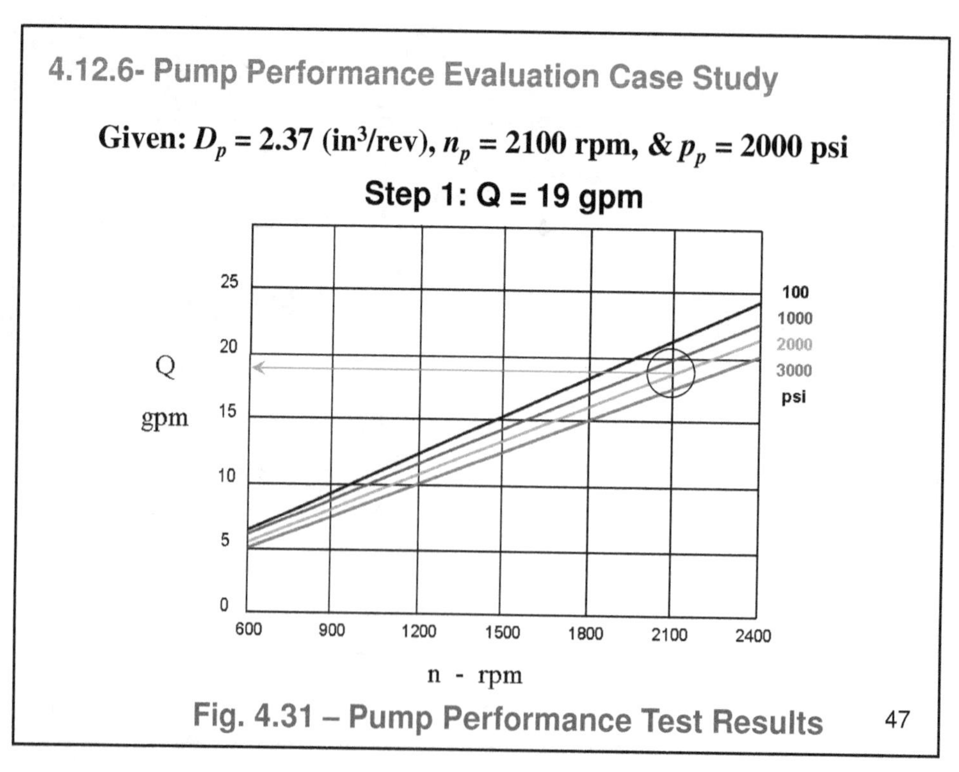

Fig. 4.31 – Pump Performance Test Results

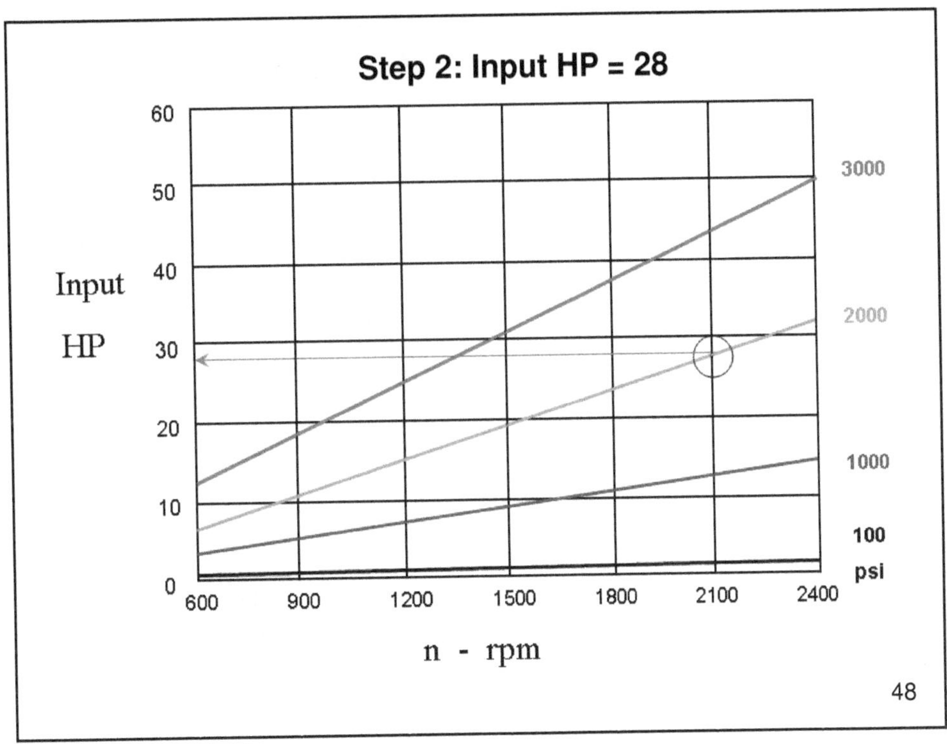

Step 3:

$$e_{vp} = \frac{231\,Q_p}{n_p D_p} = \frac{231 \times 19}{2100 \times 2.37} = 0.882 = 88.2\%$$

$$e_{tp} = \frac{Q_p p_p}{1714\,h_p} = \frac{19 \times 2000}{1714 \times 28} = 0.792 = 79.2\,\%$$

$$e_{mp} = \frac{e_{tp}}{e_{vp}} = \frac{0.792}{0.882} = 0.898 = 89.8\,\%$$

49

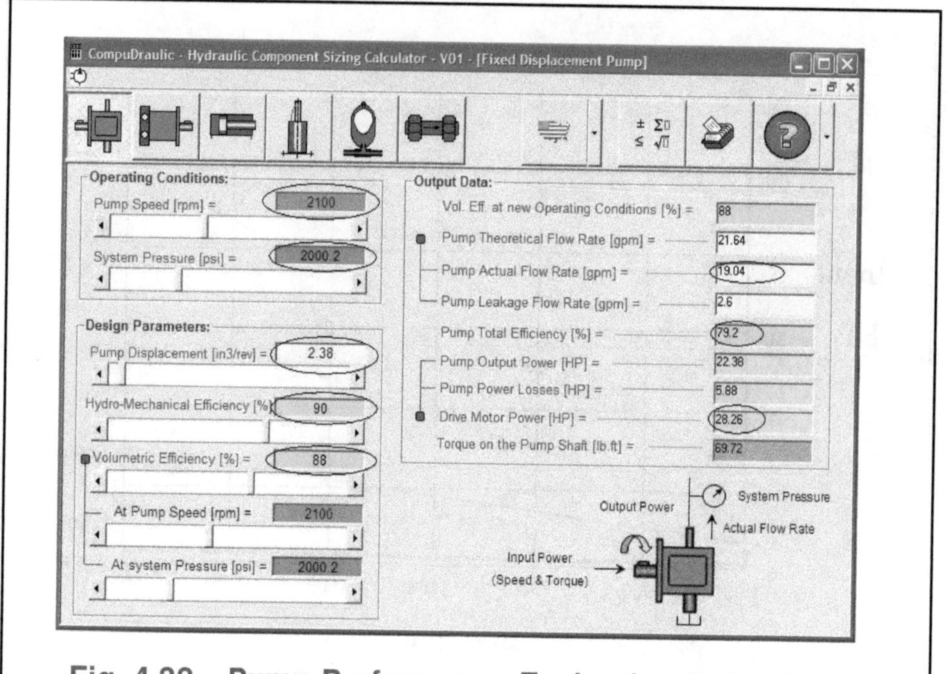

Fig. 4.32 – Pump Performance Evaluation Validation 50

4.12.7- Motor Performance Evaluation Case Study

Given: $D_m = 4$ (in^3/rev), $n_m = 1300$ rpm, and $T_m = 1000$ lb.in.

Step 1:

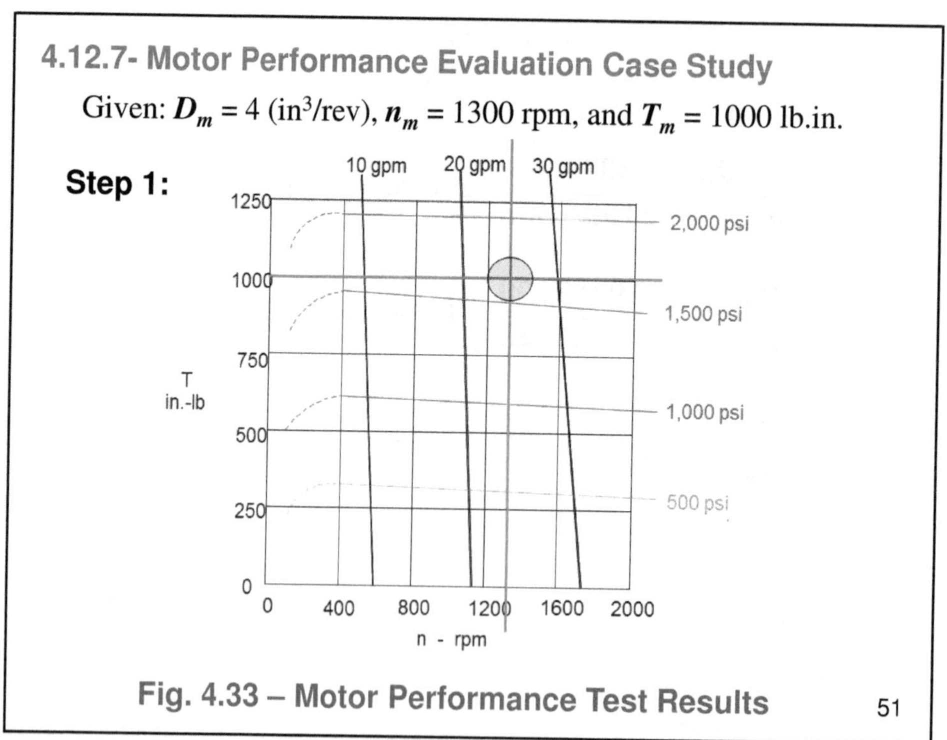

Fig. 4.33 – Motor Performance Test Results 51

Step 2:

$$e_{vm} = \frac{n_m D_m}{231\, Q_m} = \frac{4 \times 1300}{231 \times 25} = 0.9 = 90\%$$

$$e_{mm} = \frac{2\pi T_m}{D_m \times \Delta Dp_m} = \frac{2\pi \times 1000}{4 \times 1675} = 0.94 = 94\ \%$$

$$e_{tp} = e_{vp} \times e_{mp} = 0.9\ \times 0.94 = \ 0.846 = 84.6\ \%$$

52

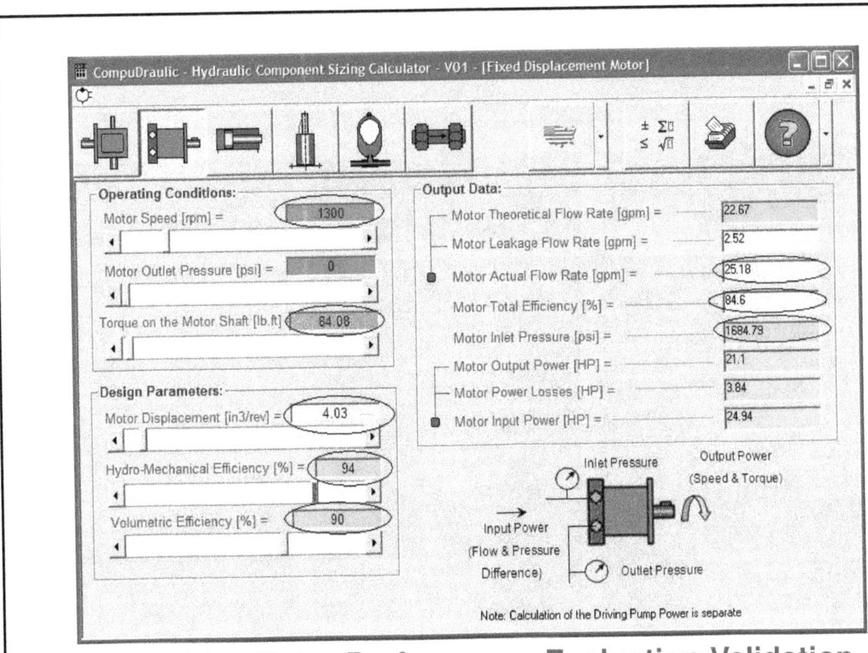

Fig. 4.34 – Motor Performance Evaluation Validation

53

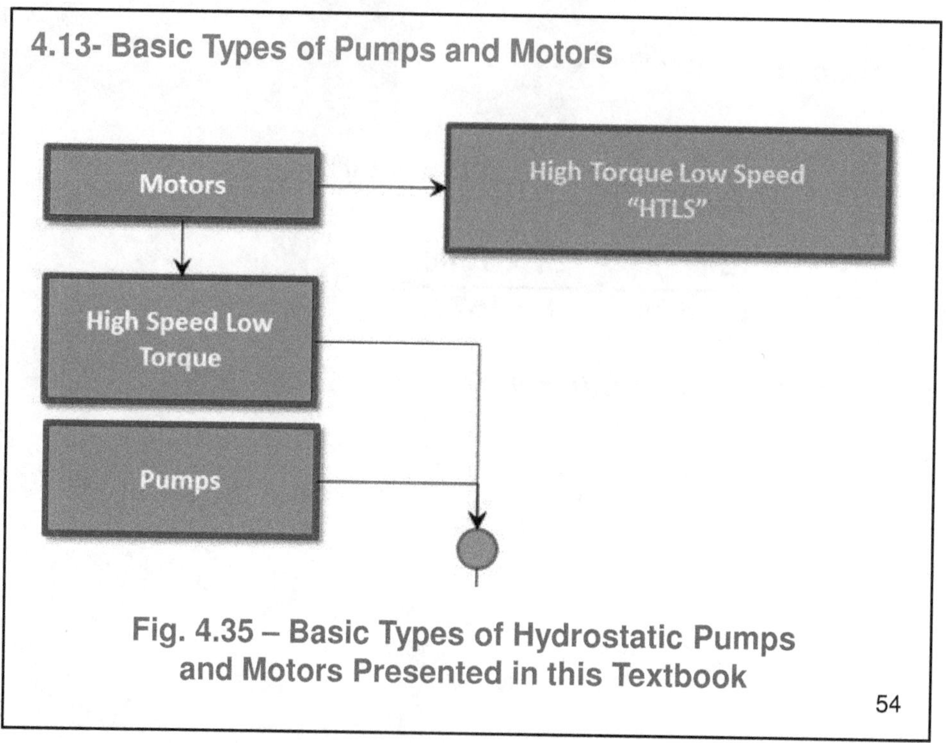

Fig. 4.35 – Basic Types of Hydrostatic Pumps and Motors Presented in this Textbook

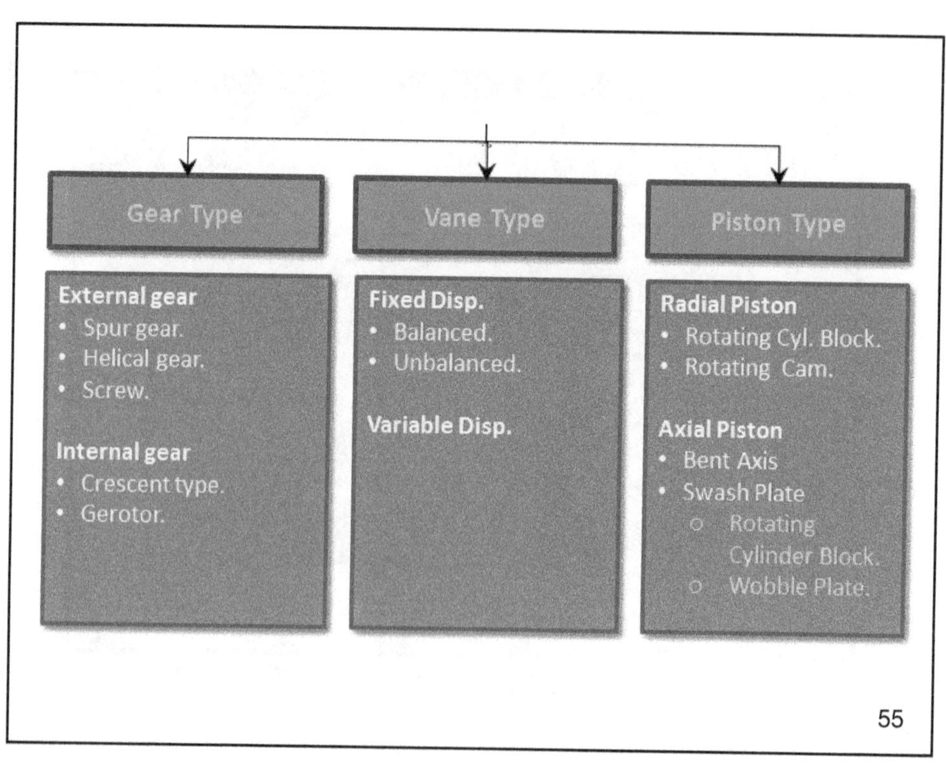

4.14- External Gear Pumps and Motors

4.14.1- External Gear Pumps

Gear pumps in general are famous and known for a very long time for the following reasons:

- **Simple in design, fabrication and assembly.**
- **Least expensive among other pumps.**
- **Tolerant to oil contamination unlike other pumps.**
- **Working pressure range up to 200 bar (3000 psi).**
- **Offers wide range of sizes up to 200 cc/rev (12 in^3/rev).**

 *Discuss suction & discharge process?*

56

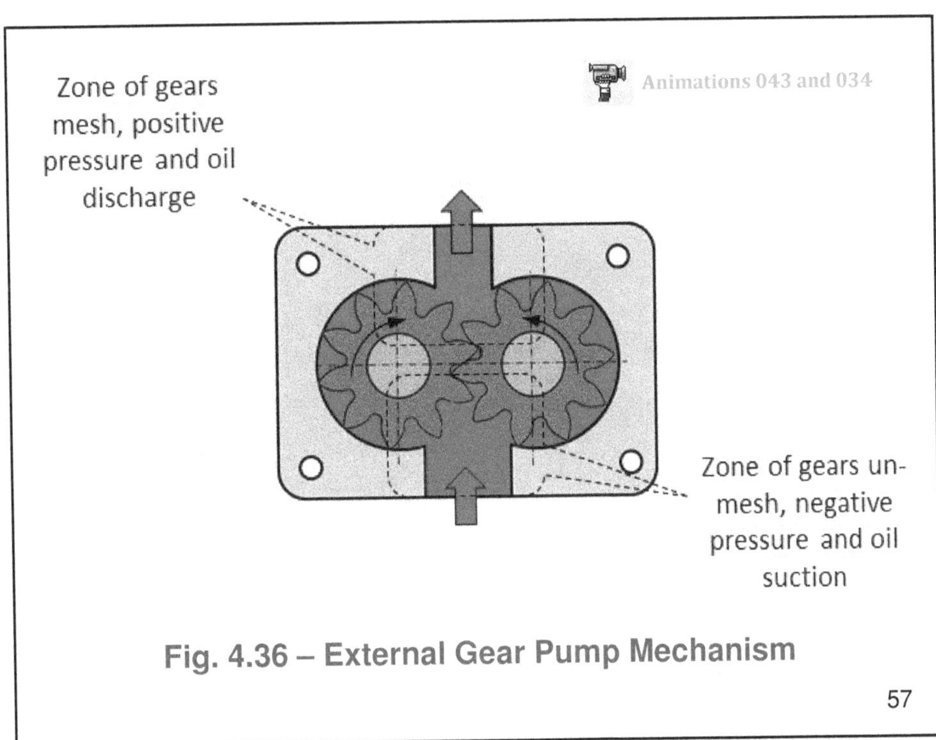

Fig. 4.36 – External Gear Pump Mechanism

57

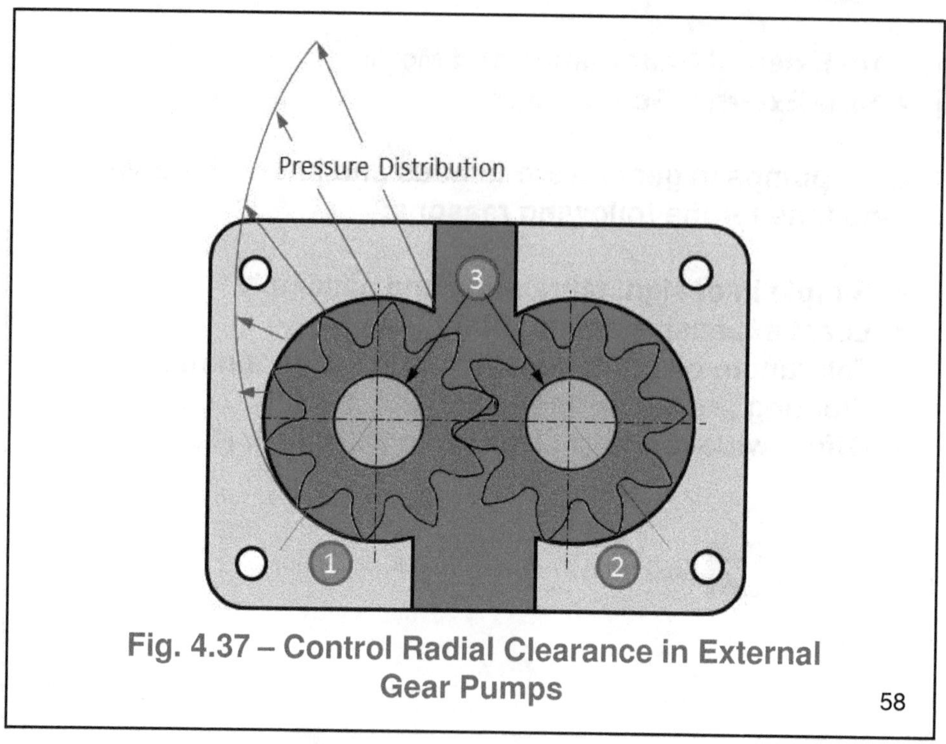

Fig. 4.37 – Control Radial Clearance in External Gear Pumps

58

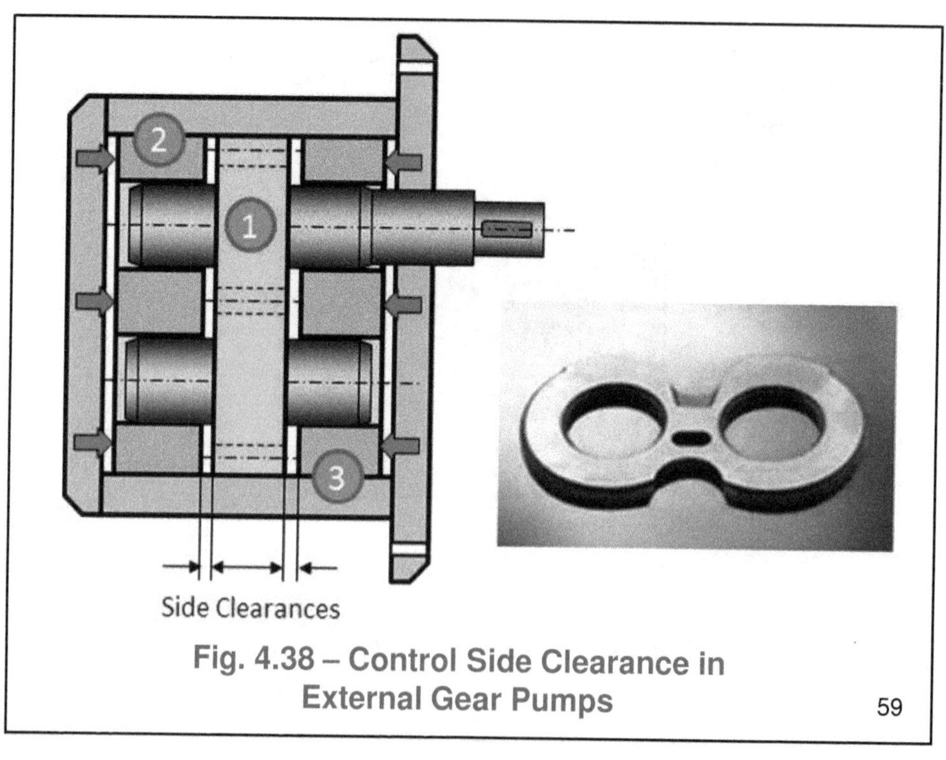

Fig. 4.38 – Control Side Clearance in External Gear Pumps

59

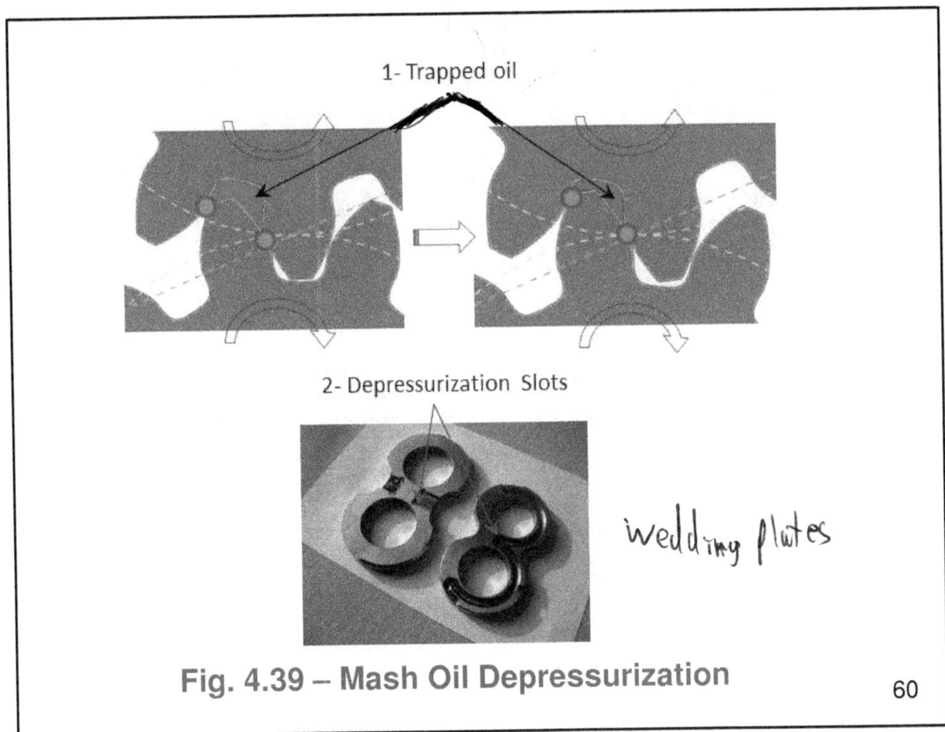

Fig. 4.39 – Mash Oil Depressurization

60

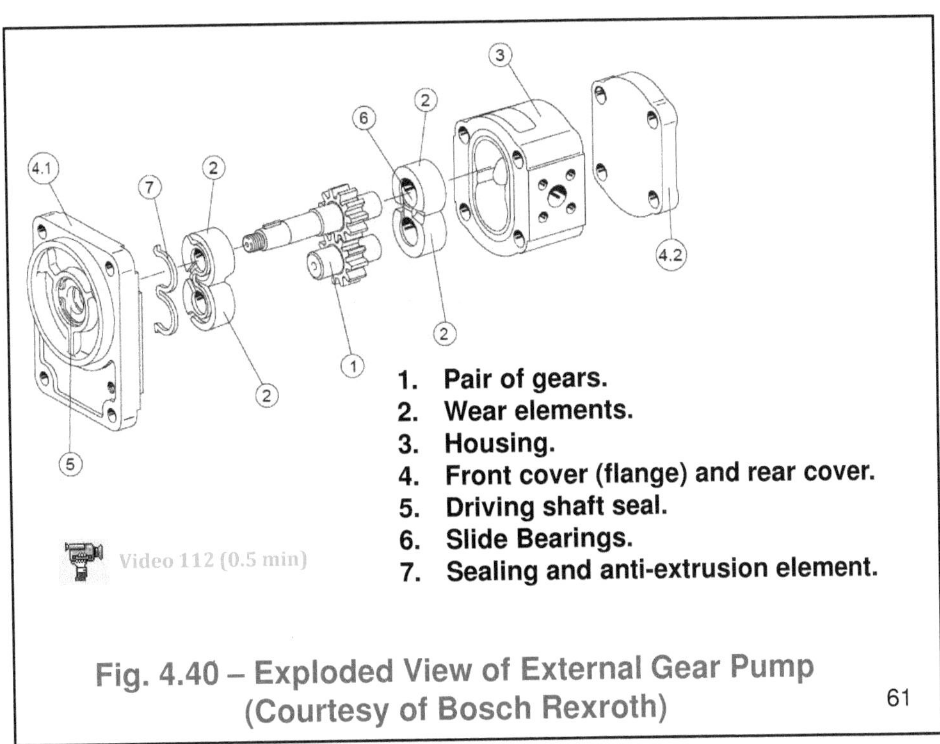

1. Pair of gears.
2. Wear elements.
3. Housing.
4. Front cover (flange) and rear cover.
5. Driving shaft seal.
6. Slide Bearings.
7. Sealing and anti-extrusion element.

Fig. 4.40 – Exploded View of External Gear Pump (Courtesy of Bosch Rexroth)

61

Fig. 4.41 – Various Pump Configurations

62

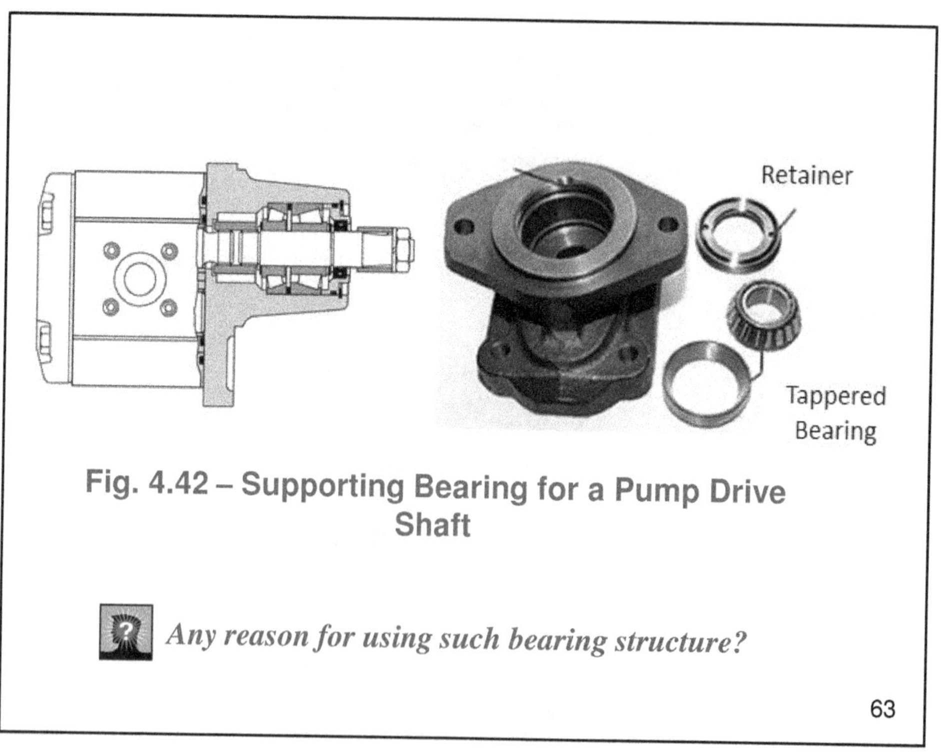

Fig. 4.42 – Supporting Bearing for a Pump Drive Shaft

Any reason for using such bearing structure?

63

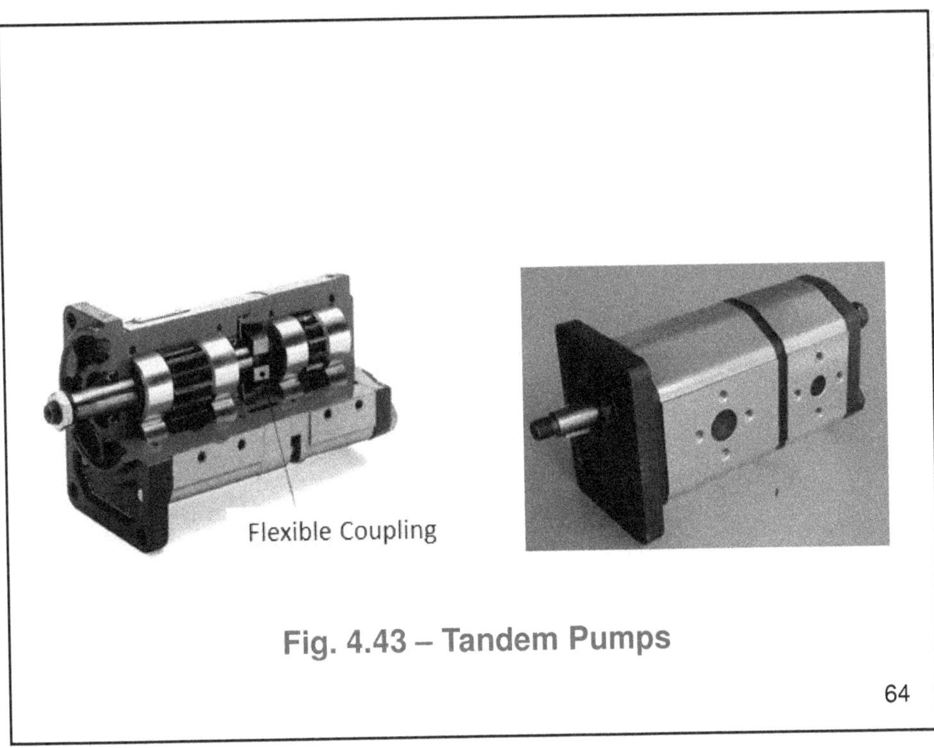

Fig. 4.43 – Tandem Pumps

64

4.14.2- External Gear Motor

Animations 014 and 015

Video 116 (0.5 min)

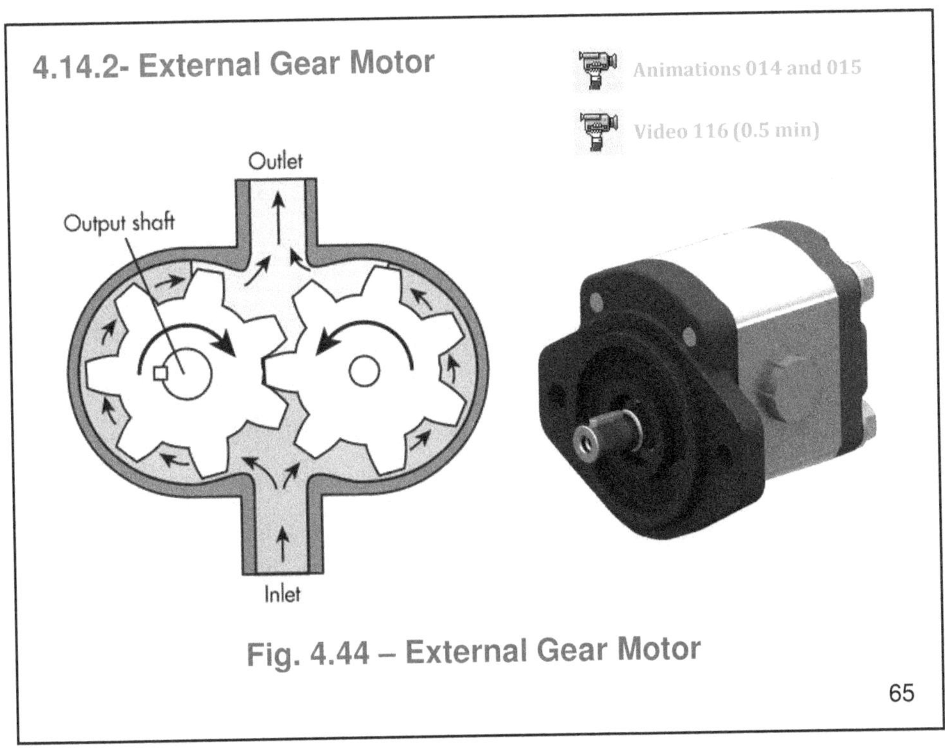

Fig. 4.44 – External Gear Motor

65

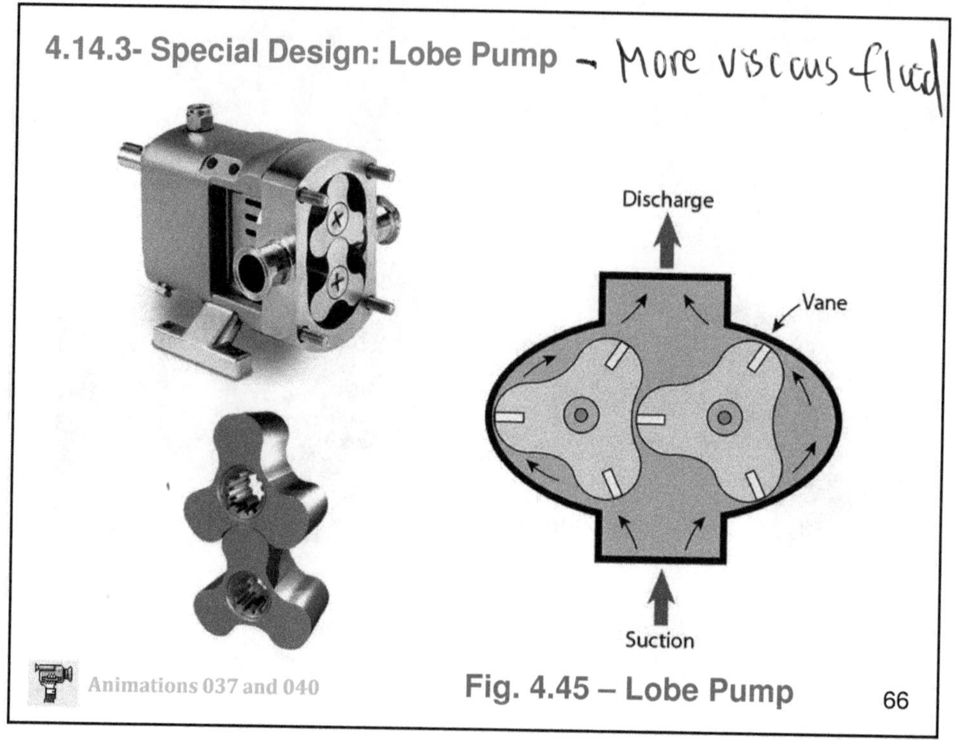

Fig. 4.45 – Lobe Pump

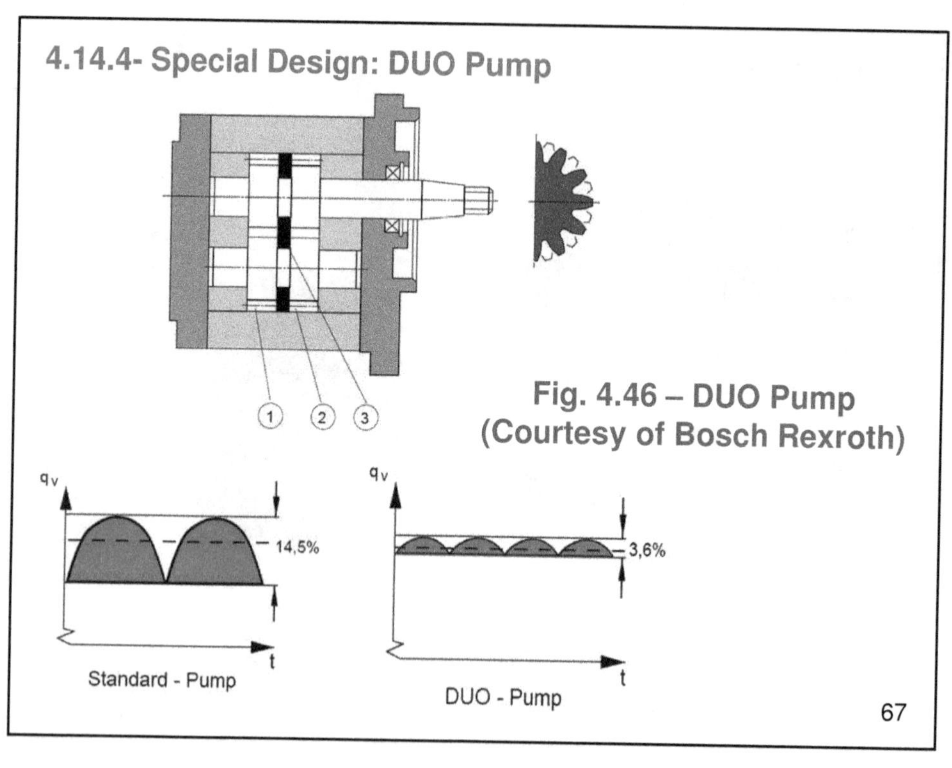

Fig. 4.46 – DUO Pump (Courtesy of Bosch Rexroth)

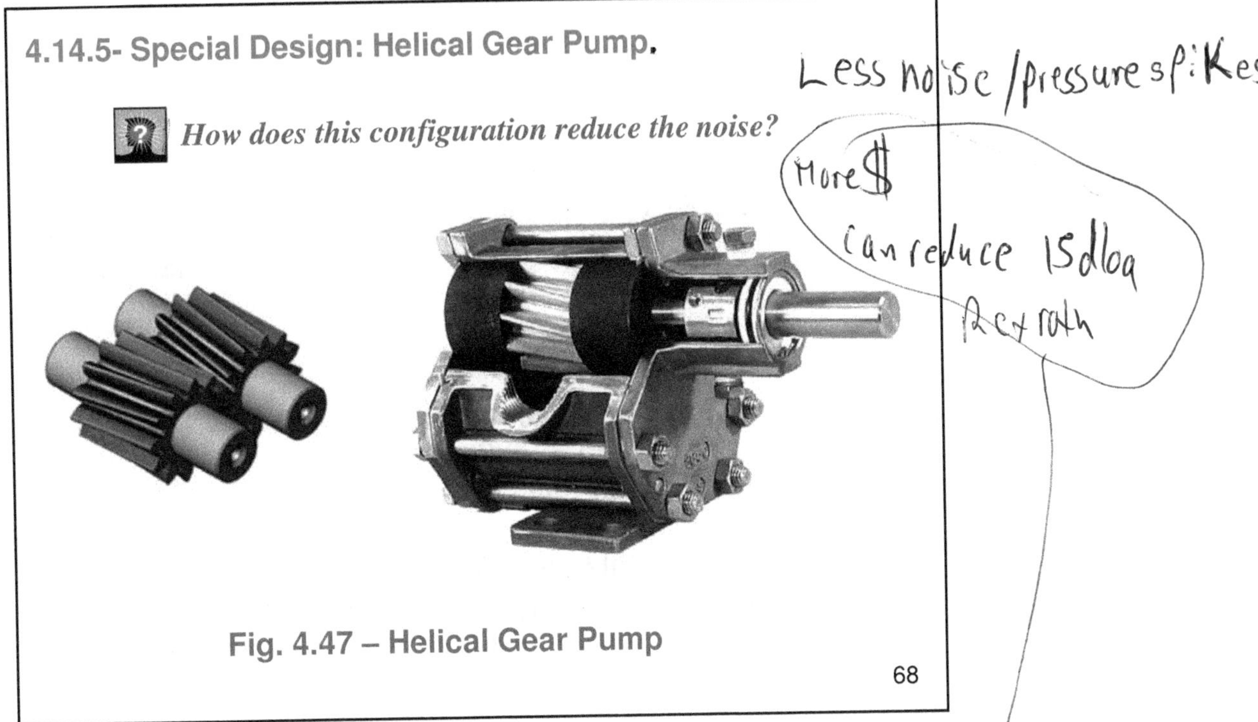
4.14.5- Special Design: Helical Gear Pump.
How does this configuration reduce the noise?
Less noise/pressure spikes
More $
Can reduce 15dba
Rexroth
Fig. 4.47 – Helical Gear Pump
68

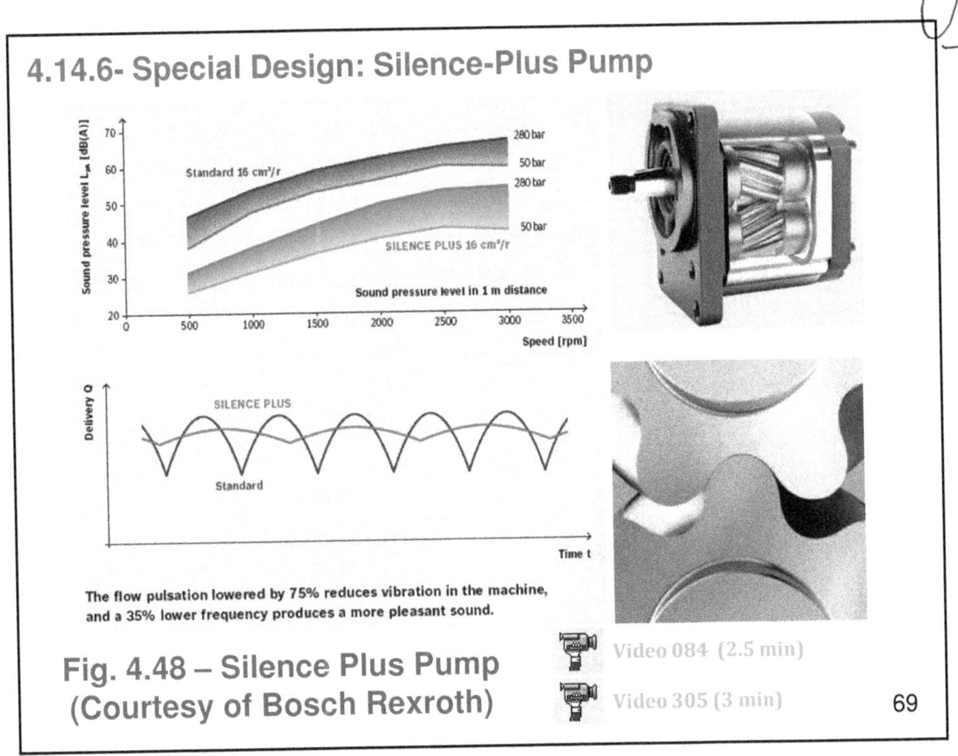
4.14.6- Special Design: Silence-Plus Pump
Sound pressure level L_{pA} [dB(A)]
70
60
50
40
30
20
0
500
1000
1500
2000
2500
3000
3500
Speed [rpm]
Standard 16 cm³/r
280 bar
50 bar
280 bar
50 bar
SILENCE PLUS 16 cm³/r
Sound pressure level in 1 m distance
Delivery Q
SILENCE PLUS
Standard
Time t
The flow pulsation lowered by 75% reduces vibration in the machine, and a 35% lower frequency produces a more pleasant sound.
Fig. 4.48 – Silence Plus Pump (Courtesy of Bosch Rexroth)
Video 084 (2.5 min)
Video 305 (3 min)
69

Shhark pump

4.14.7- Special Design: Screw Pump

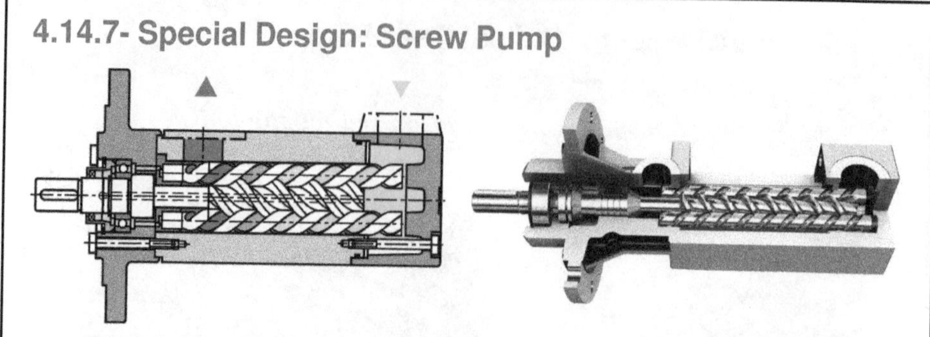

Fig. 4.49 – Screw Pump (Courtesy of Bosch Rexroth)

- **Working pressure (≅ 120bar) & peak press. (≅ 200 bar).**
- **Suction pressure up to -0.2 psig.**
- **Driving speed up to 5000 rev/min.**
- **Medium-high flow rates.**
- **Very noiseless, little vibration & almost no pulsations.**
- **Compatible with most hydraulic liquids.**
- **Overall eff. 75% under optimum working conditions.**
- **Long-lasting.**

70

4.15- Internal Gear Pumps and Motors

smaller

4.15.1- Internal Gear Pumps

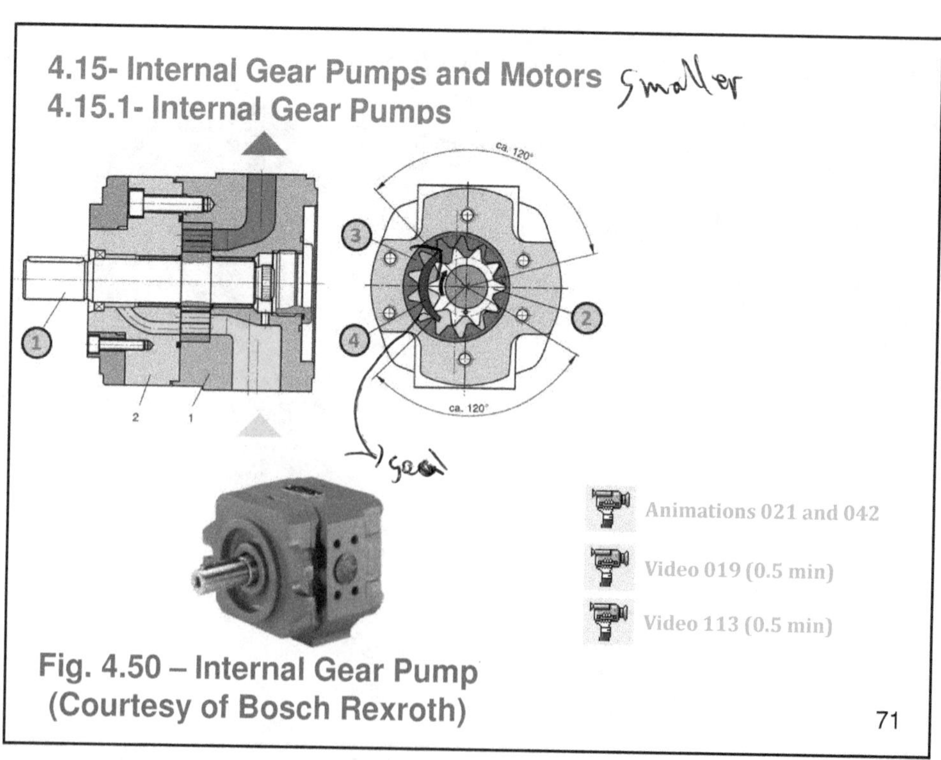

Fig. 4.50 – Internal Gear Pump (Courtesy of Bosch Rexroth)

71

4.15.2- External versus Internal Gear Pumps

External	Internal
▪ Identical gears -----------------------------	Not identical
▪ Gears running at same speed ----------	Different speeds
▪ Gears running at opposite direction--	Same Direction
▪ Gears are in external mesh -------------	Internal mesh
▪ No physical sealing element -----------	Crescent seal
▪ Four bearings ------------------------------	Two bearings
▪ Bigger size ------------------------	Compact
▪ Less expensive --------------------	Less noise (2 bearings)

72

4.15.3- Special Design: Gerotor Pumps and Motors

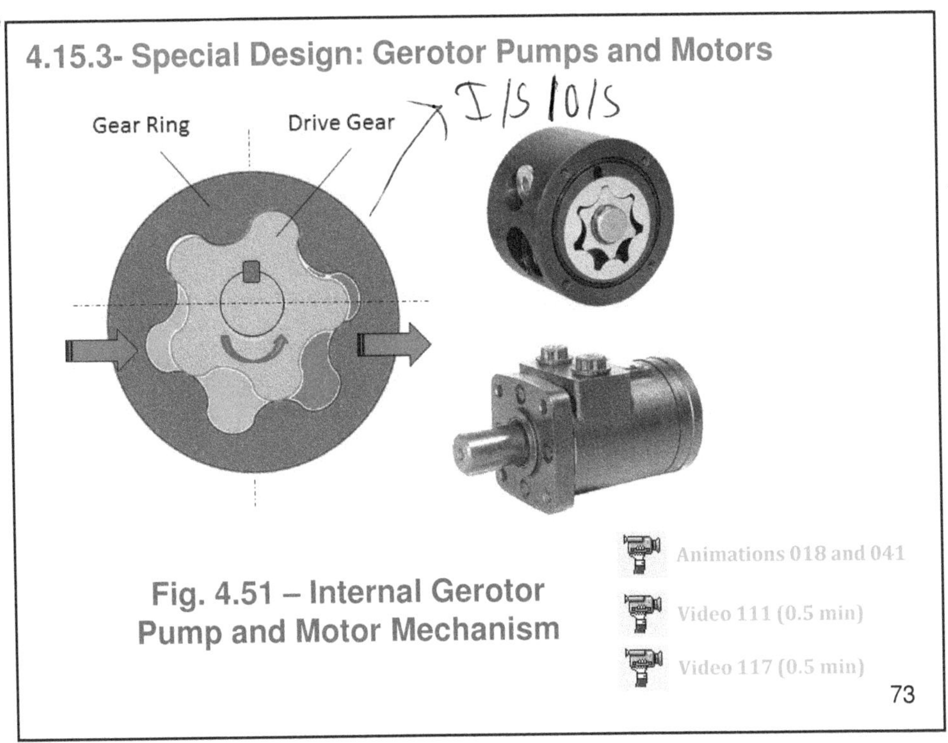

Fig. 4.51 – Internal Gerotor Pump and Motor Mechanism

Animations 018 and 041

Video 111 (0.5 min)

Video 117 (0.5 min)

73

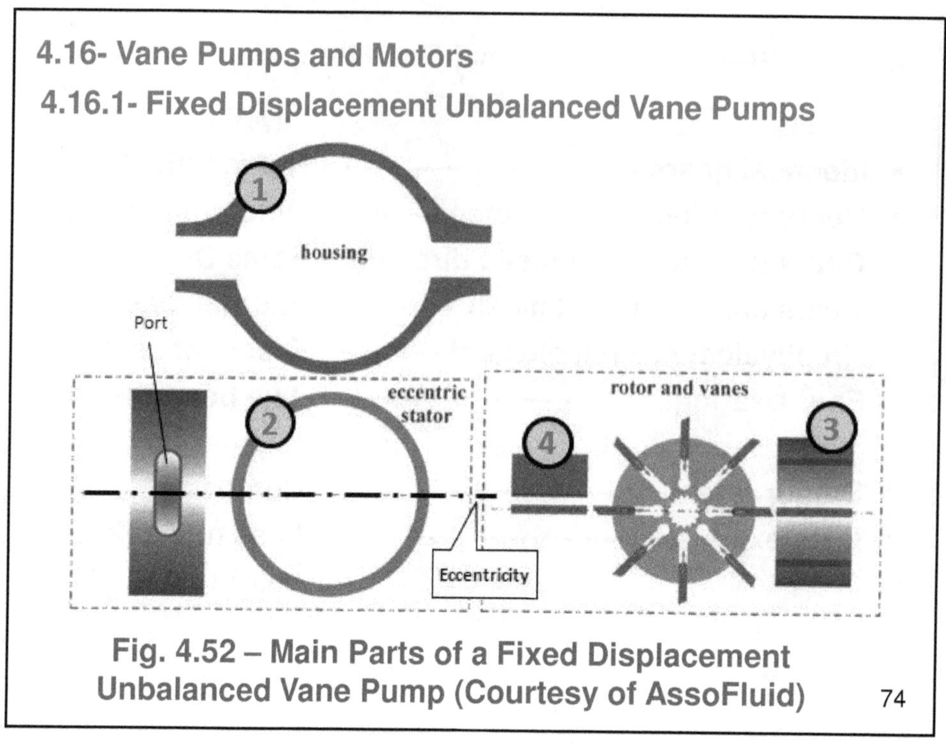

Fig. 4.52 – Main Parts of a Fixed Displacement Unbalanced Vane Pump (Courtesy of AssoFluid) 74

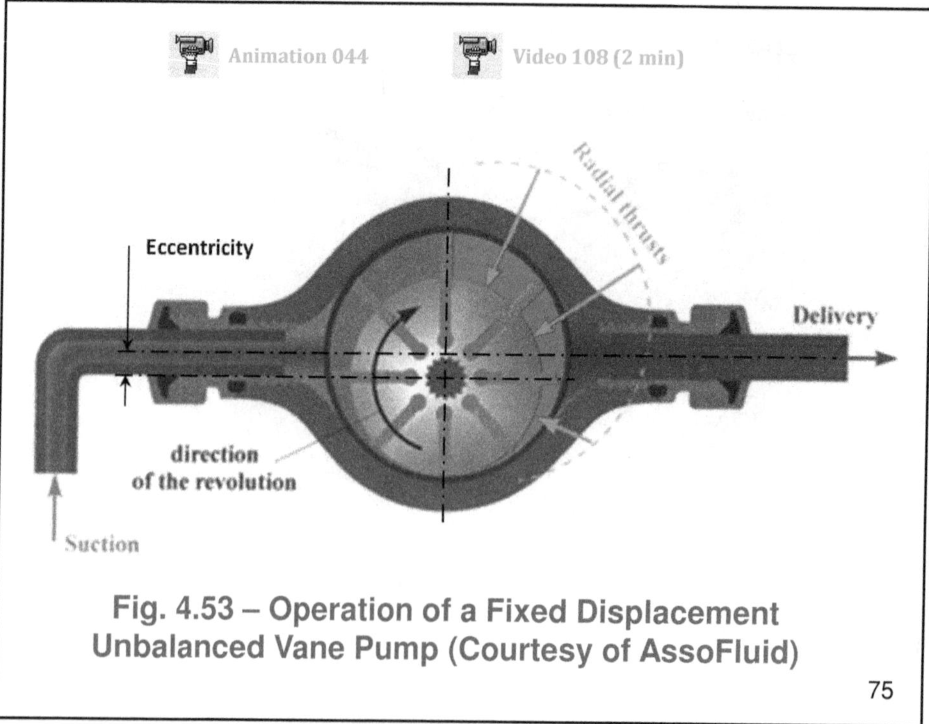

Fig. 4.53 – Operation of a Fixed Displacement Unbalanced Vane Pump (Courtesy of AssoFluid) 75

Should not be used @ high Or low speed

Fig. 4.54 – Single and Tandem Vane Pumps 76

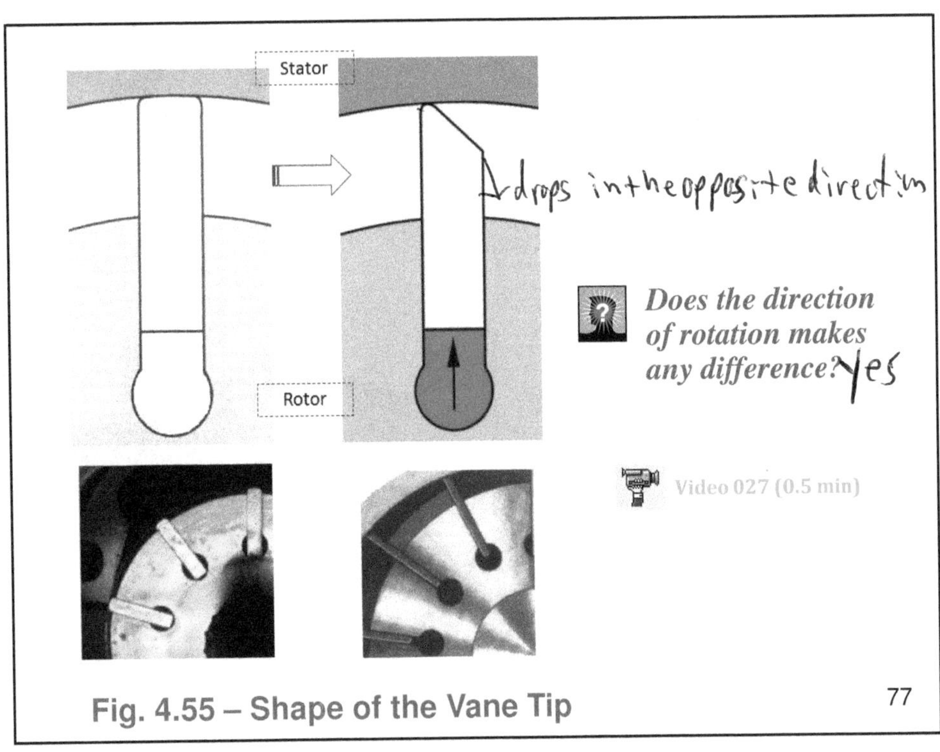

Fig. 4.55 – Shape of the Vane Tip 77

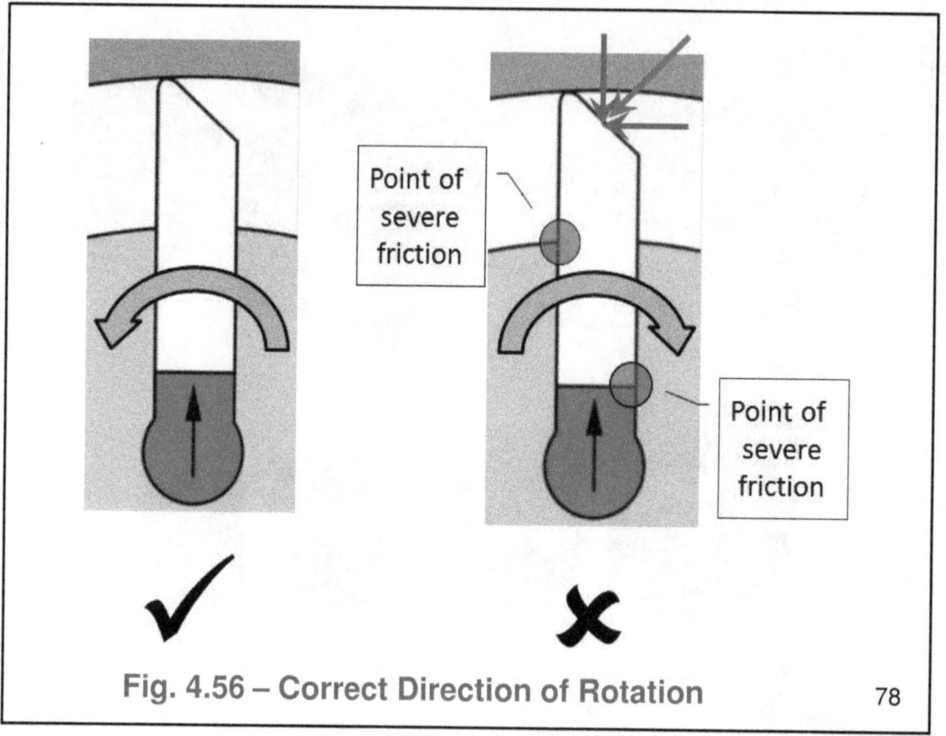

Fig. 4.56 – Correct Direction of Rotation

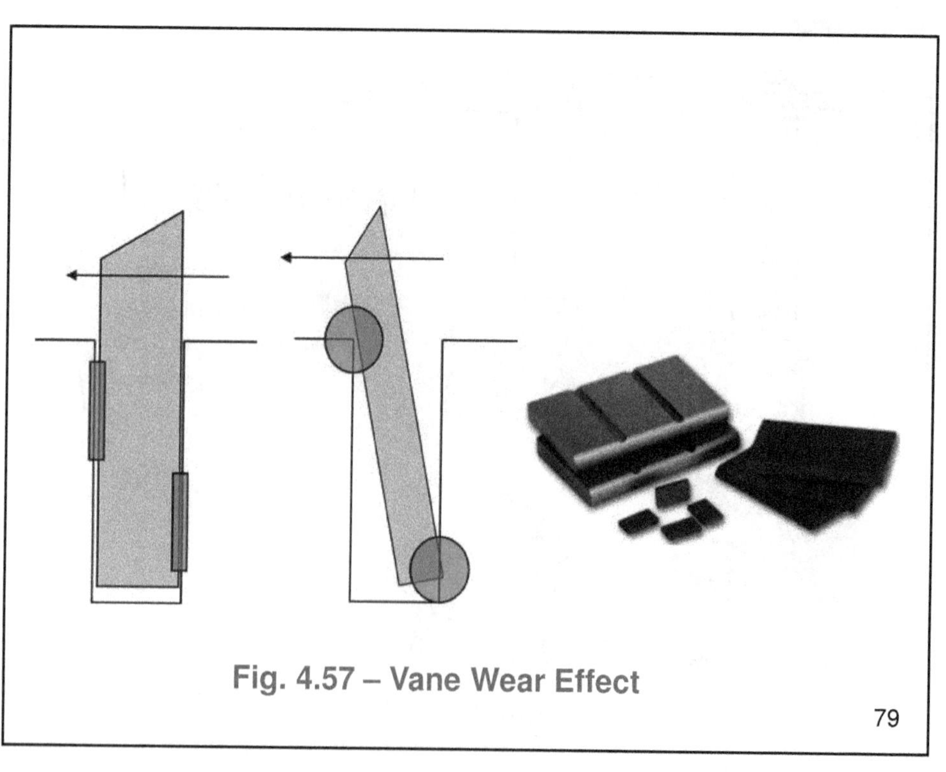

Fig. 4.57 – Vane Wear Effect

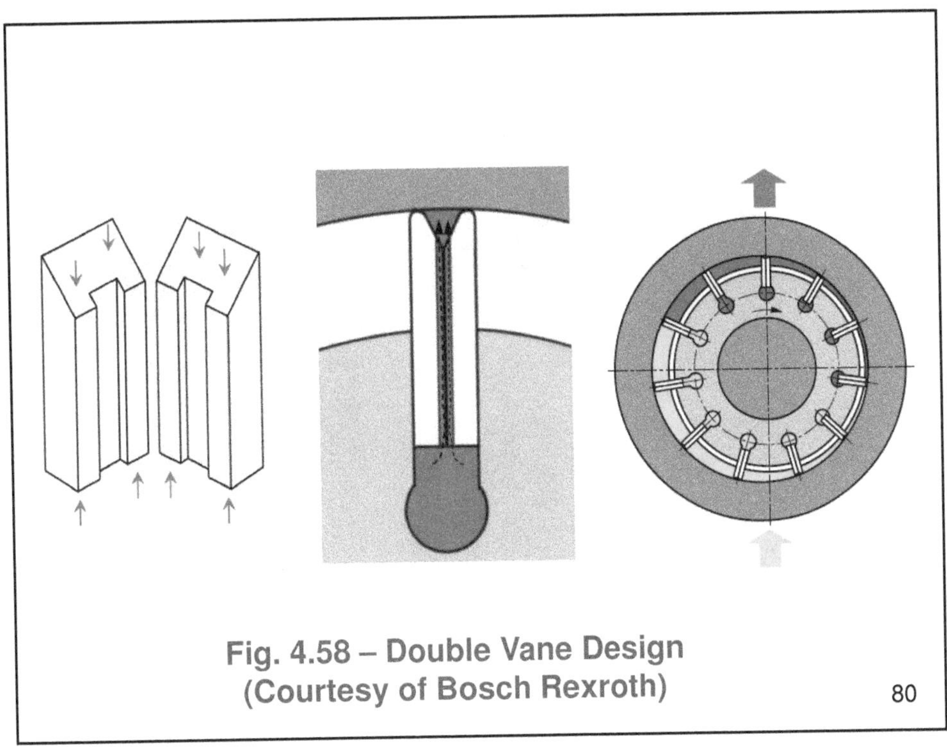

Fig. 4.58 – Double Vane Design
(Courtesy of Bosch Rexroth)

80

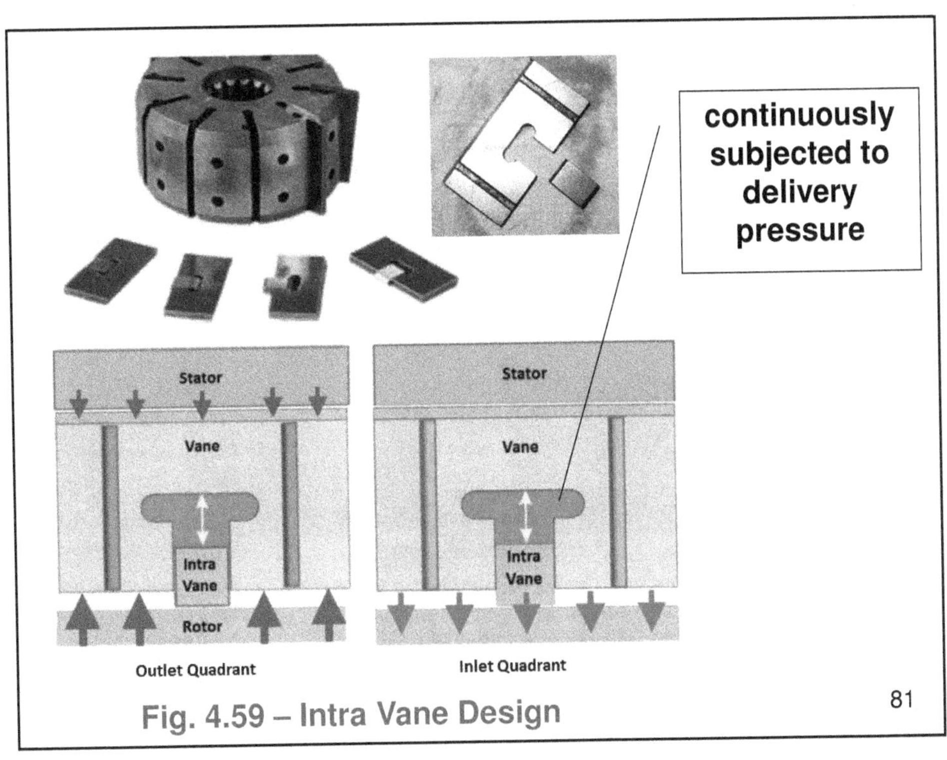

Fig. 4.59 – Intra Vane Design

81

4.16.2- Fixed Displacement Unbalanced Vane Motors

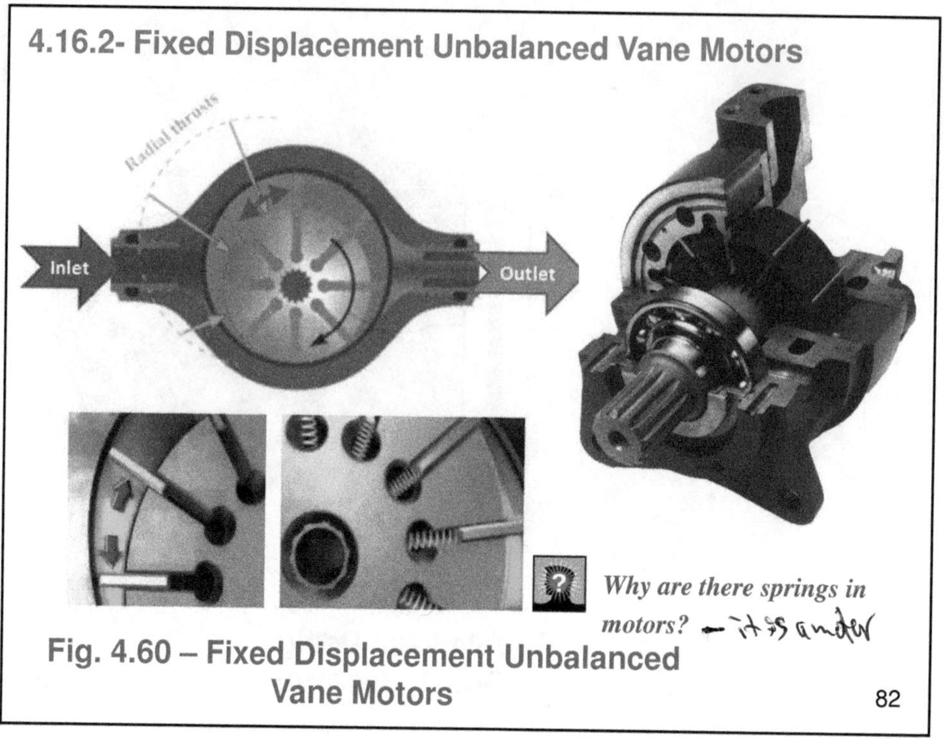

Why are there springs in motors?

Fig. 4.60 – Fixed Displacement Unbalanced Vane Motors

82

4.16.3- Fixed Displacement Balanced Vane Pumps

4.16.4- Fixed Displacement Balanced Vane Motors

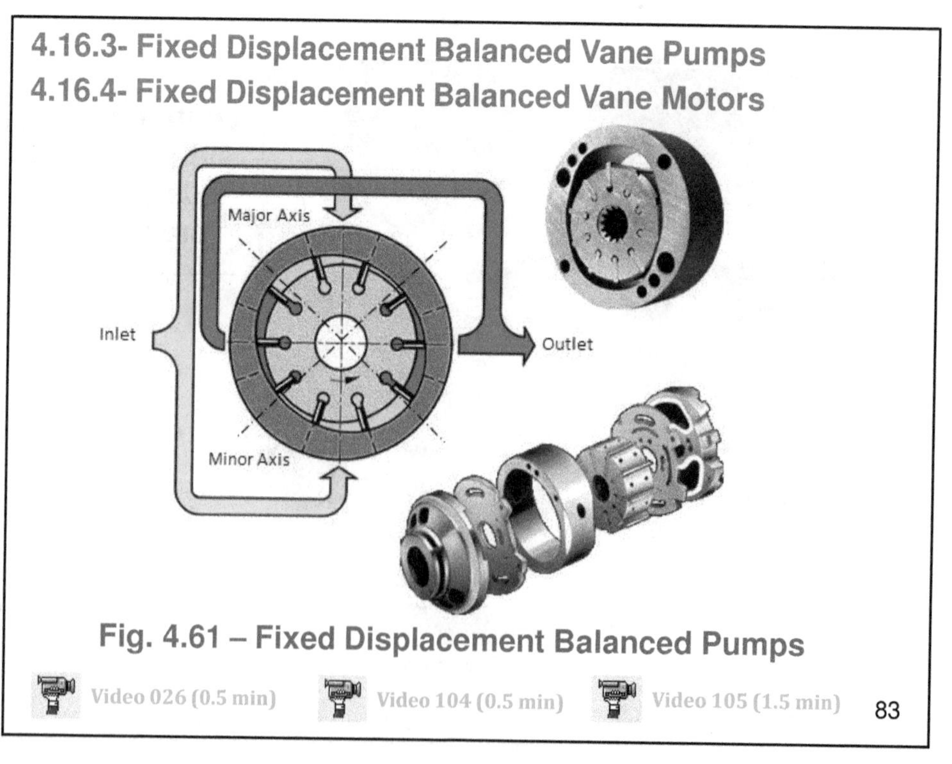

Fig. 4.61 – Fixed Displacement Balanced Pumps

Video 026 (0.5 min) Video 104 (0.5 min) Video 105 (1.5 min)

83

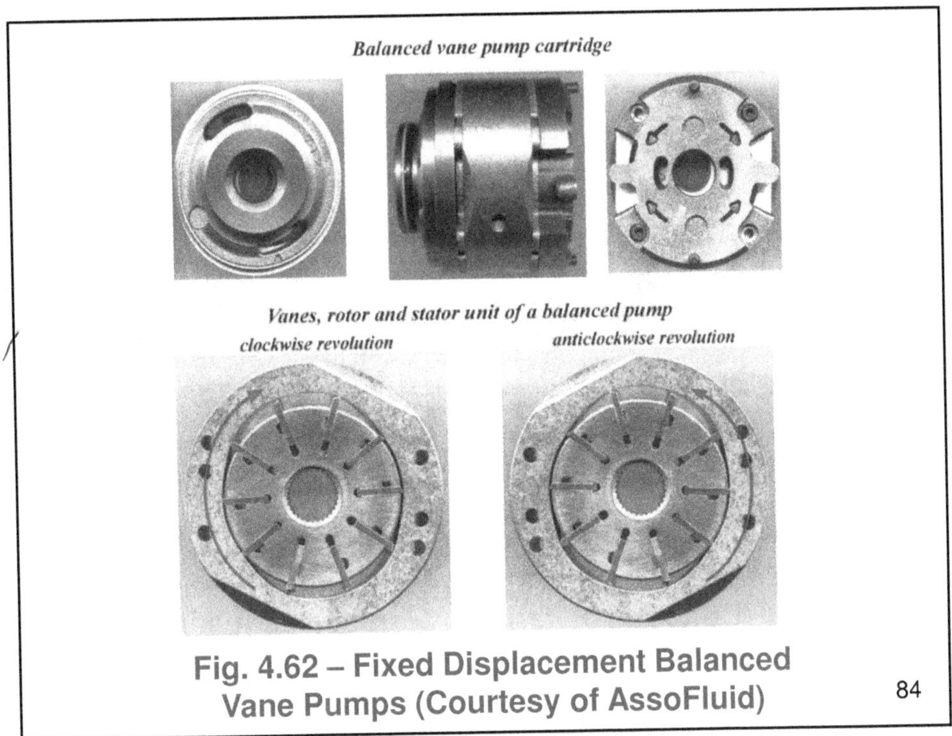

Fig. 4.62 – Fixed Displacement Balanced Vane Pumps (Courtesy of AssoFluid)

84

4.16.5- Variable Displacement Vane Pumps

unbalanced Pump

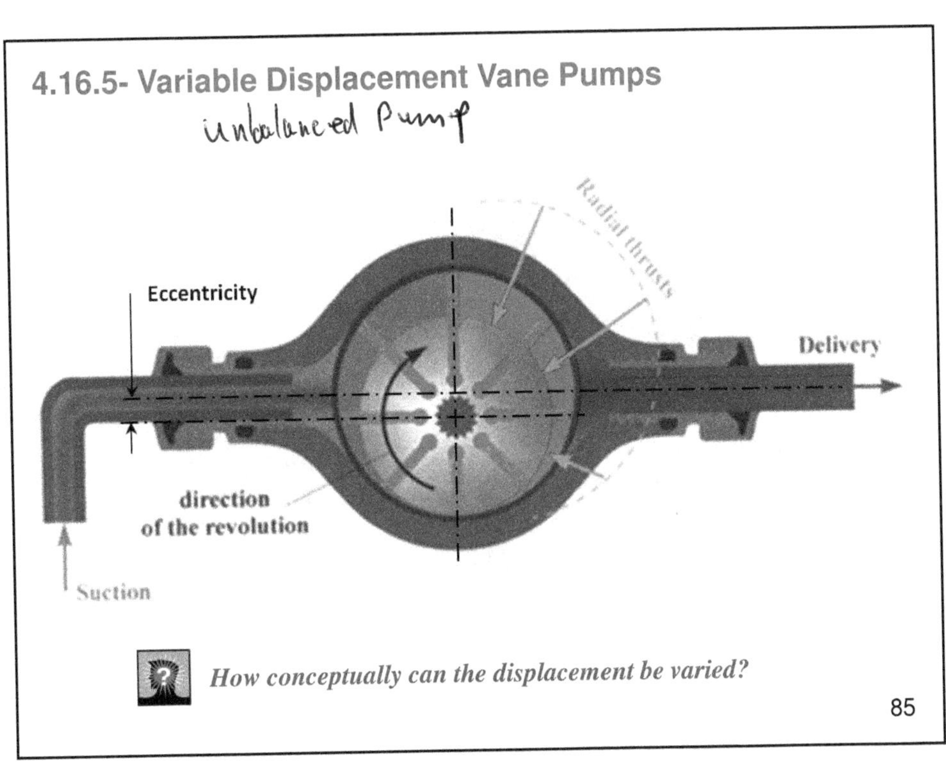

How conceptually can the displacement be varied?

85

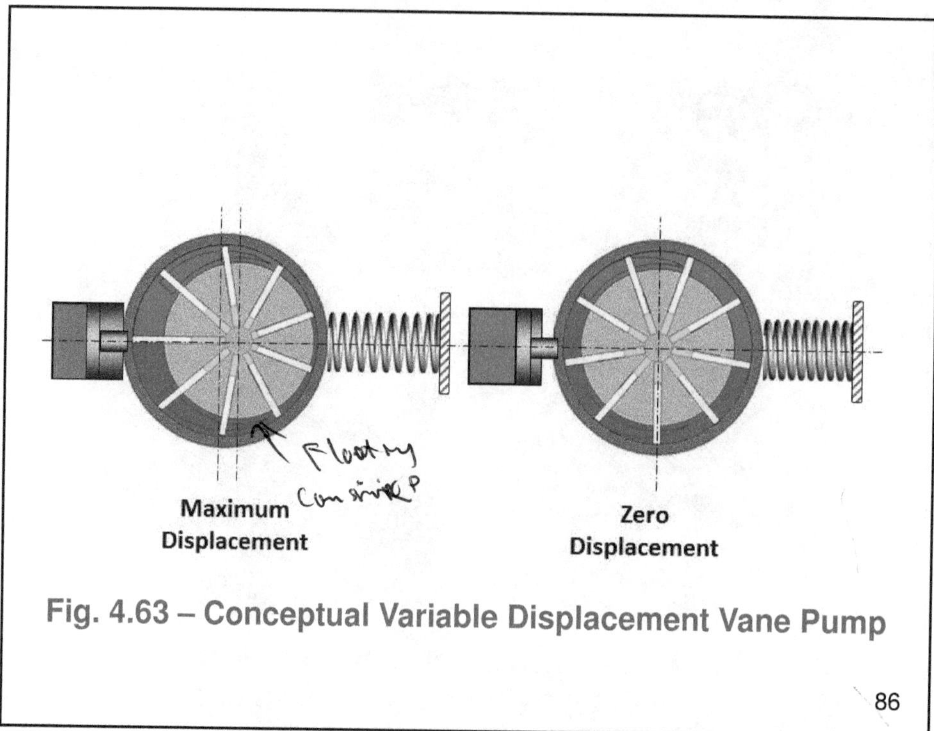

Fig. 4.63 – Conceptual Variable Displacement Vane Pump

86

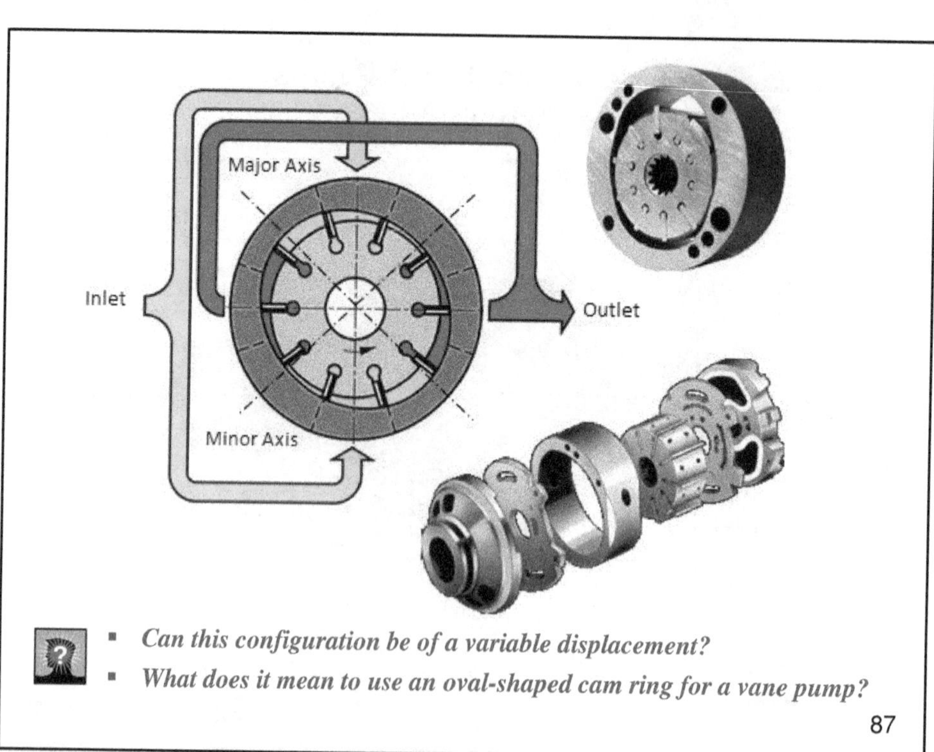

- *Can this configuration be of a variable displacement?*
- *What does it mean to use an oval-shaped cam ring for a vane pump?*

87

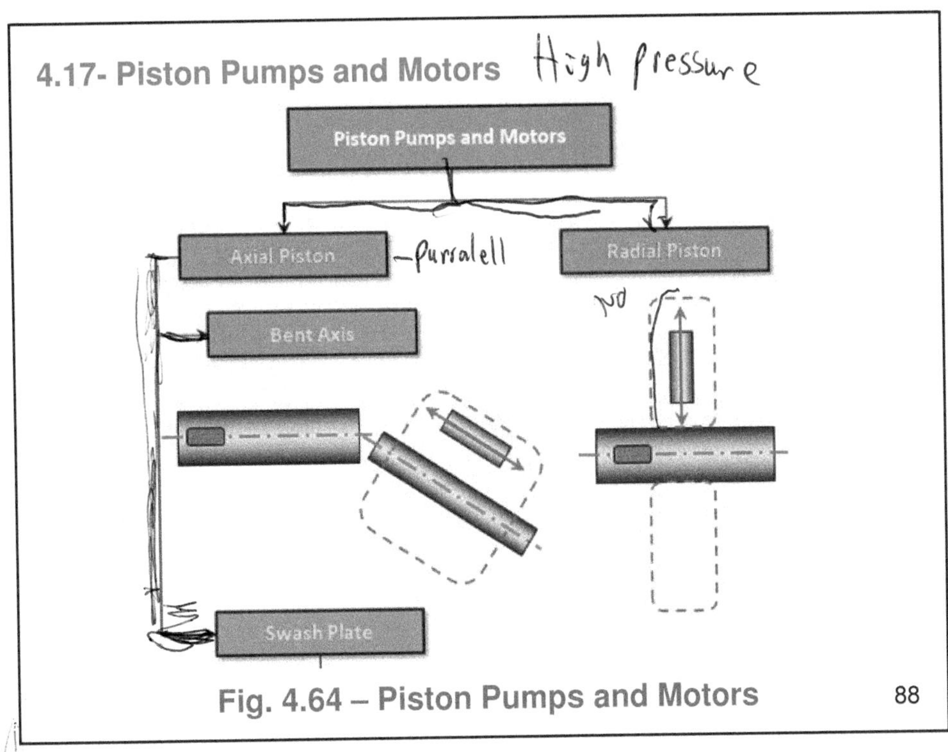

Fig. 4.64 – Piston Pumps and Motors

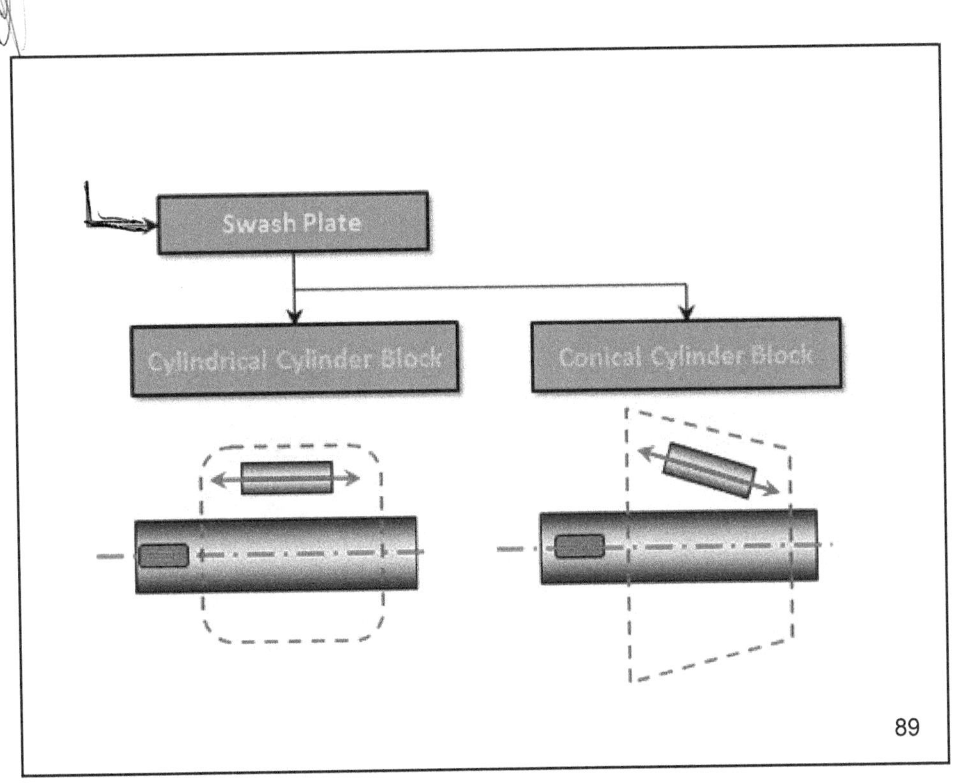

4.18- Radial Piston Pumps and Motors

4.18.1- Radial Piston Pumps

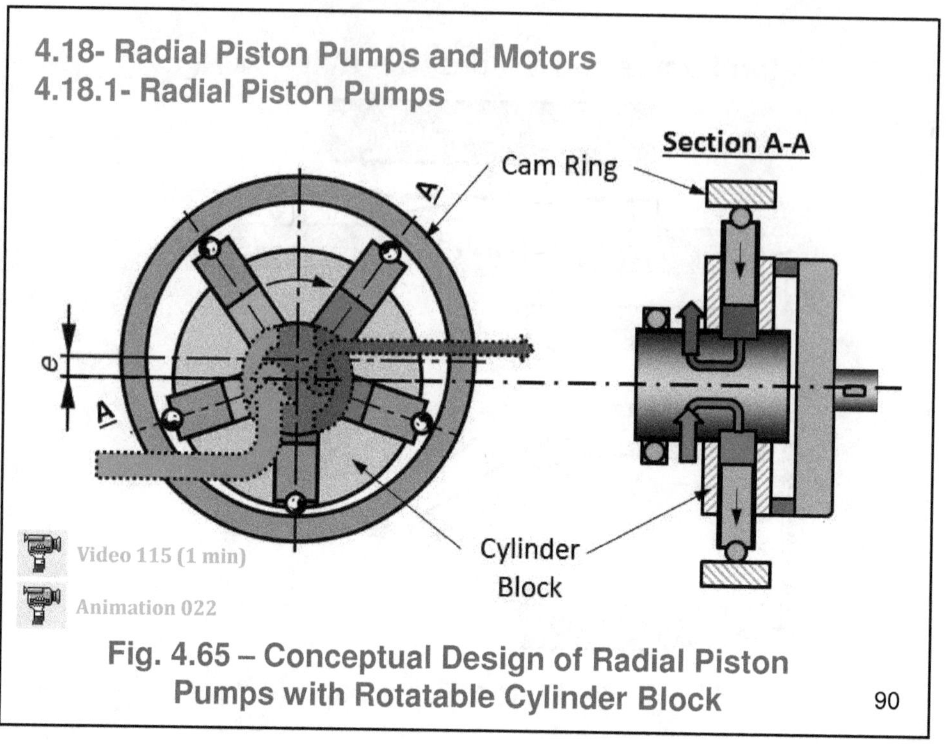

Fig. 4.65 – Conceptual Design of Radial Piston Pumps with Rotatable Cylinder Block

90

Variable Radial Piston Pump has a rotatable cylinder block

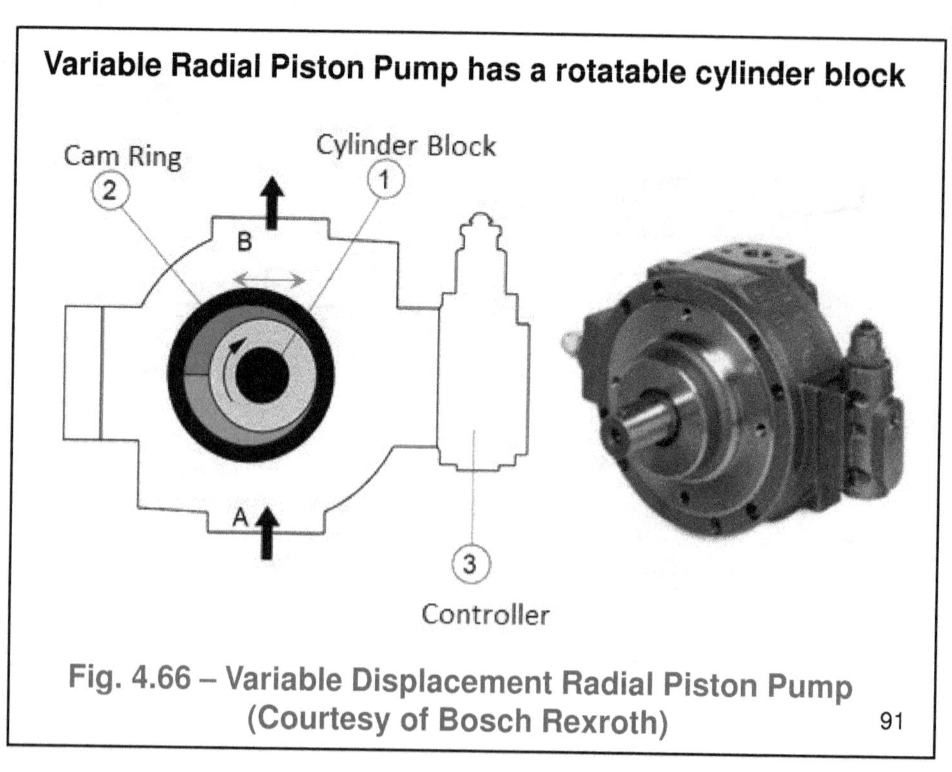

Fig. 4.66 – Variable Displacement Radial Piston Pump (Courtesy of Bosch Rexroth)

91

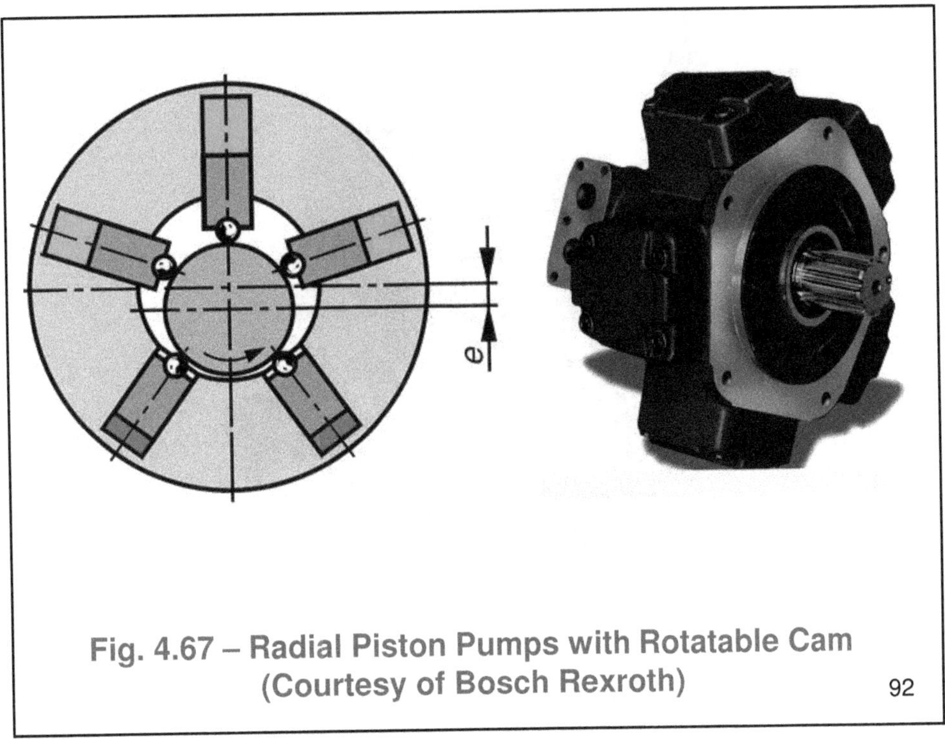

Fig. 4.67 – Radial Piston Pumps with Rotatable Cam (Courtesy of Bosch Rexroth) 92

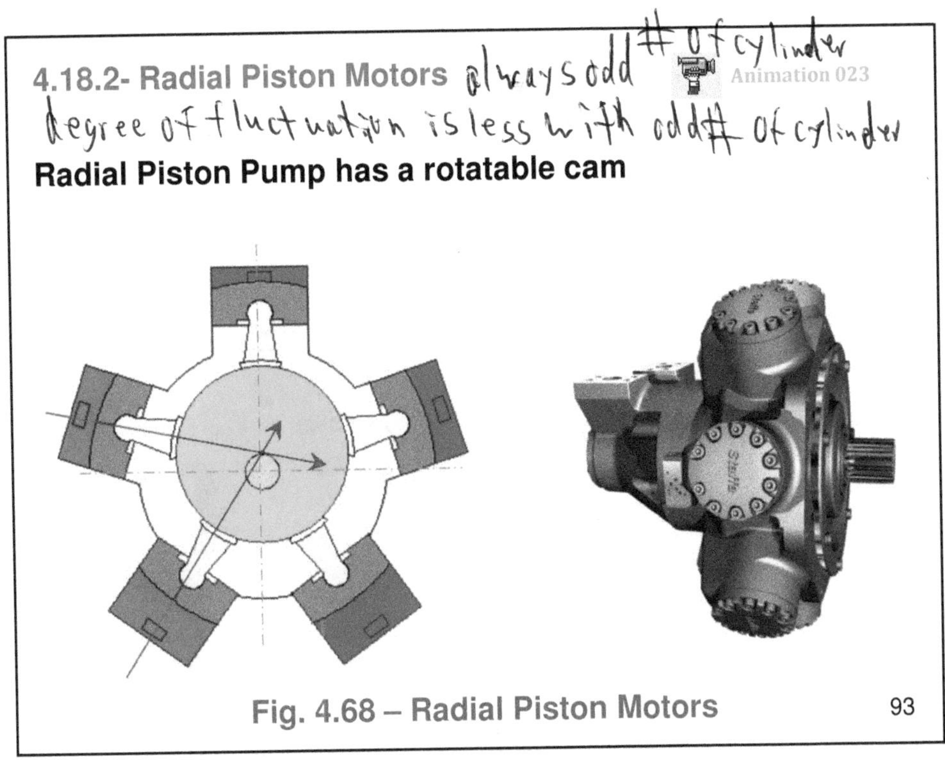

Fig. 4.68 – Radial Piston Motors 93

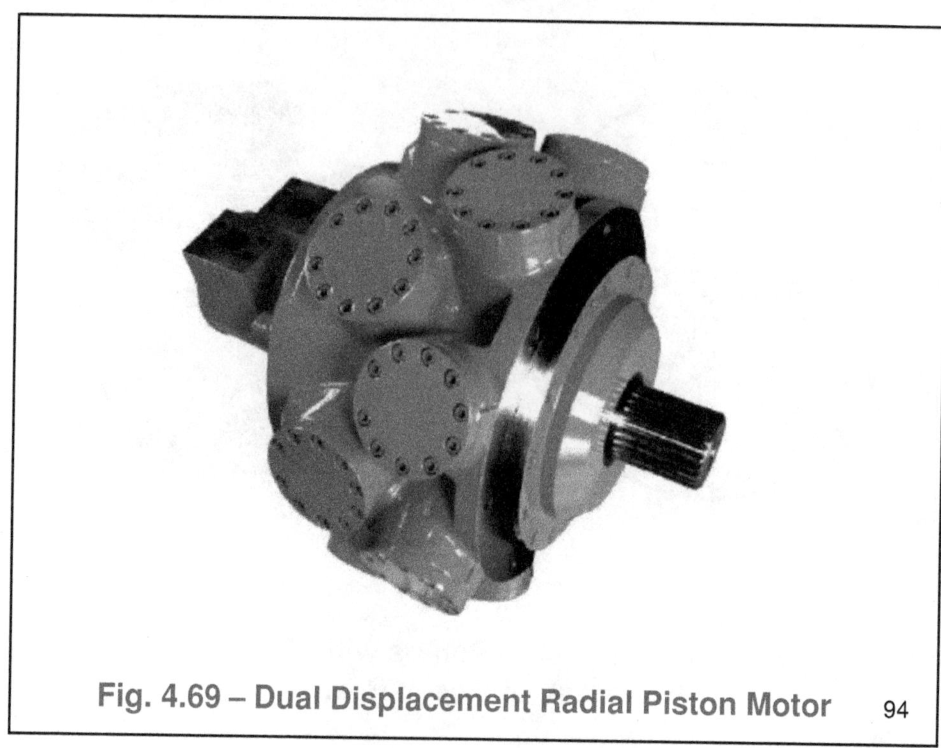

Fig. 4.69 – Dual Displacement Radial Piston Motor 94

4.19- Axial Piston Bent-Axis Pumps and Motors
4.19.1- Fixed Displacement Bent-Axis Pumps

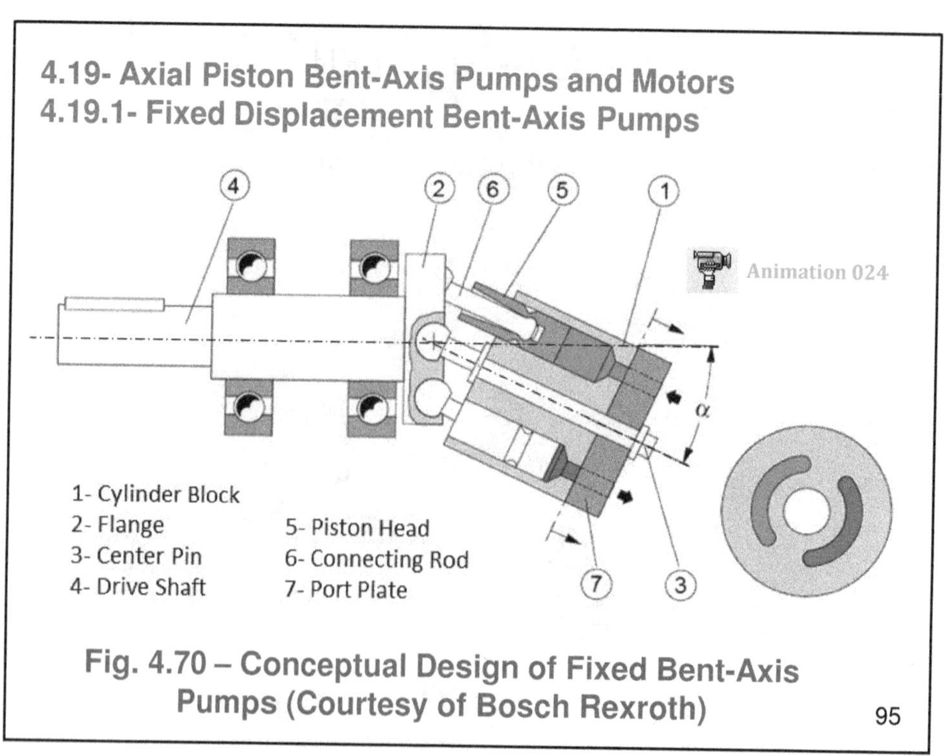

Fig. 4.70 – Conceptual Design of Fixed Bent-Axis Pumps (Courtesy of Bosch Rexroth) 95

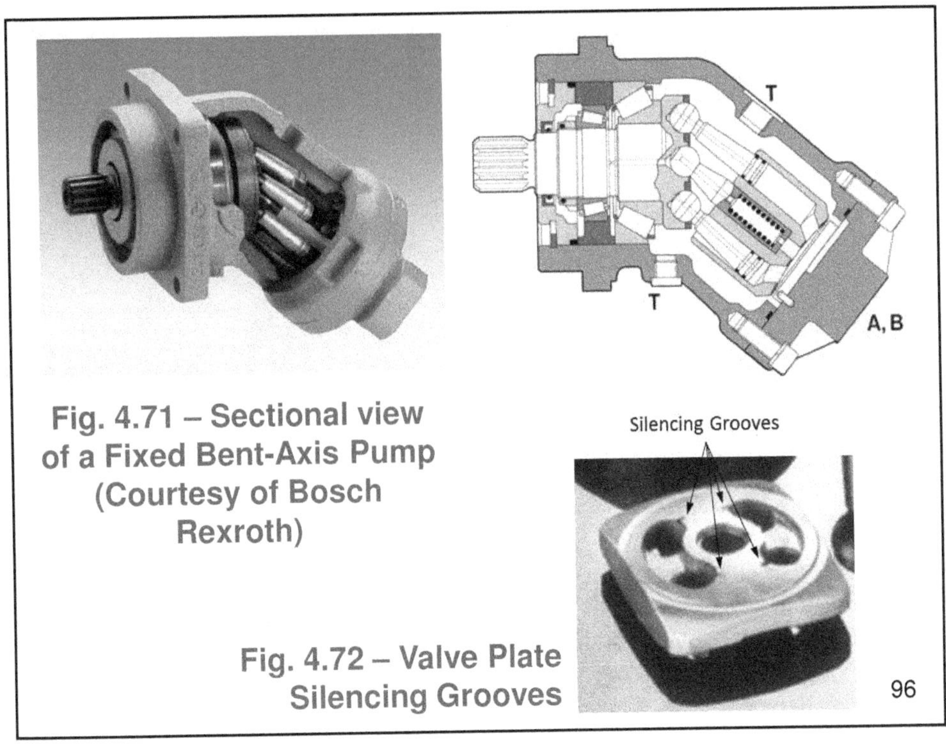

Fig. 4.71 – Sectional view of a Fixed Bent-Axis Pump (Courtesy of Bosch Rexroth)

Fig. 4.72 – Valve Plate Silencing Grooves

4.19.2- Fixed Displacement Bent-Axis Motors

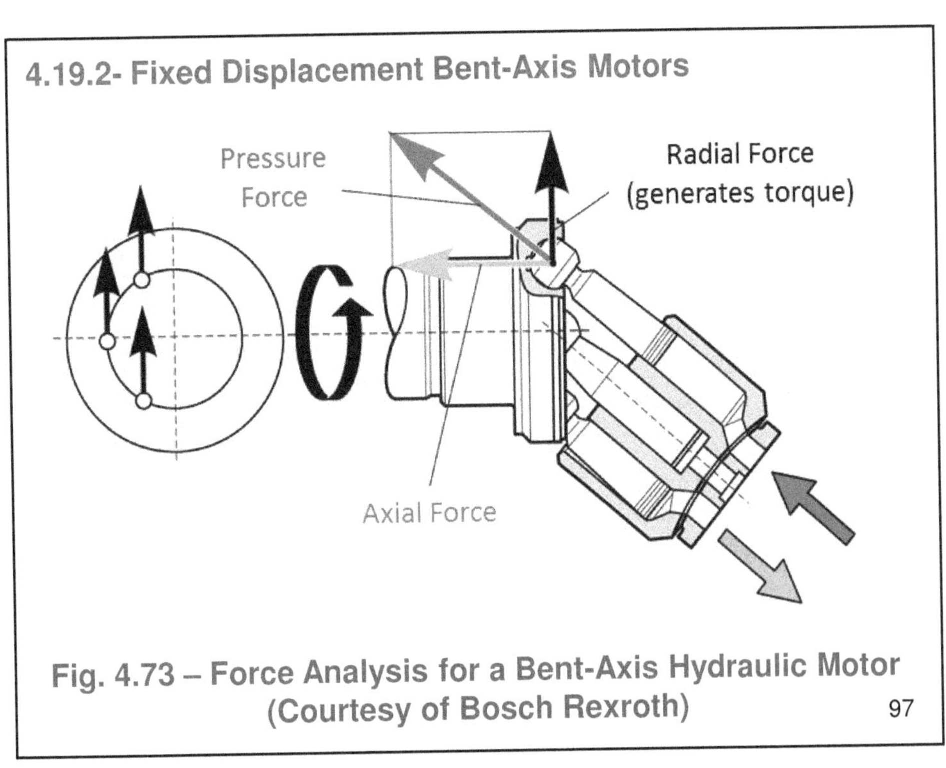

Fig. 4.73 – Force Analysis for a Bent-Axis Hydraulic Motor (Courtesy of Bosch Rexroth)

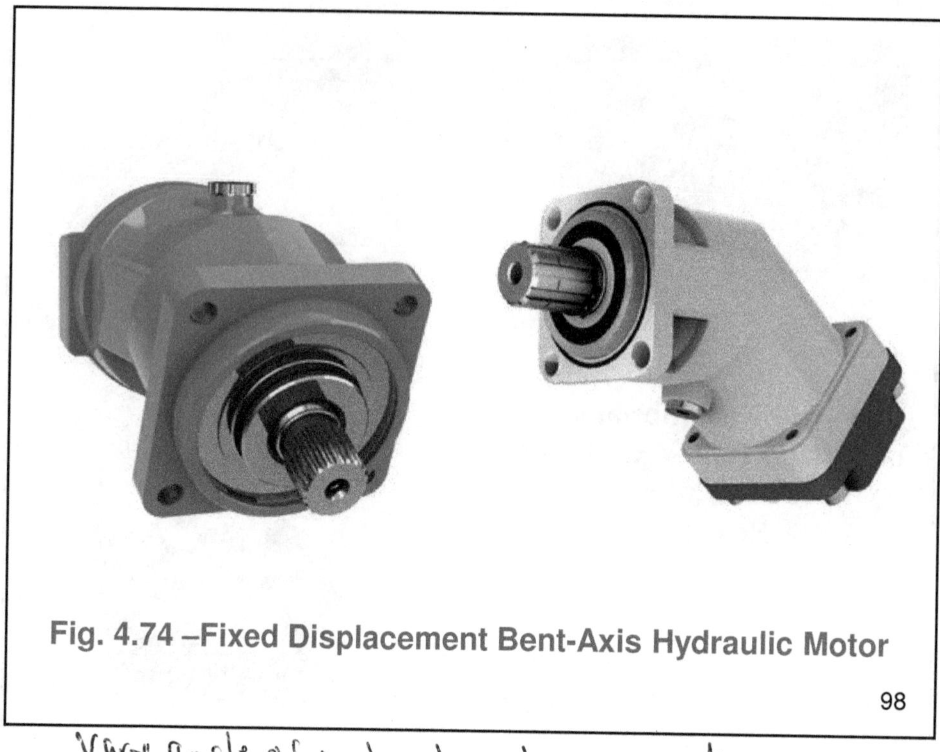

Fig. 4.74 –Fixed Displacement Bent-Axis Hydraulic Motor

98

Vary angle of inclenation to vary stroke

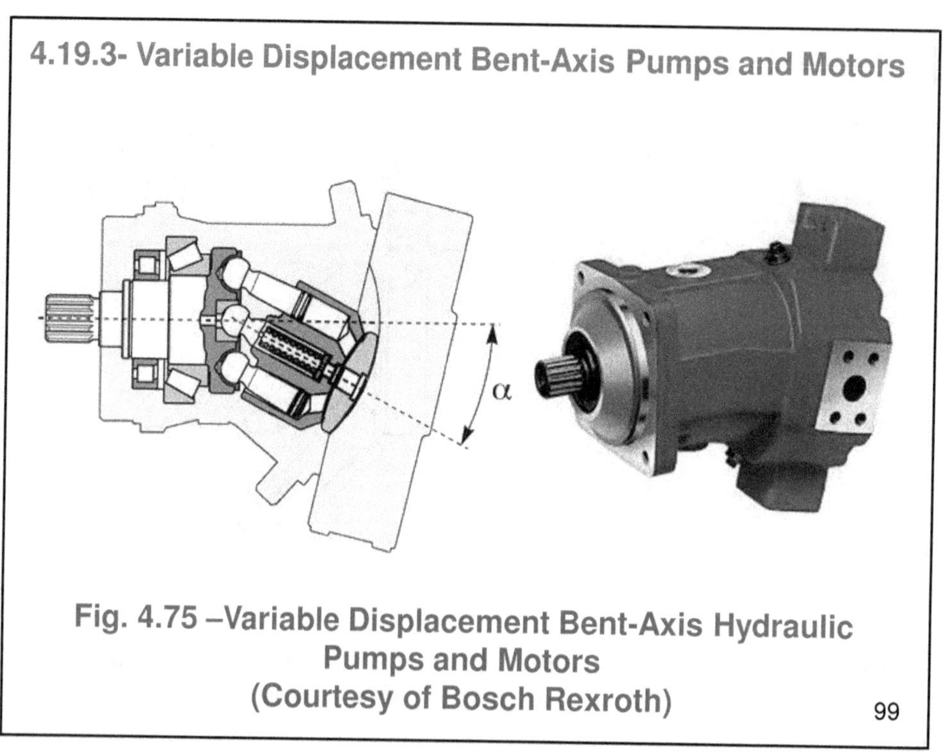

4.19.3- Variable Displacement Bent-Axis Pumps and Motors

Fig. 4.75 –Variable Displacement Bent-Axis Hydraulic Pumps and Motors (Courtesy of Bosch Rexroth)

99

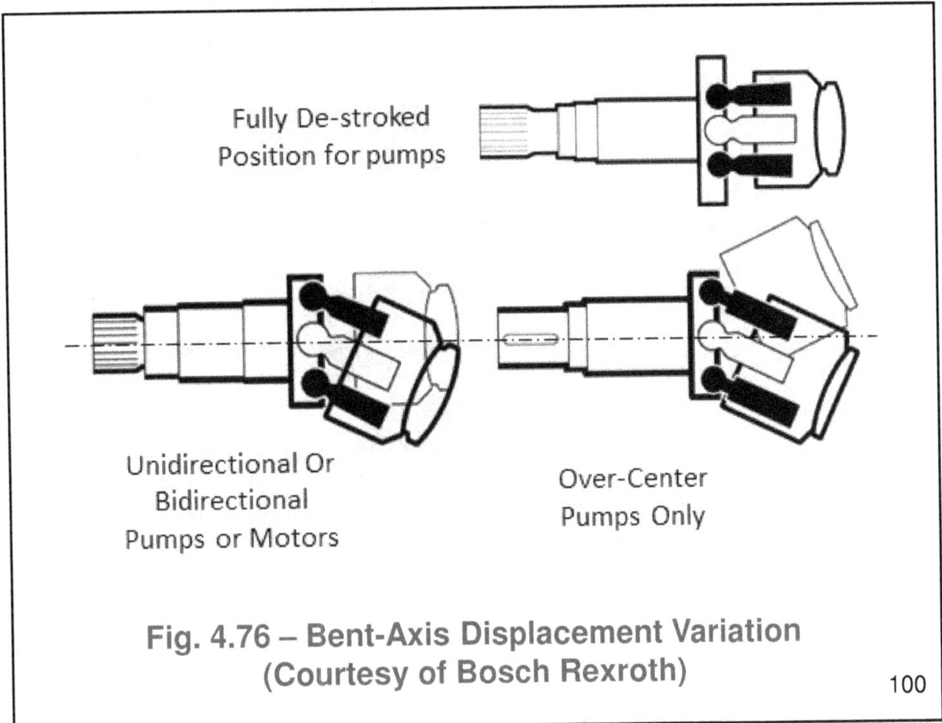

Fig. 4.76 – Bent-Axis Displacement Variation
(Courtesy of Bosch Rexroth)

not applicable for motors - motor will stall

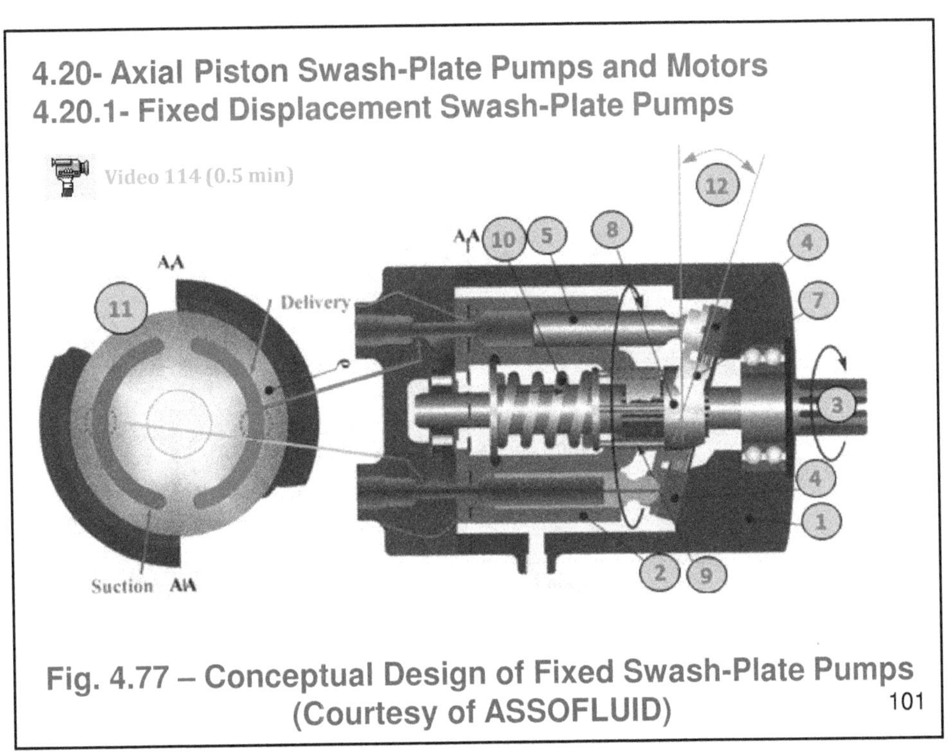

Fig. 4.77 – Conceptual Design of Fixed Swash-Plate Pumps
(Courtesy of ASSOFLUID)

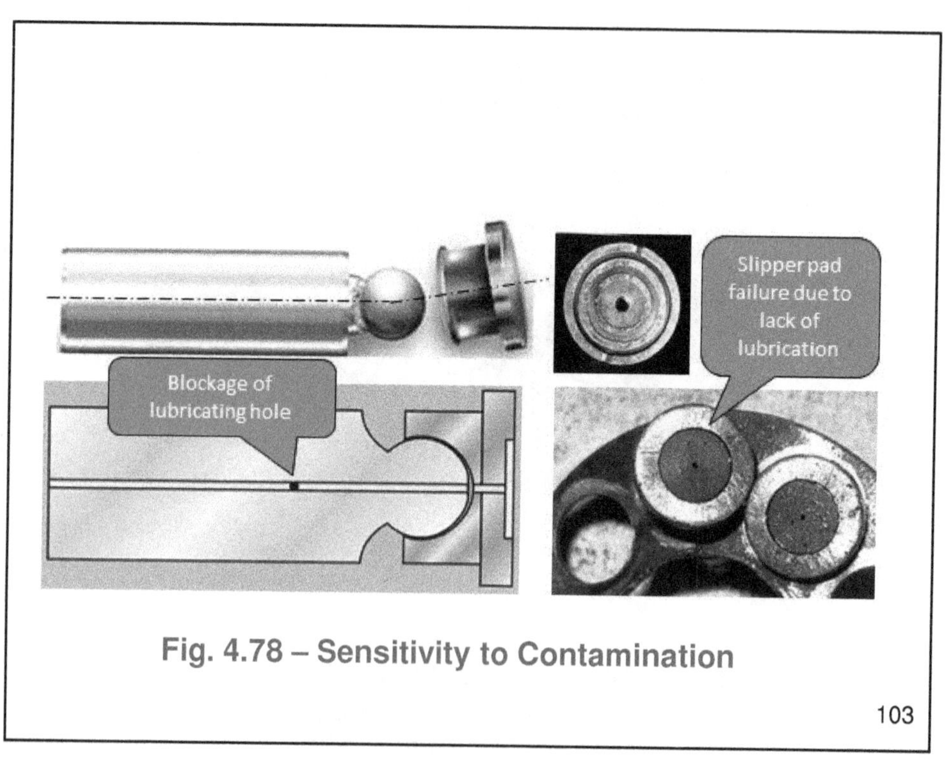

Fig. 4.78 – Sensitivity to Contamination

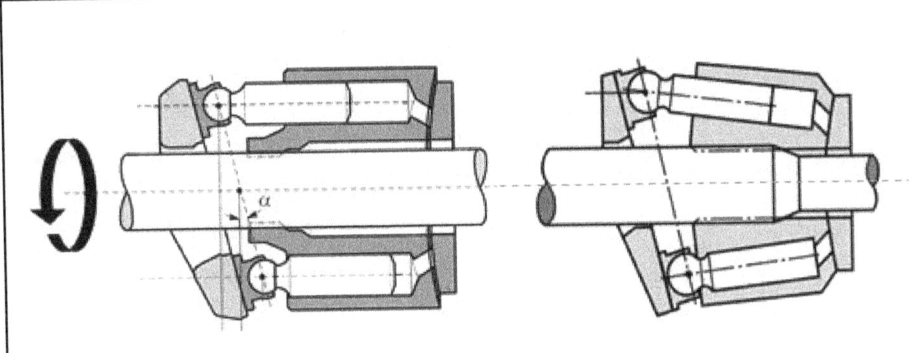

Fig. 4.79 – Cylindrical and Conical Cylinder Block (Courtesy of Bosch Rexroth)

Conical Cylinder Block:

- **Longer stroke.**
- **Detaching force between the slipper pad and the spherical head of the piston is reduced by about 15%.**

104

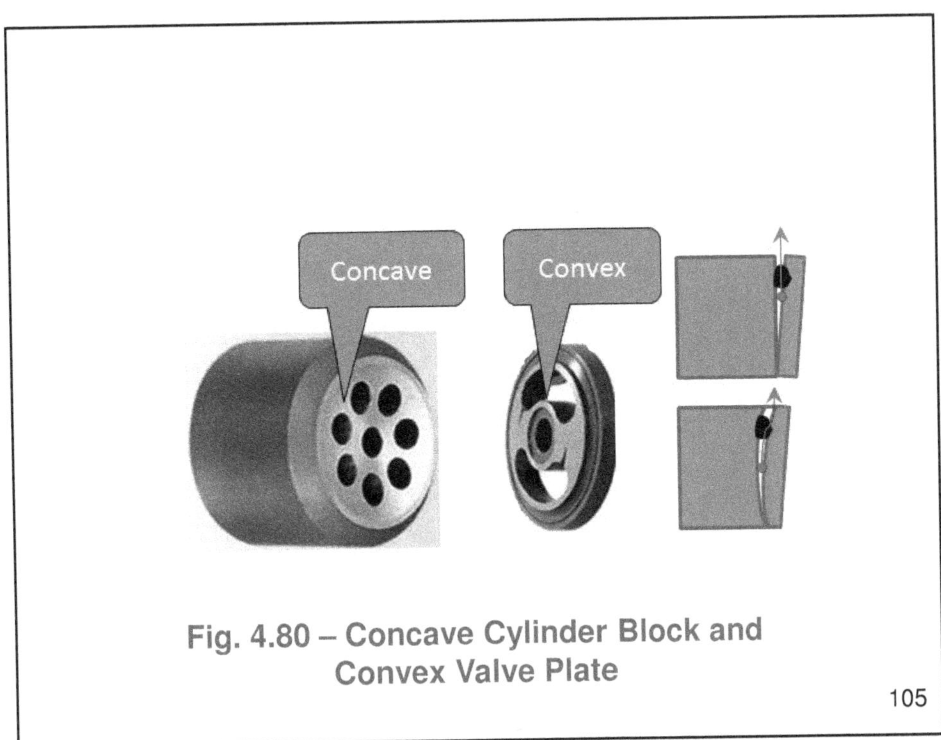

Fig. 4.80 – Concave Cylinder Block and Convex Valve Plate

105

High PSI
30,000 - 50,000

4.20.2- Special Design: Wobble Plate Pumps

Checkball Pump

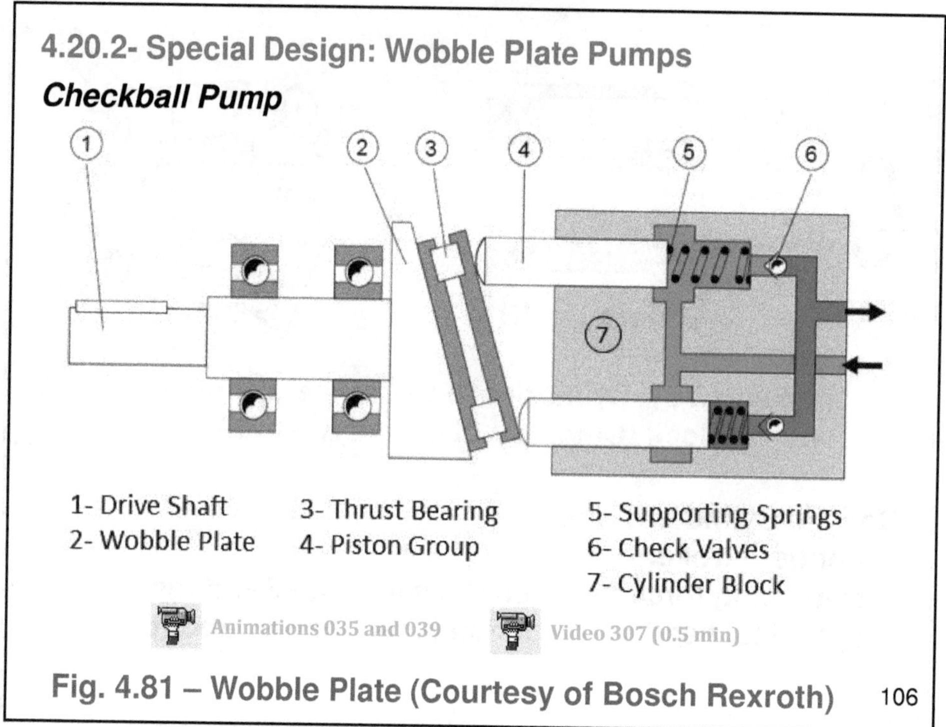

1- Drive Shaft
2- Wobble Plate
3- Thrust Bearing
4- Piston Group
5- Supporting Springs
6- Check Valves
7- Cylinder Block

Animations 035 and 039
Video 307 (0.5 min)

Fig. 4.81 – Wobble Plate (Courtesy of Bosch Rexroth) 106

4.20.3- Fixed Displacement Swash-Plate Motors

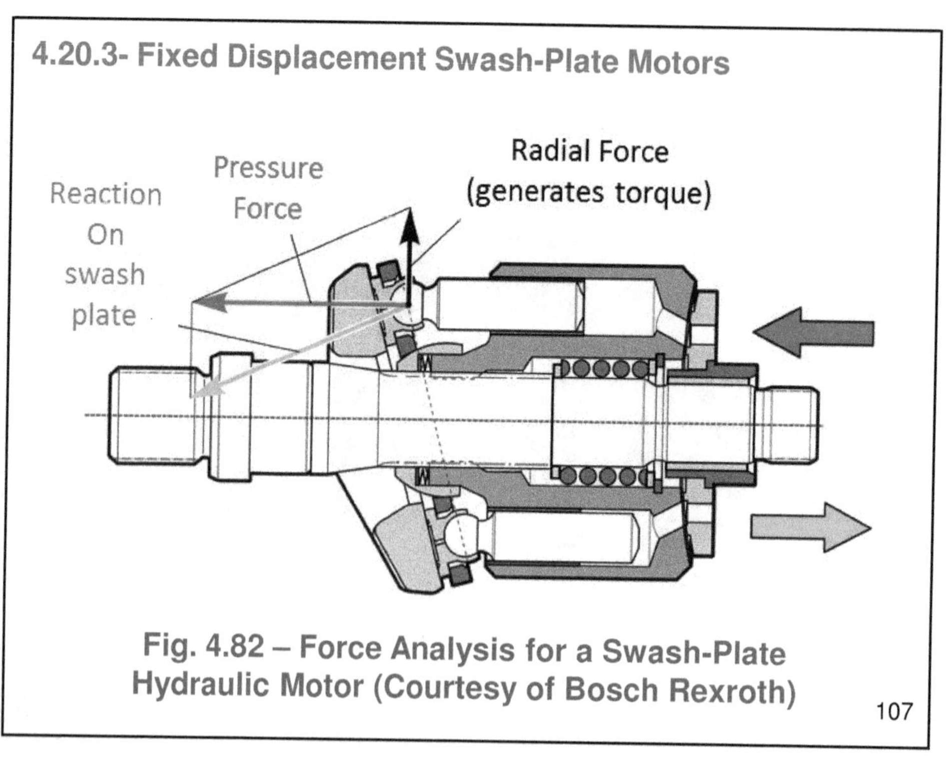

Fig. 4.82 – Force Analysis for a Swash-Plate Hydraulic Motor (Courtesy of Bosch Rexroth) 107

Fig. 4.83 –Fixed Displacement Swash-Plate Hydraulic Motor (Courtesy of Bosch Rexroth)

108

4.20.4- Variable Displacement Swash-Plate Pumps and Motors

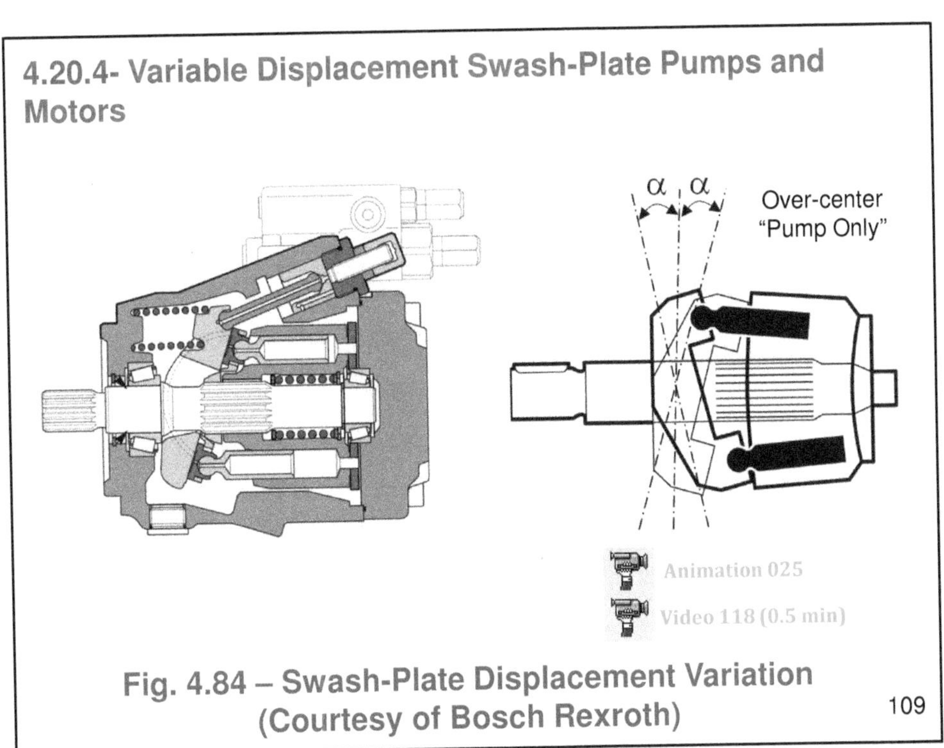

Fig. 4.84 – Swash-Plate Displacement Variation (Courtesy of Bosch Rexroth)

109

Fig. 4.85 – Swash-Plate Hydraulic Pumps and Motors (Courtesy of Bosch Rexroth)

110

4.20.5- Comparison between Swash-Plate and Bent-Axis Mechanisms

Swash-Plate Pumps and Motors	*Bent-Axis Pumps and Motors*
▪ Compact in size.	▪ Larger physical volume.
▪ Faster speed of response.	▪ Higher starting torque.
▪ Generates higher level of noise.	▪ Less frictional surfaces

Table 4.1 – Swash-Plate versus Bent-Axis Mechanisms

111

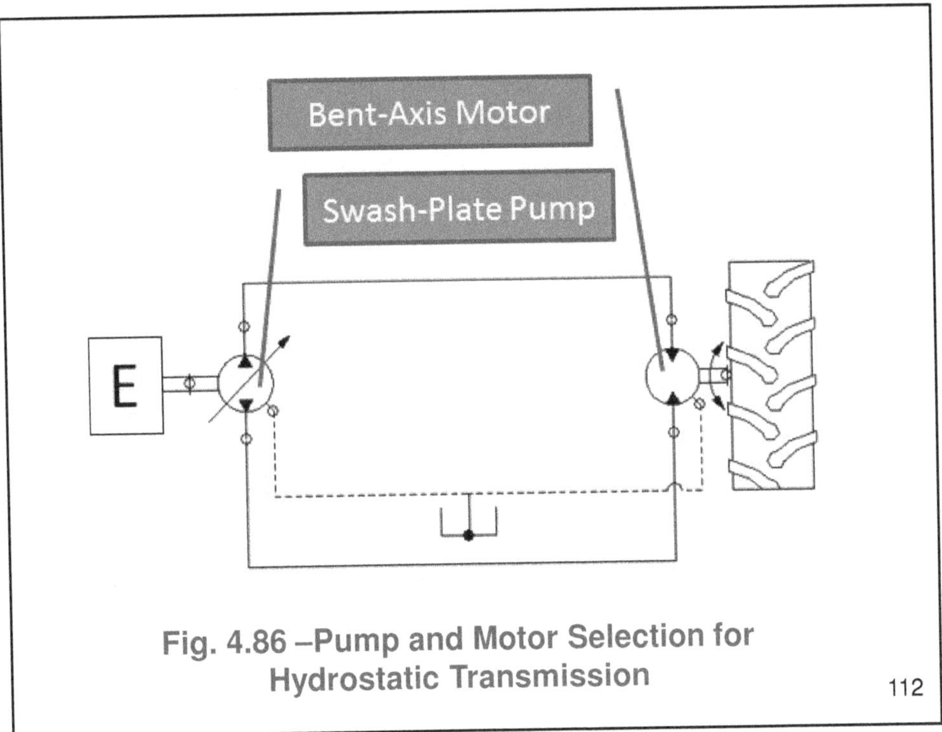

Fig. 4.86 –Pump and Motor Selection for Hydrostatic Transmission

112

4.21- Pumps and Motors Selection Criteria

- **Maximum/Optimum operating pressure.**
- **Min/Maximum/Optimum operating speed.**
- **Maximum/Optimum operating torque (for motors).**
- **Maximum overall efficiency at the optimum operating conditions.**
- **Size and displacement control requirements.**
- **Type of fluid.**
- **Contamination tolerance.**
- **Noise level.**
- **Initial cost.**
- **Approximate service life.**
- **Availability and interchangeability.**
- **Maintenance and spare parts.**
- **Physical size and weight.**

113

	Gear	*Vane*	*Piston*
Pressure	Medium	Medium	High
Flow	Medium	High	High
Cost	Low	Medium	High

Table 4.2 – The Highlighted Features of the Different Mechanisms

114

Chapter 4 Reviews

1. **Basic function of a hydraulic pump** is to**?**
   - A. Generate pressure.
   - B. Control actuator speed.
   - C. Generate flow.
   - D. Control external load.

2. What is the main **objective of using variable displacement pumps**?
   - A. Runs at low speed with high efficiency.
   - B. Generate variable flow for constant input rpm.
   - C. Build high operating pressure.
   - D. All of the above.

3. The main **objective of using variable displacement motors** is to?
   - A. Drive a rotational load with variable speed even with constant supply flow.
   - B. Generate variable flow for constant input rpm.
   - C. Build high operating pressure.
   - D. All of the above.

4. Which statement is **False?**
   - A. Displacement of a variable pump can be reduced to zero.
   - B. Displacement of a variable motor can be reduced to zero.
   - C. Cylinders require cushioning if the speed is above 10 cm/s.
   - D. Gear pumps are only of a fixed displacement.

5. Which statement is **False?**
   - A. Vane pumps could be of a fixed or variable displacement.
   - B. Gear pumps could be of a fixed or variable displacement.
   - C. Piston pumps could be of a fixed or variable displacement.
   - D. Vane pumps could be of a balanced or unbalanced design.

6. What is the correct **definition of pump displacement?**
   - A. Flowrate discharged by the pump per unit time.
   - B. Flowrate discharged by the pump per unit revolution.
   - C. Volume of oil displaced by the pump per unit revolution.
   - D. Volume of oil displaced by the pump per unit time.

7. **Leakage losses** in a hydraulic pump are quantified by?
   - A. Mechanical efficiency.
   - B. Overall efficiency.
   - C. Volumetric efficiency.
   - D. All of the above.

8. In your experience, why does the increase in working pressure improve the **mechanical efficiency** of a hydraulic motor?
    A. Because at high pressure, lubrication oil film thickness is increased.
    B. Because at high pressure, oil viscosity is increased, which results in better lubricity.
    C. Because at high pressure, clearances between mating surfaces are filled with more oil.
    D. All of the above.

9. The **overall efficiency of a hydraulic motor** is defined as:
    A. Ratio between the motor's Output energy and input energy (during the same time period).
    B. The product of the motor's volumetric efficiency and the motor's mechanical efficiency.
    C. Ratio between the motor's mechanical output power and its hydraulic input power.
    D. All of the above.

10. A pump is working at a certain pressure and driven at a certain speed. If both the working pressure and the drive speed are increased by 50%, assuming linear pump characteristics, what is your **prediction for the pump's volumetric efficiency** at the new working conditions?
    A. Increased by 50%.
    B. Decreased by 50%.
    C. Be the same.
    D. None of the above.

11. A hydraulic motor is replaced with one that has half of the displacement of the first one. Assuming constant flowrate is supplied to the motor, **what is your expectation of the new motor's speed as compared to the first motor?**
    A. Decreased to half.
    B. Motor speed is doubled.
    C. Decreased to one quarter.
    D. No change in the motor speed because the supplied flow is the same.

12. A hydraulic pump is working at a pressure of 60 bar and discharge 100 L/min. It has an overall efficiency of 80%. What is **the power loss?**
    A. 8 kW.
    B. 8 HP.
    C. 2 kW.
    D. None of the above.

13. The four versions that are classified as gear pumps or special versions of it are**?**
    A. External gear pump - screw pump - internal gear pump - swash plate pump.
    B. External gear pump - bent axis pump - internal gear pump - gerotor.
    C. External gear pump - radial piston - internal gear pump - gerotor.
    D. External gear pump - screw pump - internal gear pump - gerotor.

14. In the unidirectional **external gear pump** shown below, which port is the discharge port**?**

A. Upper port.
B. Lower port.
C. Upper or lower port.
D. None of the above.

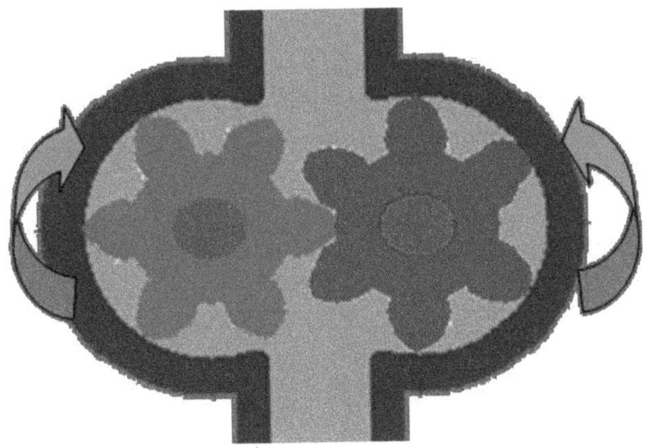

15. Which of the following pump mechanisms could be of a **balanced or unbalanced design?**

A. Vane pump.
B. Gear pump.
C. Piston pump.
D. Centrifugal pump.

16. For the hydraulic vane pump shown below, **what is the correct direction of rotation?**

A. Clockwise.
B. Counter clockwise.
C. Both directions.
D. None of the above.

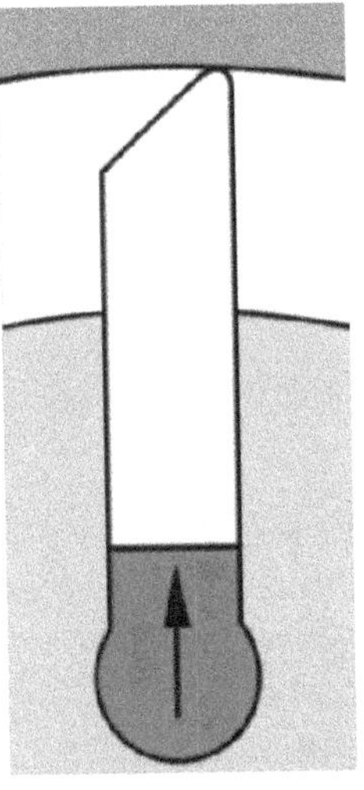

17. The **oval shaped cam surface in a vane pump** is used for**?**
    A. Variable vane pump.
    B. Fixed vane pump..
    C. Balanced vane pump.
    D. Fixed and balanced vane pump.

18. The **balanced vane pump design** changes its displacement by**?**
    A. Changing the swash plate inclination angle.
    B. Changing the bent axis inclination angle.
    C. Changing the eccentricity between the rotor and cam surface.
    D. None of the above.

19. Which of the following pumps listed below can **generate relatively large flowrates and work at higher working pressures?**
    A. Axial piston pump.
    B. Gear pump.
    C. Vane pump.
    D. Centrifugal pump.

20. The **pumping mechanism** shown below is for?
    A. Bent axis axial piston pump.
    B. Swash plate axial piston pump.
    C. Radial piston pump.
    D. Vane pump.

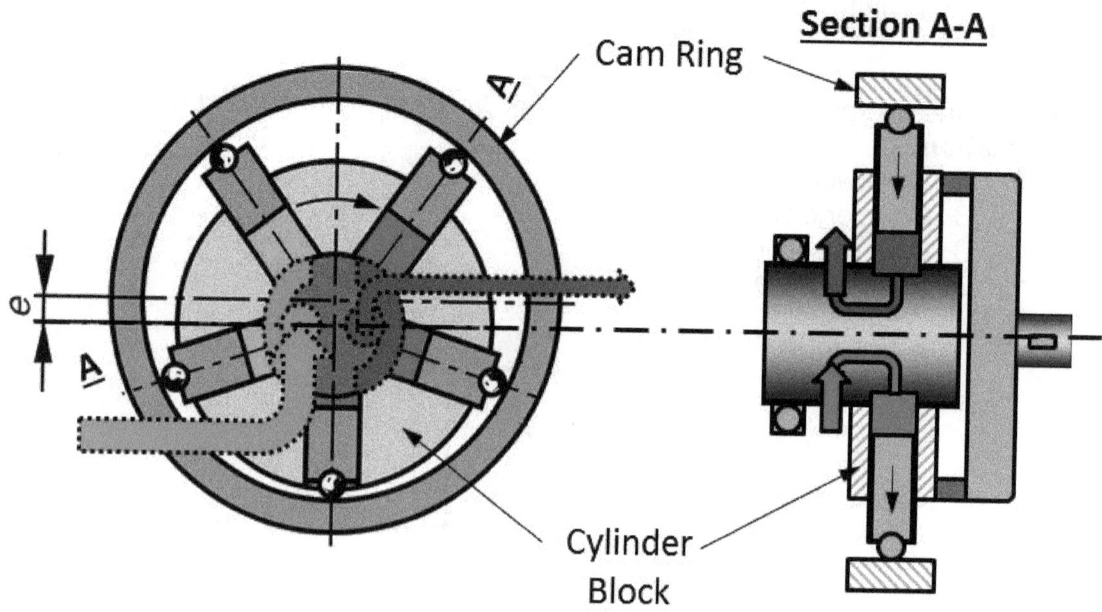

21. What is the main reason that **axial piston pumps** have an odd number of pistons?
    A. To reduce flow fluctuations and pressure ripple.
    B. To increase the pump flow rate.
    C. To increase the pump pressure.
    D. To increase the pump efficiency.

22- Why is the variable **bent axis design** preferred for a motor rather than a pump?

A. Work at high pressure.
B. Discharge large flow.
C. Has a better starting torque.
D. Easy and fast to control and compact.

23. A **swash plate design** piston pump changes its flowrate by?

A. Changing the swash plate inclination angle.
B. Increasing the pump efficiency.
C. Changing the bent-axis inclination angle.
D. Changing the pump eccentricity.

24. **Over-center design pumps** are used to?

A. Discharge large amount of flow rate.
B. Work under high pressure.
C. Vary the actuator's direction of motion without the need of a directional valve.
D. Provide power enough to drive the load.

25. Which of the following pumps are most suitable for a commercial **cost-effective** application that requires 10 GPM constant flow and 1000 psi pressure?

A. Swash plate axial piston variable displacement pump.
B. Bent axis axial piston variable displacement pump.
C. Variable displacement vane pump.
D. External gear pump.

Chapter 4 Assignment

Student Name: -- Student ID: ------------------

Date: -- Score: ------------------------

Given: The shown below test data are for a pump that has 2.37 cubic inches size.

Required: Evaluate the performance of the pump if it runs at 1800 rpm against 3000 psi system pressure.

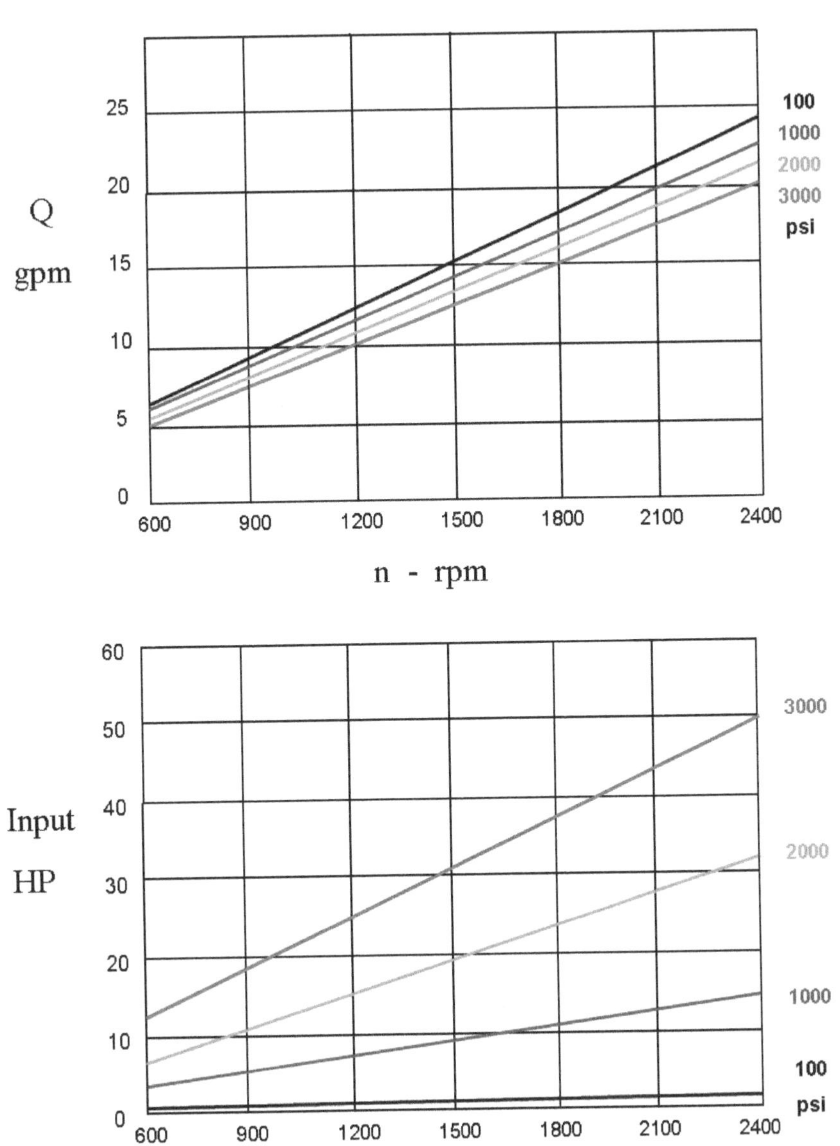

Objectives:
This chapter presents the construction and operating principles of the three families of control valves: namely pressure, directional, and flow control valves. The chapter presents the reasons why one of these valves should be selected over the other from a practical point of view. Valve symbols and comparison tables are presented.

Brief Contents:
5.1- Classification of Hydraulic Valves
5.2- Pressure Control Valves – load
5.3- Directional Control Valves –
5.4- Flow Control Valves – Speed Control

0

5.1- Classification of Hydraulic Valves

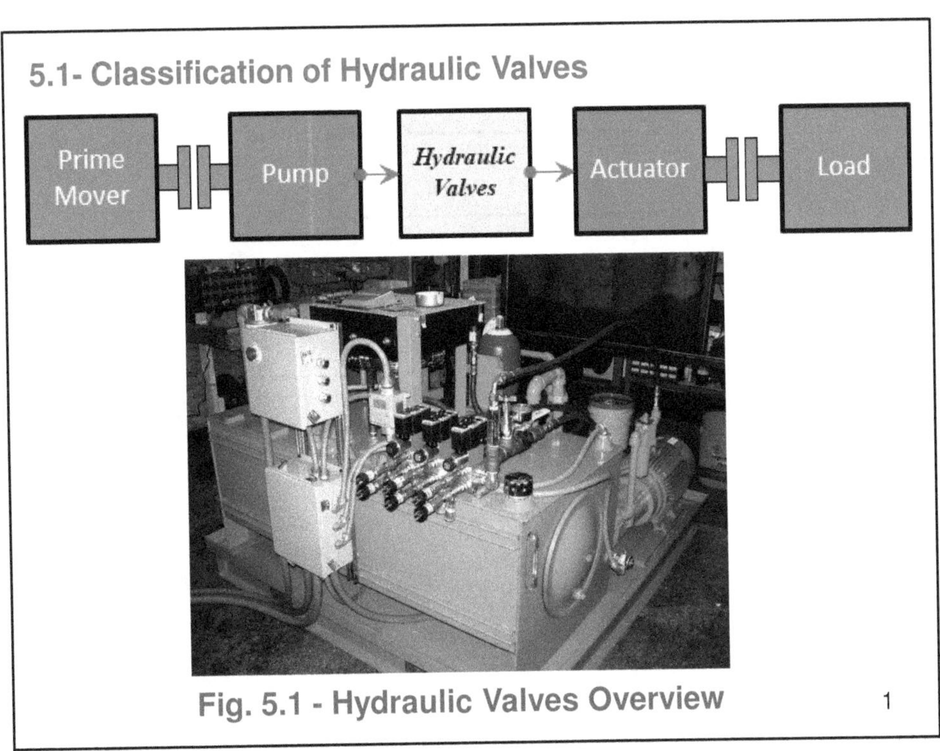

Fig. 5.1 - Hydraulic Valves Overview

1

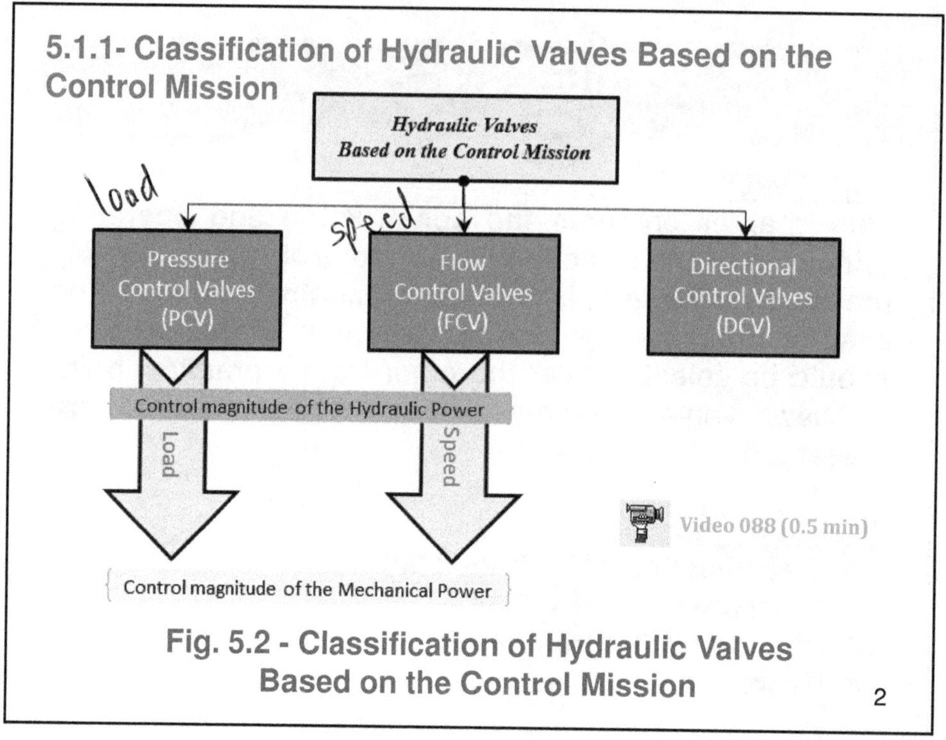

Fig. 5.2 - Classification of Hydraulic Valves Based on the Control Mission

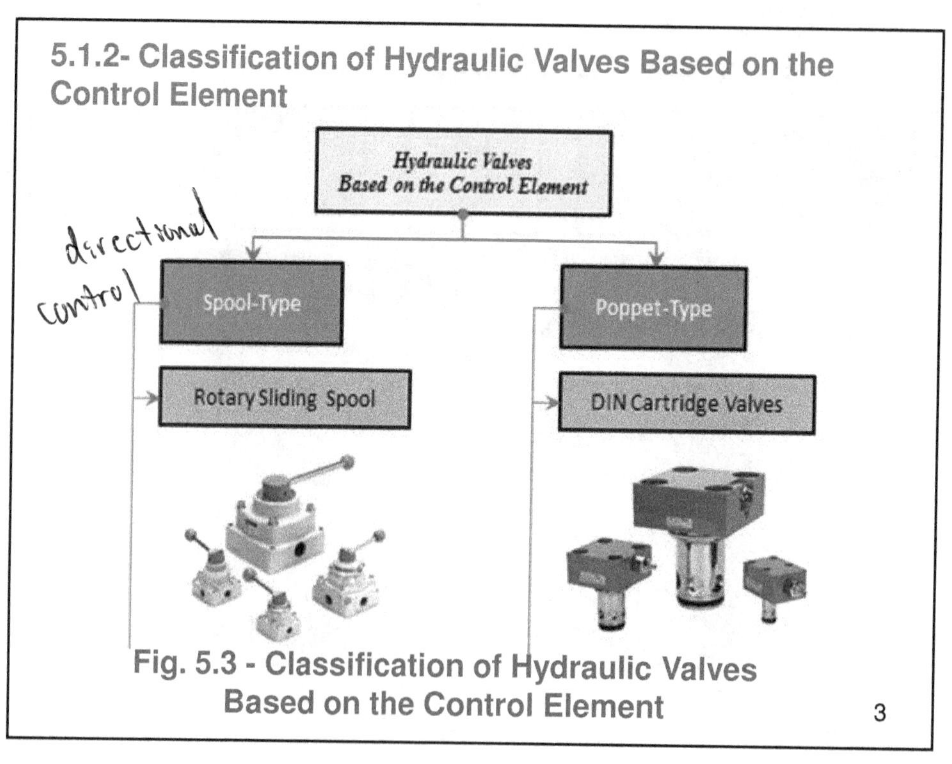

Fig. 5.3 - Classification of Hydraulic Valves Based on the Control Element

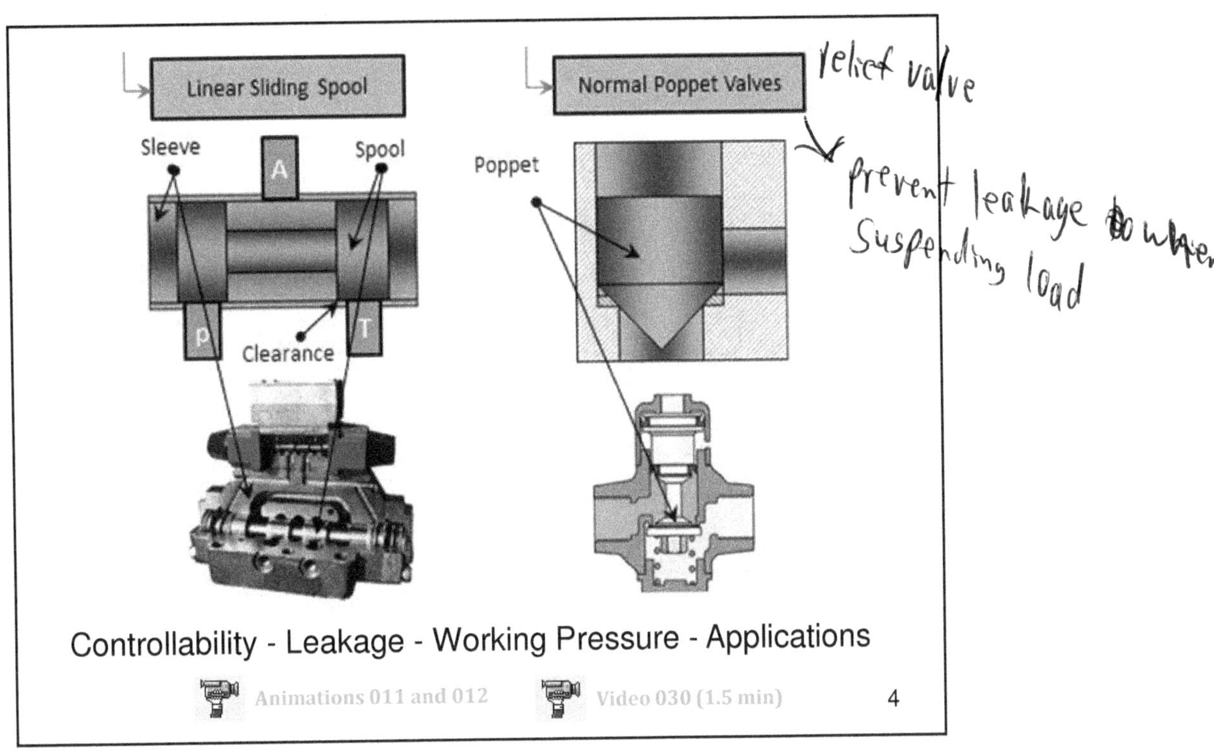

5.1.3- Classification of Hydraulic Valves Based on the Valve Actuation

Manual Actuation		Mechanical Actuation	
▪ Manual		▪ Roller	
▪ Push Button / Pull-out Knob		▪ Lever with roller	
▪ Lever		▪ Plunger	
▪ Pedal		▪ Plunger with Adjustable Stroke	

Table 5.1- Classification of Hydraulic Valves Based on the Valve Actuation

5

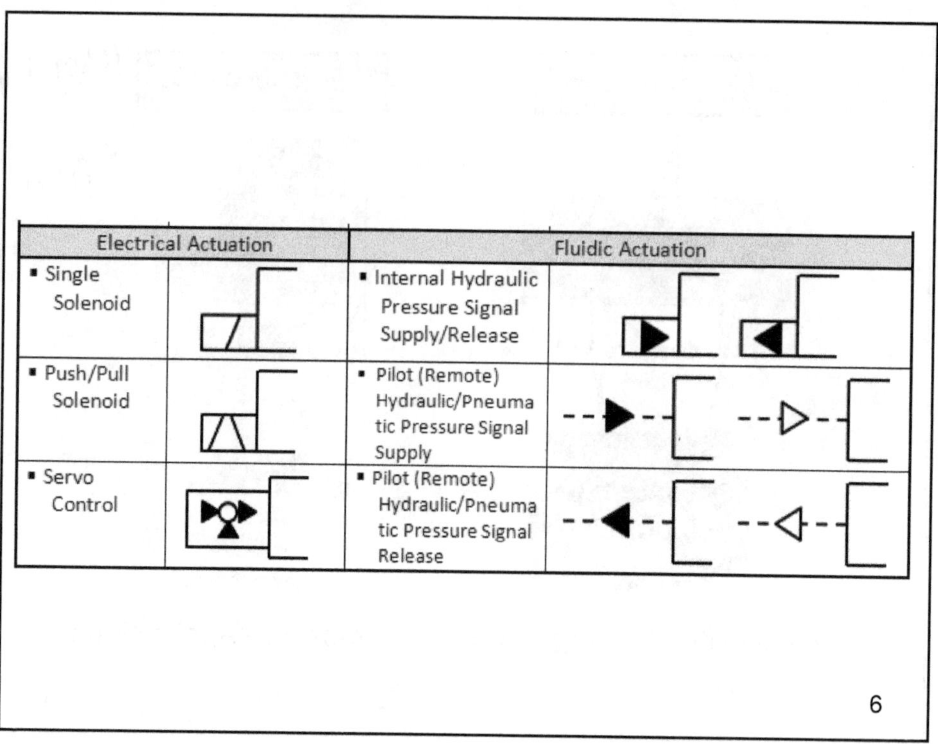

Electrical Actuation		Fluidic Actuation		
▪ Single Solenoid		▪ Internal Hydraulic Pressure Signal Supply/Release		
▪ Push/Pull Solenoid		▪ Pilot (Remote) Hydraulic/Pneumatic Pressure Signal Supply		
▪ Servo Control		▪ Pilot (Remote) Hydraulic/Pneumatic Pressure Signal Release		

6

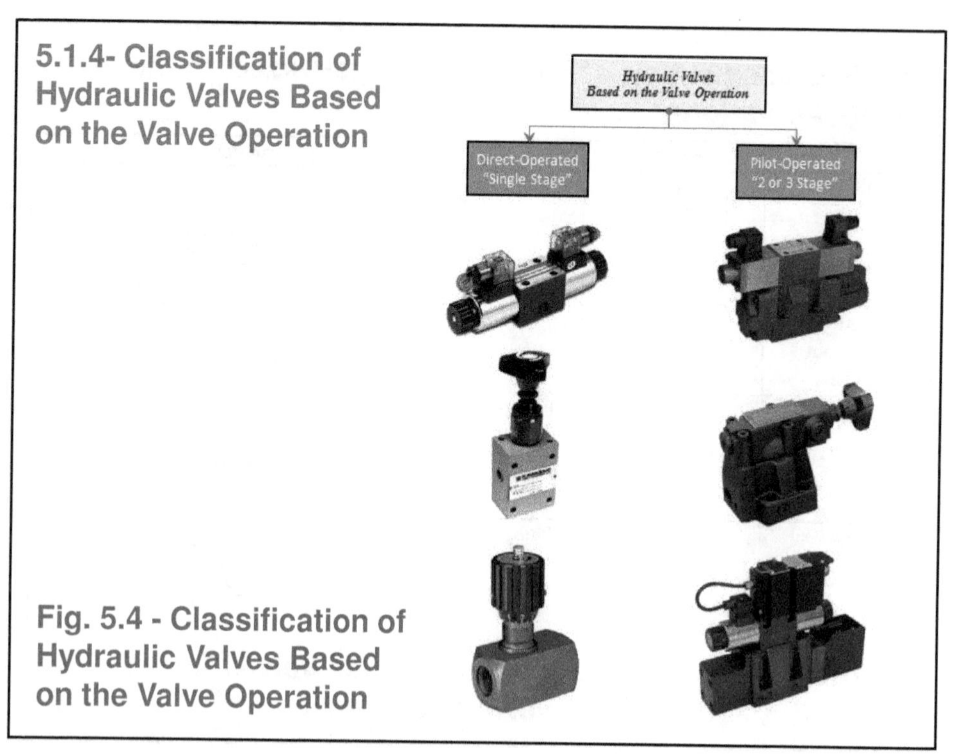

Fig. 5.4 - Classification of Hydraulic Valves Based on the Valve Operation

5.1.5- Classification of Hydraulic Valves Based on the Control Pressure Source

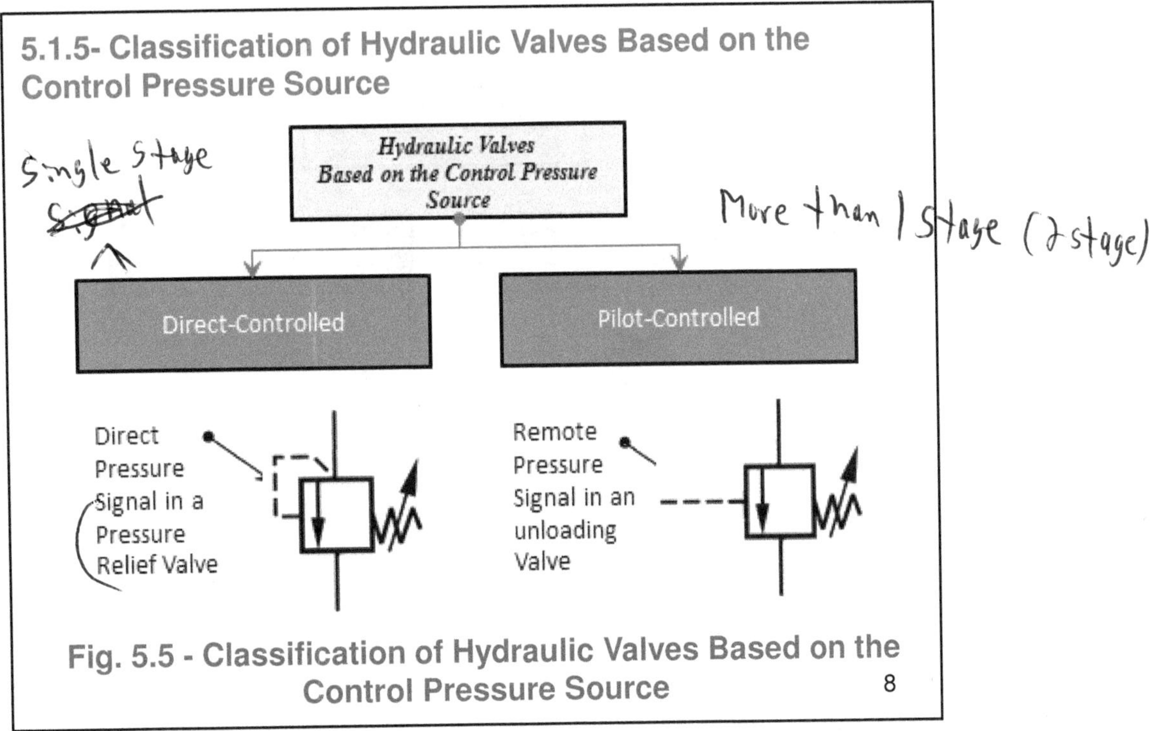

Fig. 5.5 - Classification of Hydraulic Valves Based on the Control Pressure Source 8

5.1.6- Classification of Hydraulic Valves Based on the Valve Control Level

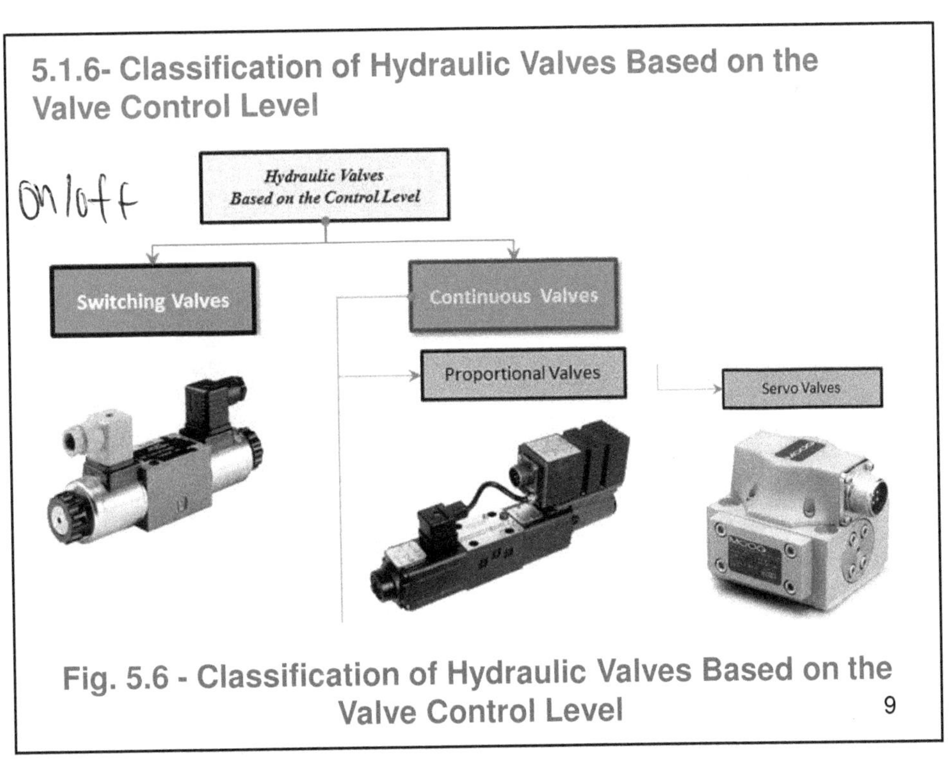

Fig. 5.6 - Classification of Hydraulic Valves Based on the Valve Control Level 9

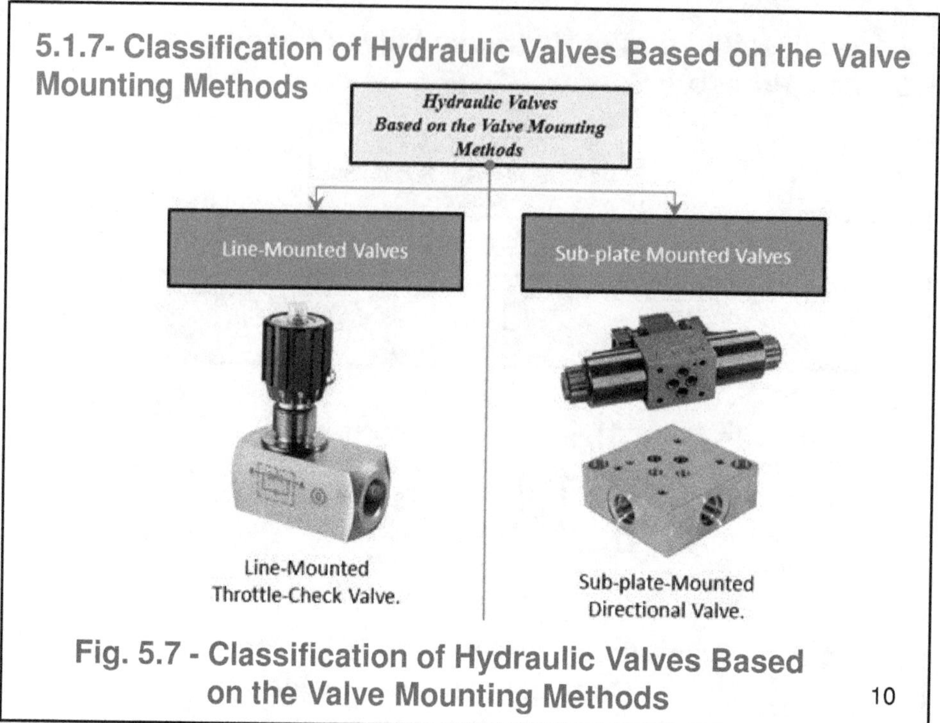

Fig. 5.7 - Classification of Hydraulic Valves Based on the Valve Mounting Methods

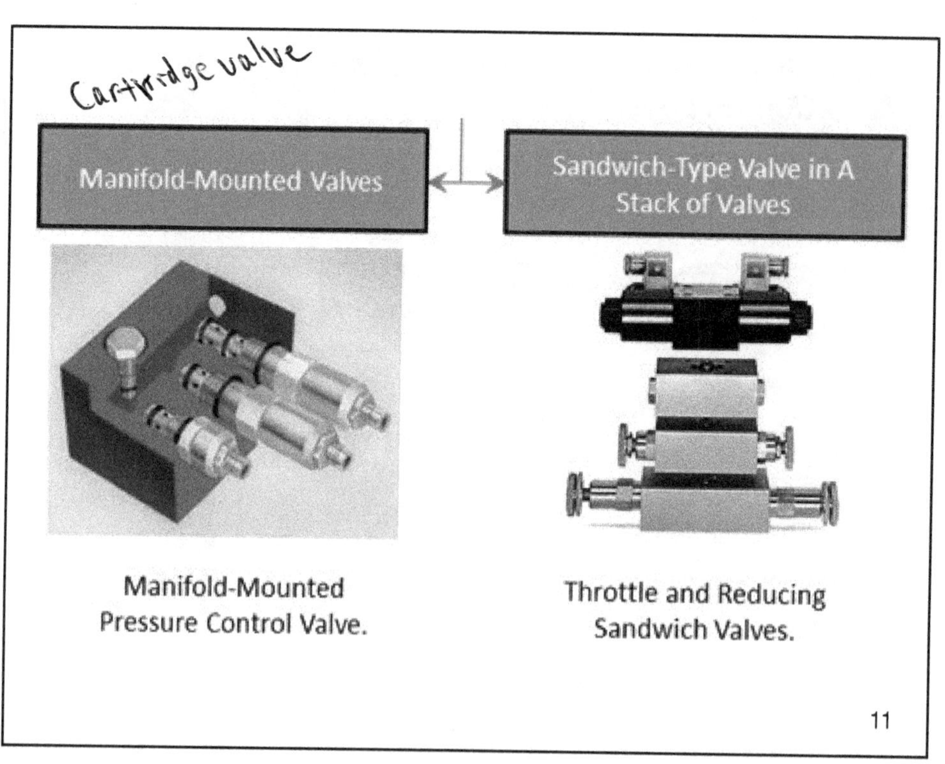

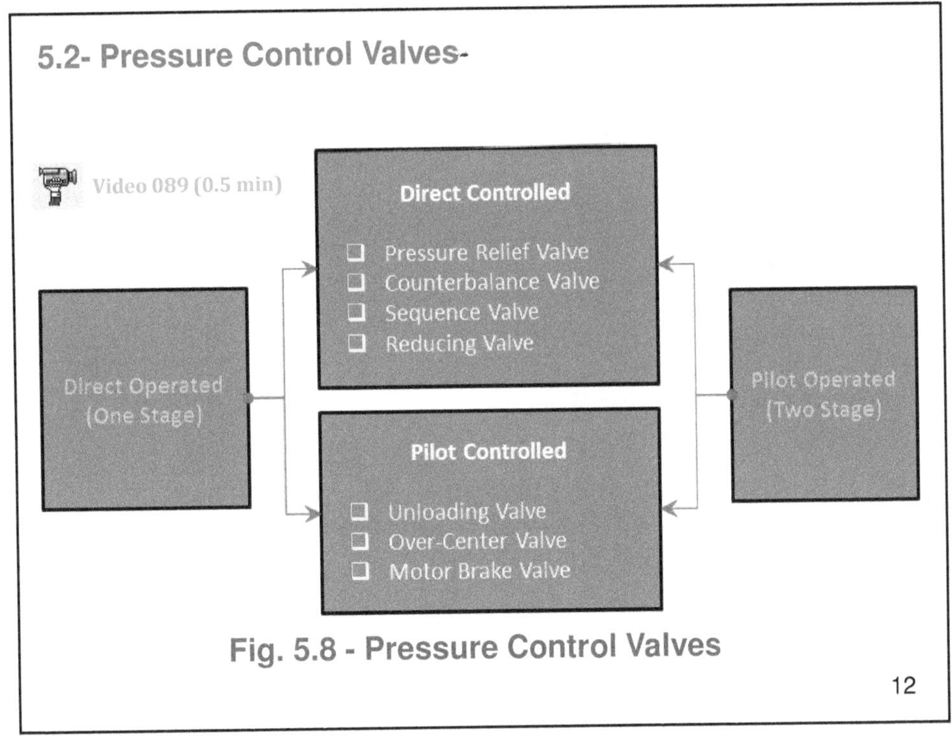

Fig. 5.8 - Pressure Control Valves

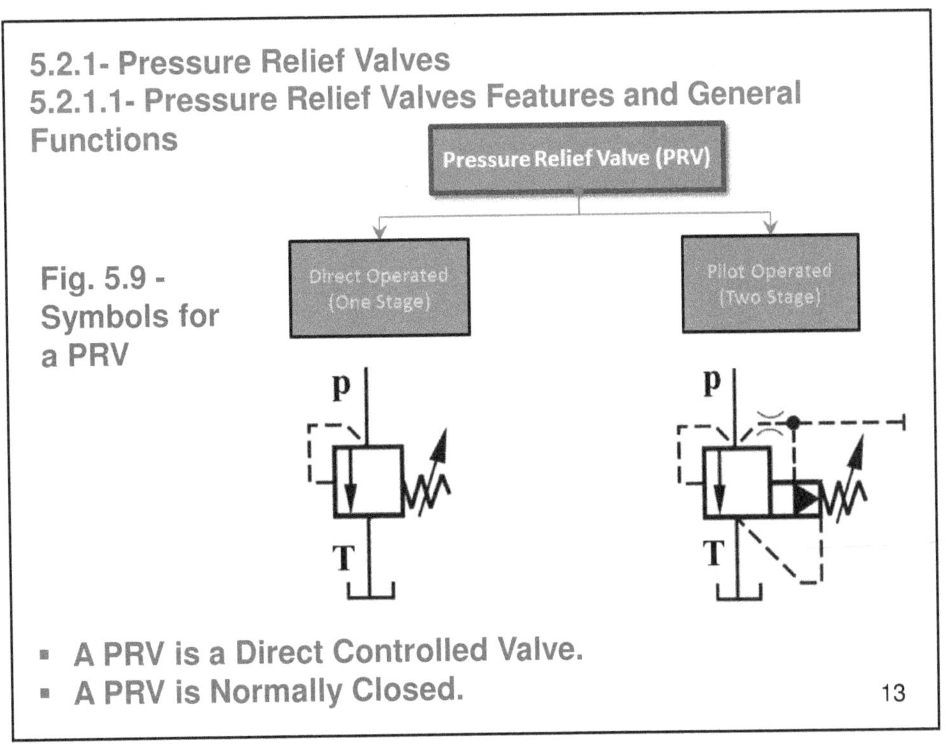

Fig. 5.9 - Symbols for a PRV

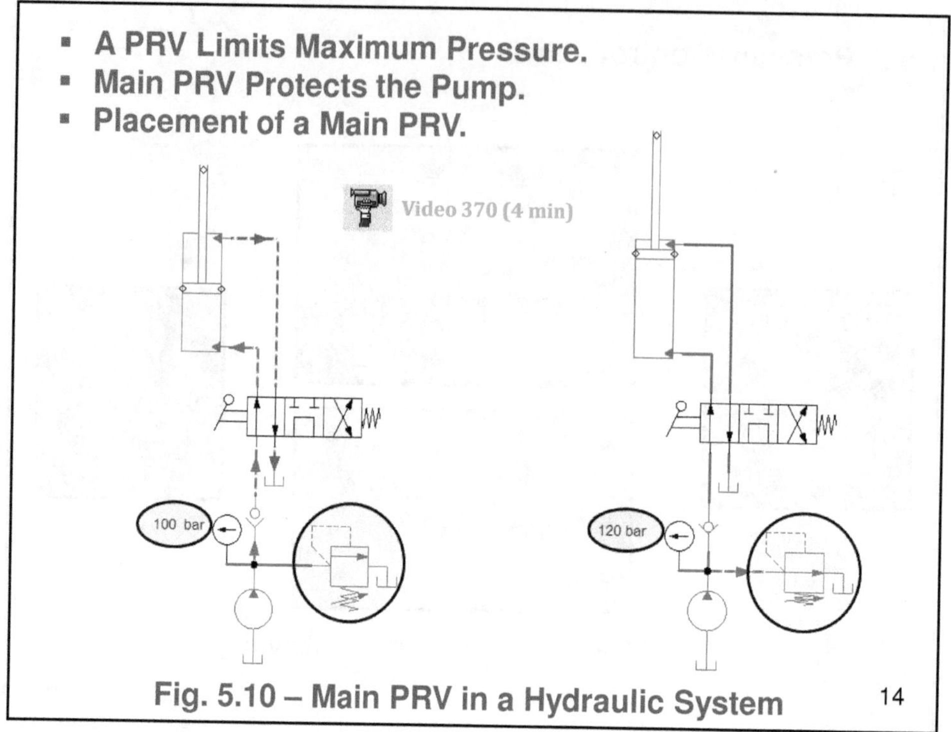

Fig. 5.10 – Main PRV in a Hydraulic System

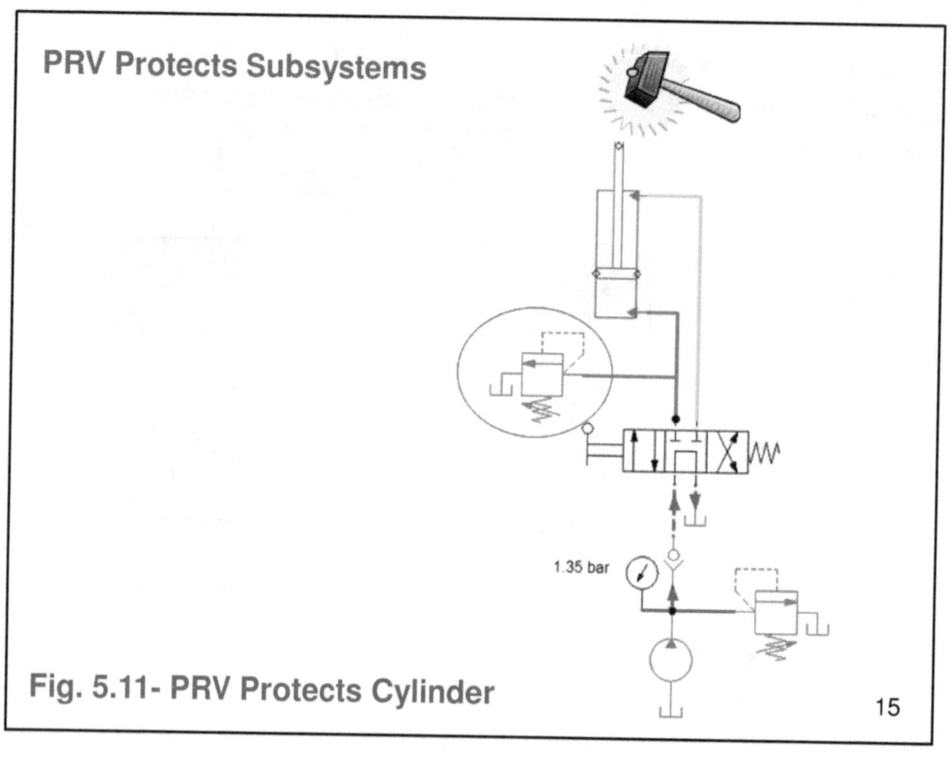

Fig. 5.11- PRV Protects Cylinder

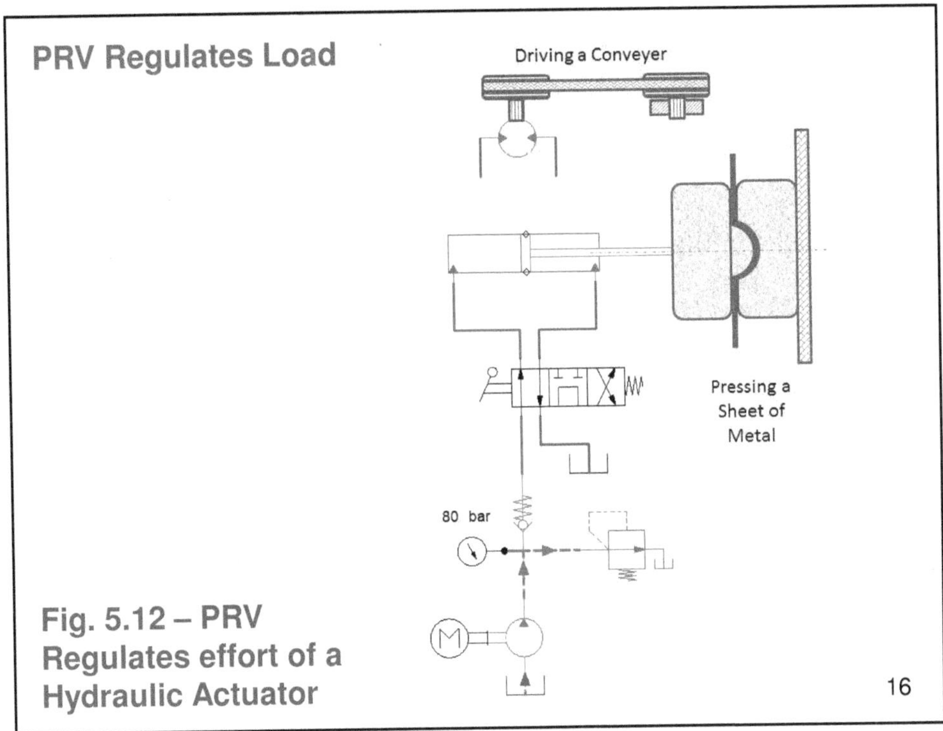

Fig. 5.12 – PRV Regulates effort of a Hydraulic Actuator

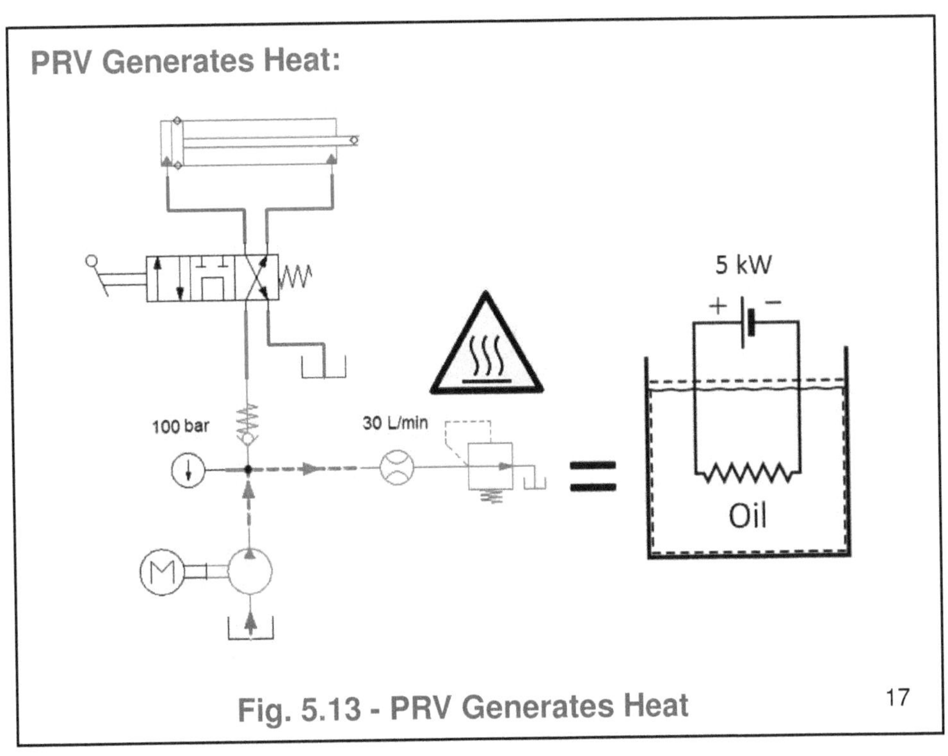

Fig. 5.13 - PRV Generates Heat

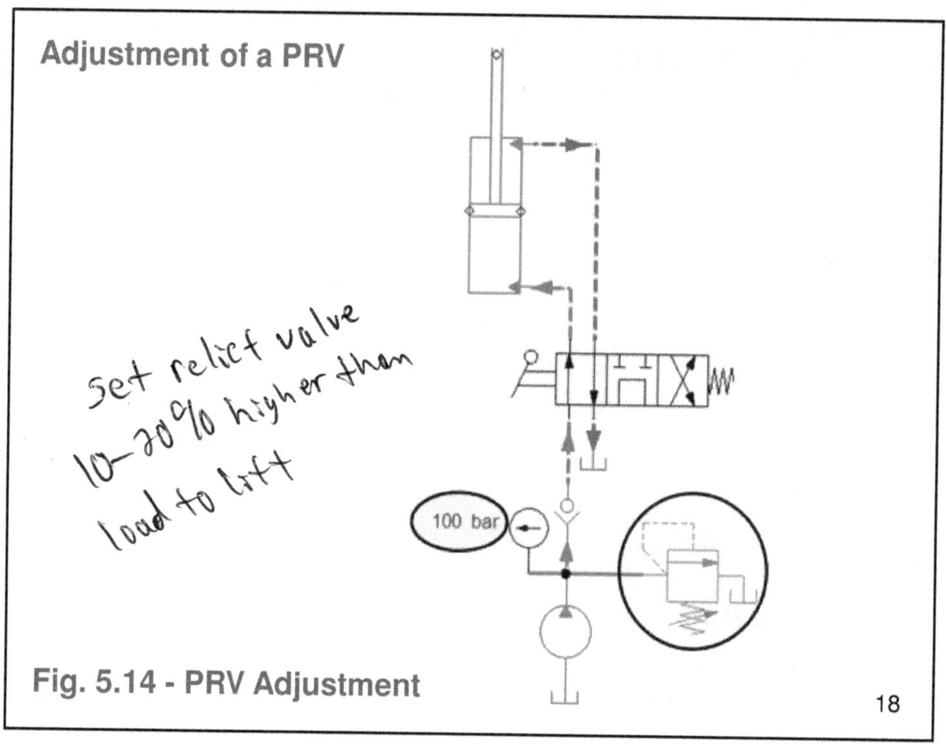

Fig. 5.14 - PRV Adjustment

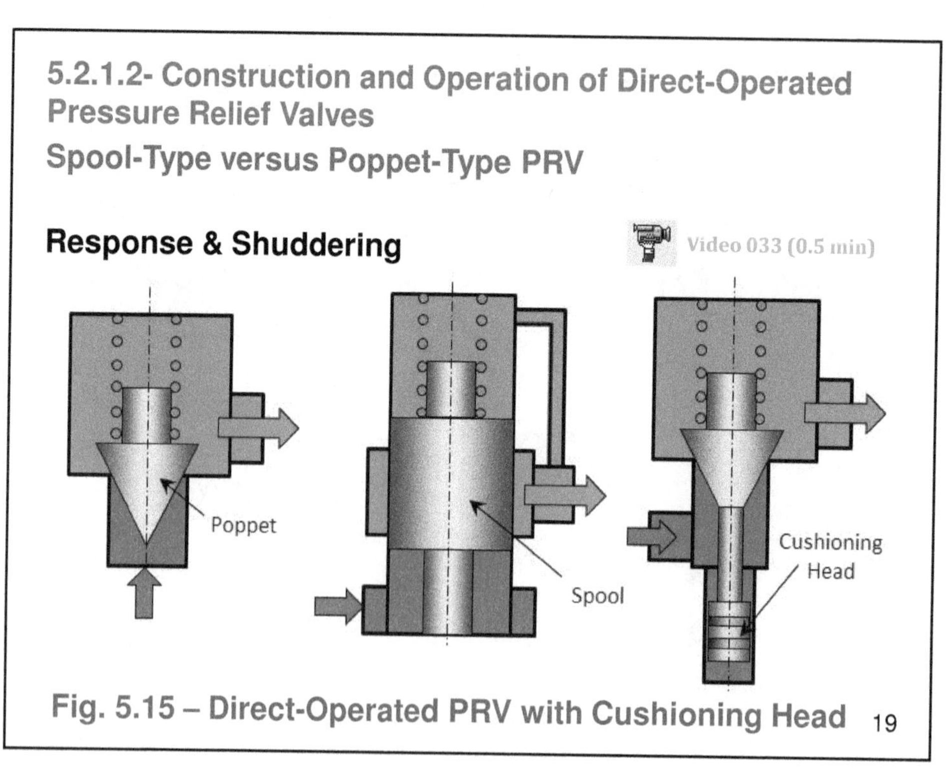

Fig. 5.15 – Direct-Operated PRV with Cushioning Head

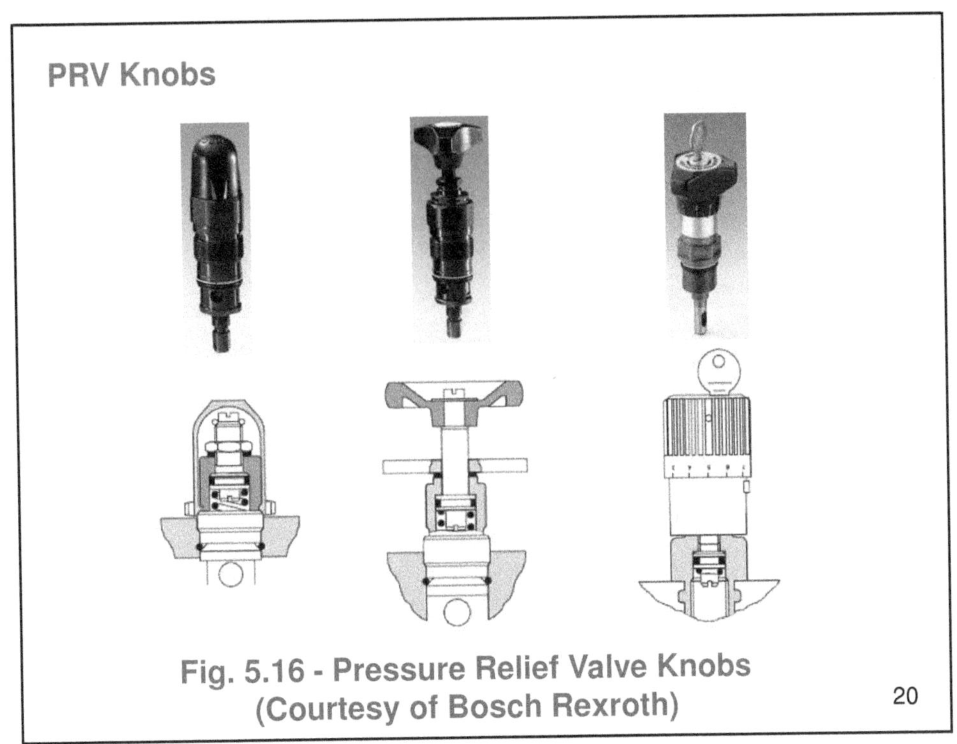

Fig. 5.16 - Pressure Relief Valve Knobs (Courtesy of Bosch Rexroth)

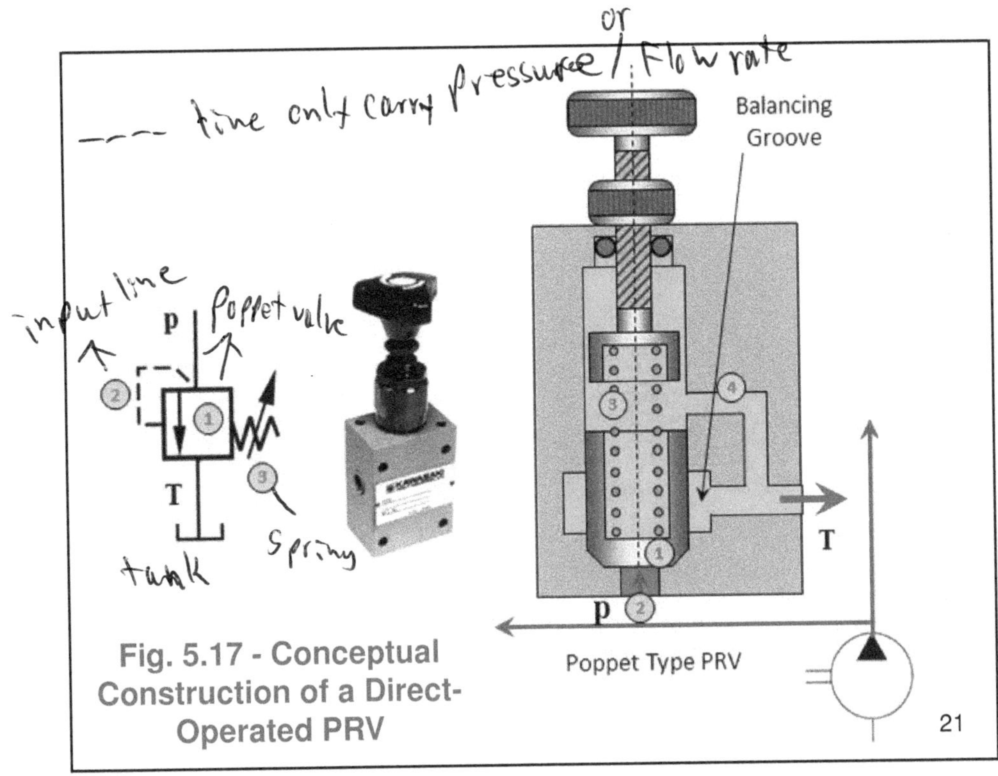

Fig. 5.17 - Conceptual Construction of a Direct-Operated PRV

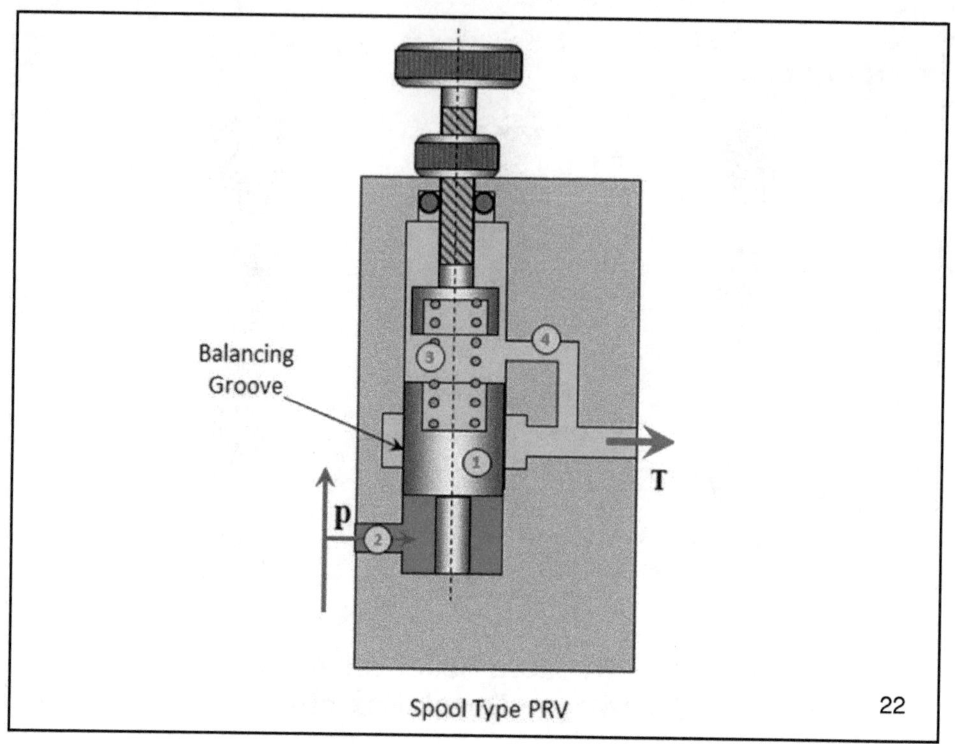

up to 20 gpm – single stage

5.2.1.3- Pilot-Operated Pressure Relief Valves

Case A	Case B	Case C
3000 psi	3000 psi	3000 psi
10 gpm	20 gpm	30 gpm
(40 lit/min)	(80 lit/min)	(120 lit/min)

P → T

Fig. 5.18 - Why Pilot-Operated Valves are used 23

5.2.1.4- Construction and Operation of Pilot-Operated Pressure Relief Valves

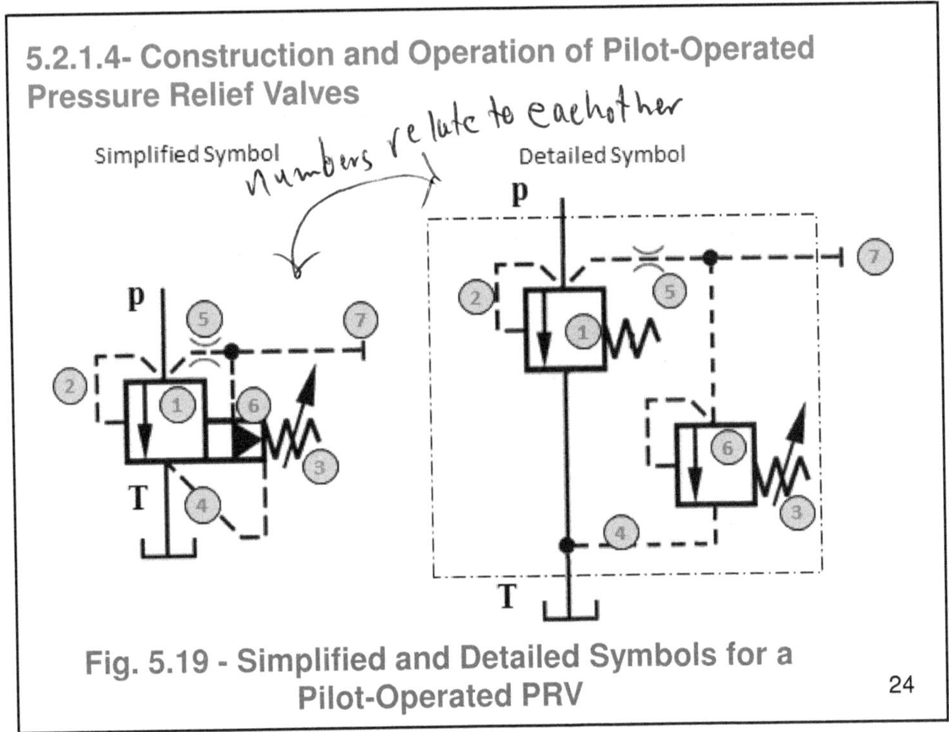

Fig. 5.19 - Simplified and Detailed Symbols for a Pilot-Operated PRV

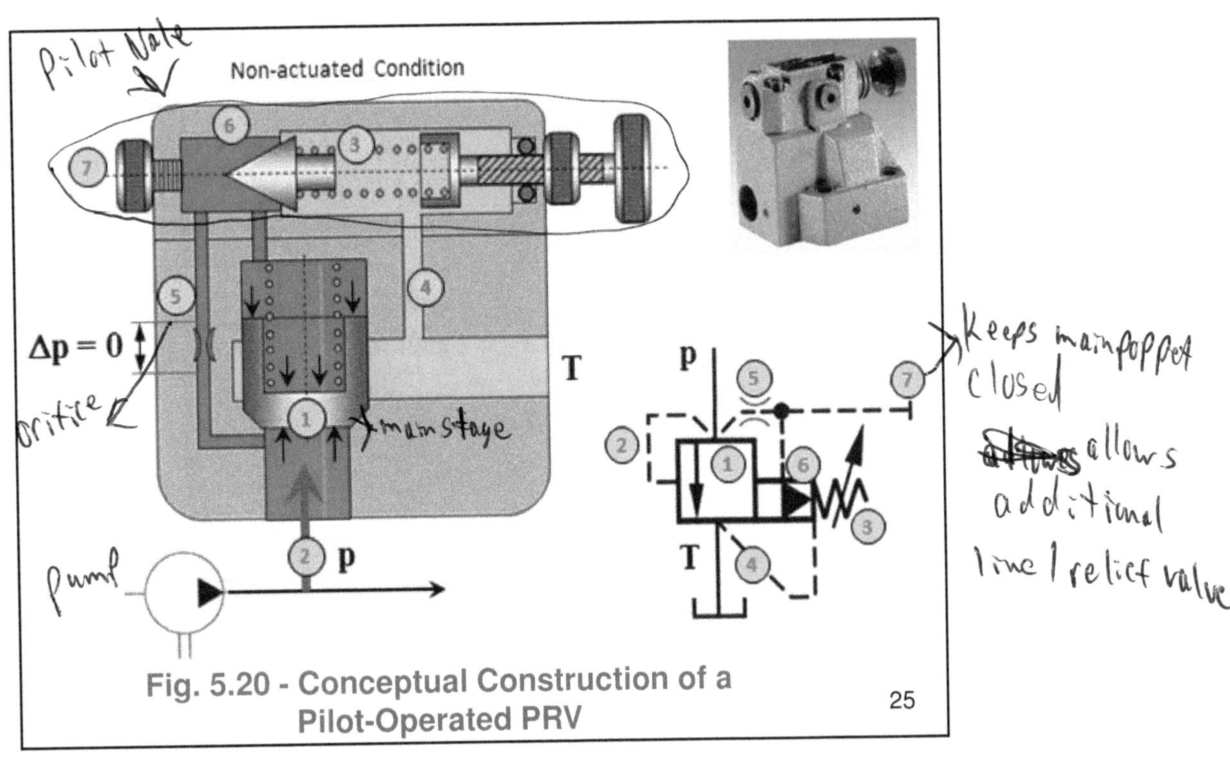

Fig. 5.20 - Conceptual Construction of a Pilot-Operated PRV

used for 100-1200 require 2 stage valves

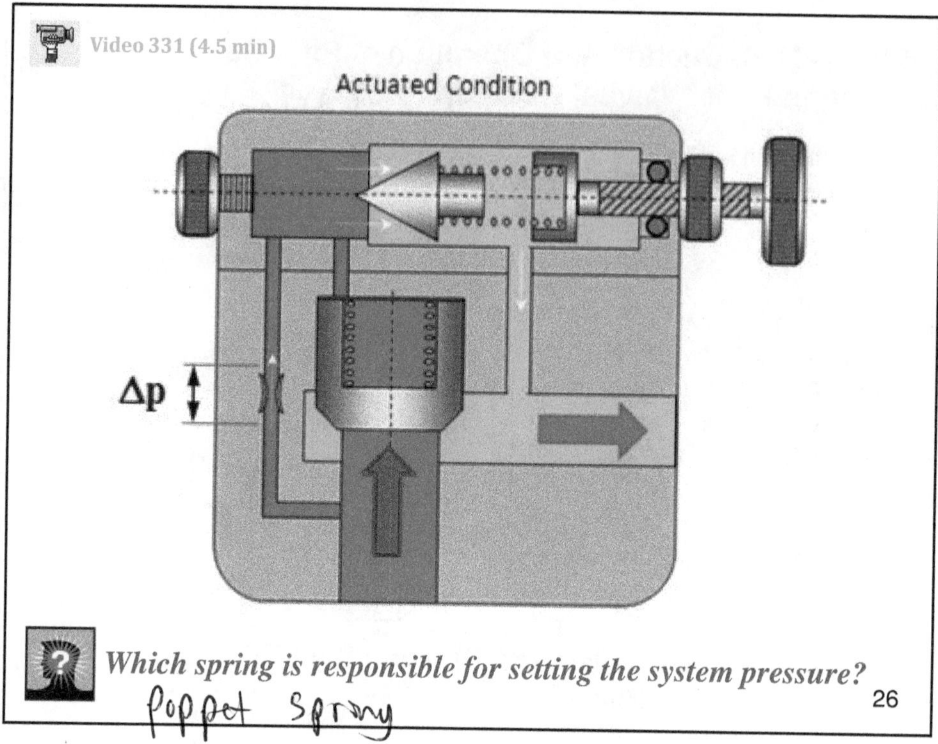

Which spring is responsible for setting the system pressure?

Poppet Spring

26

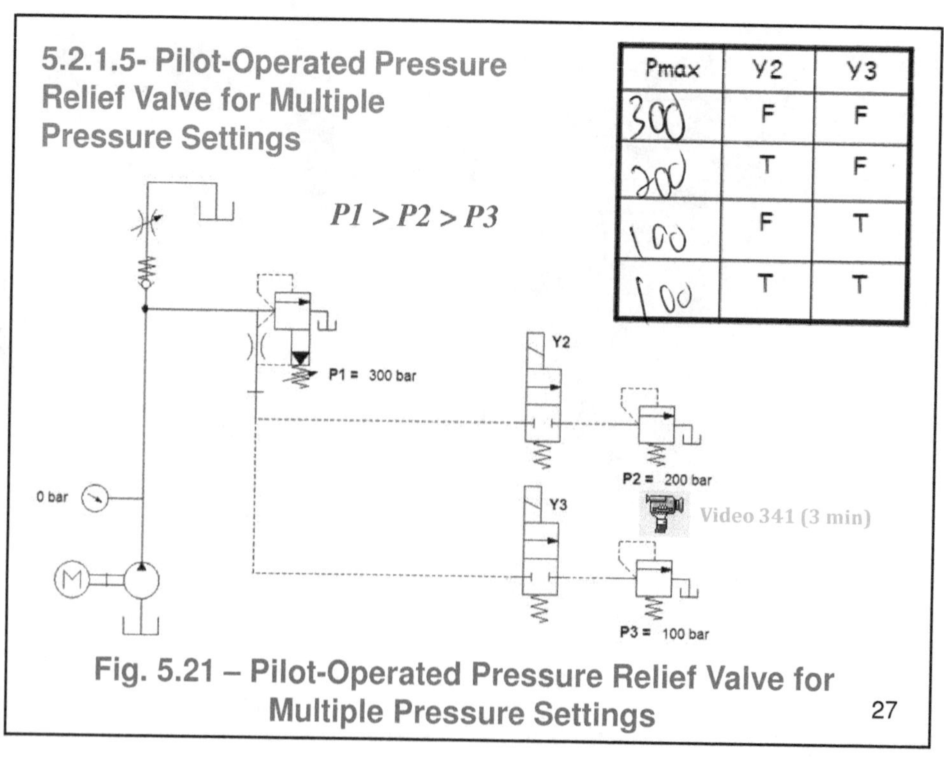

Pmax	Y2	Y3
300	F	F
200	T	F
100	F	T
100	T	T

Fig. 5.21 – Pilot-Operated Pressure Relief Valve for Multiple Pressure Settings

27

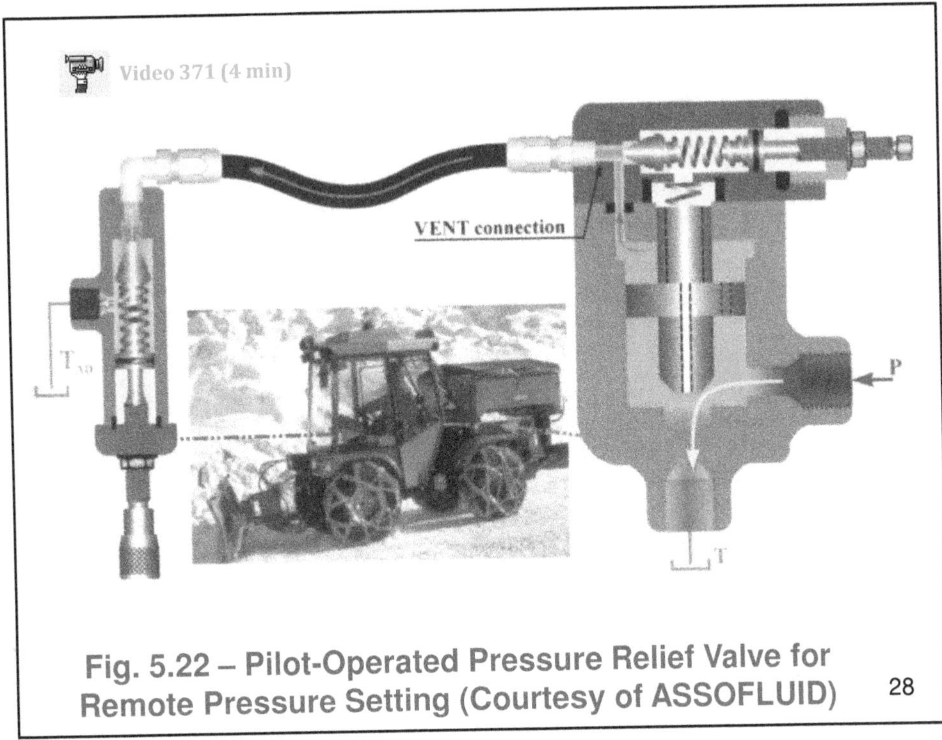

Fig. 5.22 – Pilot-Operated Pressure Relief Valve for Remote Pressure Setting (Courtesy of ASSOFLUID) 28

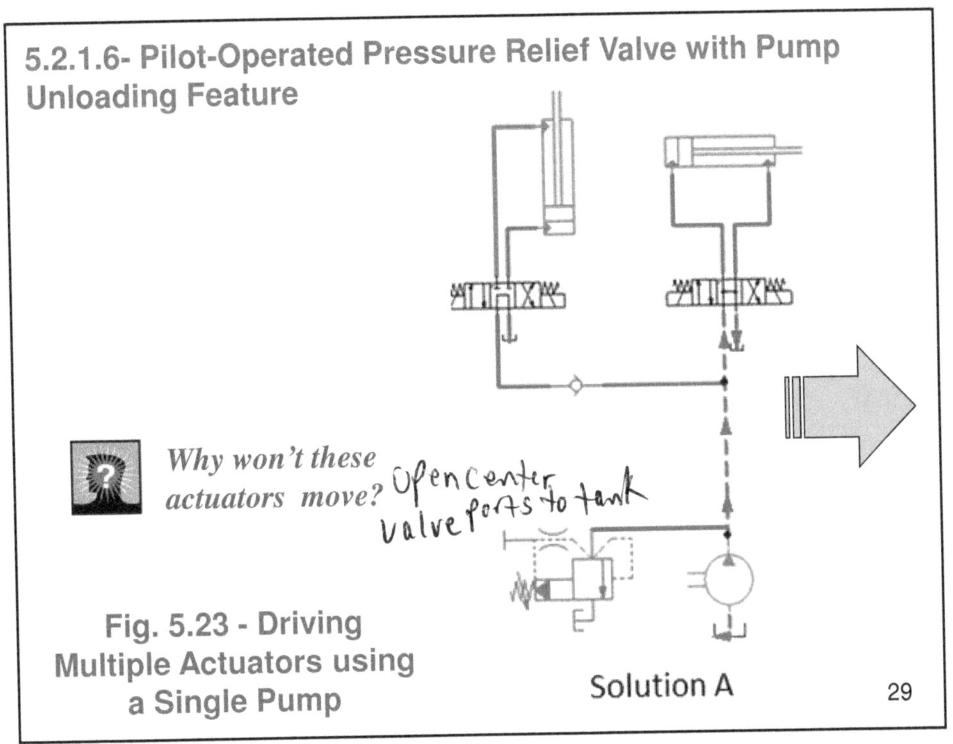

Fig. 5.23 - Driving Multiple Actuators using a Single Pump 29

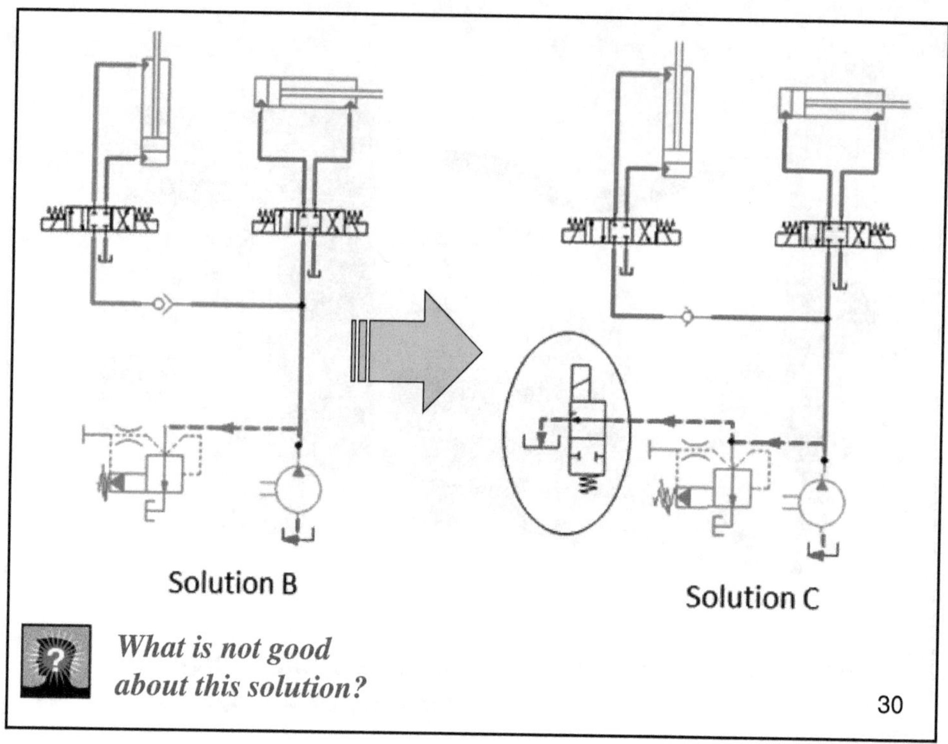

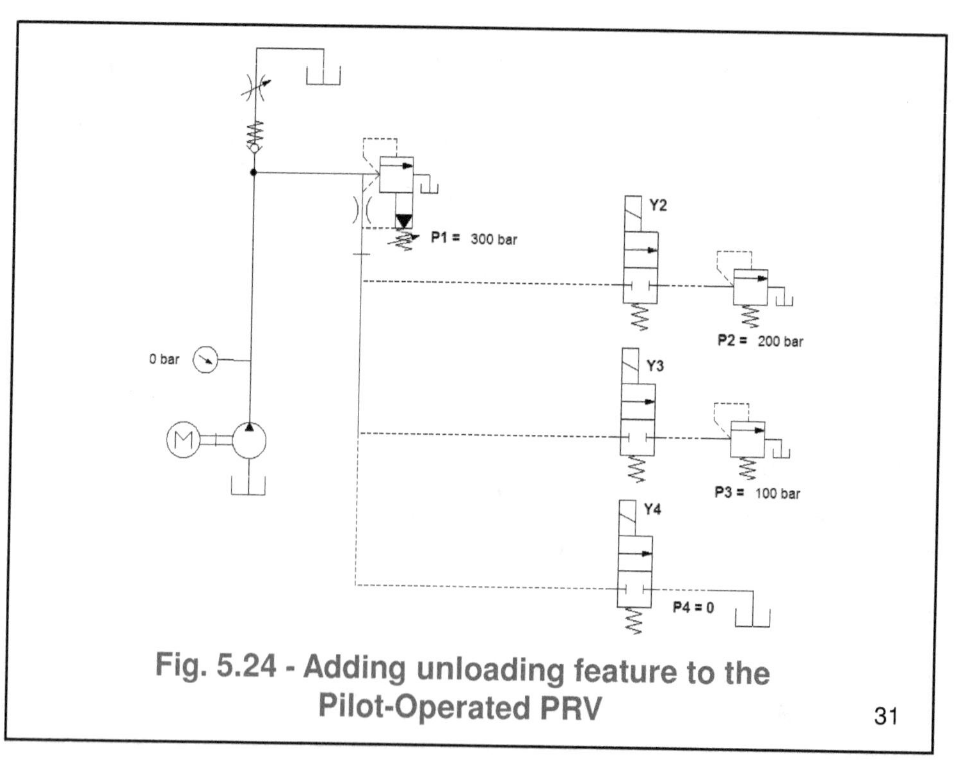

Fig. 5.24 - Adding unloading feature to the Pilot-Operated PRV

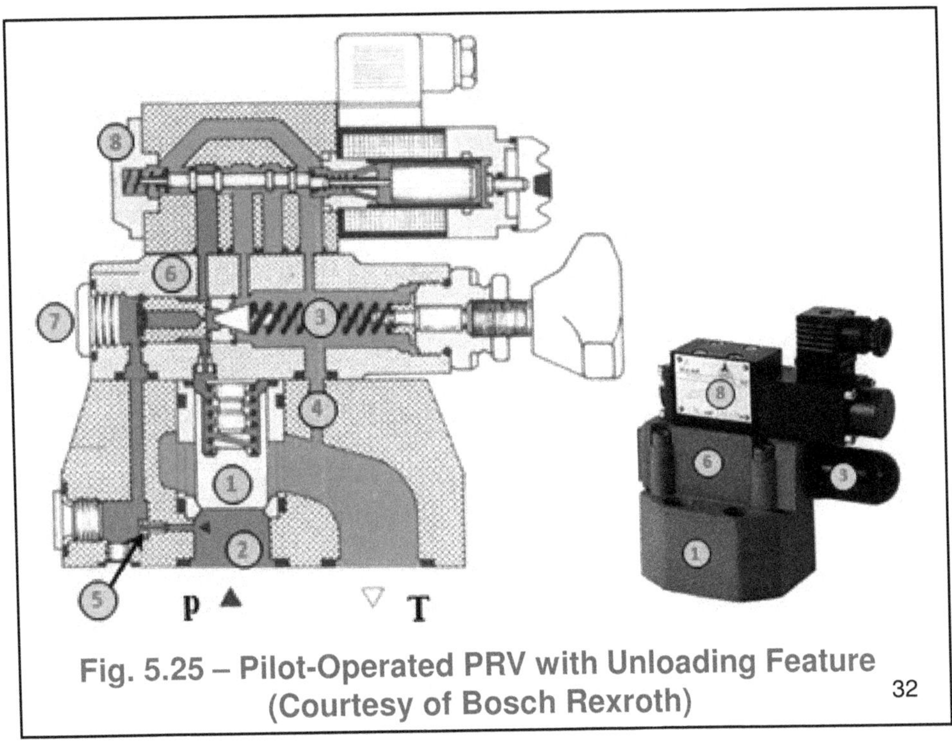

Fig. 5.25 – Pilot-Operated PRV with Unloading Feature
(Courtesy of Bosch Rexroth)

32

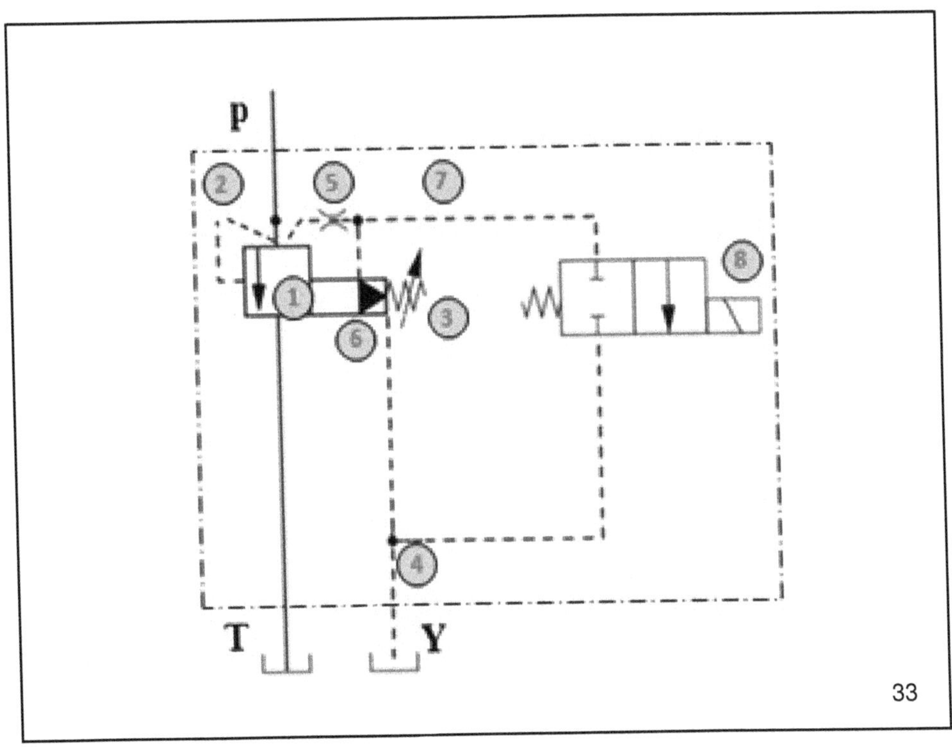

33

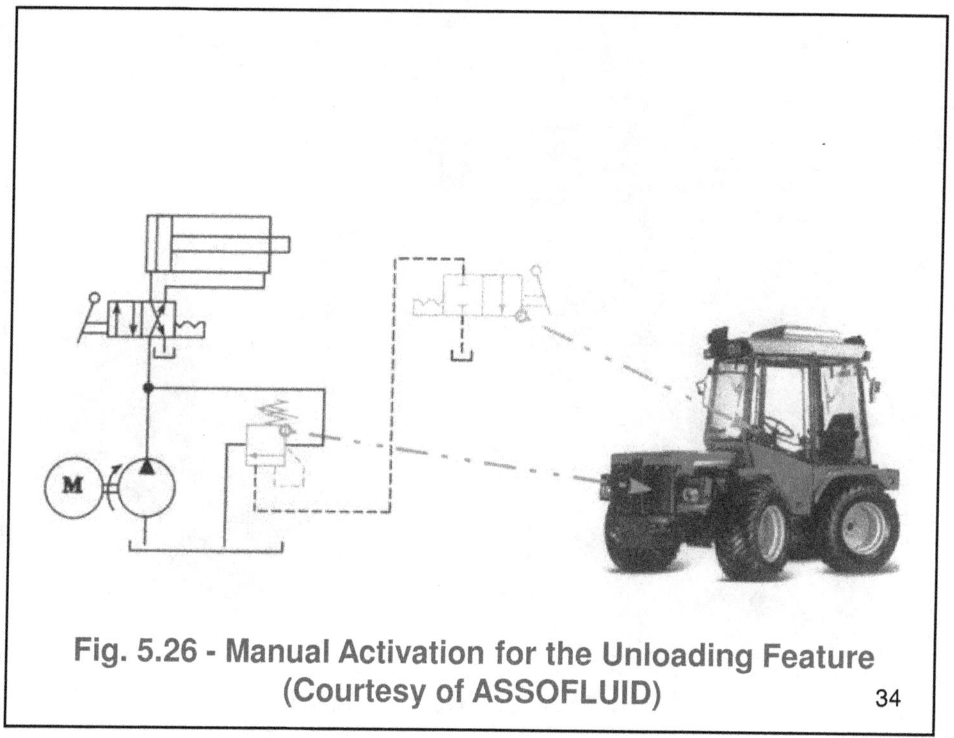

Fig. 5.26 - Manual Activation for the Unloading Feature (Courtesy of ASSOFLUID) 34

cut off psi – flow from pump can only go to relief valve

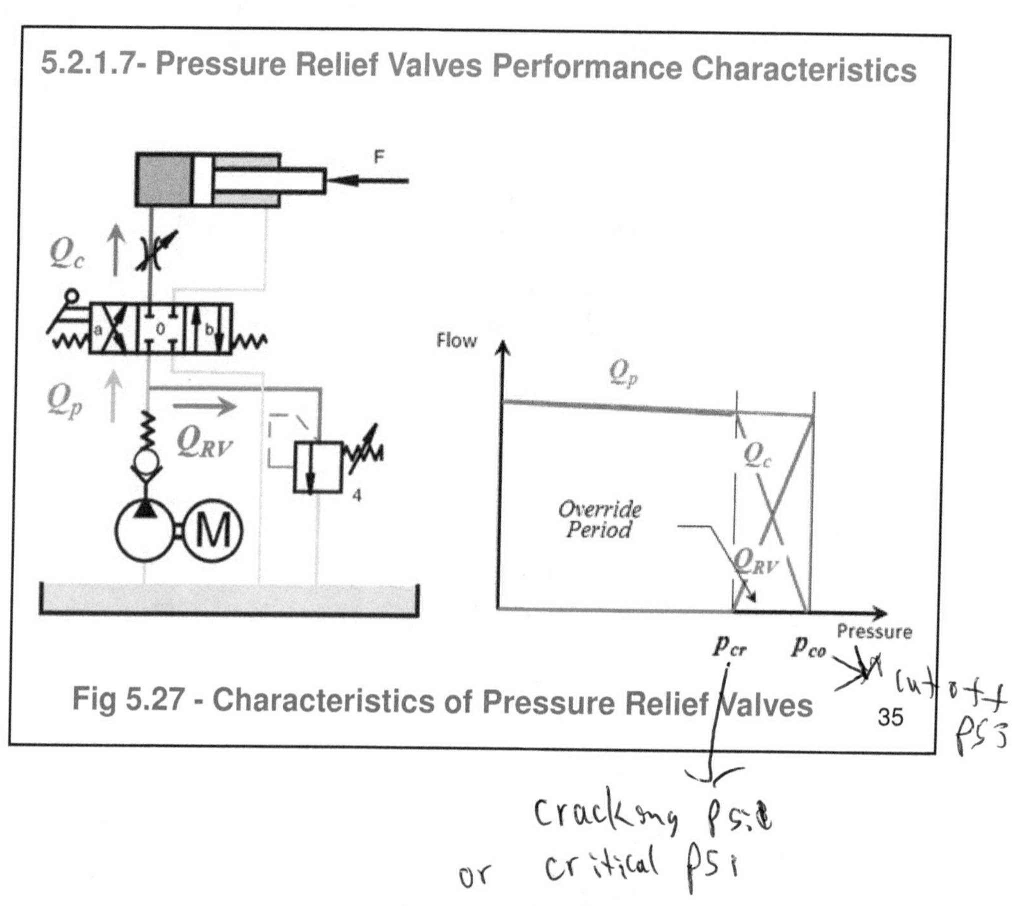

Fig 5.27 - Characteristics of Pressure Relief Valves 35

cracking psi
or critical psi

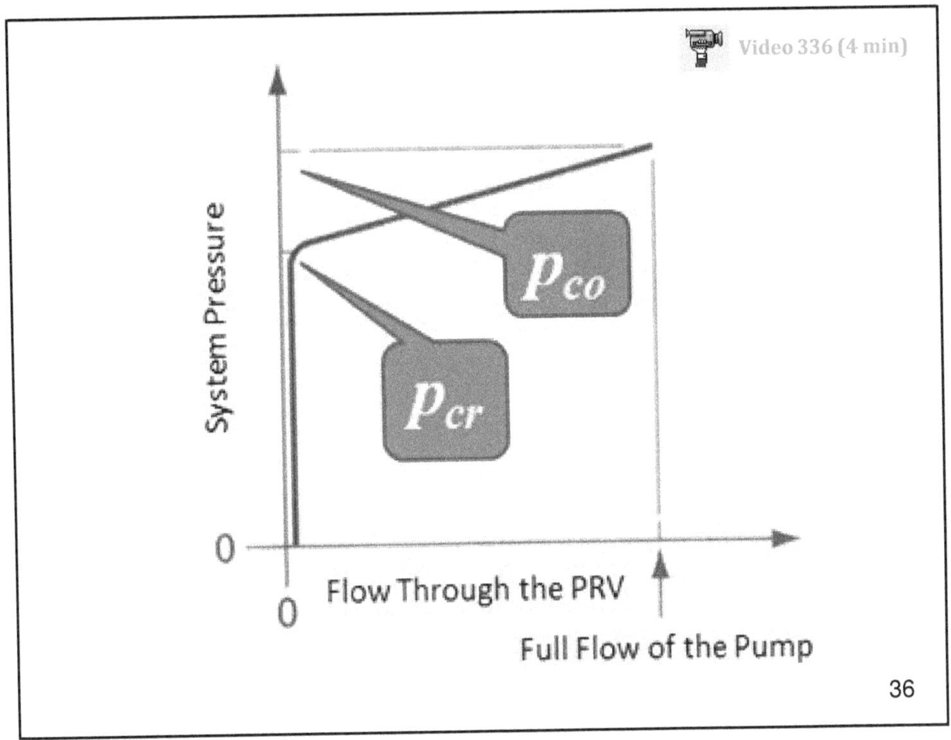

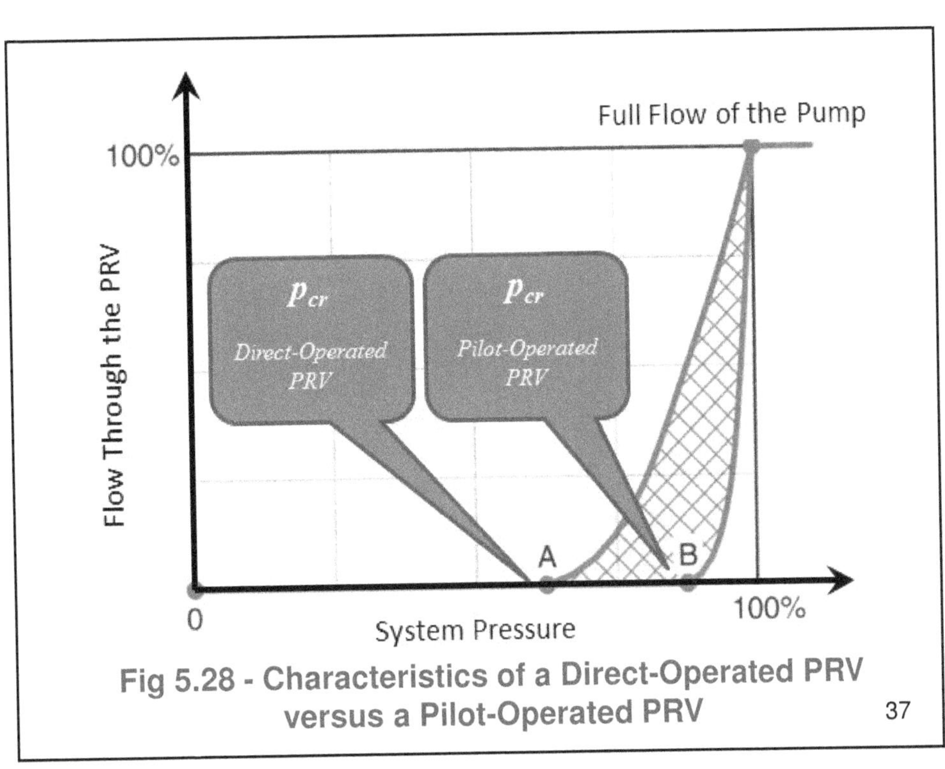

Fig 5.28 - Characteristics of a Direct-Operated PRV versus a Pilot-Operated PRV

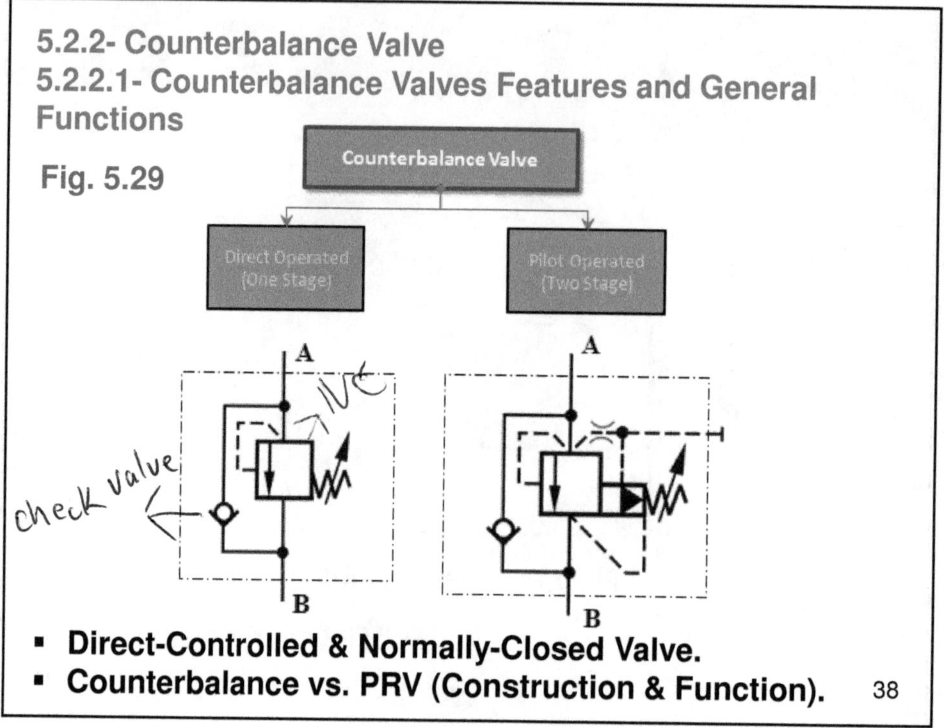

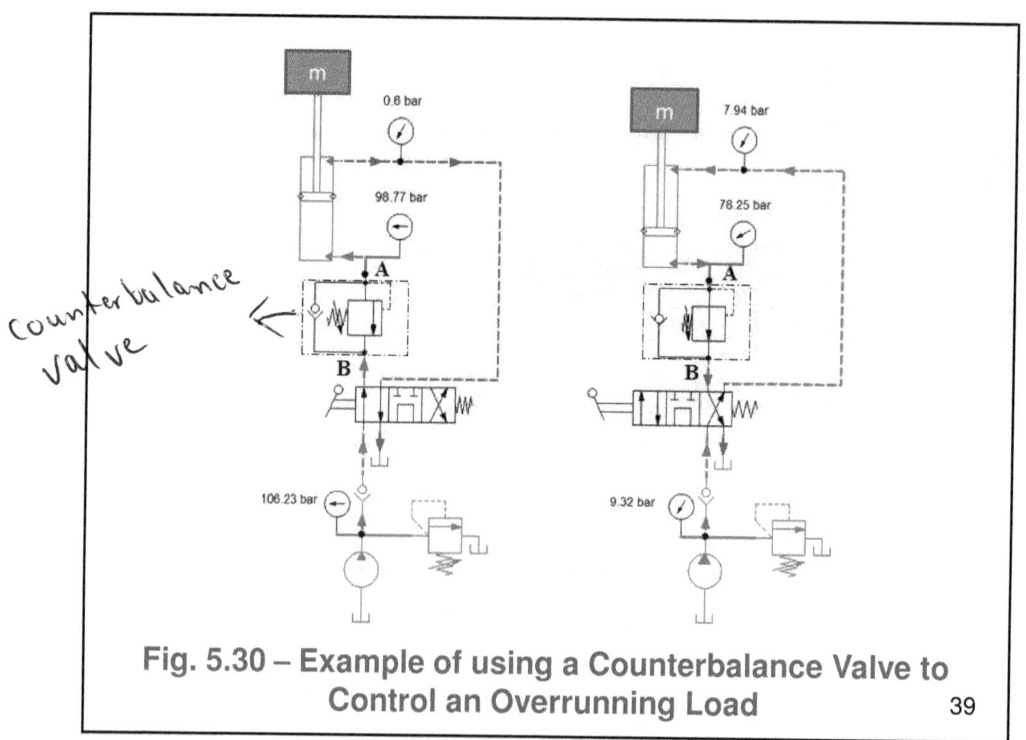

Fig. 5.30 – Example of using a Counterbalance Valve to Control an Overrunning Load

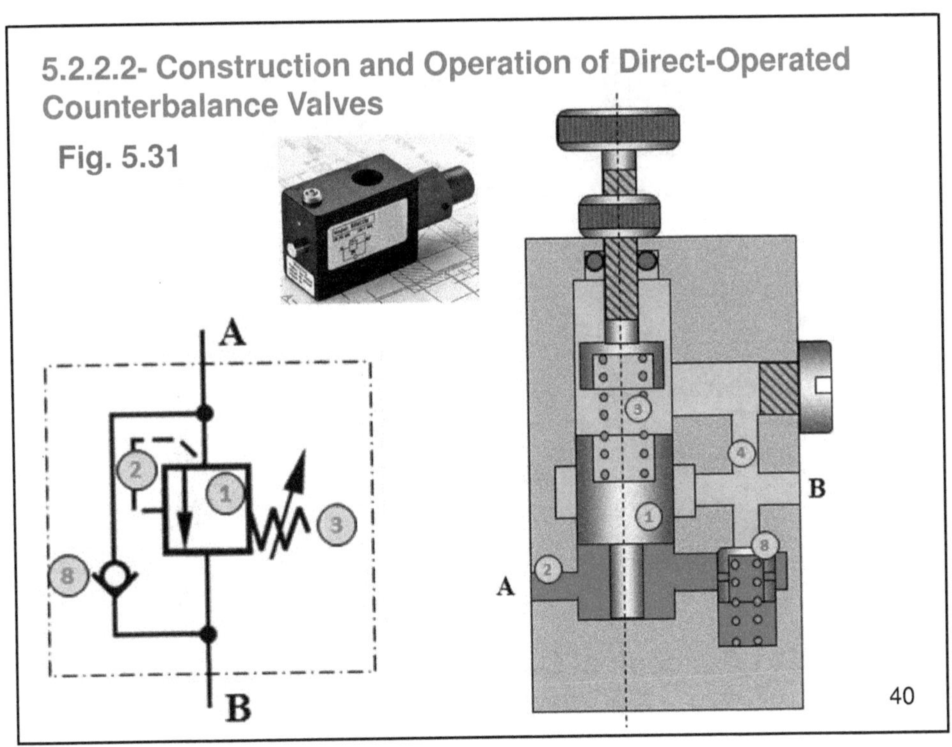

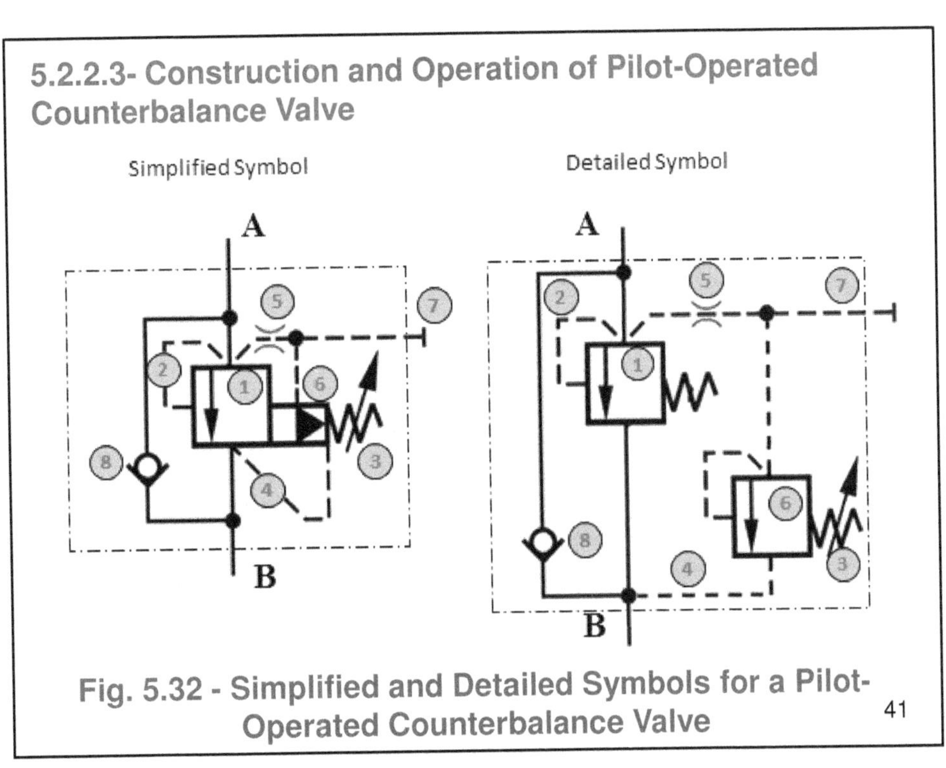

Fig. 5.32 - Simplified and Detailed Symbols for a Pilot-Operated Counterbalance Valve

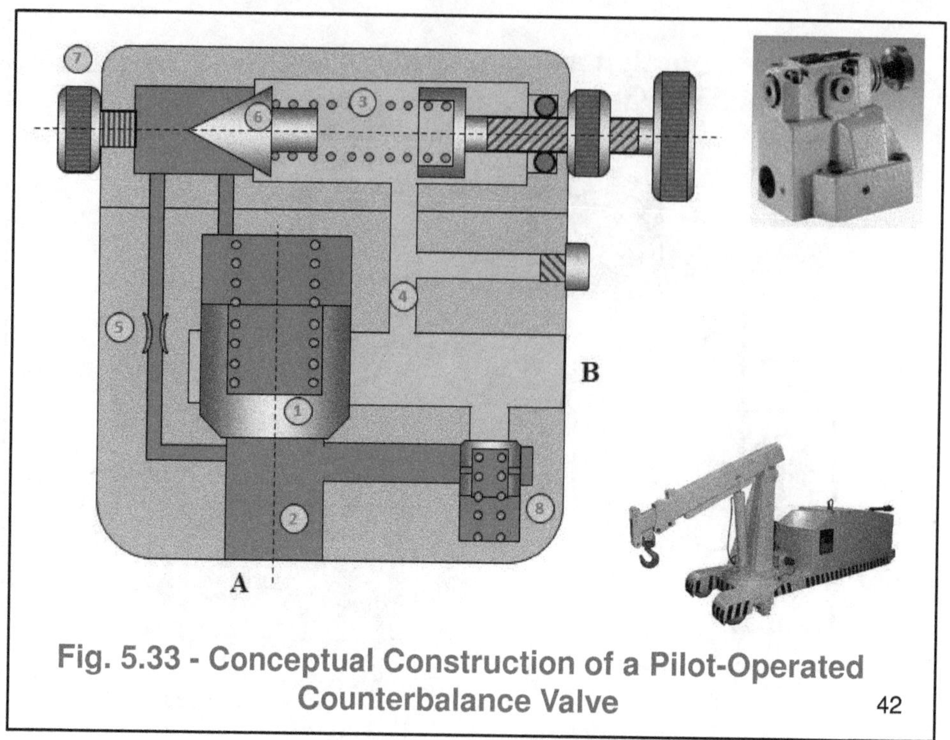

Fig. 5.33 - Conceptual Construction of a Pilot-Operated Counterbalance Valve

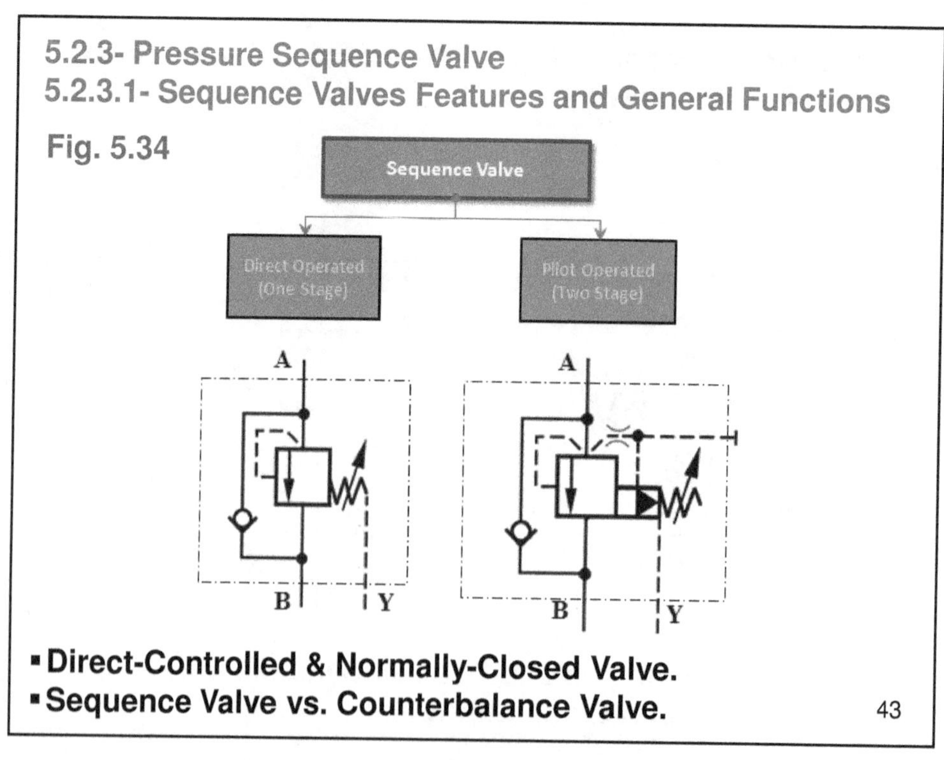

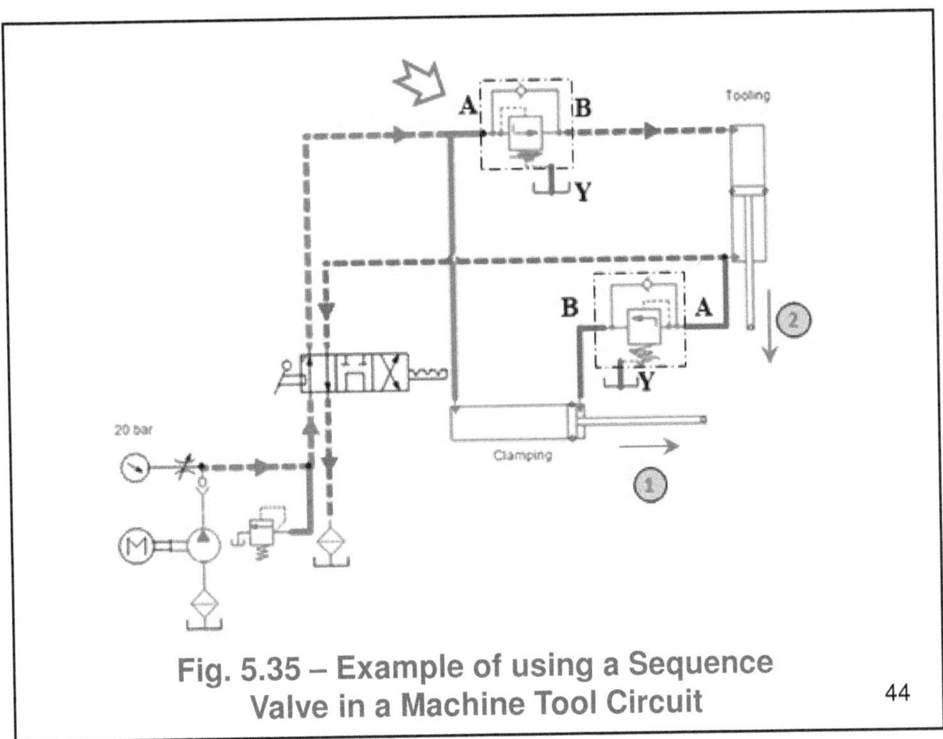

Fig. 5.35 – Example of using a Sequence Valve in a Machine Tool Circuit

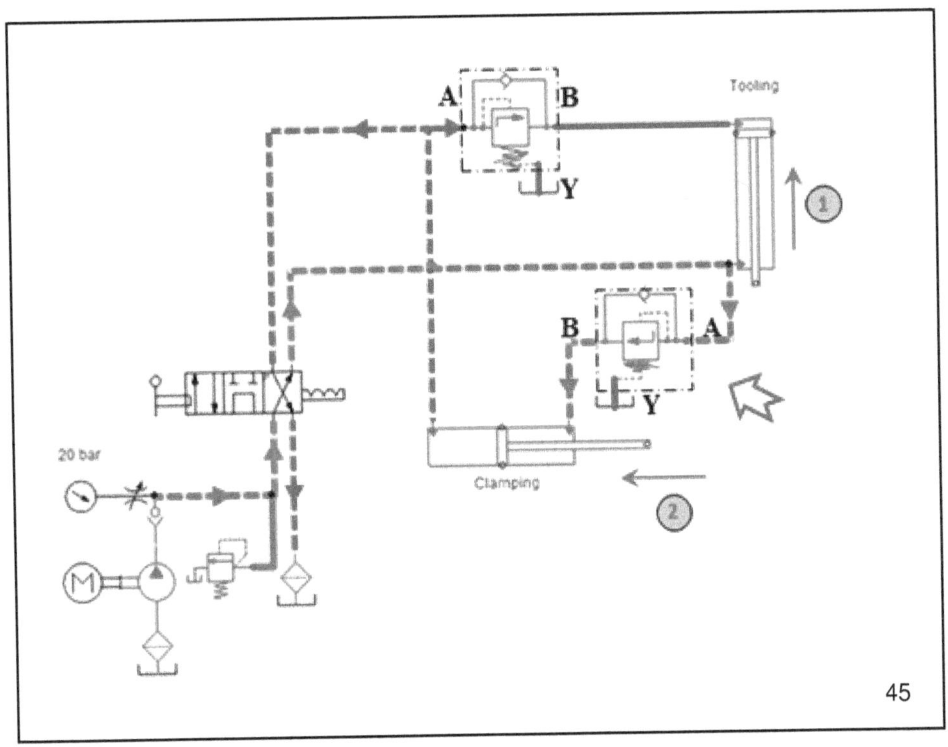

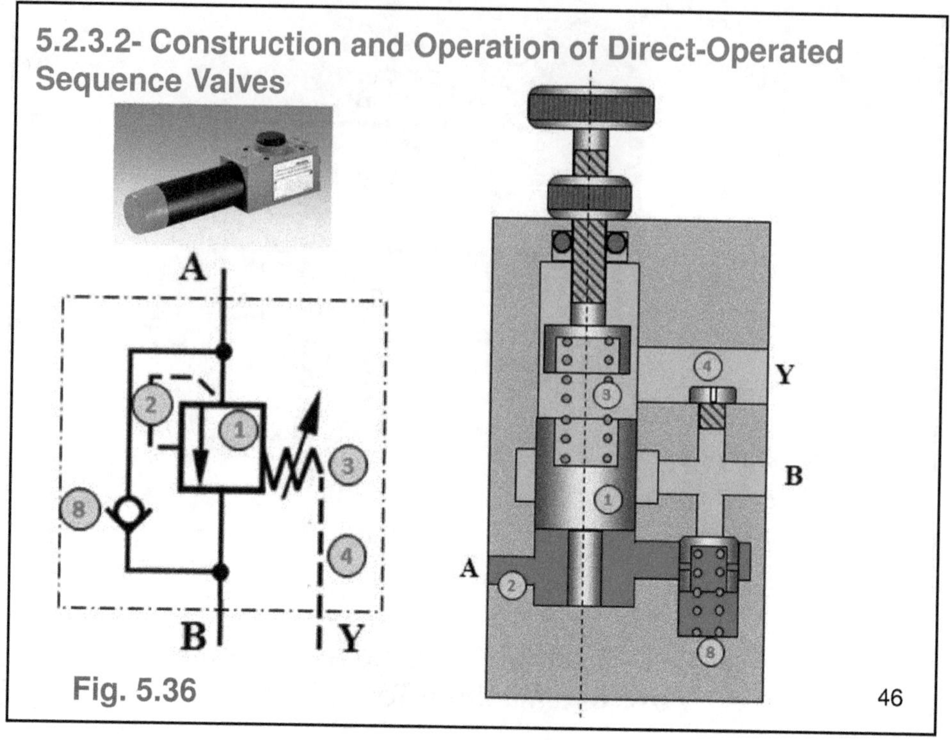

Fig. 5.36

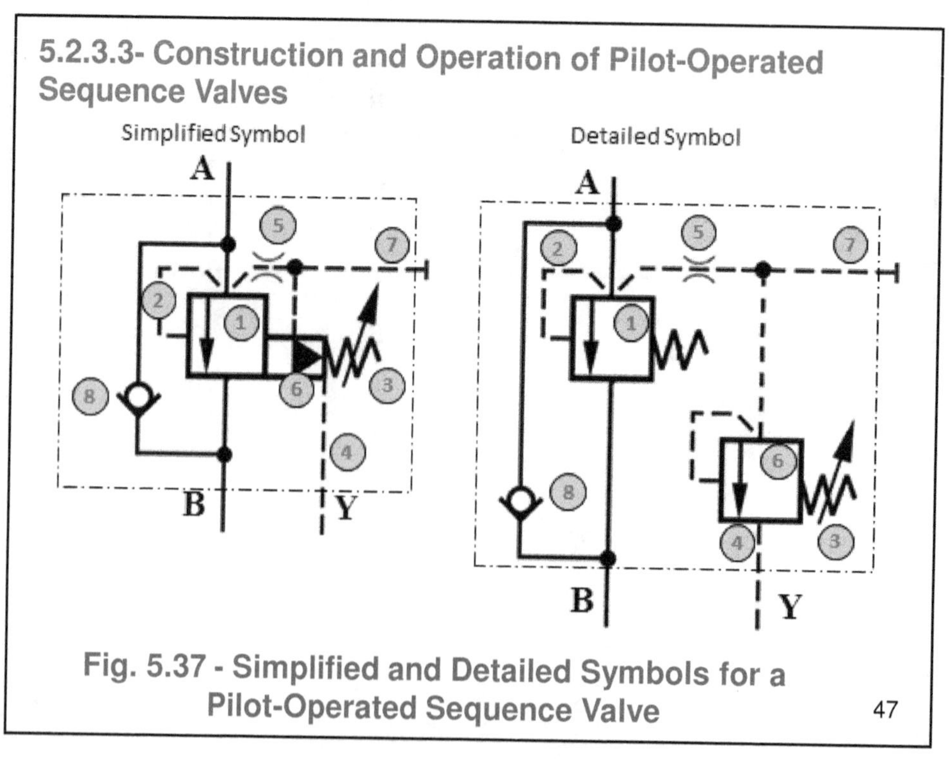

Fig. 5.37 - Simplified and Detailed Symbols for a Pilot-Operated Sequence Valve

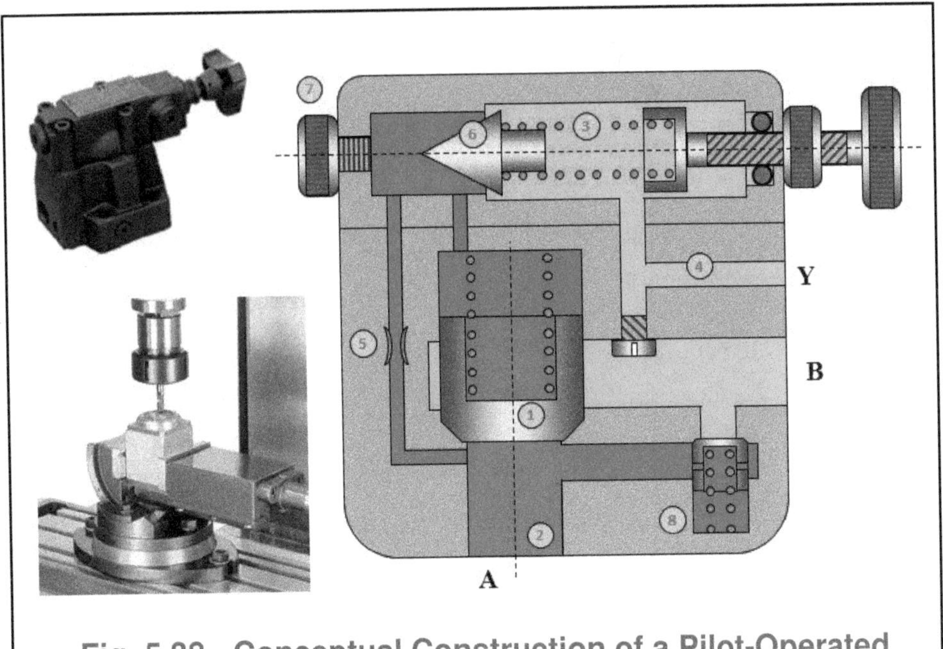

Fig. 5.38 - Conceptual Construction of a Pilot-Operated Sequence Valve

48

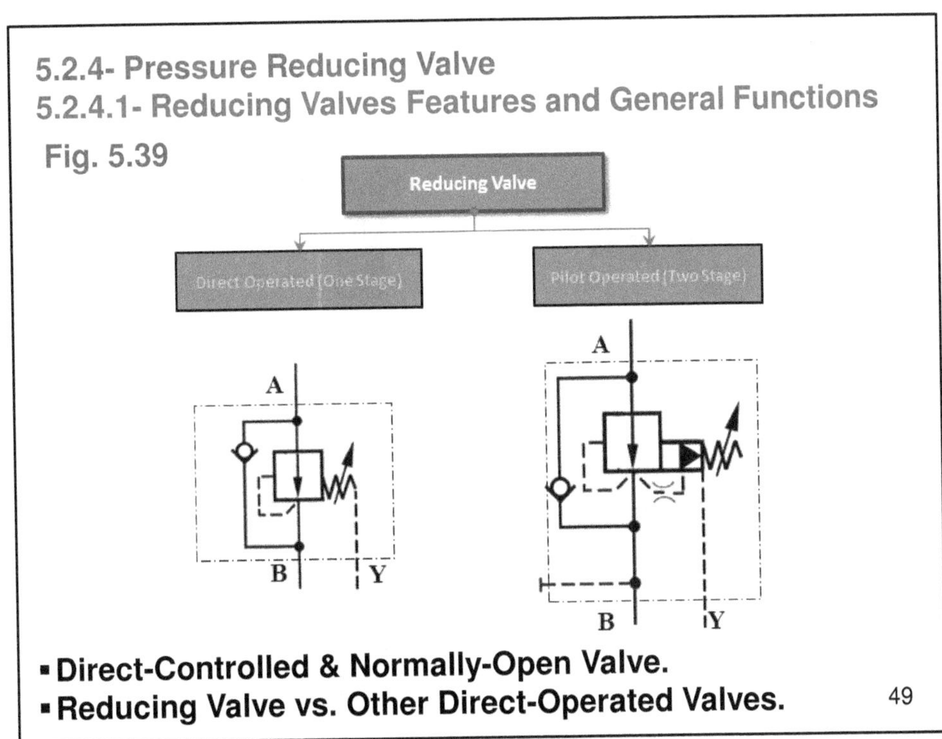

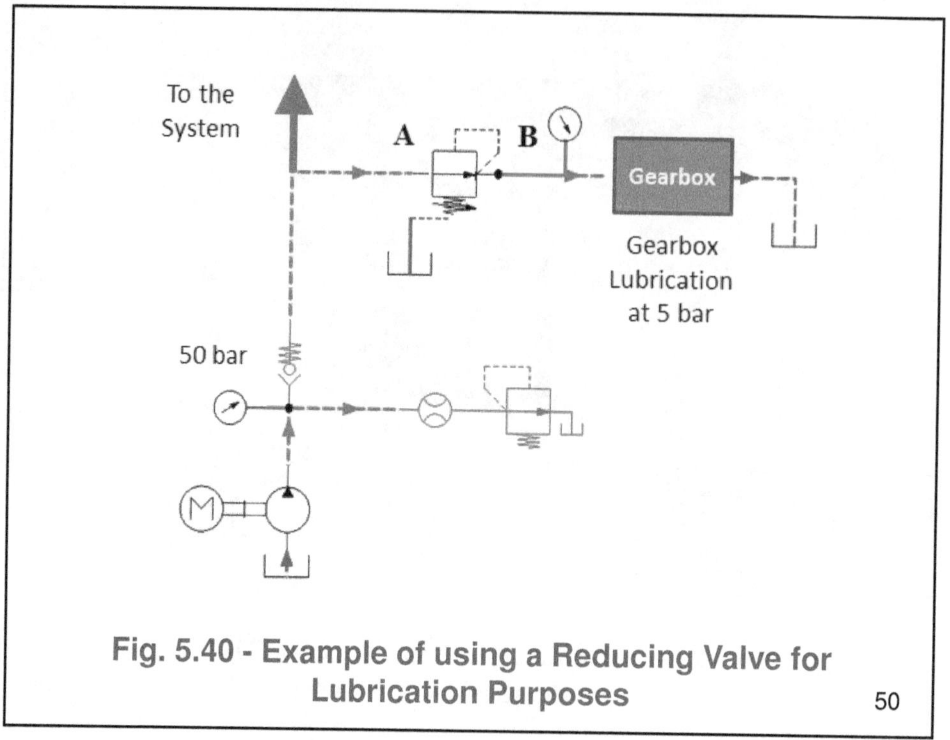

Fig. 5.40 - Example of using a Reducing Valve for Lubrication Purposes

50

5.2.4.2- Construction and Operation of Direct-Operated Pressure Reducing Valves

Fig. 5.41

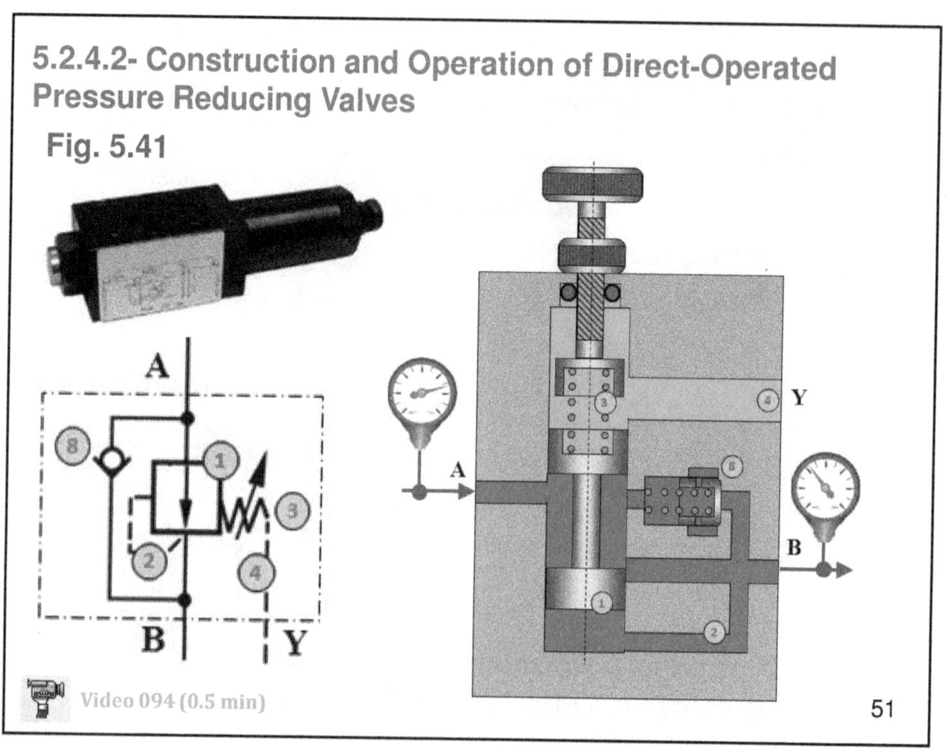

Video 094 (0.5 min)

51

5.2.4.3- Construction and Operation of Pilot-Operated Pressure Reducing Valves

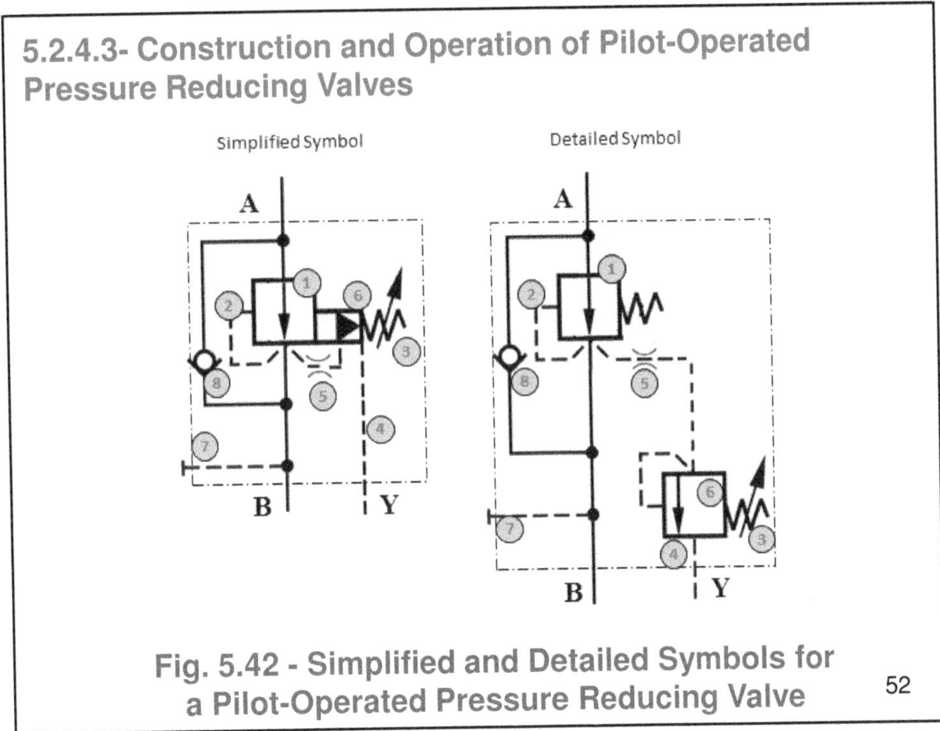

Fig. 5.42 - Simplified and Detailed Symbols for a Pilot-Operated Pressure Reducing Valve

- Closing the knob → Spool 1 moves down → $p_B \uparrow$
- Opening the knob → Spool 1 moves up → $p_B \downarrow$
- At any settings: $p_B \uparrow$ → Spool 1 moves up and vise versa.

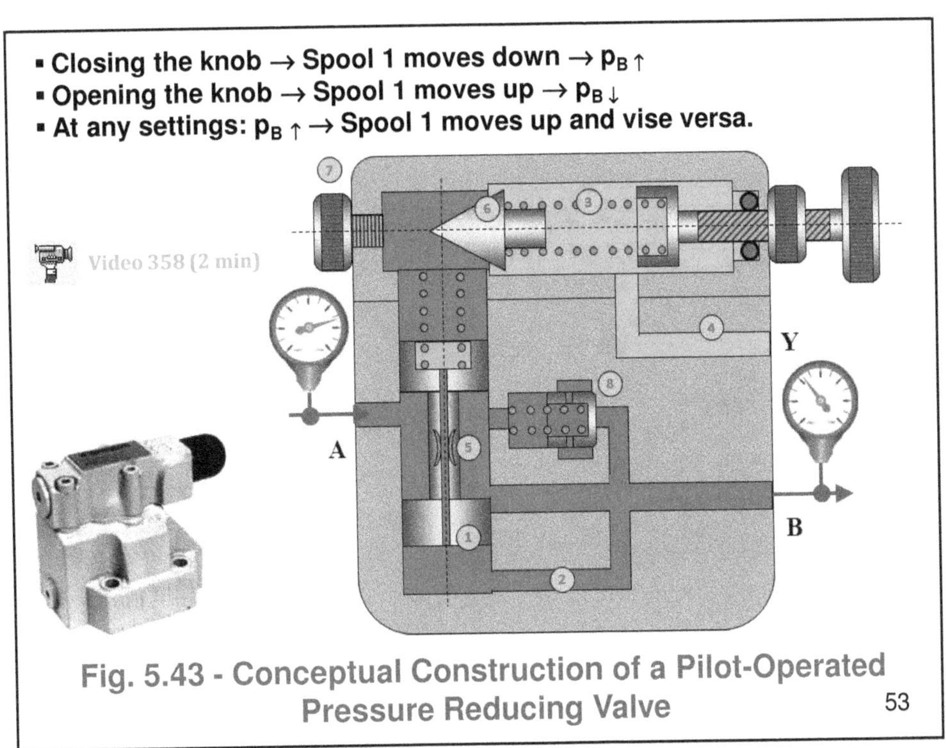

Fig. 5.43 - Conceptual Construction of a Pilot-Operated Pressure Reducing Valve

5.2.5- Unloading Valve

5.2.5.1- Unloading Valves Features and General Functions

Fig. 5.44

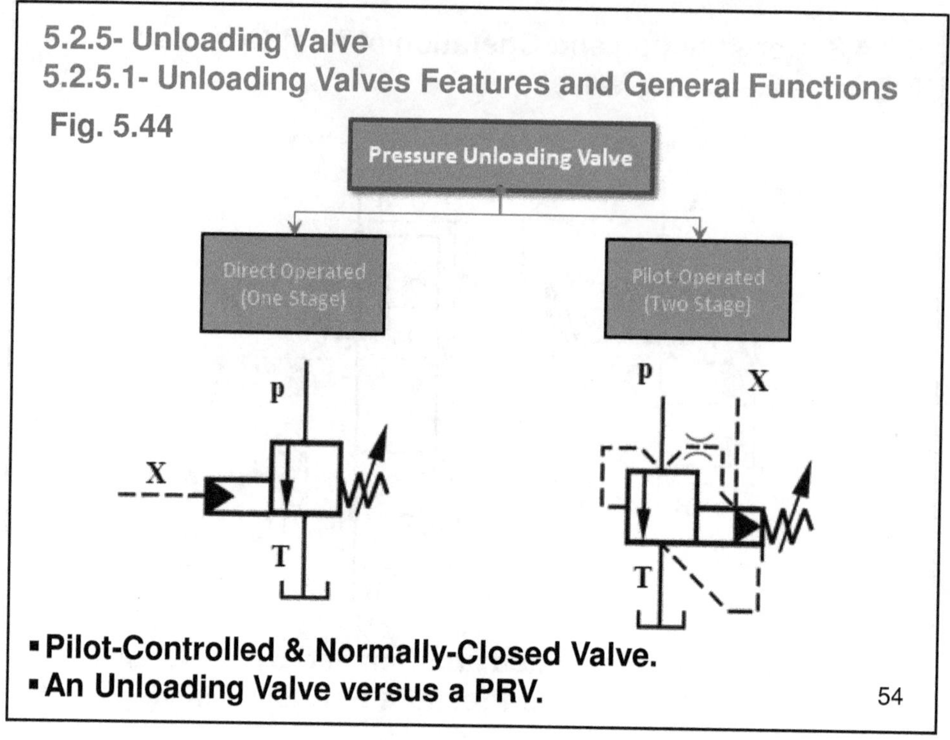

- **Pilot-Controlled & Normally-Closed Valve.**
- **An Unloading Valve versus a PRV.**

54

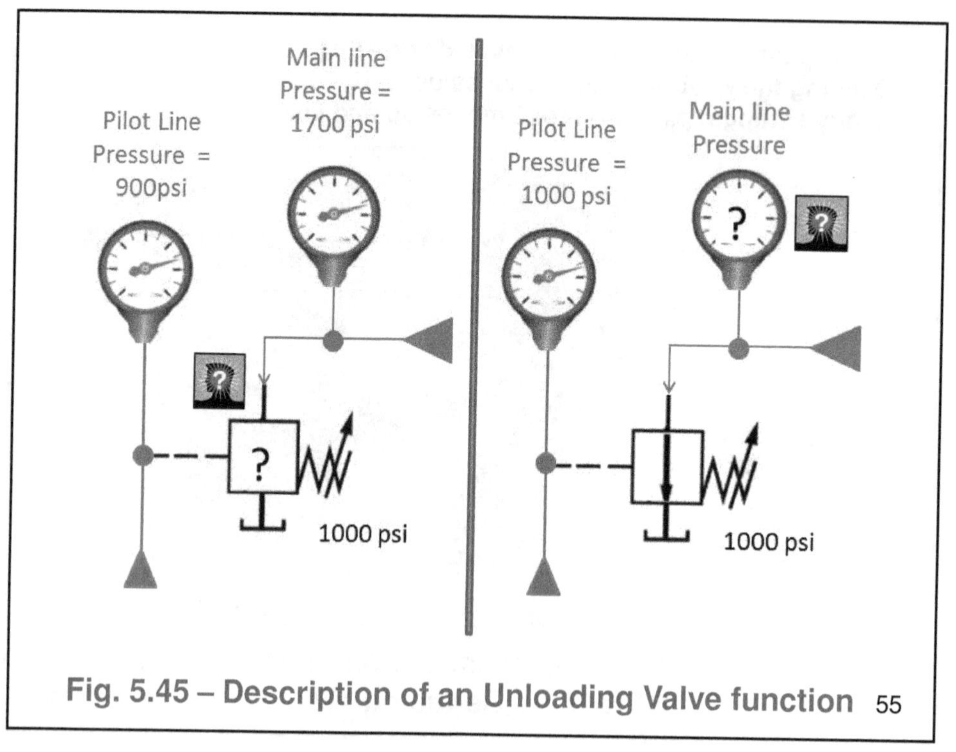

Fig. 5.45 – Description of an Unloading Valve function 55

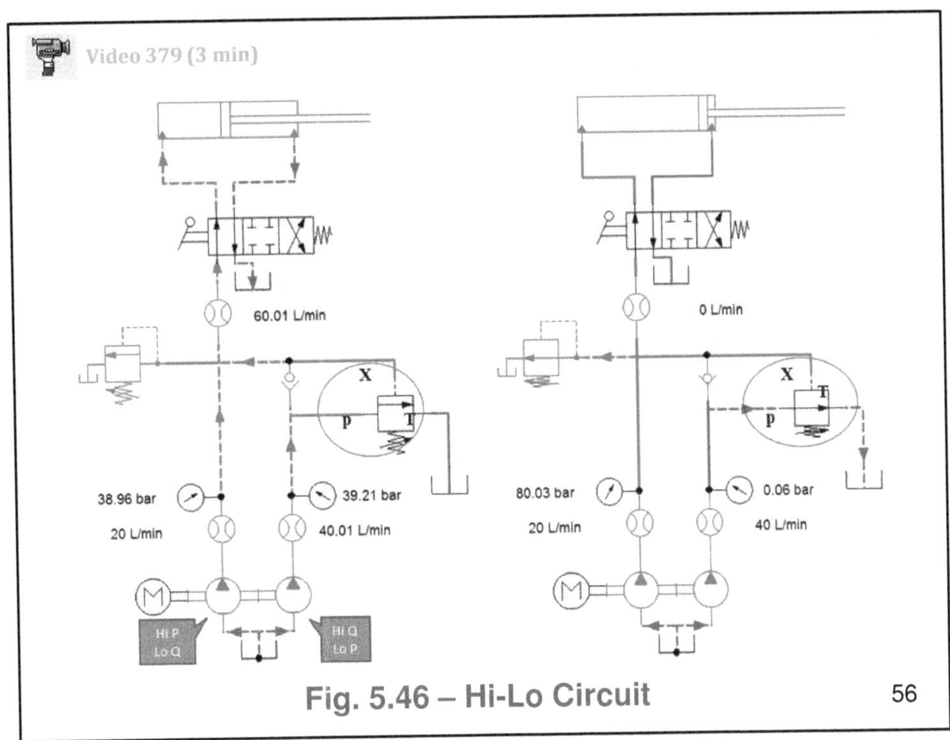

Fig. 5.46 – Hi-Lo Circuit

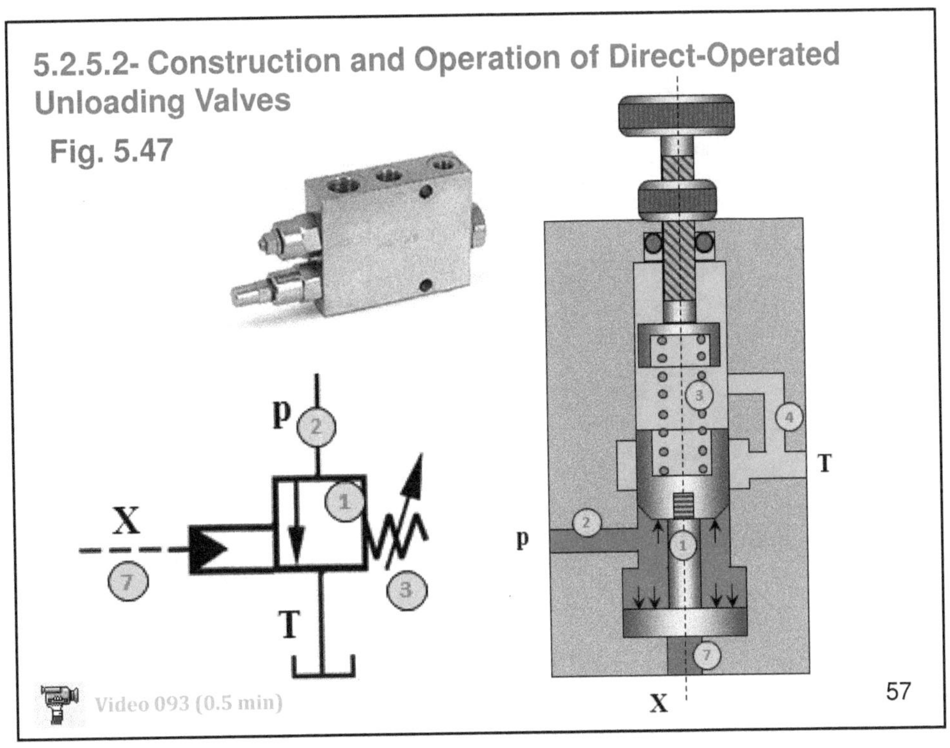

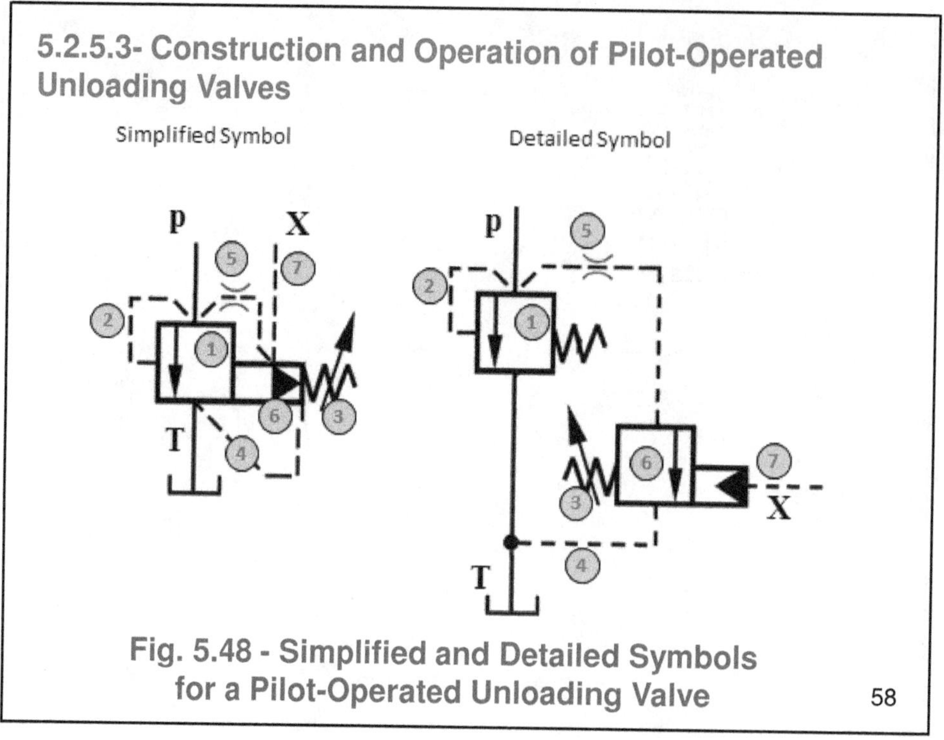

Fig. 5.48 - Simplified and Detailed Symbols for a Pilot-Operated Unloading Valve

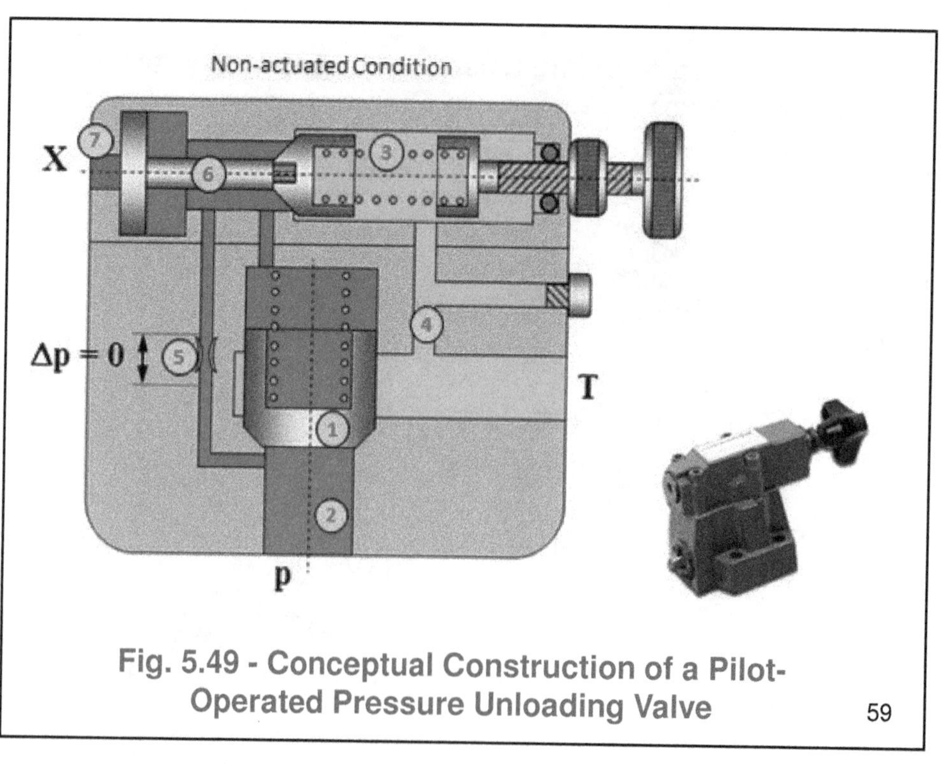

Fig. 5.49 - Conceptual Construction of a Pilot-Operated Pressure Unloading Valve

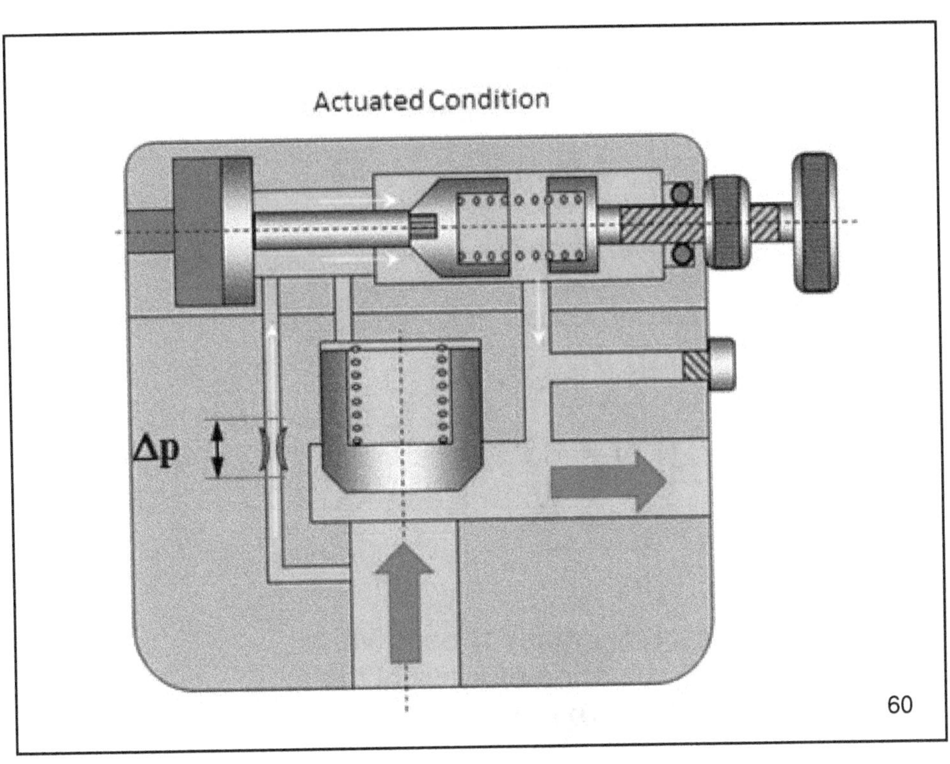
Actuated Condition
Δp
60

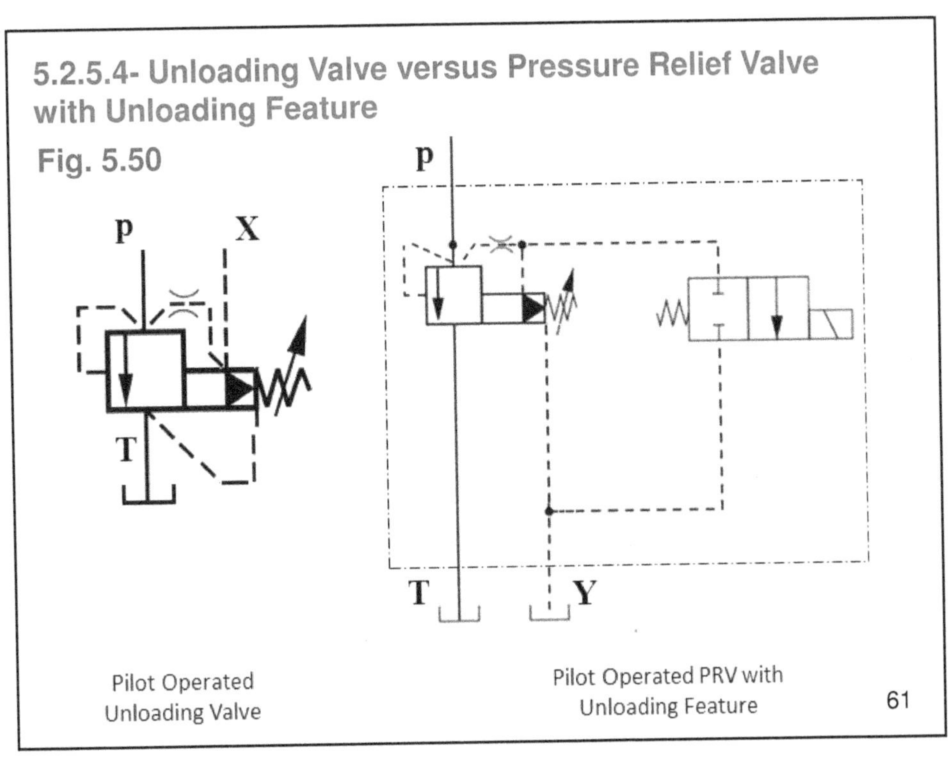
5.2.5.4- Unloading Valve versus Pressure Relief Valve with Unloading Feature
Fig. 5.50
p
X
T
p
T
Y
Pilot Operated Unloading Valve
Pilot Operated PRV with Unloading Feature
61

5.2.6- Over-Center Valve

5.2.6.1- Over-Center Valves Features and General Functions

Fig. 5.51

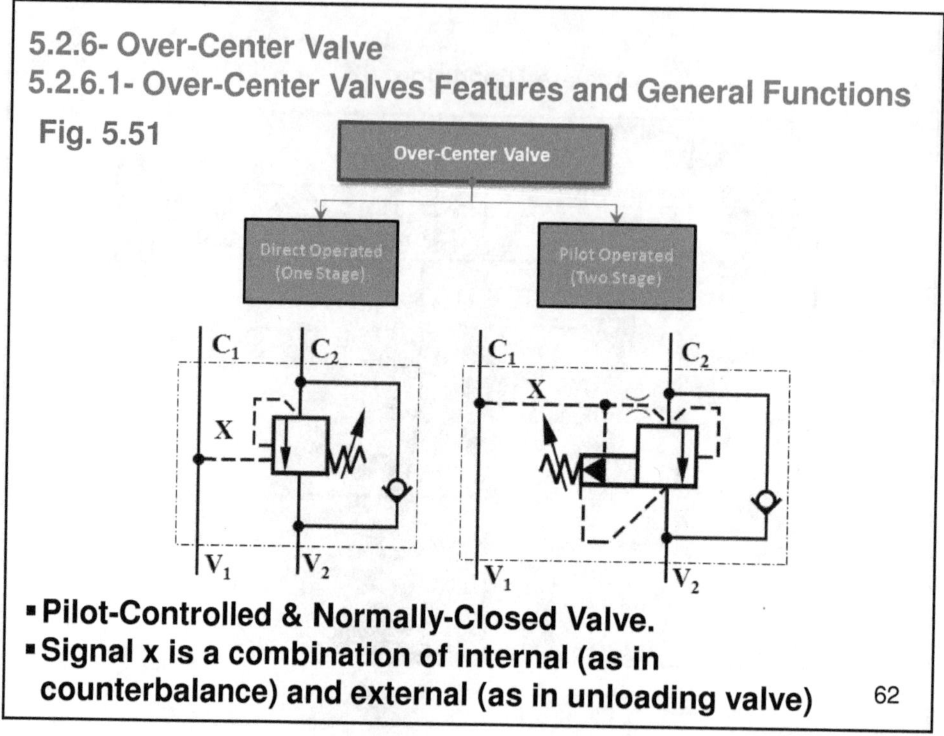

- Pilot-Controlled & Normally-Closed Valve.
- Signal x is a combination of internal (as in counterbalance) and external (as in unloading valve)

62

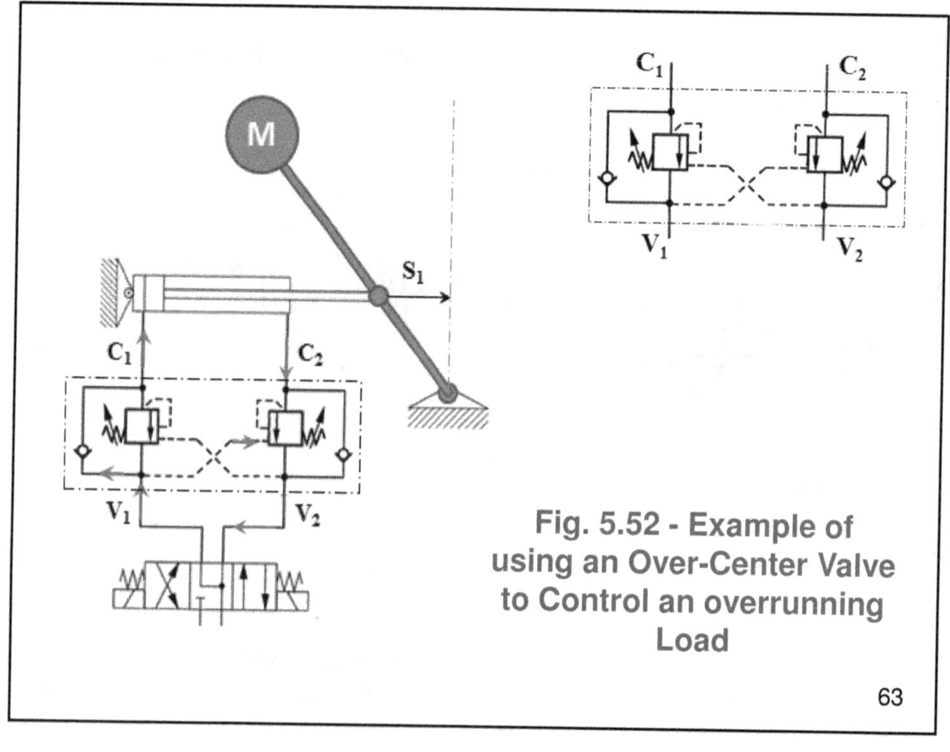

Fig. 5.52 - Example of using an Over-Center Valve to Control an overrunning Load

63

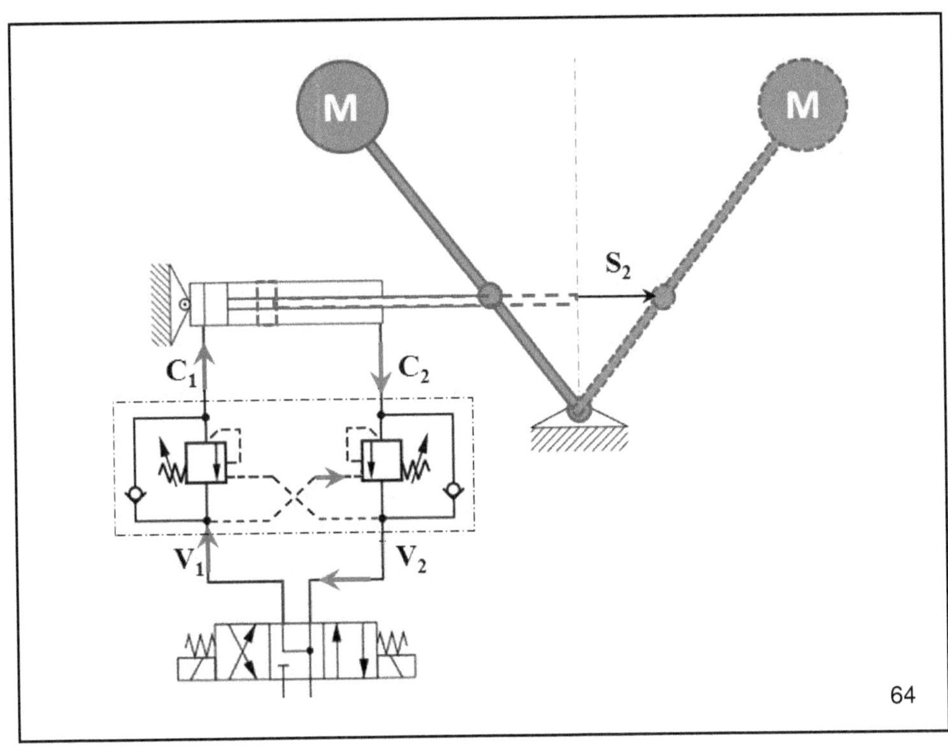

5.2.6.2- Construction and Operation of Direct-Operated Over-Center Valve

Fig. 5.53

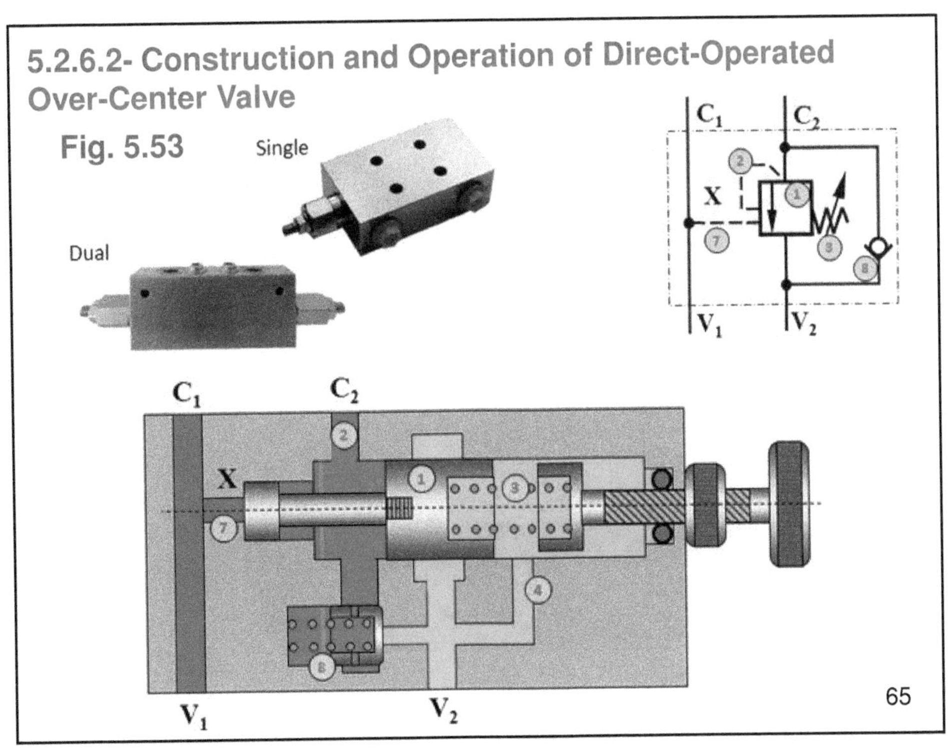

5.2.6.3- Construction and Operation of Pilot-Operated Over-Center Valve

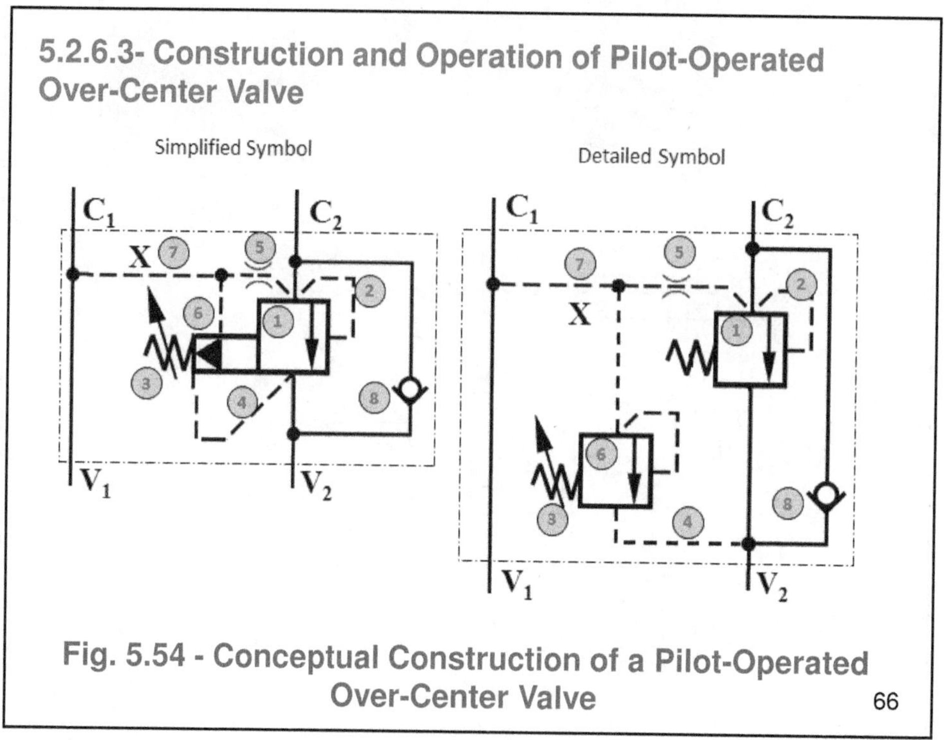

Fig. 5.54 - Conceptual Construction of a Pilot-Operated Over-Center Valve

66

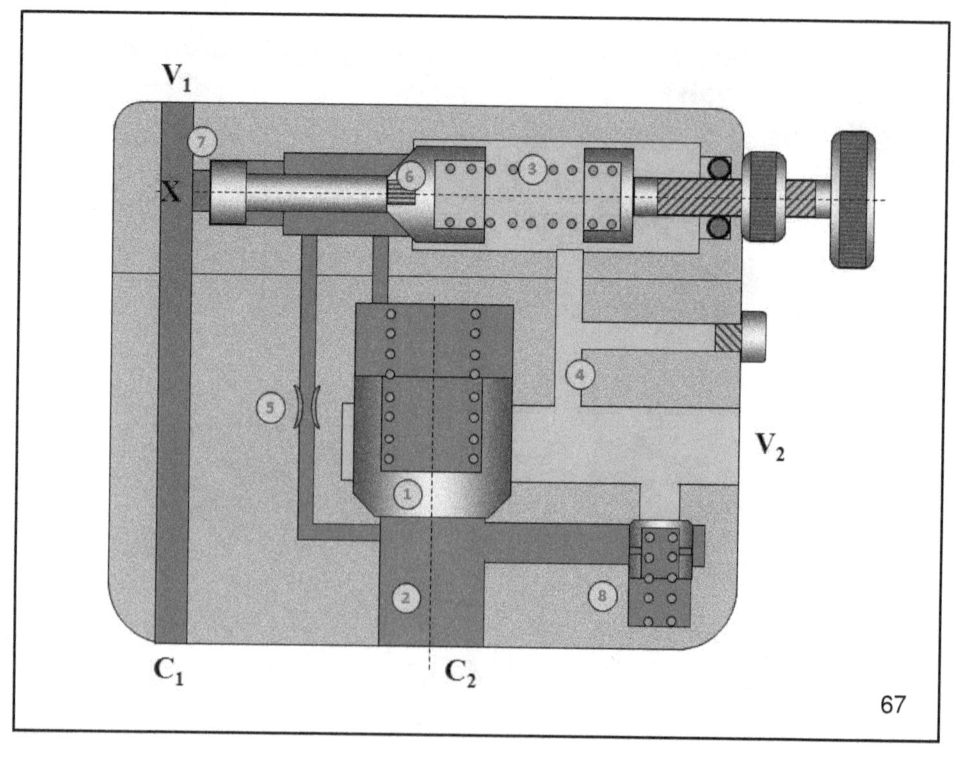

67

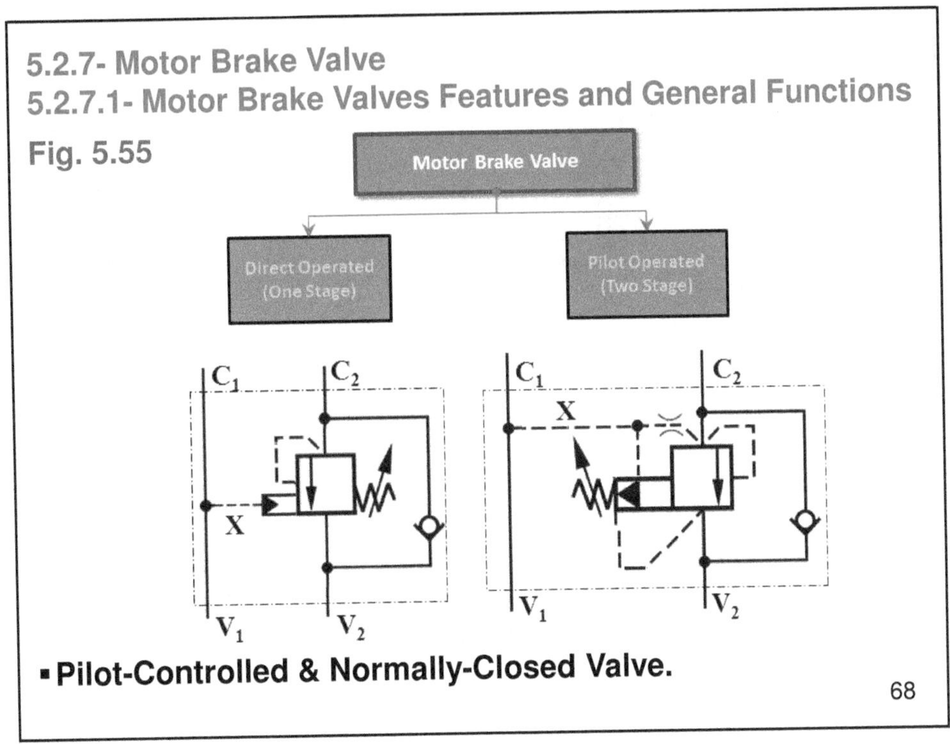
5.2.7- Motor Brake Valve
5.2.7.1- Motor Brake Valves Features and General Functions
Fig. 5.55
Motor Brake Valve
Direct Operated (One Stage)
Pilot Operated (Two Stage)
C1
C2
X
V1
V2
▪ Pilot-Controlled & Normally-Closed Valve.
68

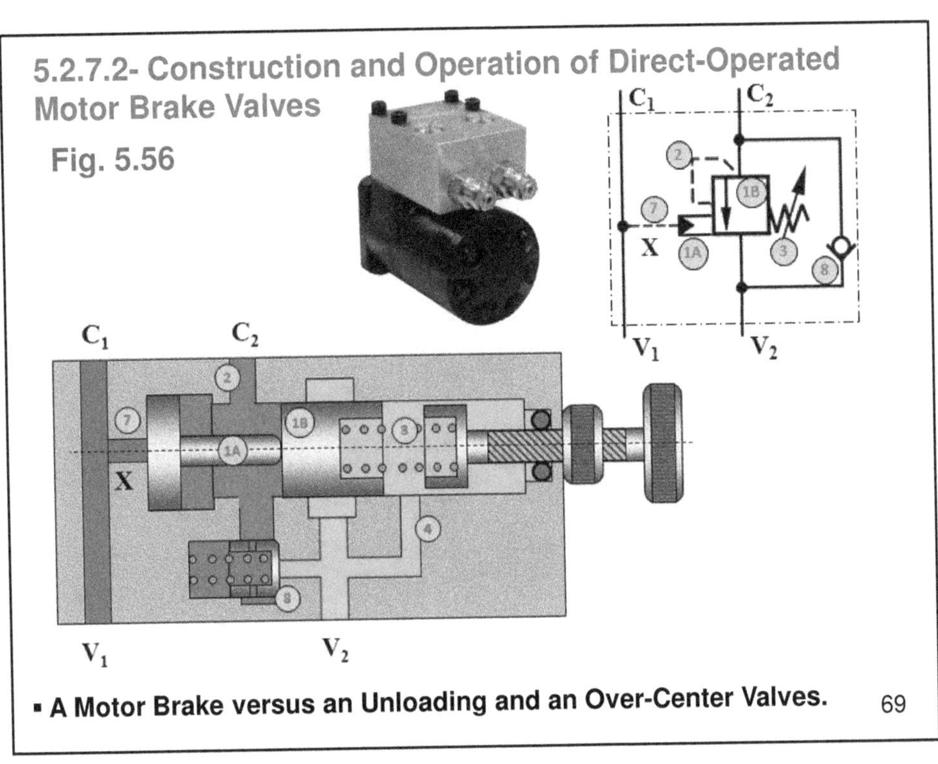
5.2.7.2- Construction and Operation of Direct-Operated Motor Brake Valves
Fig. 5.56
C1
C2
X
V1
V2
1A
1B
2
3
4
7
8
▪ A Motor Brake versus an Unloading and an Over-Center Valves.
69

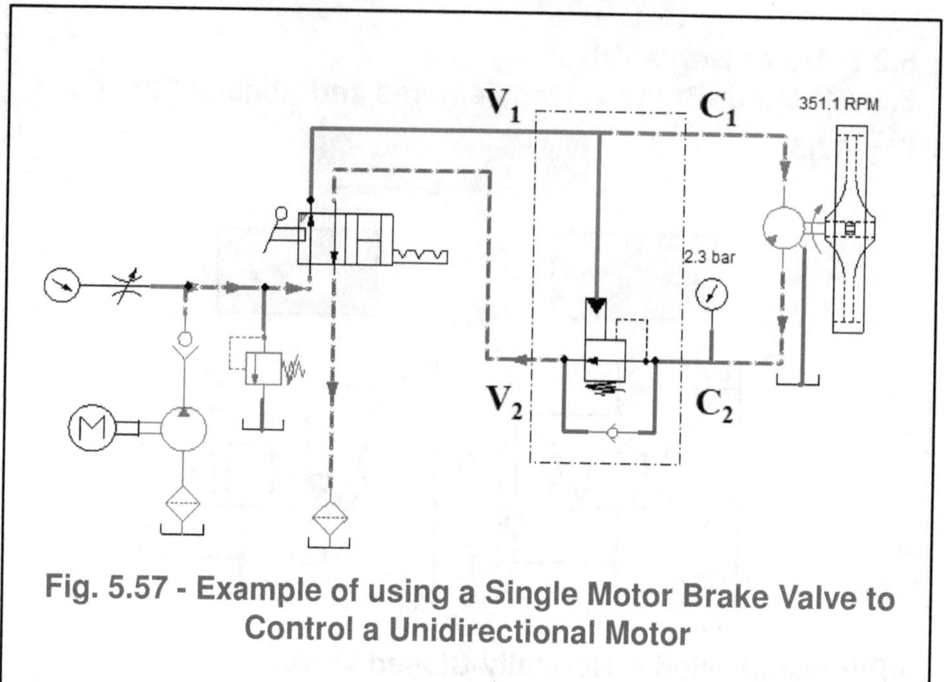

Fig. 5.57 - Example of using a Single Motor Brake Valve to Control a Unidirectional Motor

70

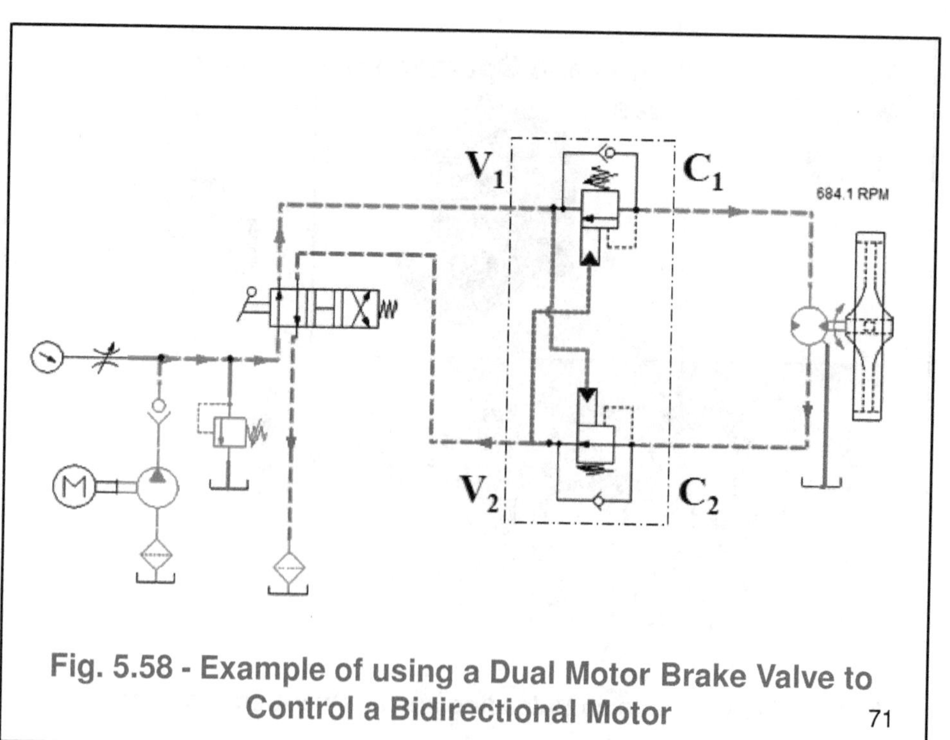

Fig. 5.58 - Example of using a Dual Motor Brake Valve to Control a Bidirectional Motor

71

5.2.7.3- Construction and Operation of Pilot-Operated Motor Brake Valves

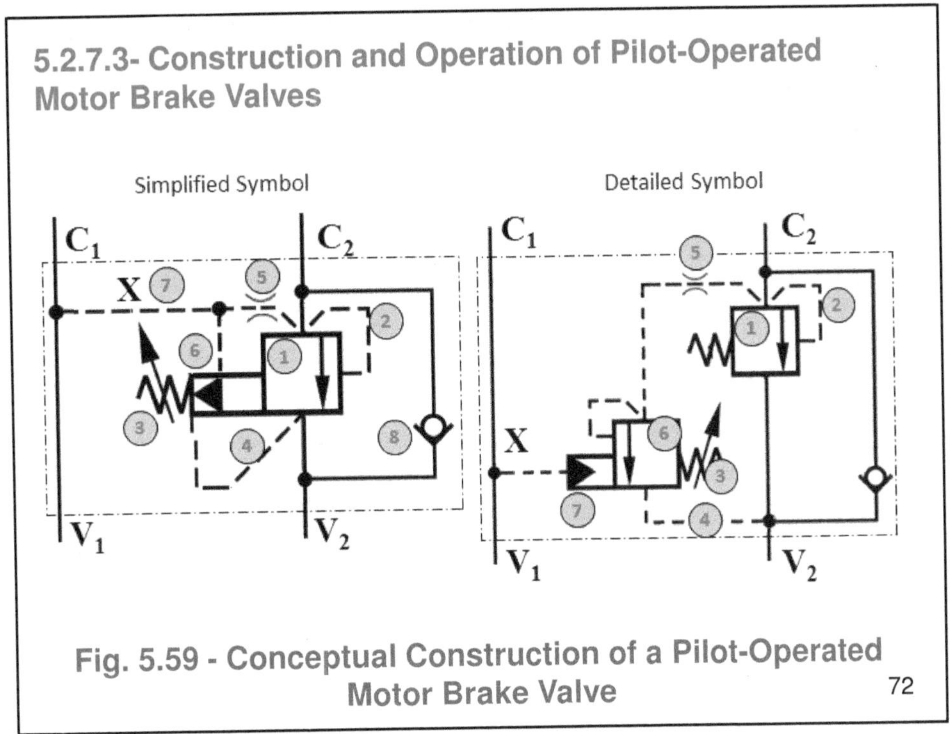

Fig. 5.59 - Conceptual Construction of a Pilot-Operated Motor Brake Valve

72

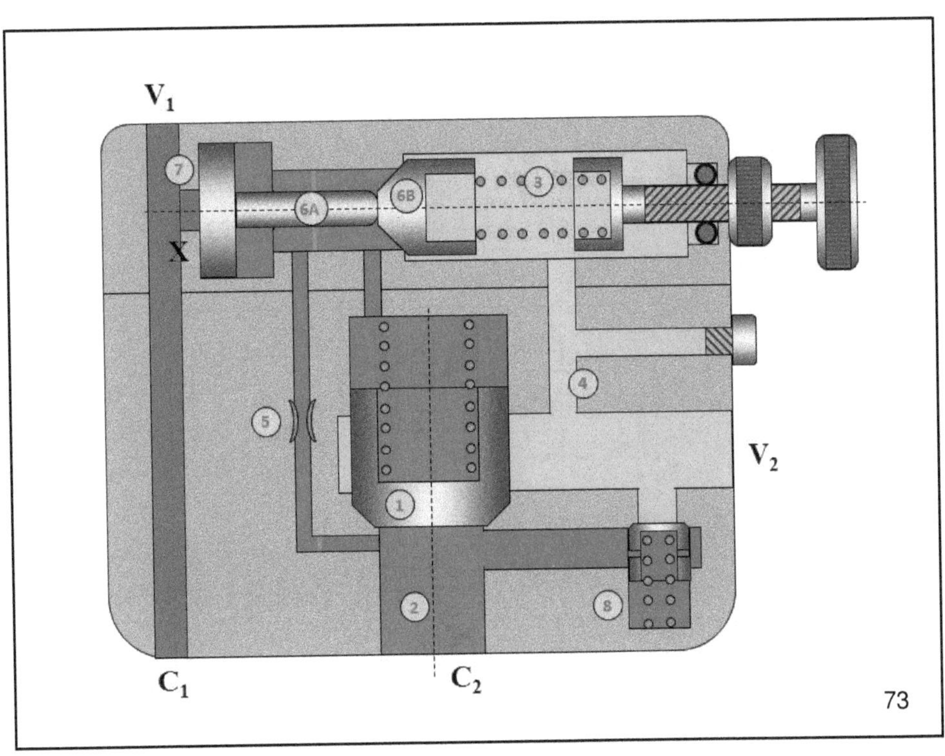

73

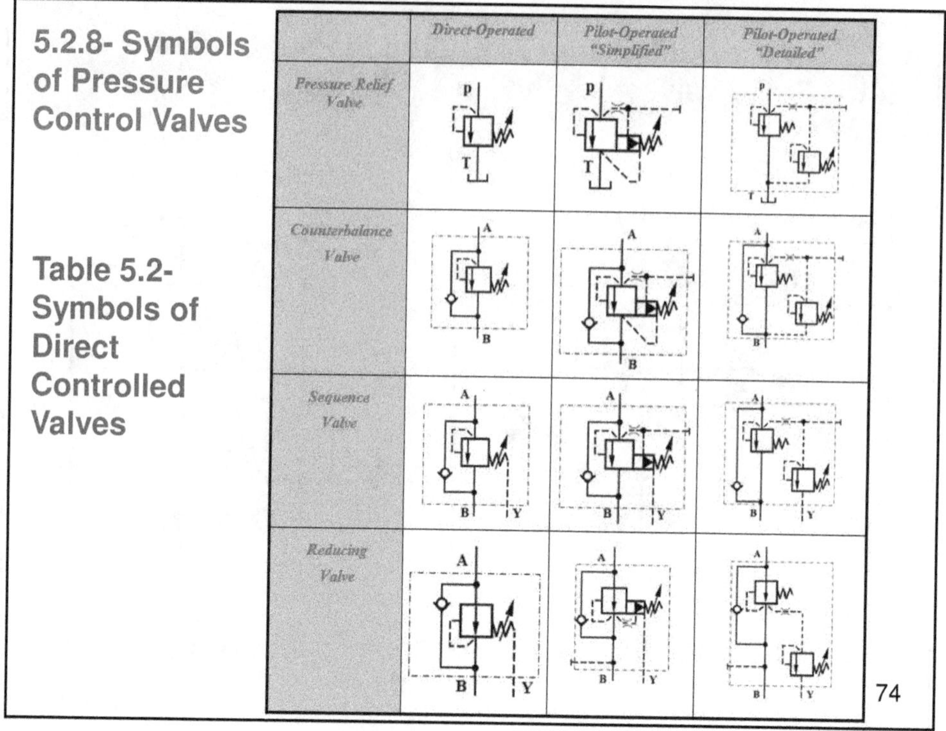

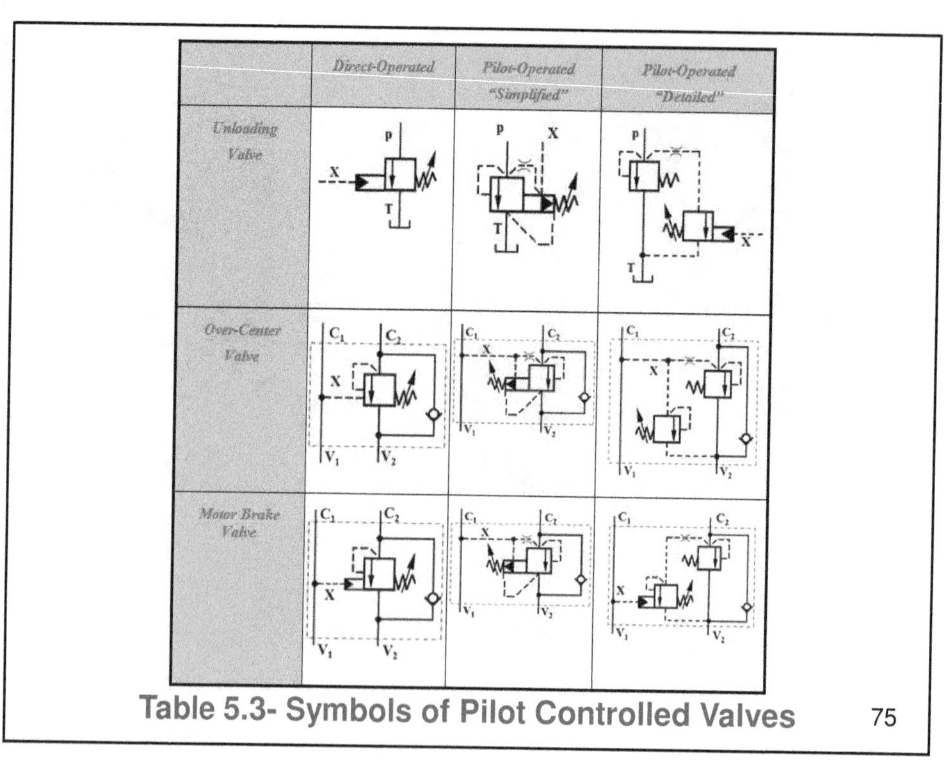

Table 5.3- Symbols of Pilot Controlled Valves

5.2.9- Pressure Control Valves Function Identification

	Normal Status	*Control Pressure Supply*	*Spring Chamber Drain*	*Built-in Check Valve*
Pressure Relief Valve	Closed	Internal	Internal	NO
Counterbalance Valve	Closed	Internal	Optional	YES
Sequence Valve	Closed	Internal	External	YES
Reducing Valve	Open	Internal	External	Optional
Unloading Valve	Closed	External	Internal	No
Over-Center Valve	Closed	Both	Internal	YES
Motor Brake Valve	Closed	Both	Internal	Optional

Table 5.4-Pressure Control Valves Function Identification[76]

5.3- Directional Control Valves

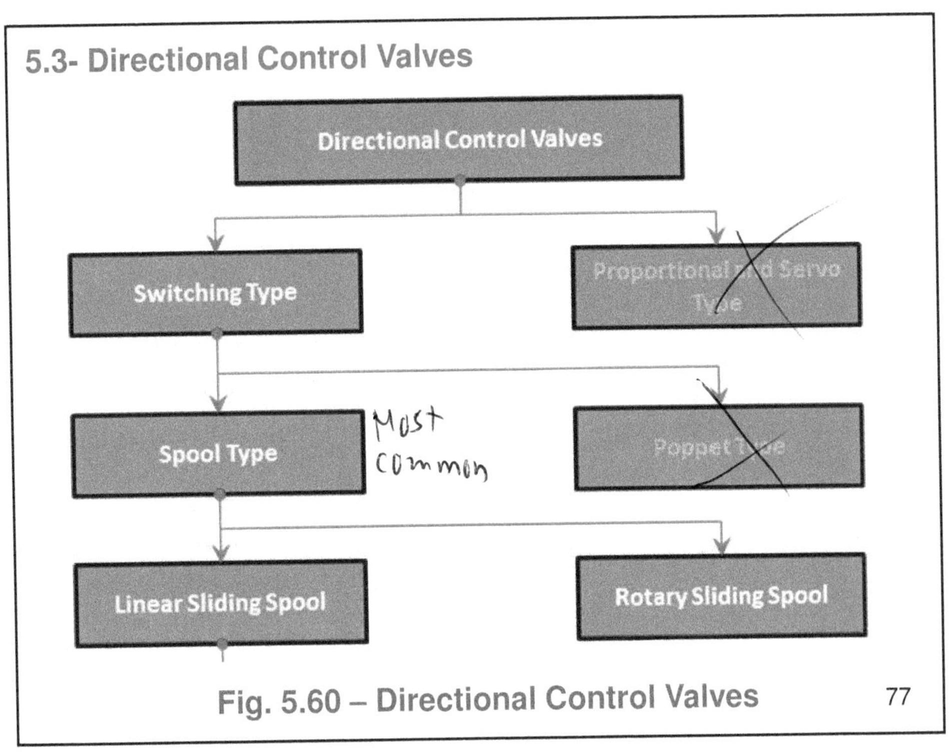

Fig. 5.60 – Directional Control Valves [77]

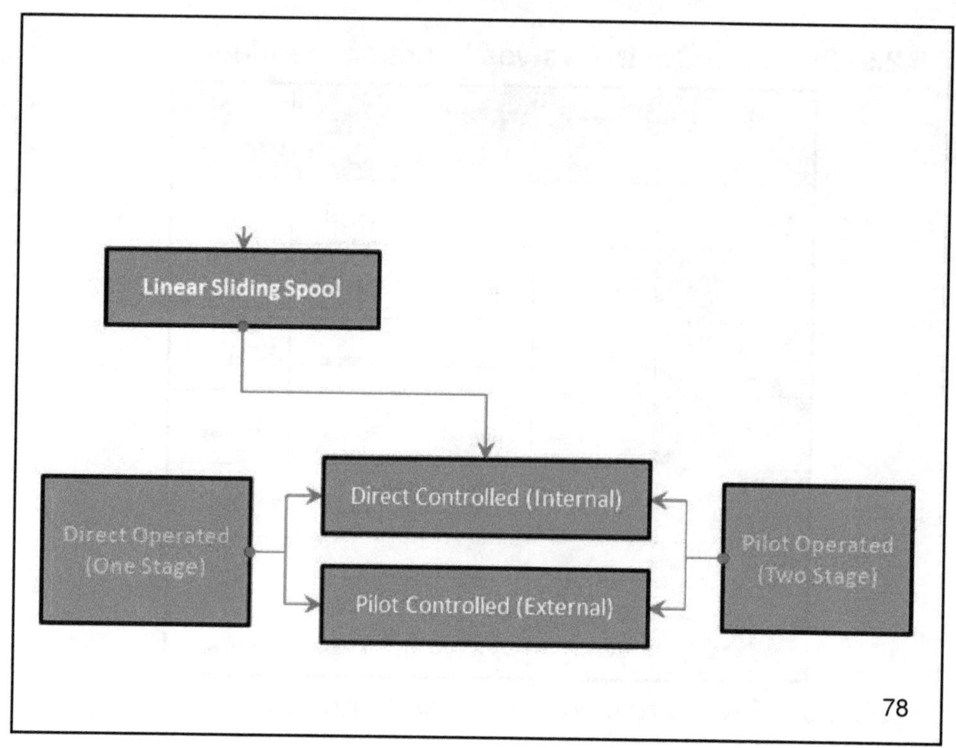

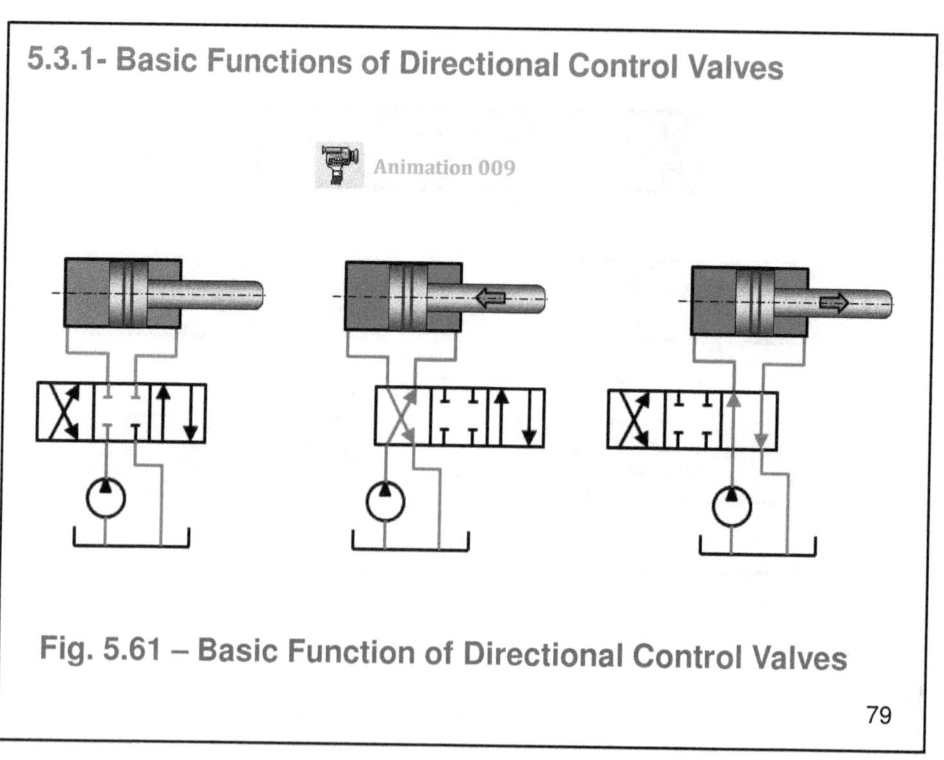

Fig. 5.61 – Basic Function of Directional Control Valves

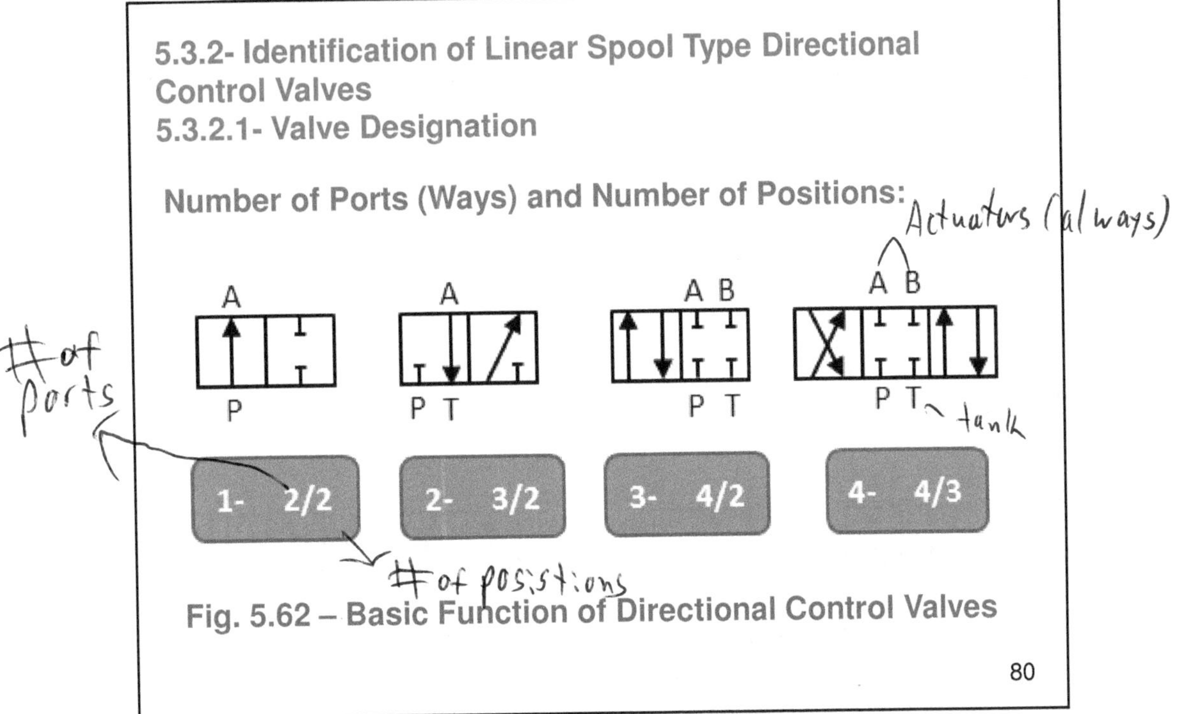

Fig. 5.62 – Basic Function of Directional Control Valves

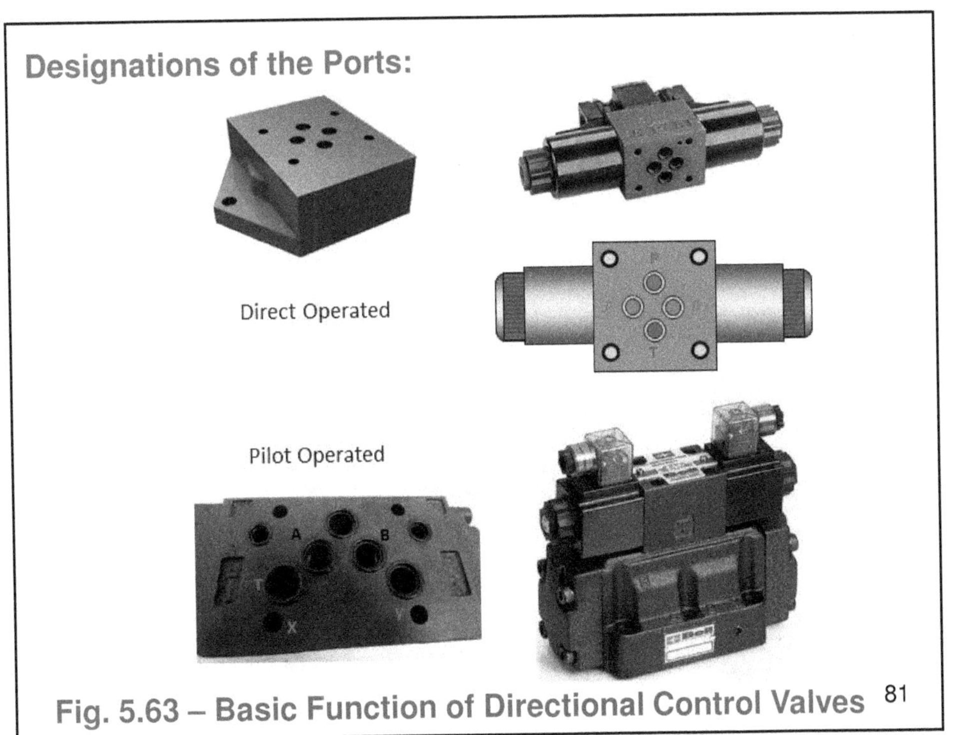

Fig. 5.63 – Basic Function of Directional Control Valves

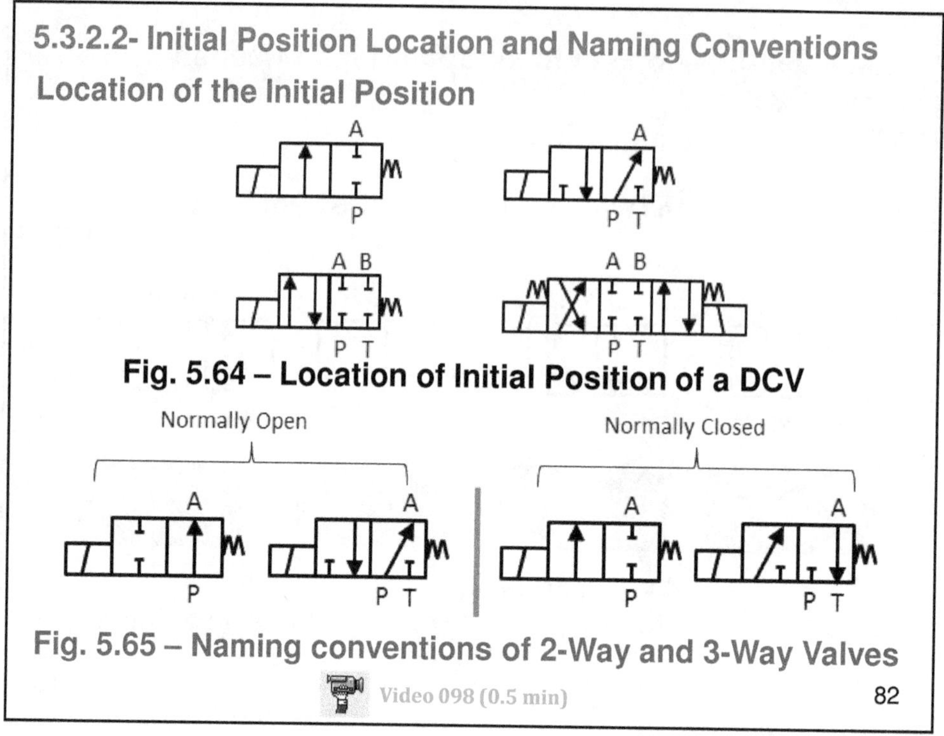

Fig. 5.64 – Location of Initial Position of a DCV

Fig. 5.65 – Naming conventions of 2-Way and 3-Way Valves

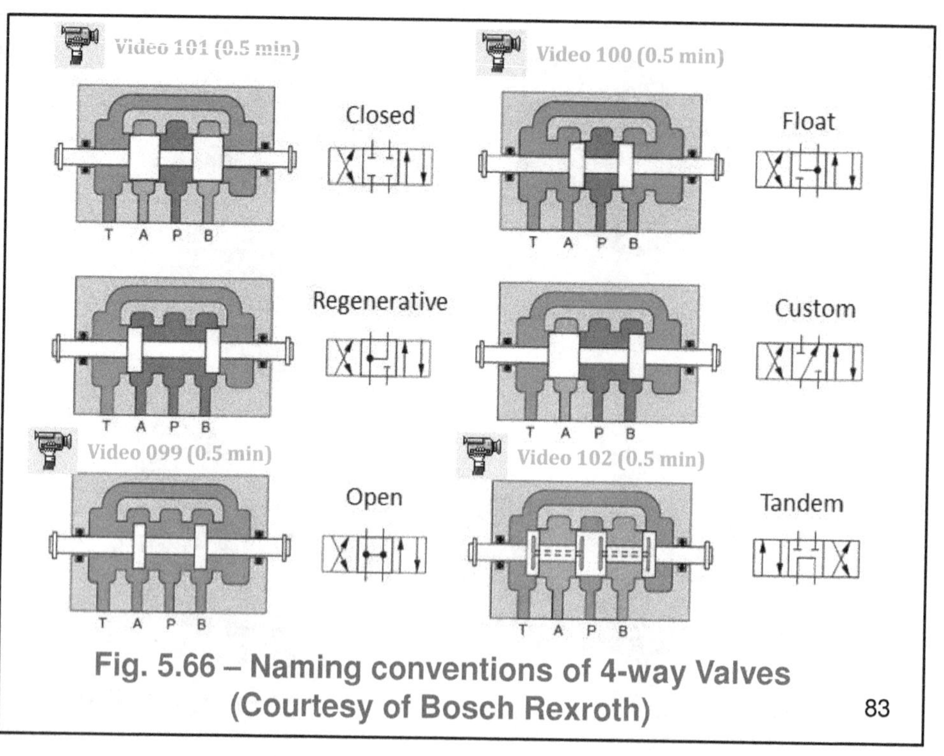

Fig. 5.66 – Naming conventions of 4-way Valves (Courtesy of Bosch Rexroth)

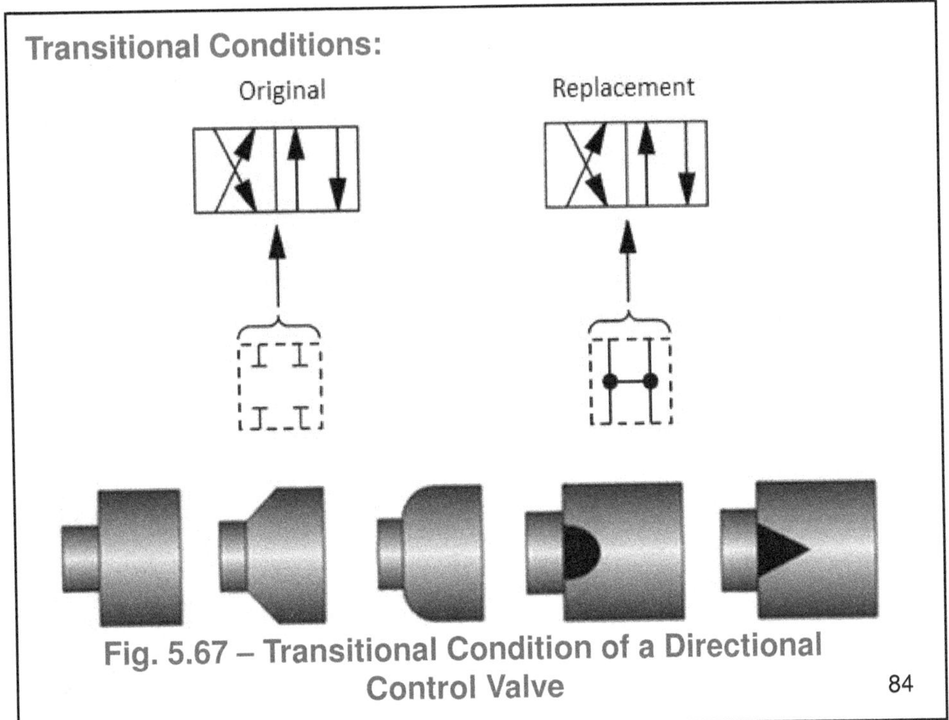

Fig. 5.67 – Transitional Condition of a Directional Control Valve

84

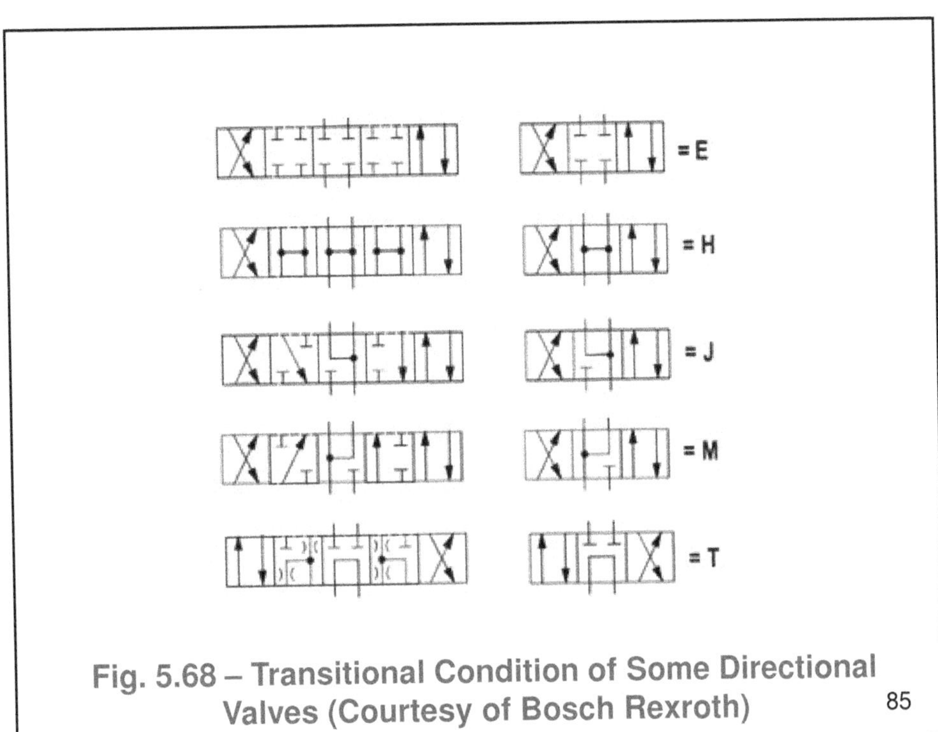

Fig. 5.68 – Transitional Condition of Some Directional Valves (Courtesy of Bosch Rexroth)

85

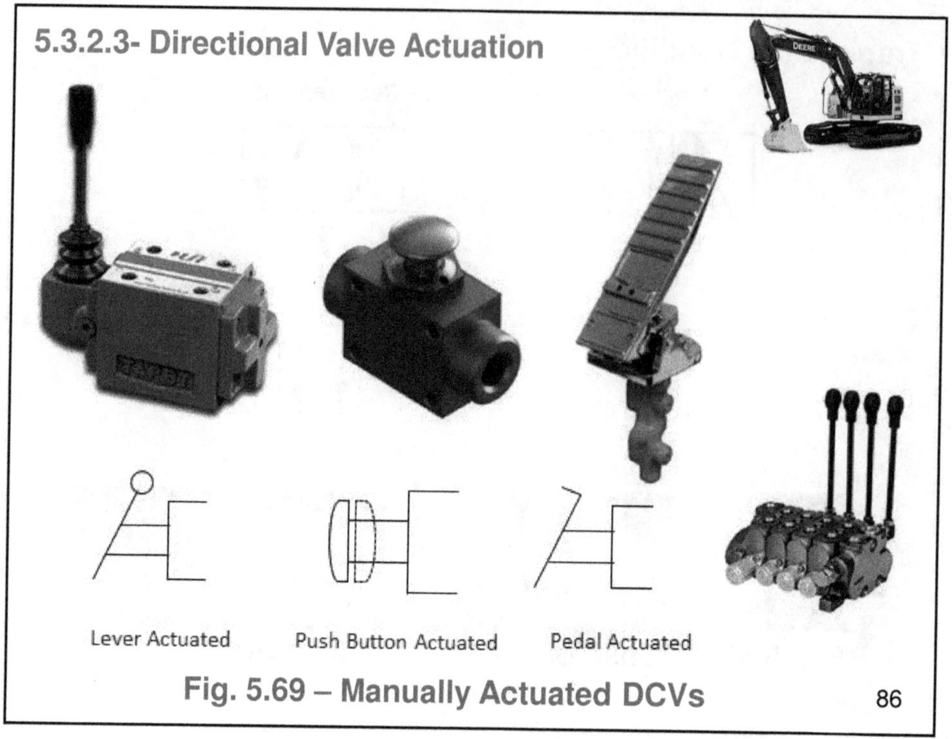

Fig. 5.69 – Manually Actuated DCVs

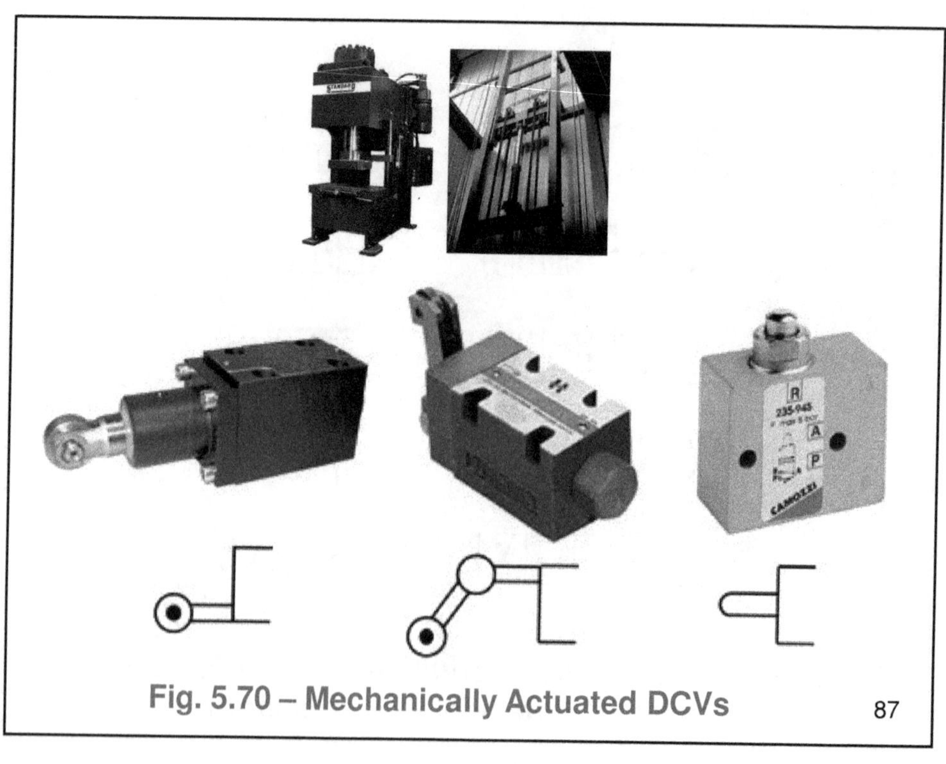

Fig. 5.70 – Mechanically Actuated DCVs

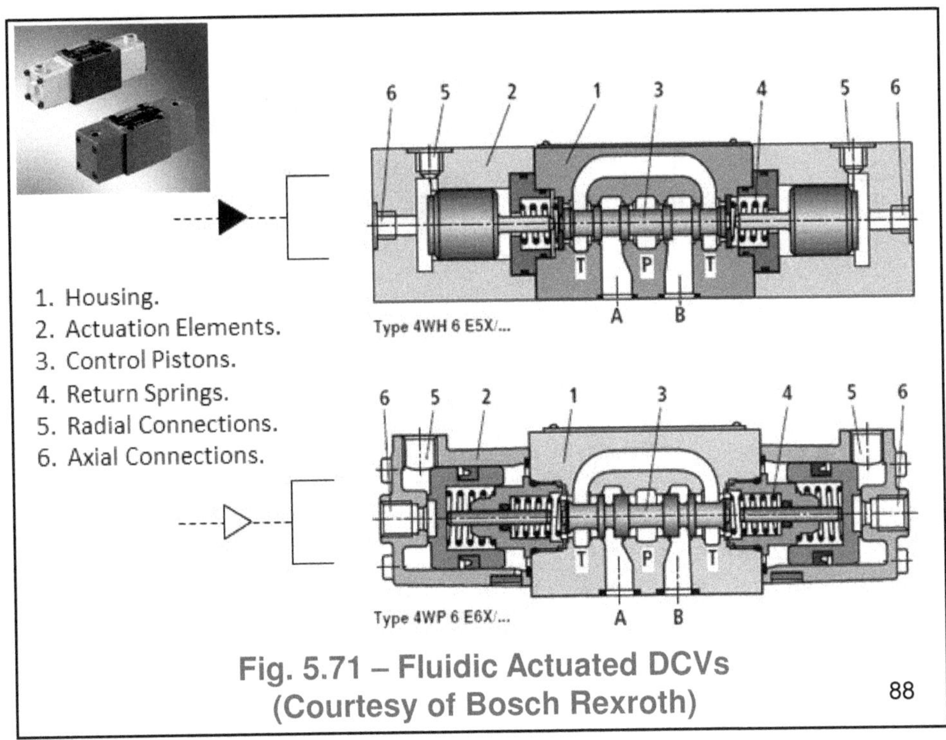

Fig. 5.71 – Fluidic Actuated DCVs
(Courtesy of Bosch Rexroth)

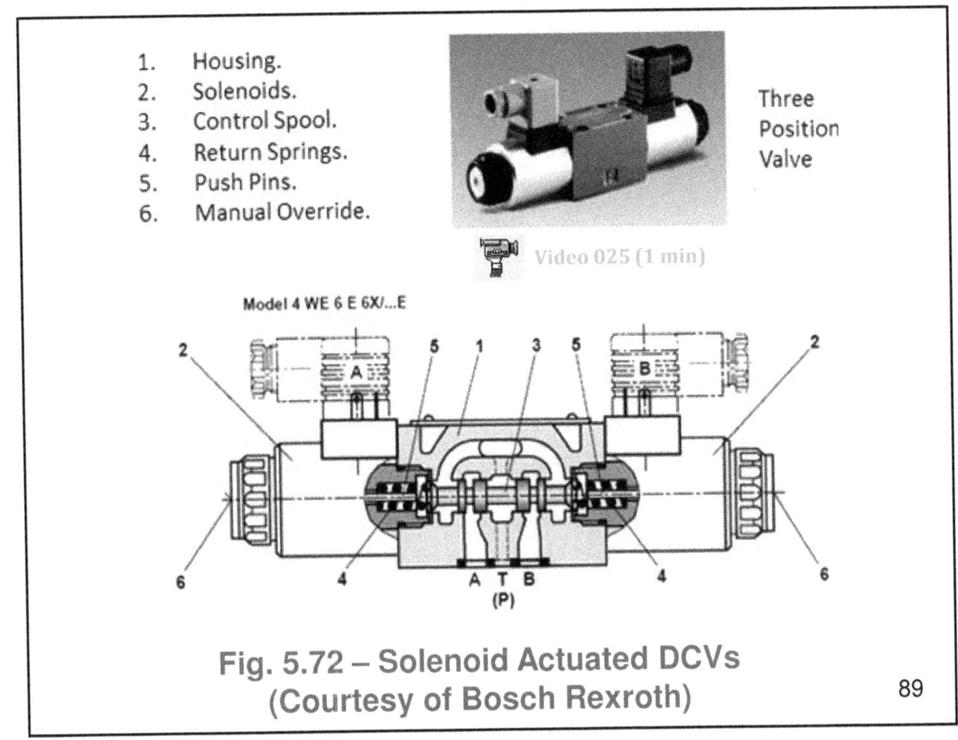

Fig. 5.72 – Solenoid Actuated DCVs
(Courtesy of Bosch Rexroth)

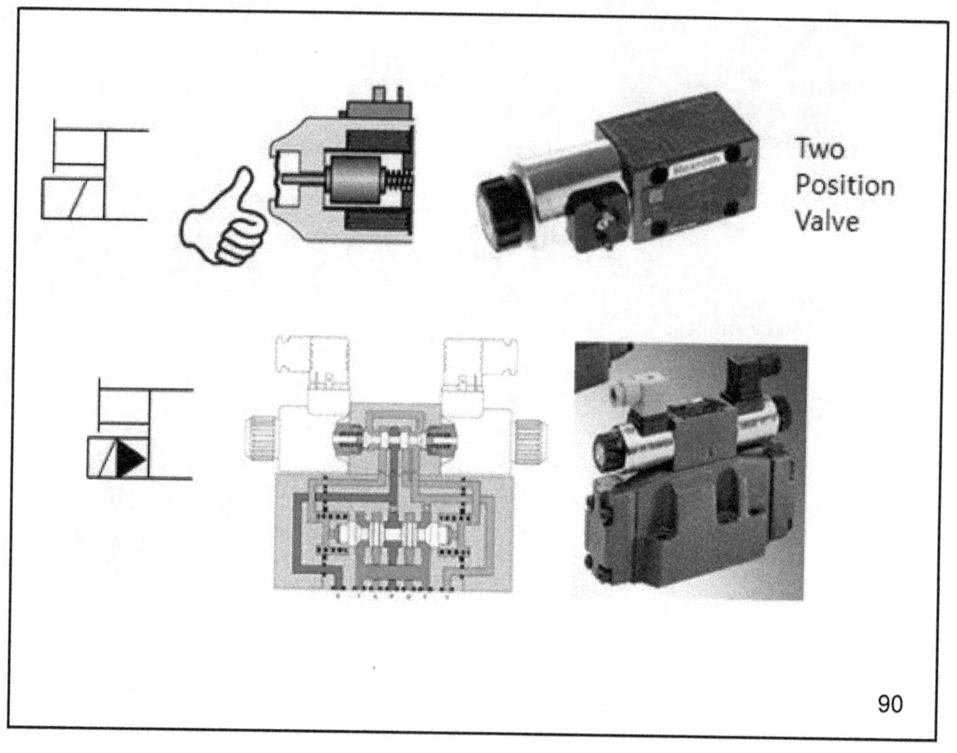

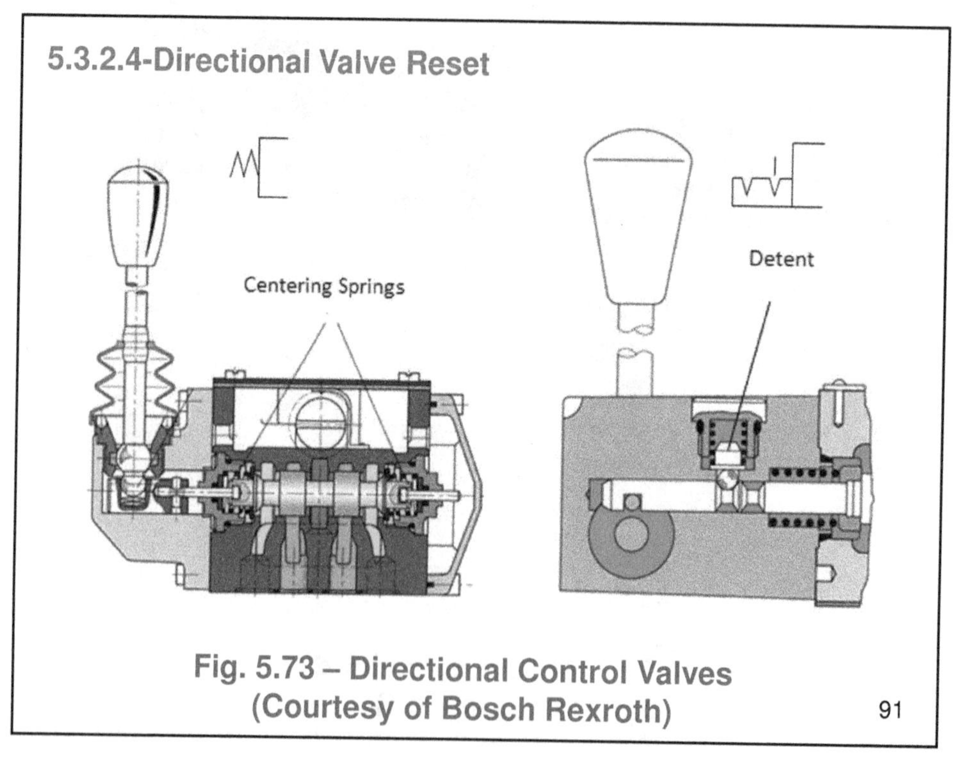

Fig. 5.73 – Directional Control Valves
(Courtesy of Bosch Rexroth)

5.3.2.5- Examples of Directional Valve Identification

Fig. 5.74

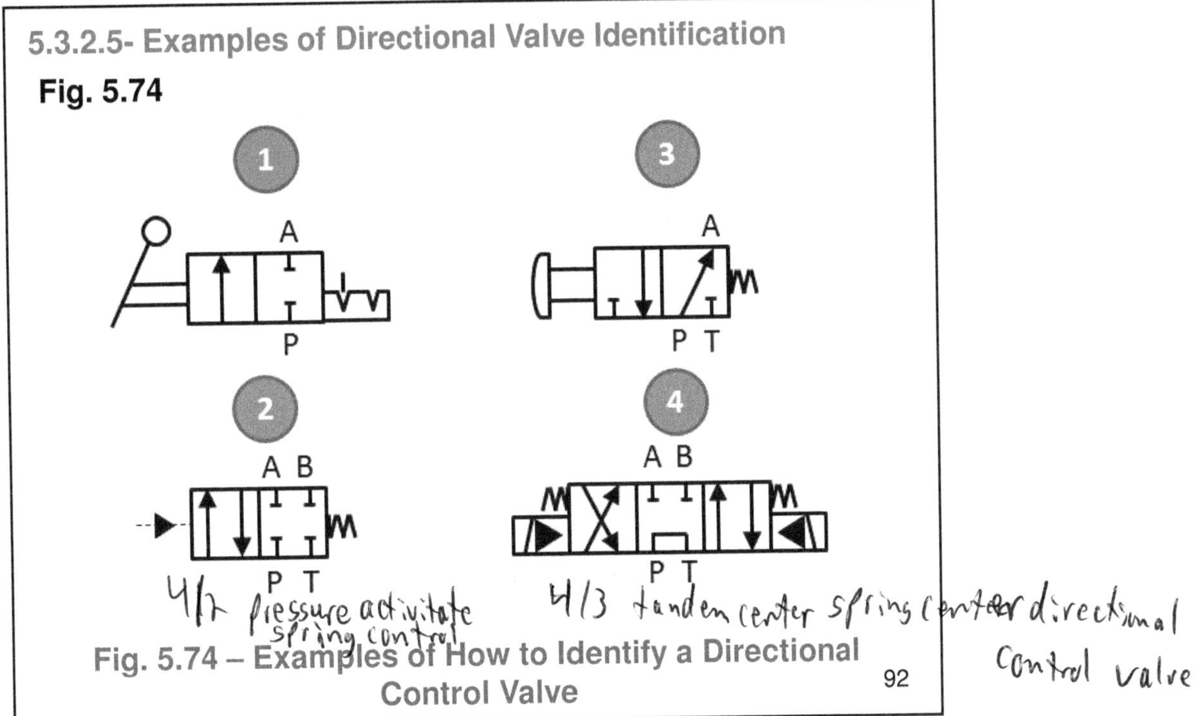

Fig. 5.74 – Examples of How to Identify a Directional Control Valve

92

5.3.3- Direct-Operated Linear Spool Directional Valves

5.3.3.1- Operation Principle of a 2/2 DCV

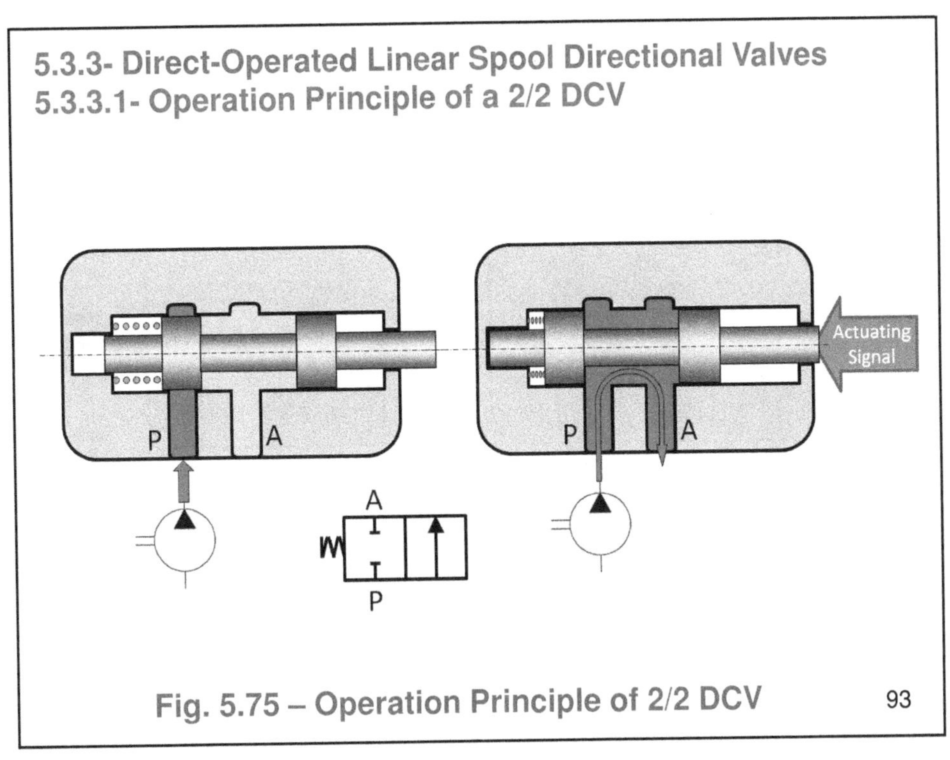

Fig. 5.75 – Operation Principle of 2/2 DCV

93

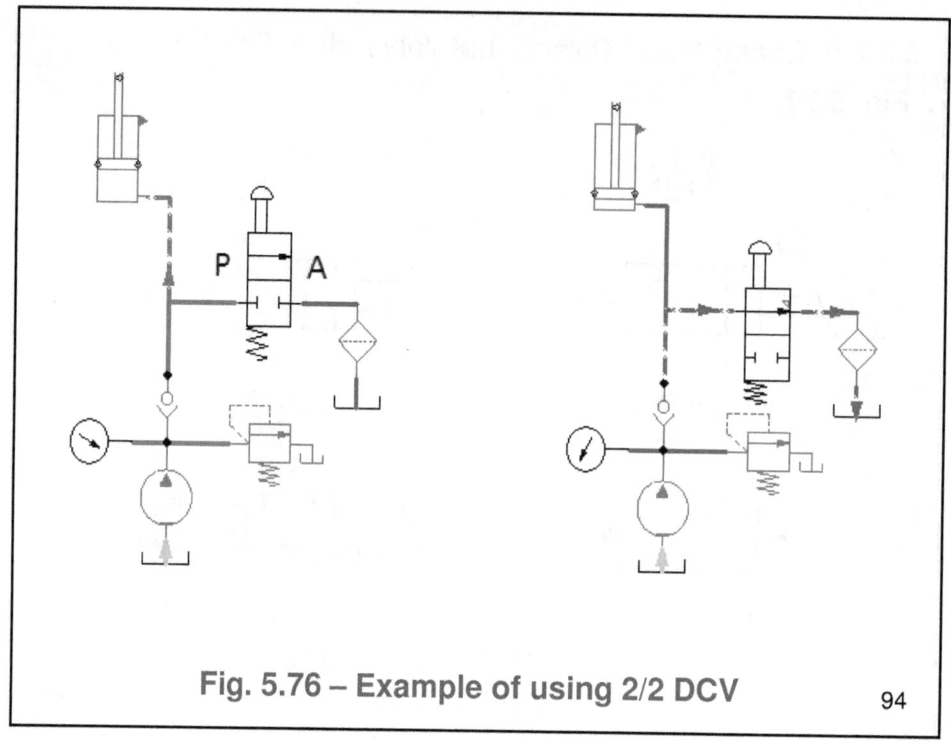

Fig. 5.76 – Example of using 2/2 DCV

94

5.3.3.2- Operation Principle of a 3/2 DCV

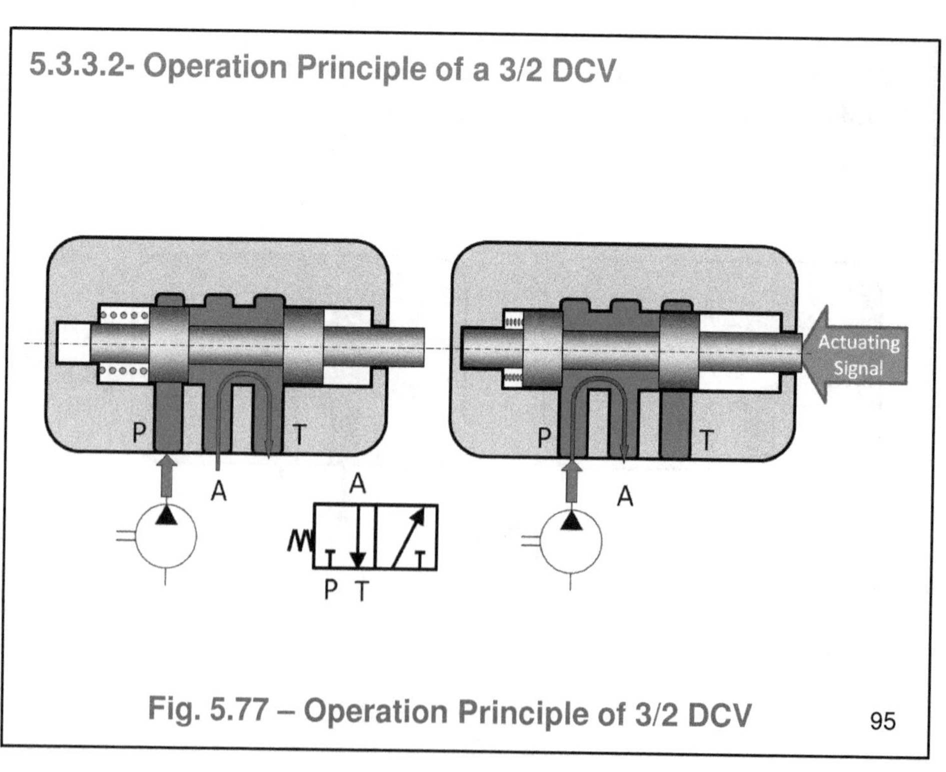

Fig. 5.77 – Operation Principle of 3/2 DCV

95

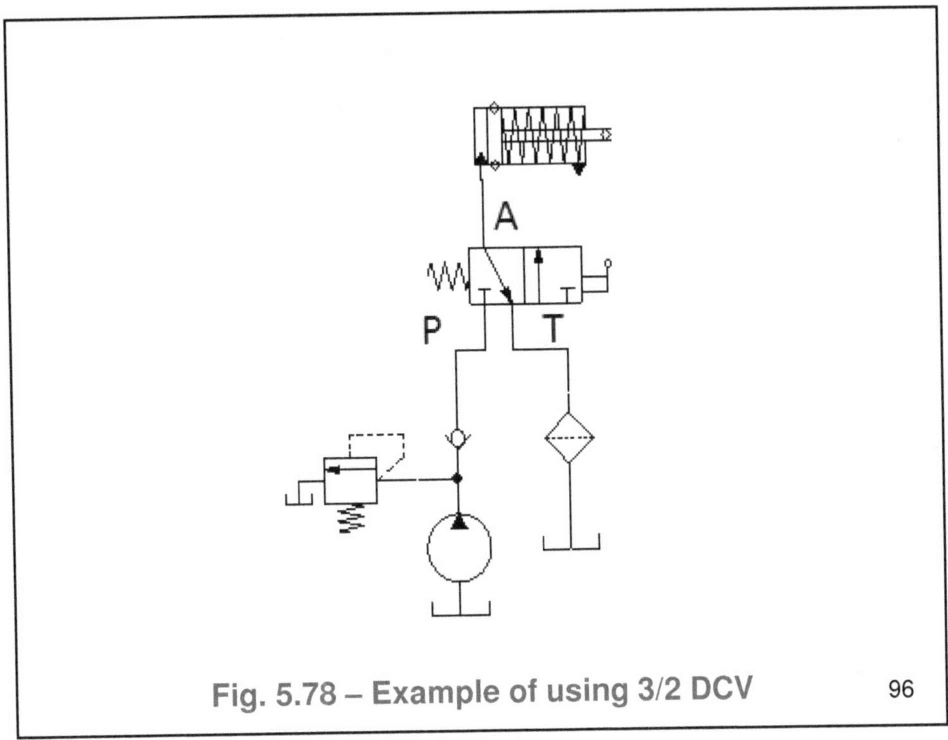

Fig. 5.78 – Example of using 3/2 DCV 96

5.3.3.3- Operation Principle of a 4/2 DCV

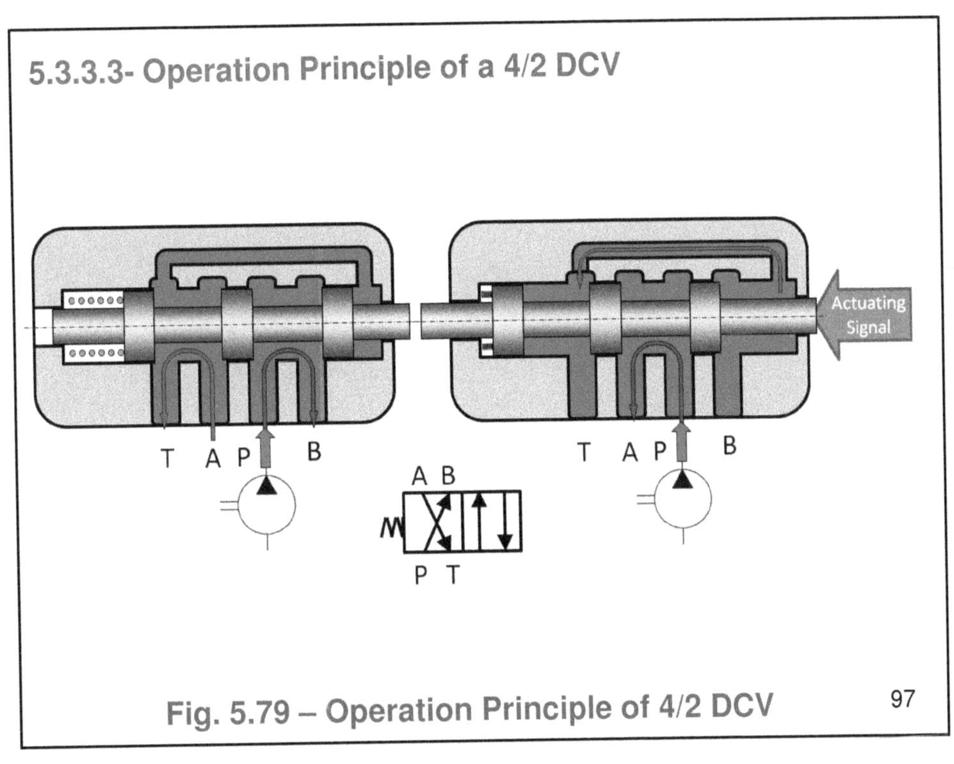

Fig. 5.79 – Operation Principle of 4/2 DCV 97

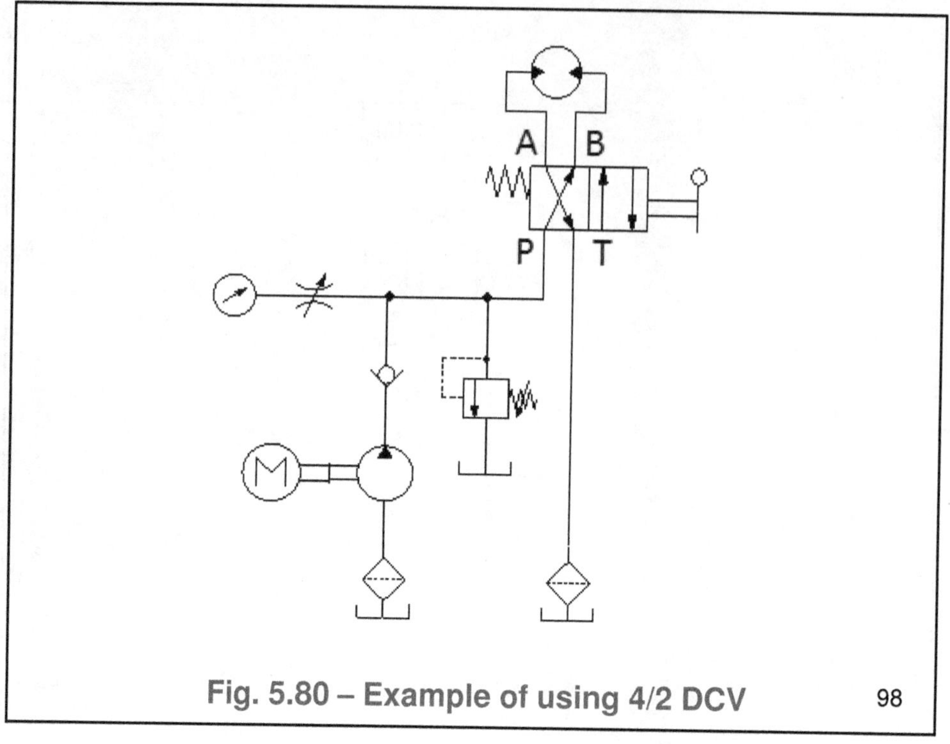

Fig. 5.80 – Example of using 4/2 DCV 98

5.3.3.4- Operation Principle of a 4/3 DCV

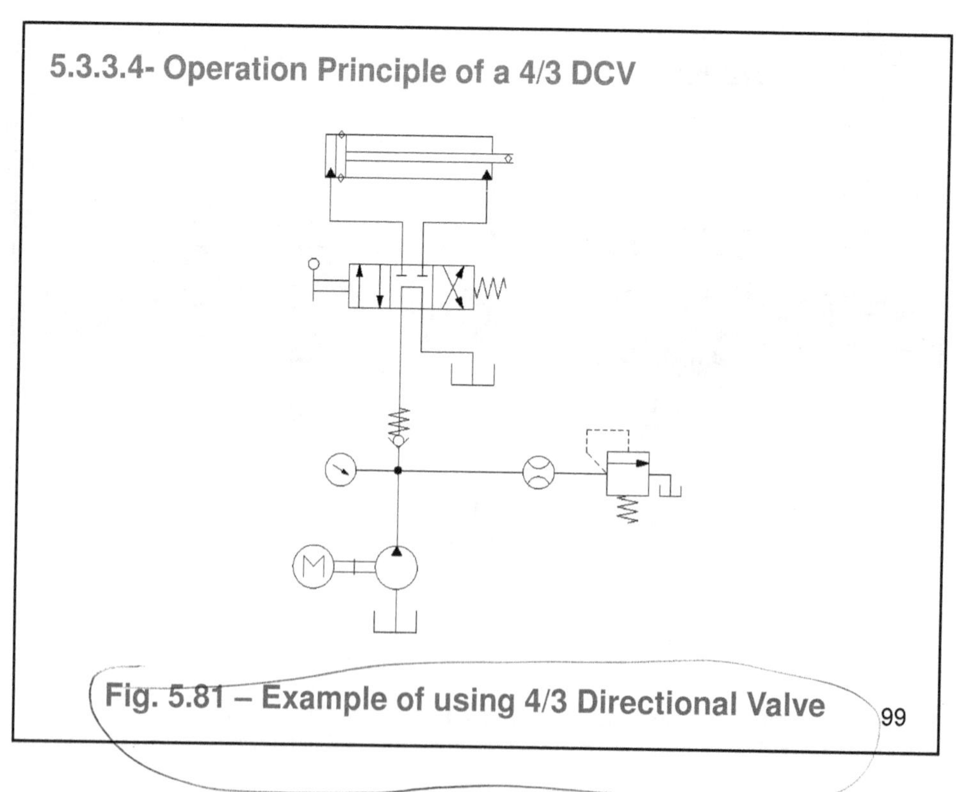

Fig. 5.81 – Example of using 4/3 Directional Valve 99

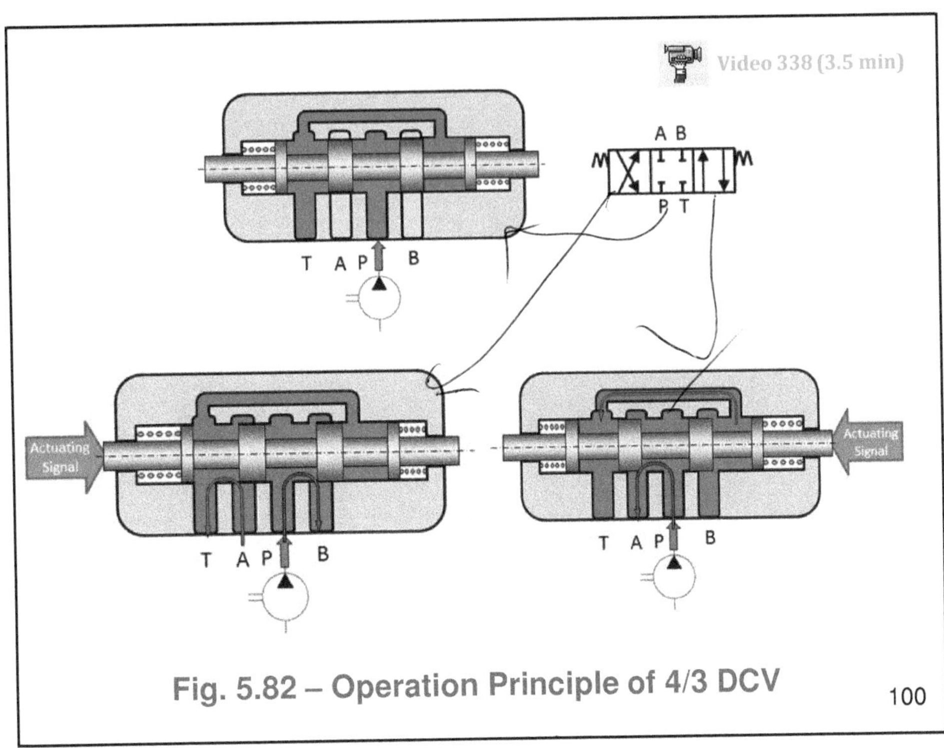

Fig. 5.82 – Operation Principle of 4/3 DCV

100

5.3.4- Pilot-Operated Linear Spool Directional Valves
5.3.4.1- Why Pilot-Operated Directional Control Valve?

$Q = V \cdot A \Rightarrow V = Q/A$

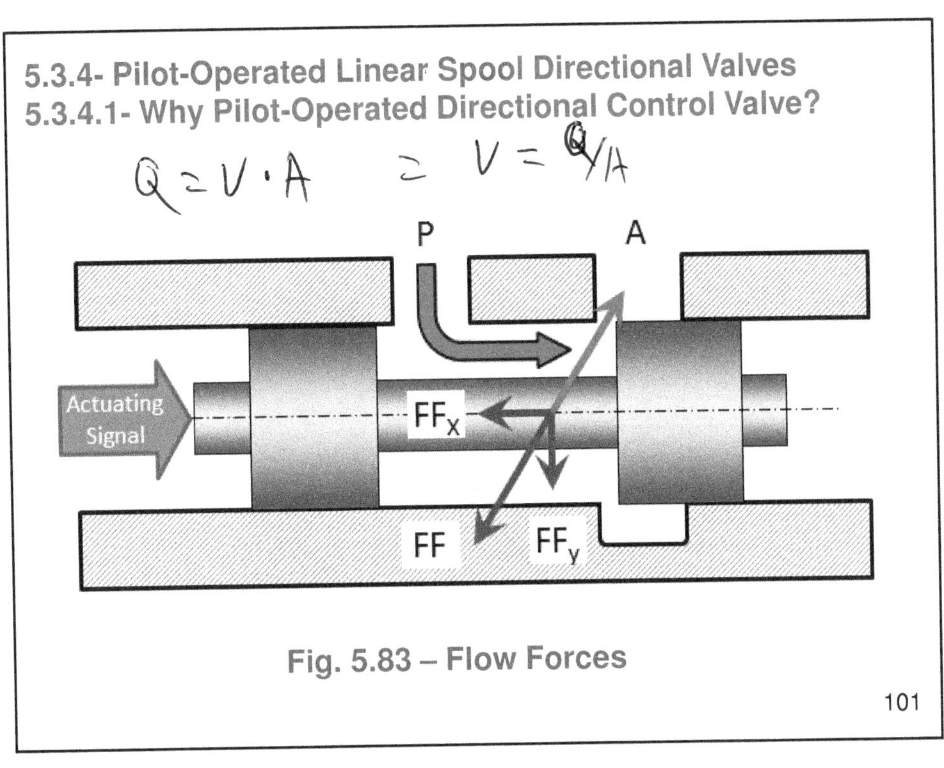

Fig. 5.83 – Flow Forces

101

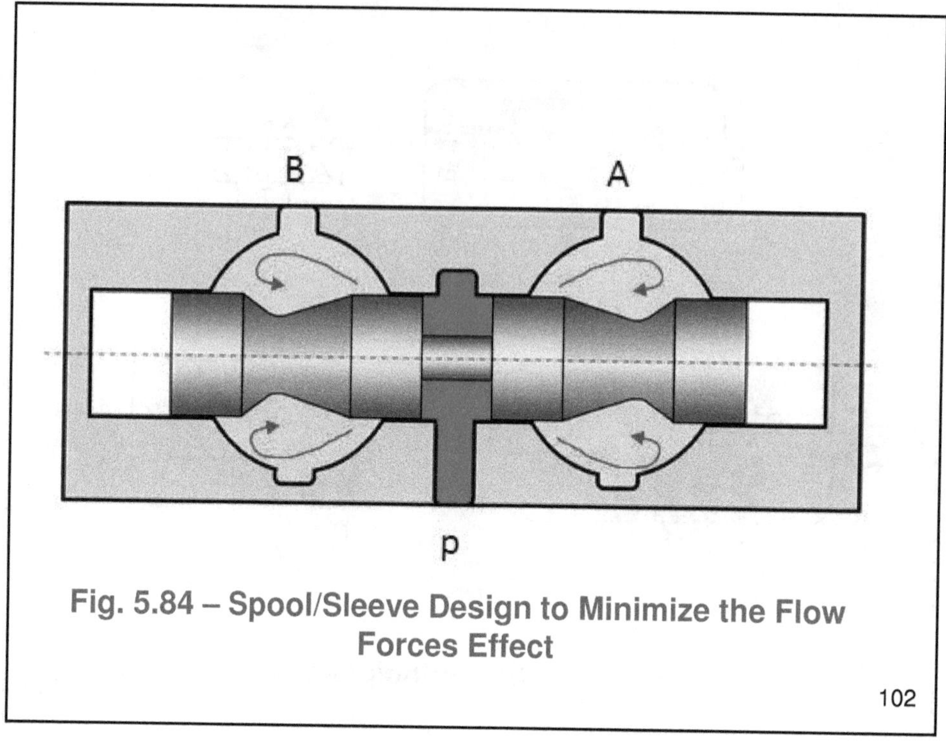

Fig. 5.84 – Spool/Sleeve Design to Minimize the Flow Forces Effect

102

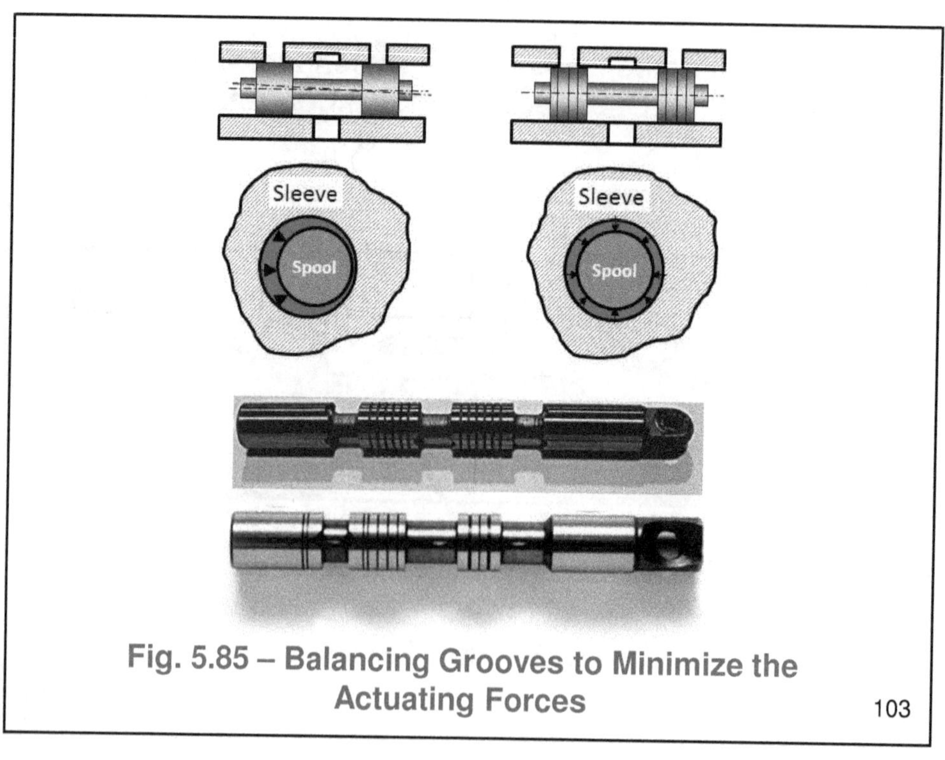

Fig. 5.85 – Balancing Grooves to Minimize the Actuating Forces

103

5.3.4.2- Construction and Operation Principle of Pilot-Operated Directional Valves

Video 103 (0.5 min)

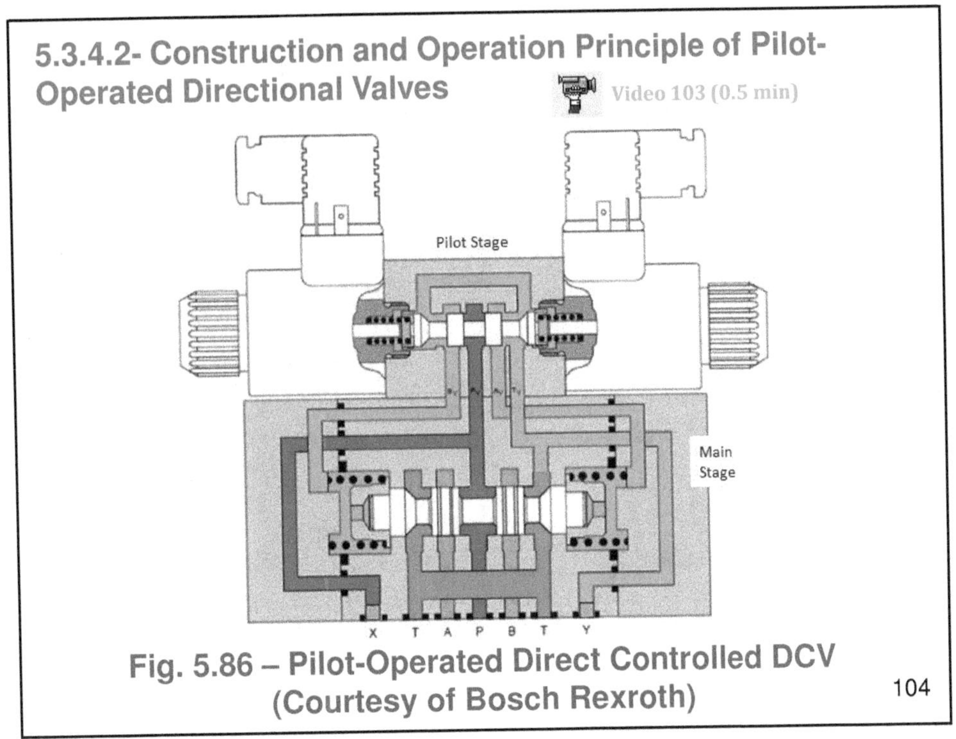

Fig. 5.86 – Pilot-Operated Direct Controlled DCV (Courtesy of Bosch Rexroth)

104

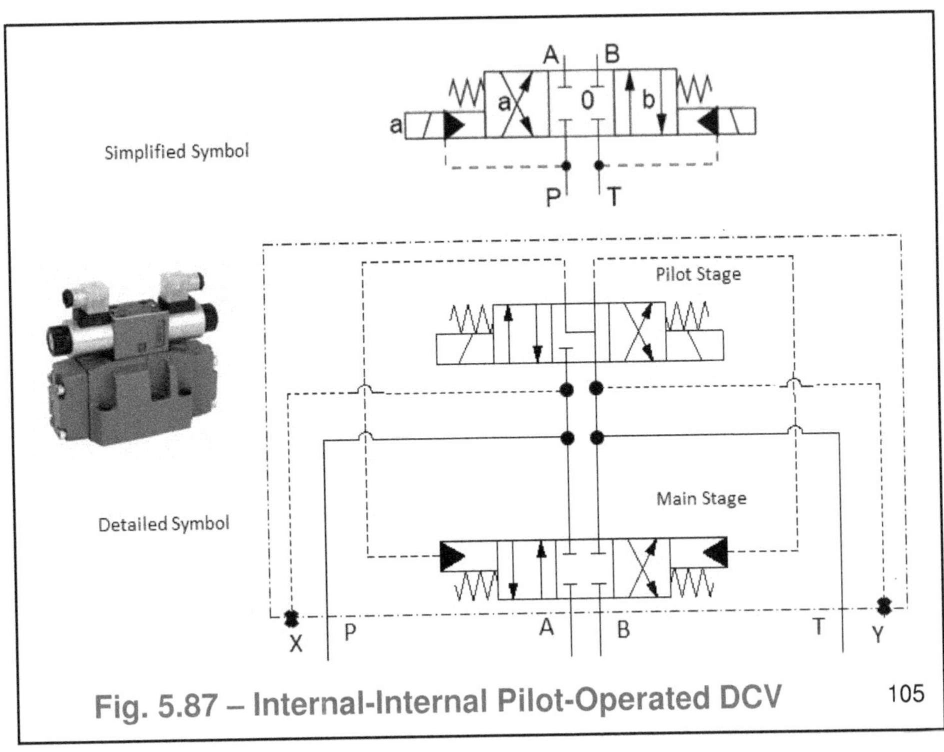

Fig. 5.87 – Internal-Internal Pilot-Operated DCV

105

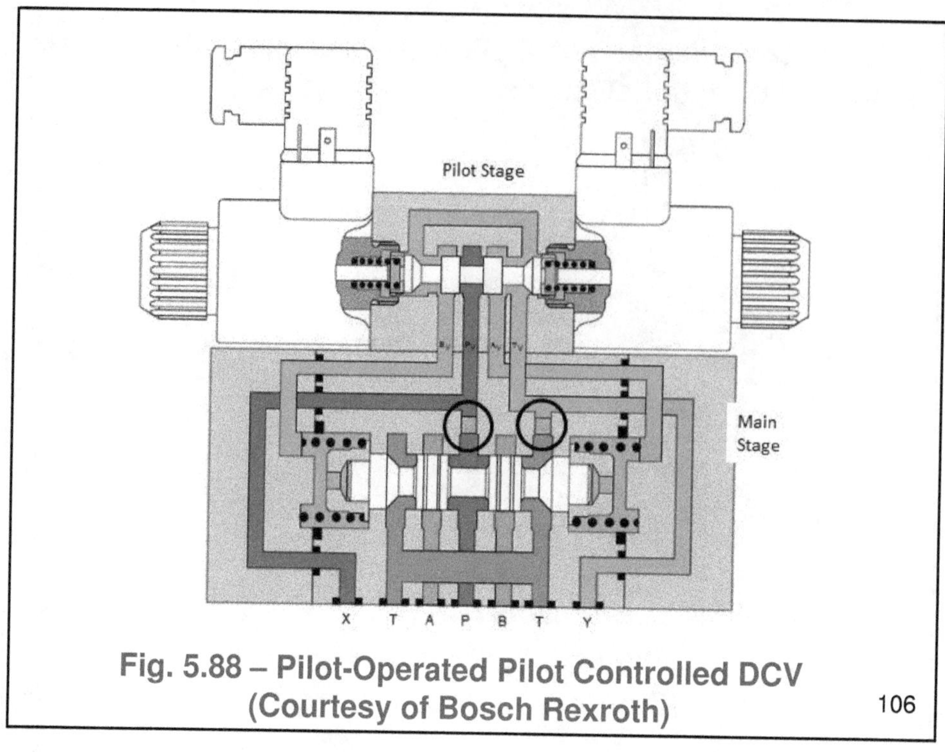

Fig. 5.88 – Pilot-Operated Pilot Controlled DCV (Courtesy of Bosch Rexroth)

106

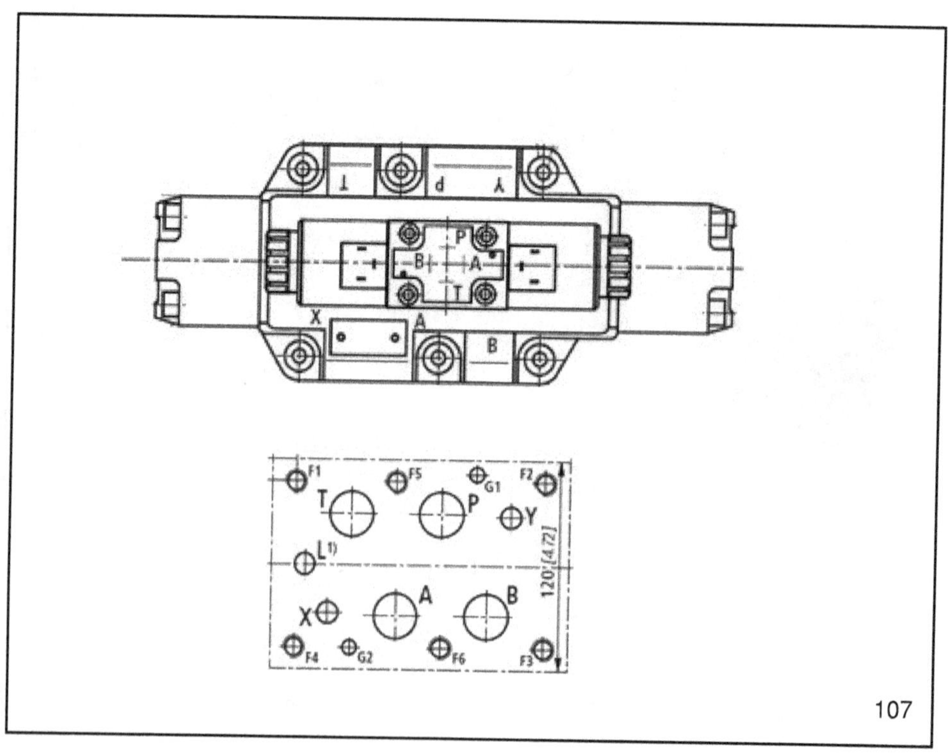

107

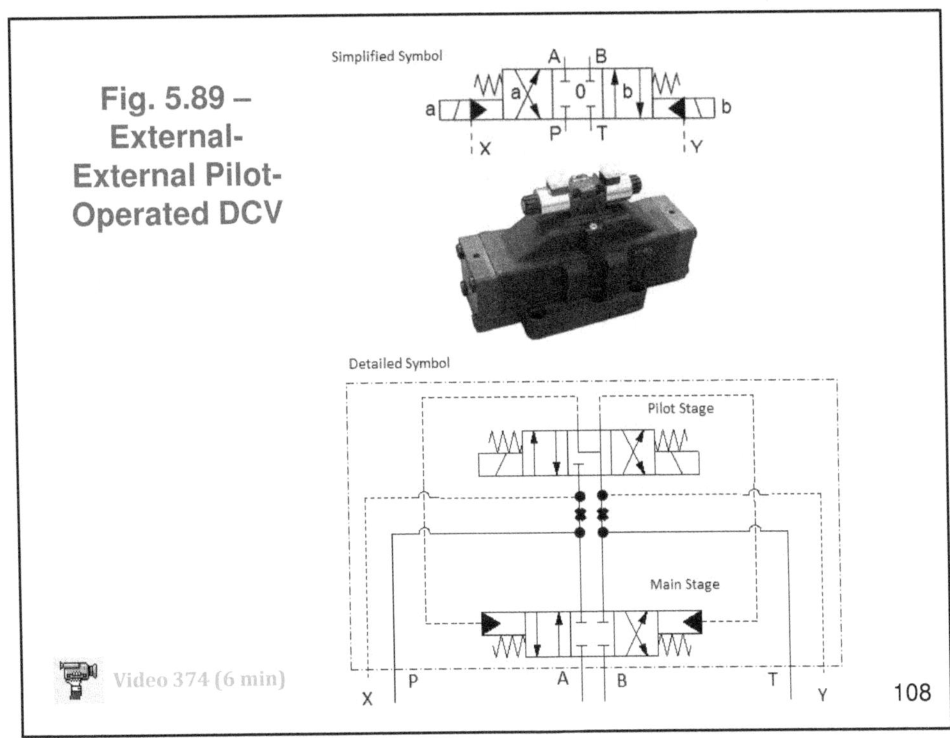

Fig. 5.89 – External-External Pilot-Operated DCV

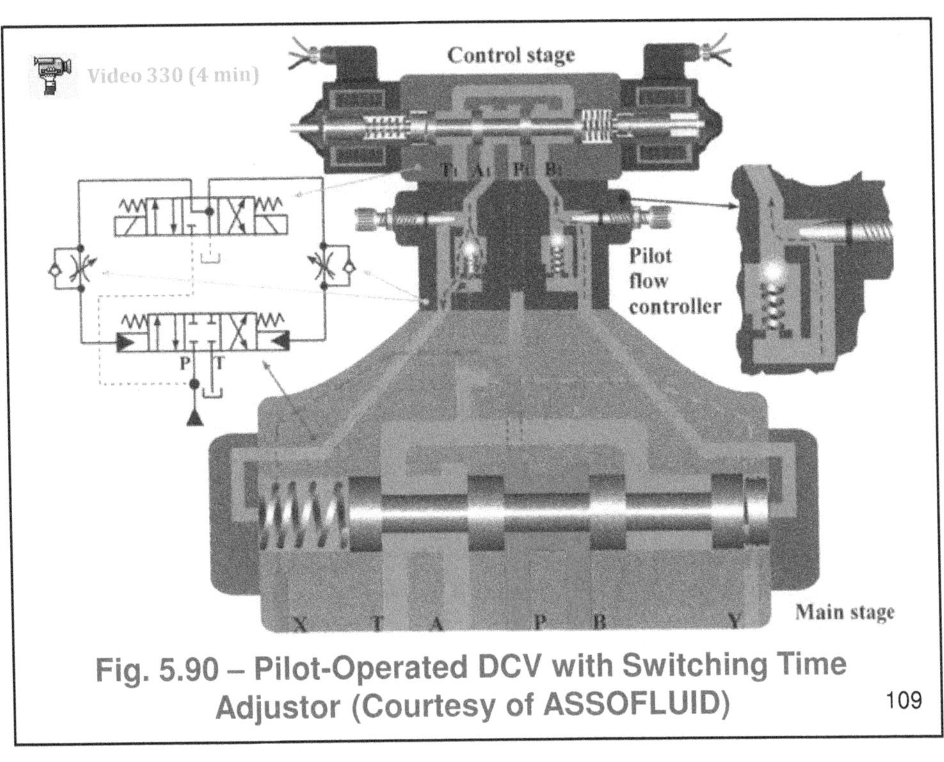

Fig. 5.90 – Pilot-Operated DCV with Switching Time Adjustor (Courtesy of ASSOFLUID)

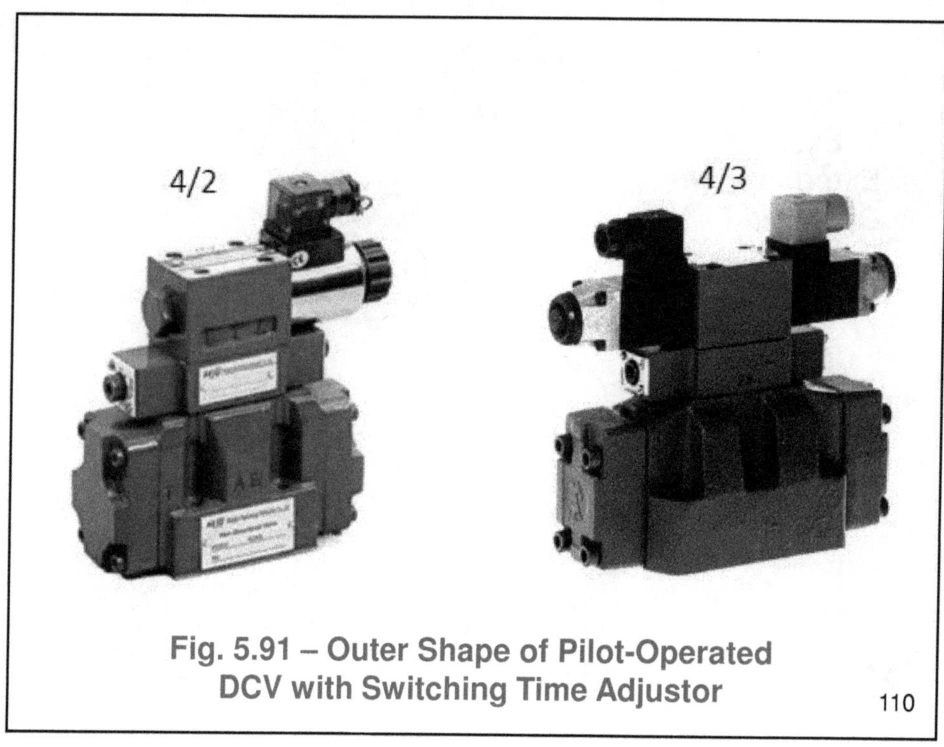

Fig. 5.91 – Outer Shape of Pilot-Operated DCV with Switching Time Adjustor

110

Special Cases:

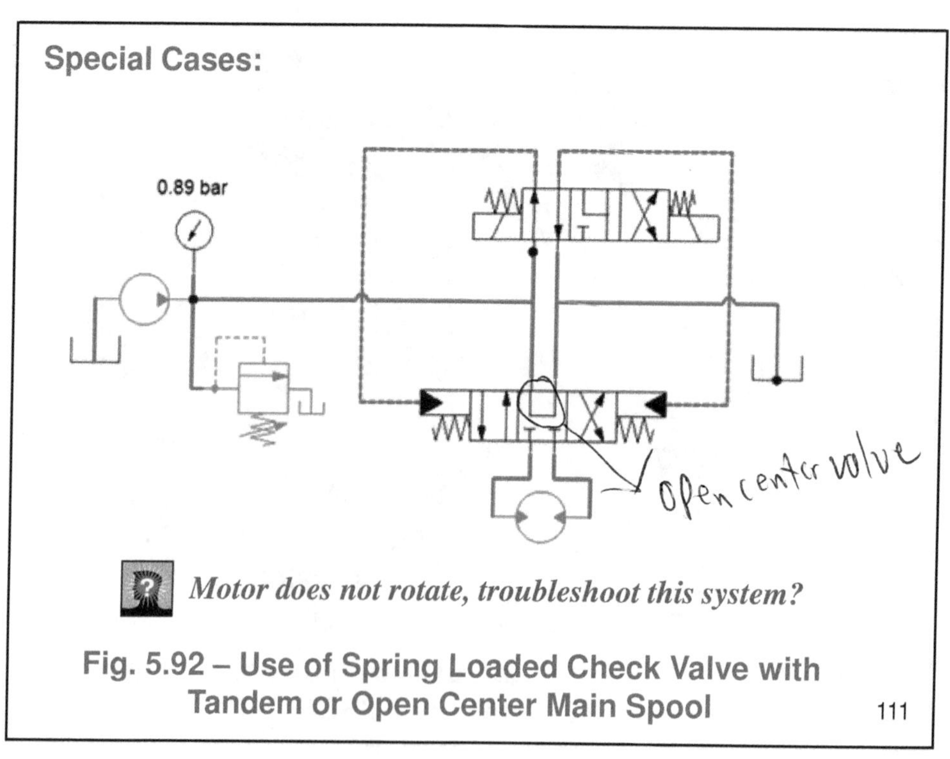

Motor does not rotate, troubleshoot this system?

Fig. 5.92 – Use of Spring Loaded Check Valve with Tandem or Open Center Main Spool

111

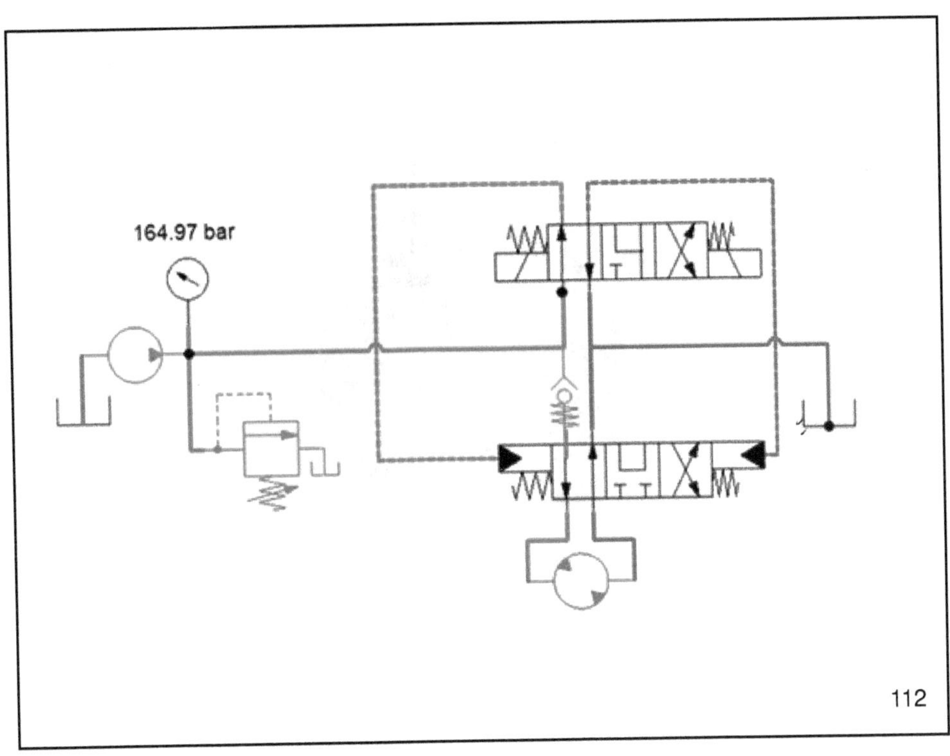

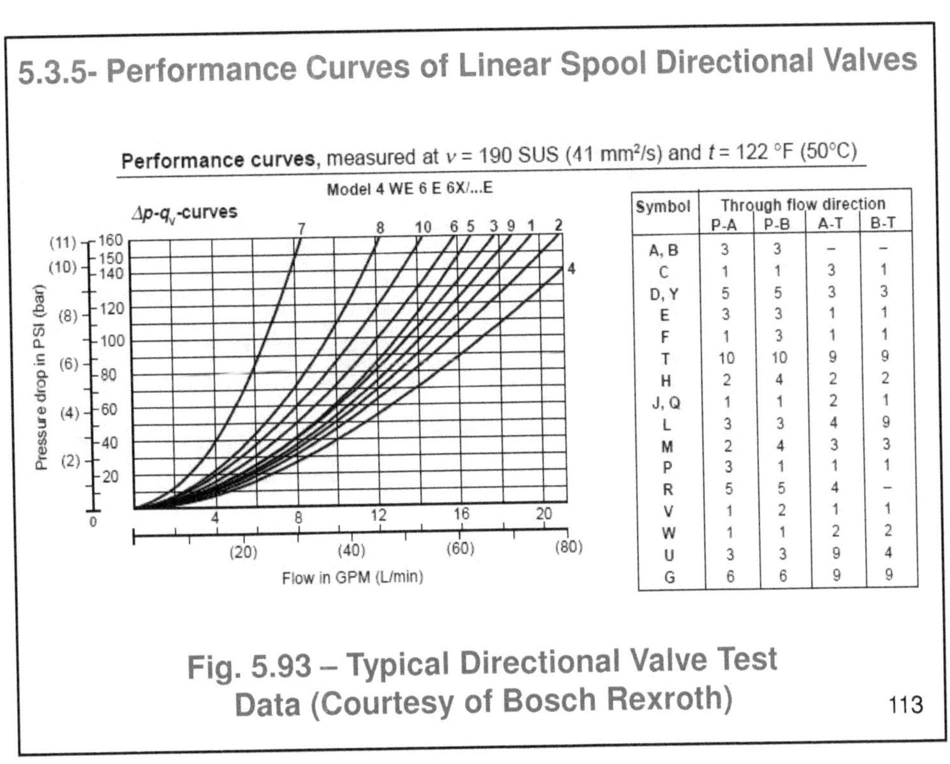

Symbol	Through flow direction			
	P-A	P-B	A-T	B-T
A, B	3	3	–	–
C	1	1	3	1
D, Y	5	5	3	3
E	3	3	1	1
F	1	3	1	1
T	10	10	9	9
H	2	4	2	2
J, Q	1	1	2	1
L	3	3	4	9
M	2	4	3	3
P	3	1	1	1
R	5	5	4	–
V	1	2	1	1
W	1	1	2	2
U	3	3	9	4
G	6	6	9	9

Fig. 5.93 – Typical Directional Valve Test Data (Courtesy of Bosch Rexroth)

Ex.3

114

5.3.6- Mobile Control Blocks

Fig. 5.94 – Mono Blocks for Mobile Machines

115

Fig. 5.95 – Modular Control Blocks for Mobile Machines

116

Mono Block Design	*Modular (Sectional) Design*
▪ One casting with spools inside.	▪ Multiple sections assembled side by side.
▪ Fixed design.	▪ Flexible design allows optional functions.
▪ Unrepairable.	▪ Only failed section is replaced.
▪ Rigid and resists twisting forces.	▪ Twisting forces will move the sections relatively to each other causing leakage.
▪ Compact.	▪ Relatively larger.

Table 5.5- Control Blocks in Mobile Machines

117

5.3.7- Operation Principle of Rotary Sliding Spool Directional Valves

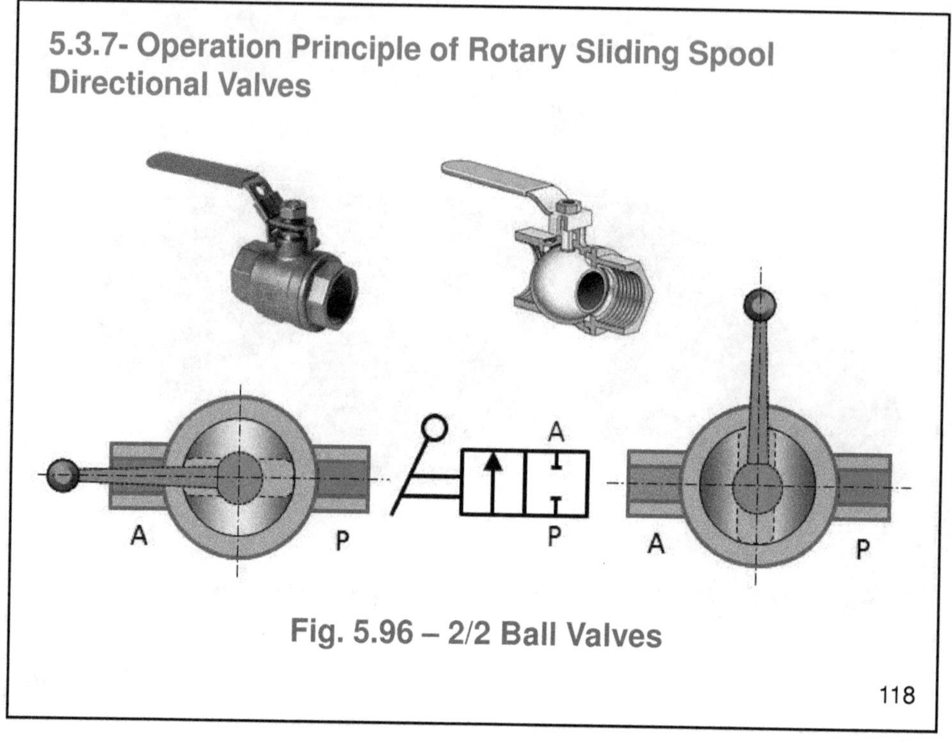

Fig. 5.96 – 2/2 Ball Valves

118

Animation 038

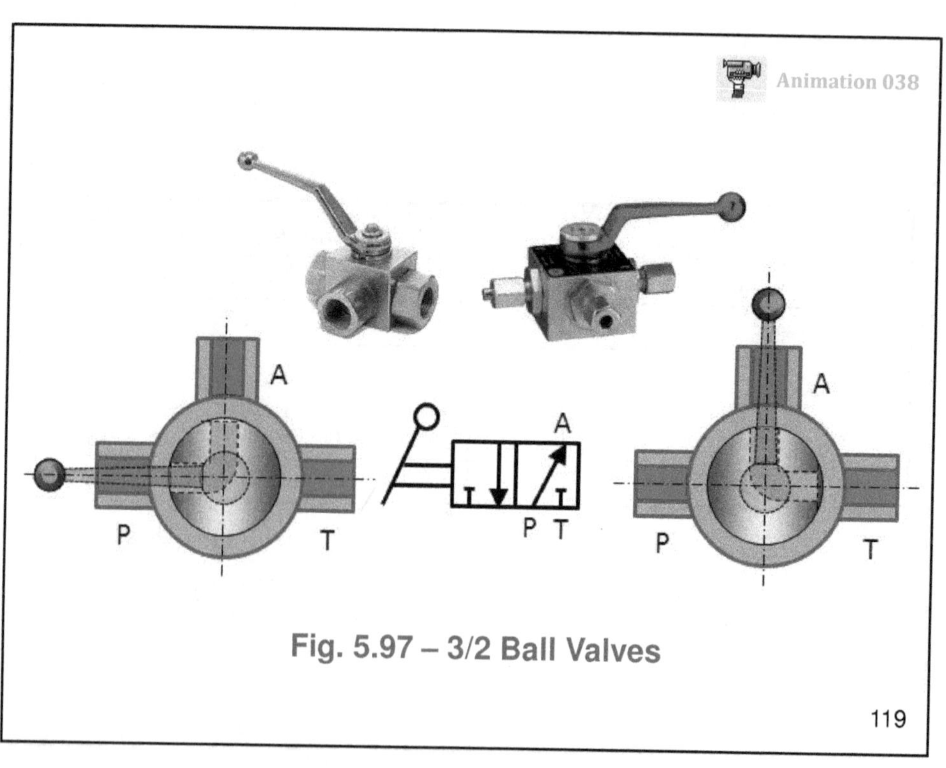

Fig. 5.97 – 3/2 Ball Valves

119

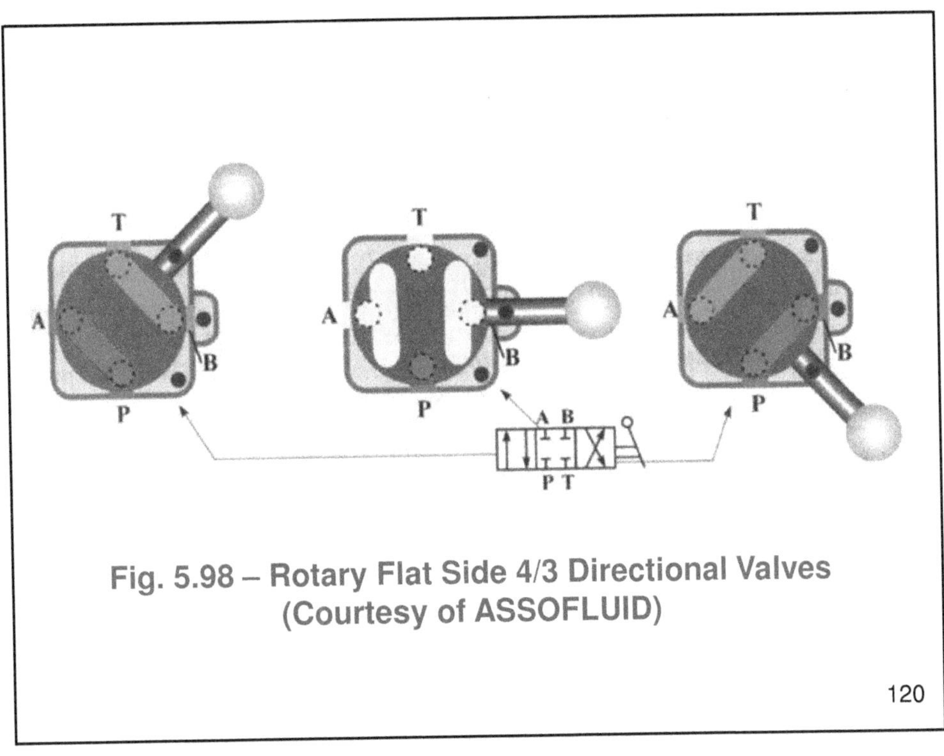

Fig. 5.98 – Rotary Flat Side 4/3 Directional Valves
(Courtesy of ASSOFLUID)

120

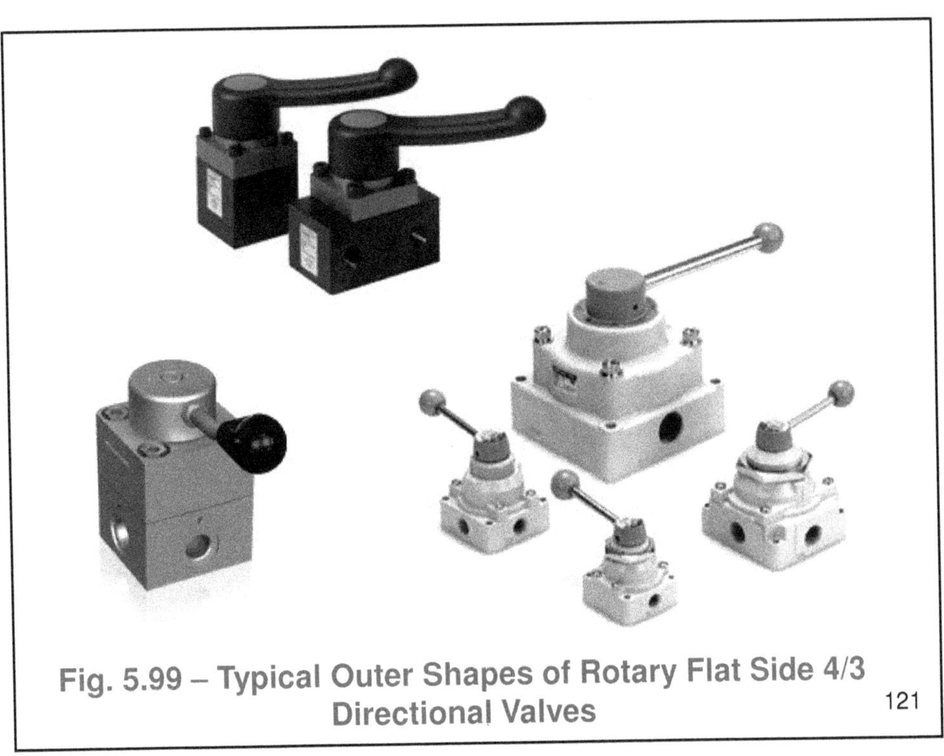

Fig. 5.99 – Typical Outer Shapes of Rotary Flat Side 4/3 Directional Valves

121

5.3.8- Check Valves

5.3.8.1- Basic Function of Check Valves

Fig. 5.100 – Check Valve Basic Function

122

5.3.8.2- Applications of Check Valves

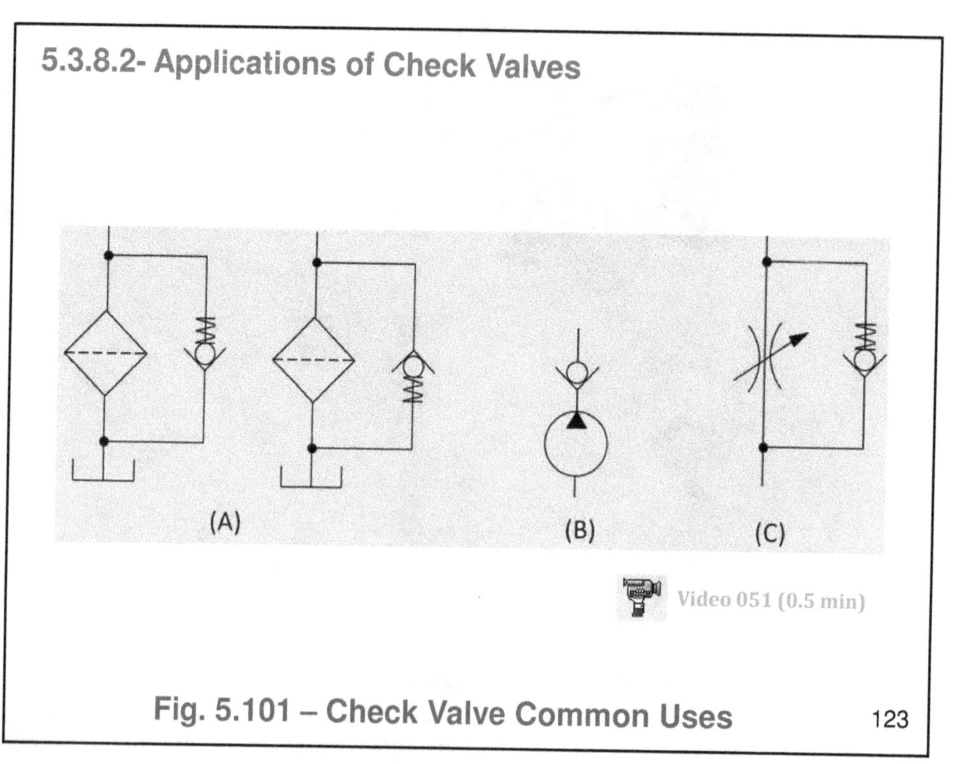

Video 051 (0.5 min)

Fig. 5.101 – Check Valve Common Uses

123

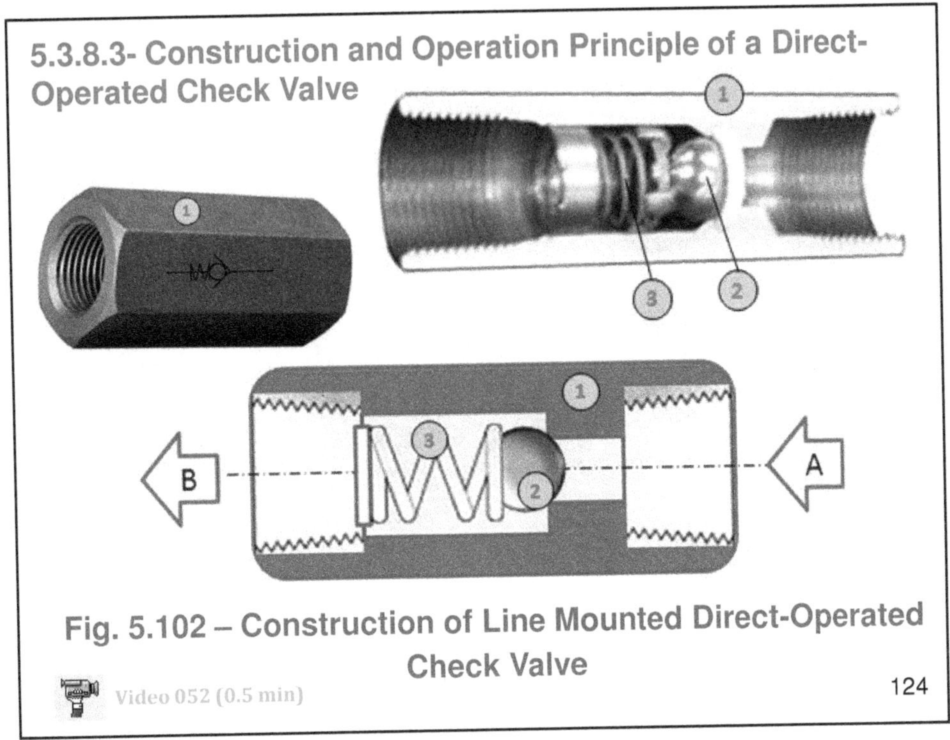

Fig. 5.102 – Construction of Line Mounted Direct-Operated Check Valve

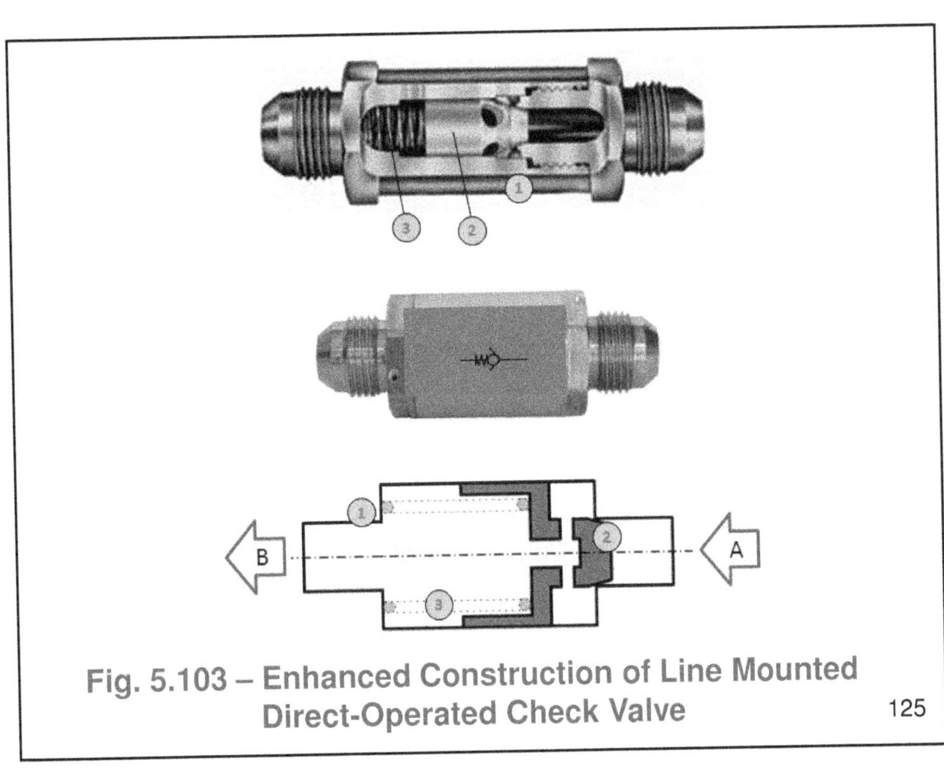

Fig. 5.103 – Enhanced Construction of Line Mounted Direct-Operated Check Valve

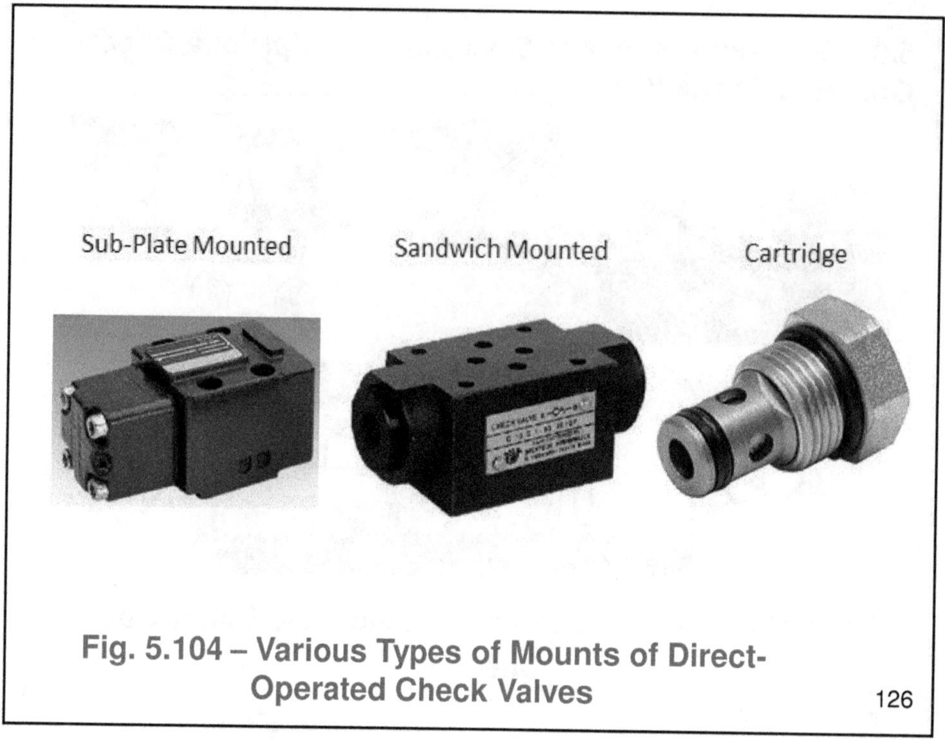

Fig. 5.104 – Various Types of Mounts of Direct-Operated Check Valves

126

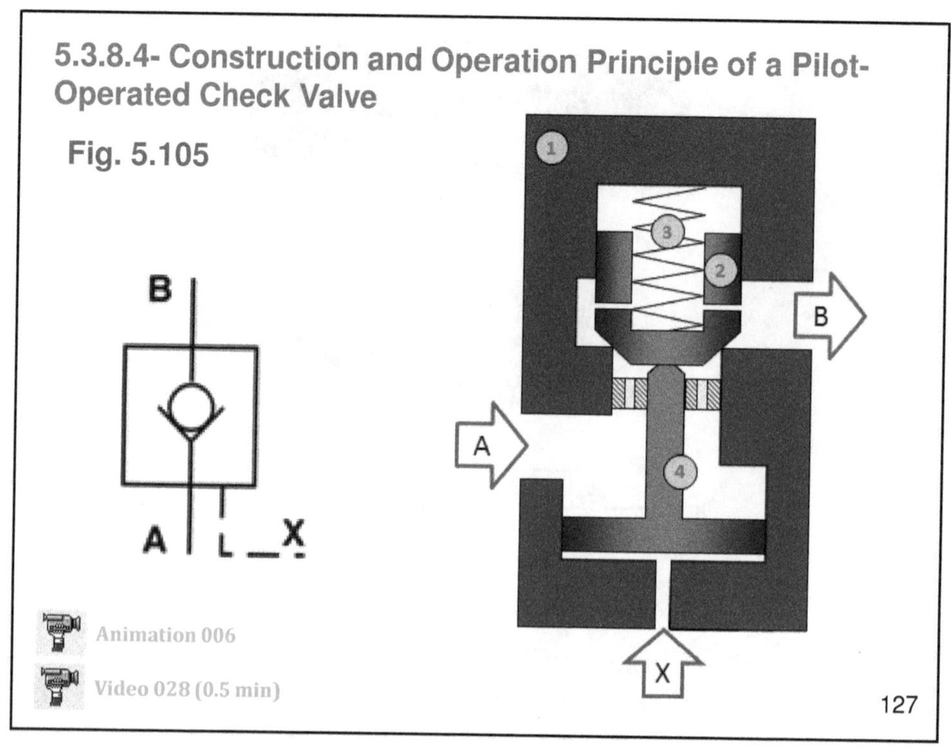

127

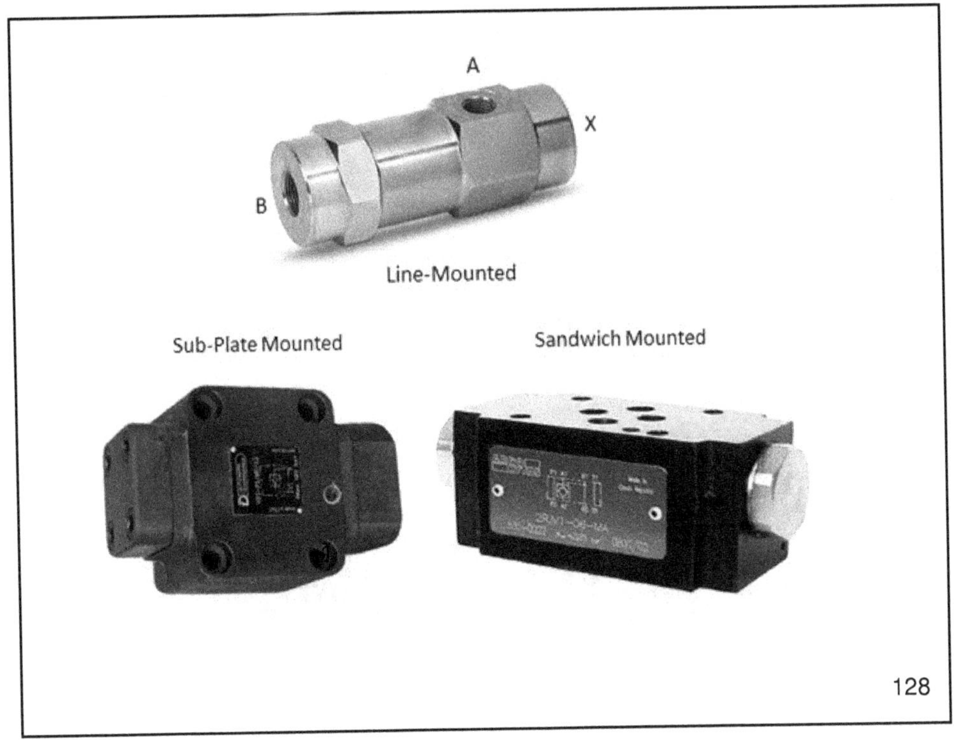

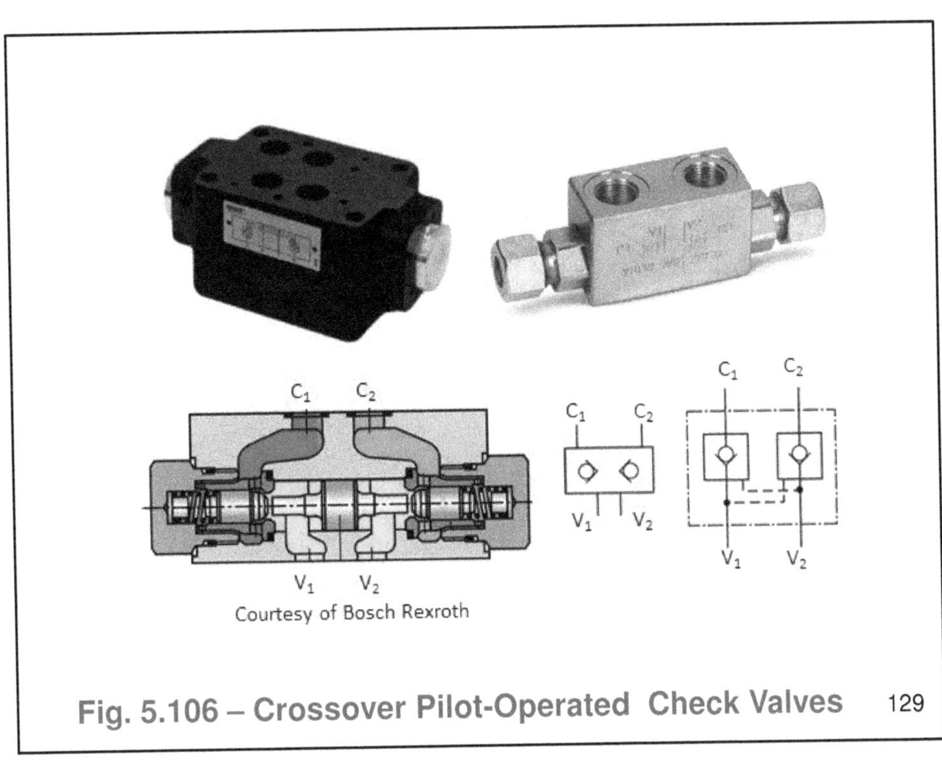

Fig. 5.106 – Crossover Pilot-Operated Check Valves

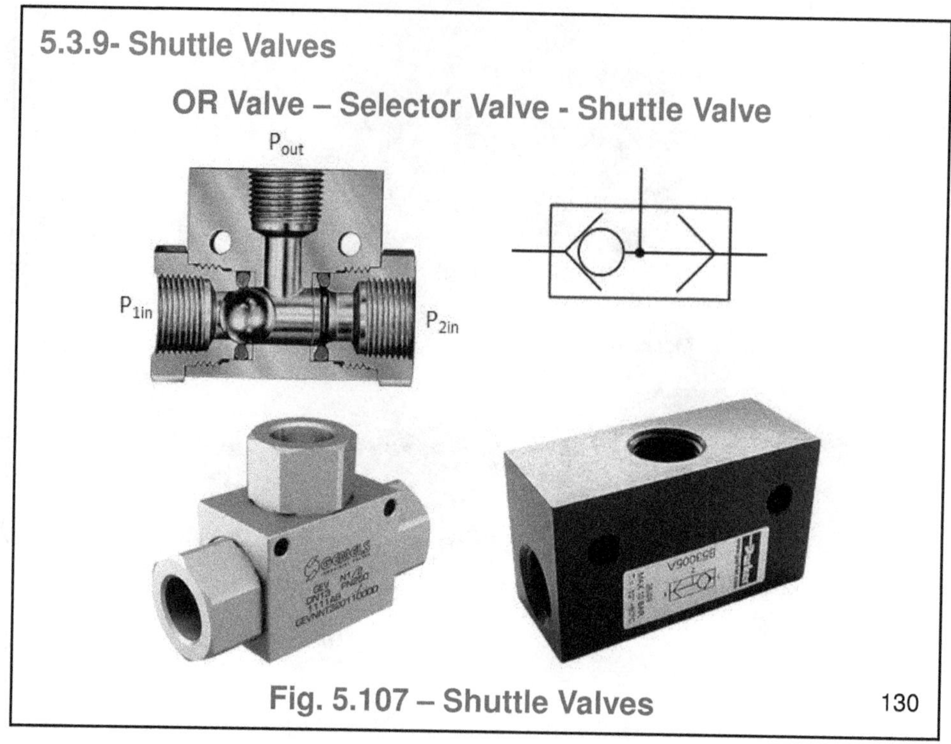

Fig. 5.107 – Shuttle Valves

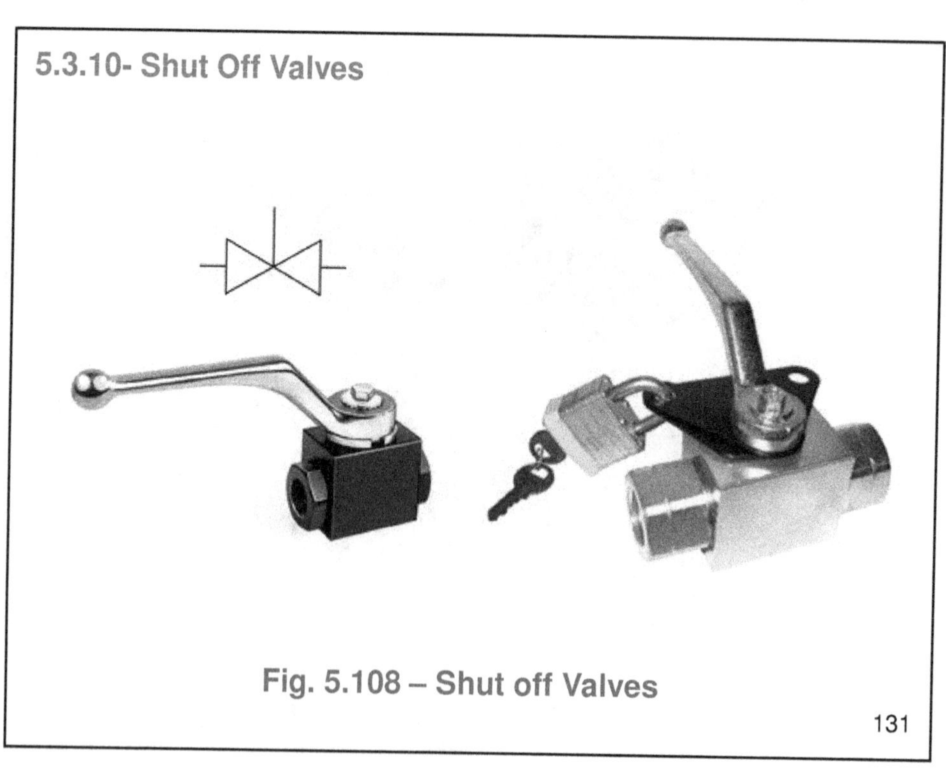

Fig. 5.108 – Shut off Valves

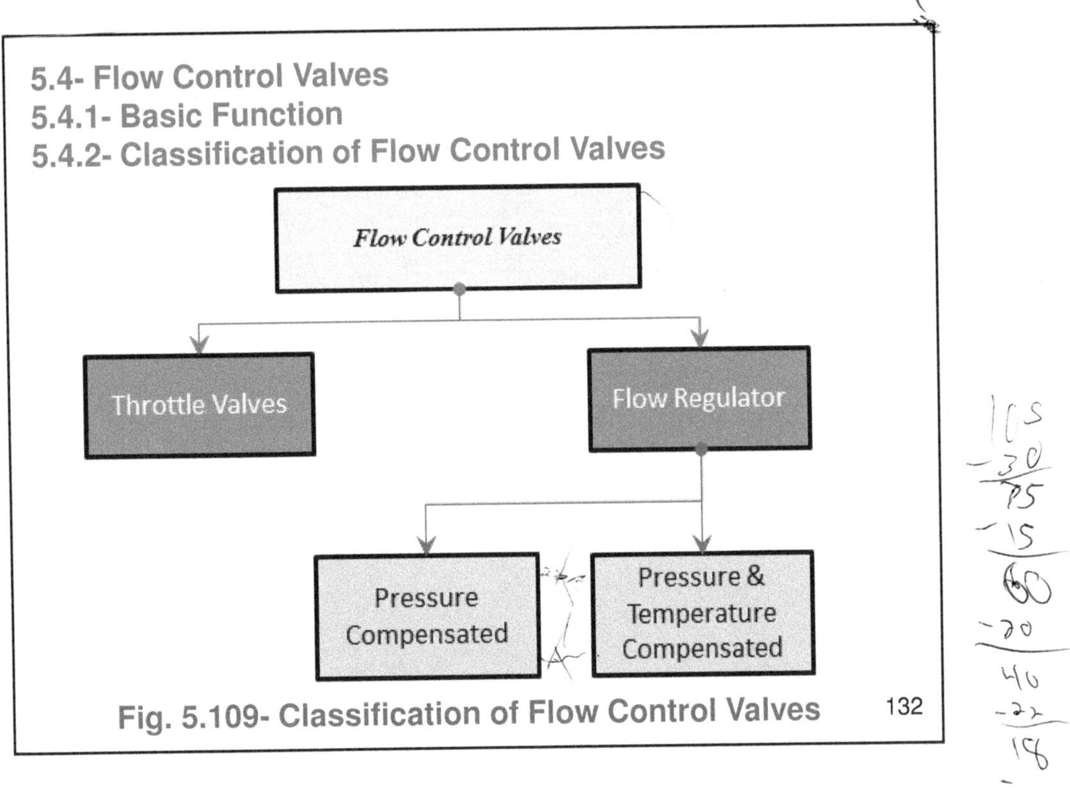

Fig. 5.109- Classification of Flow Control Valves 132

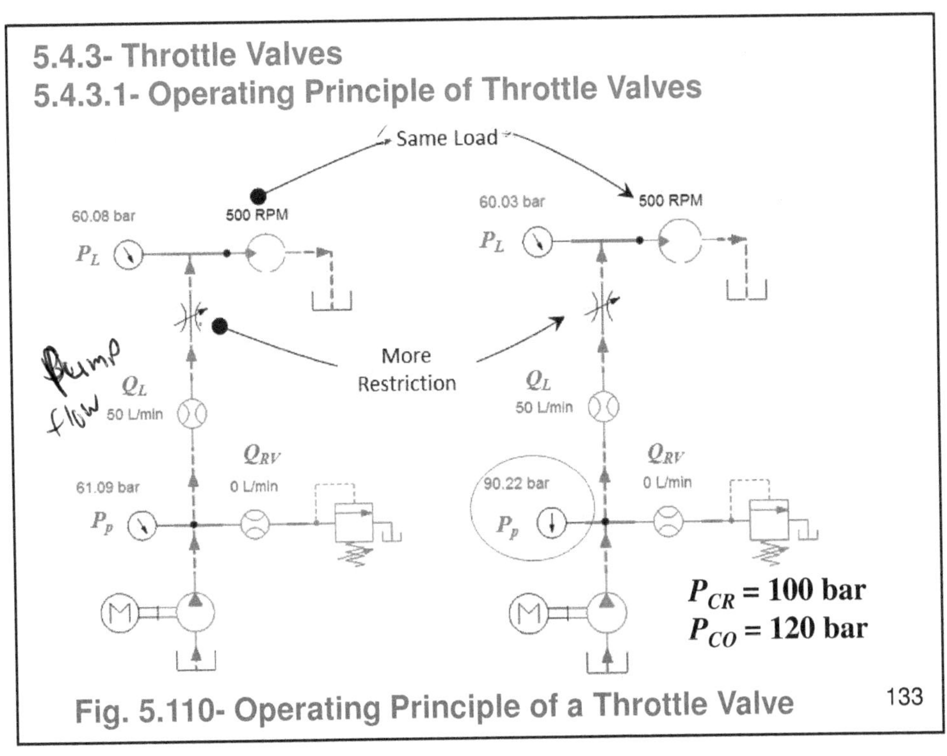

Fig. 5.110- Operating Principle of a Throttle Valve 133

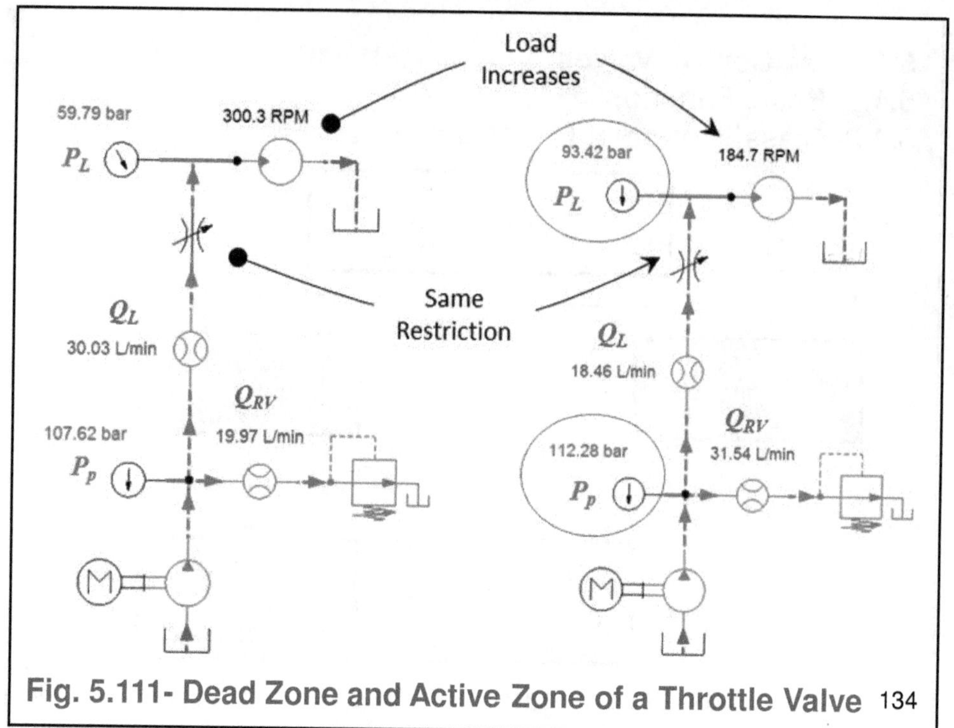

Fig. 5.111- Dead Zone and Active Zone of a Throttle Valve 134

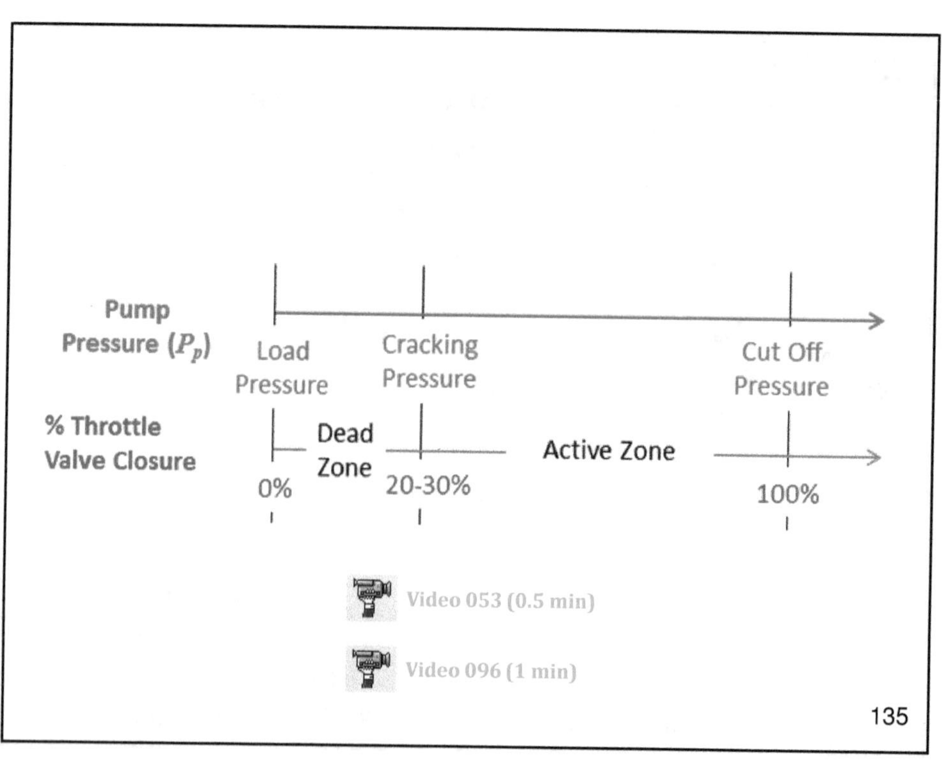

135

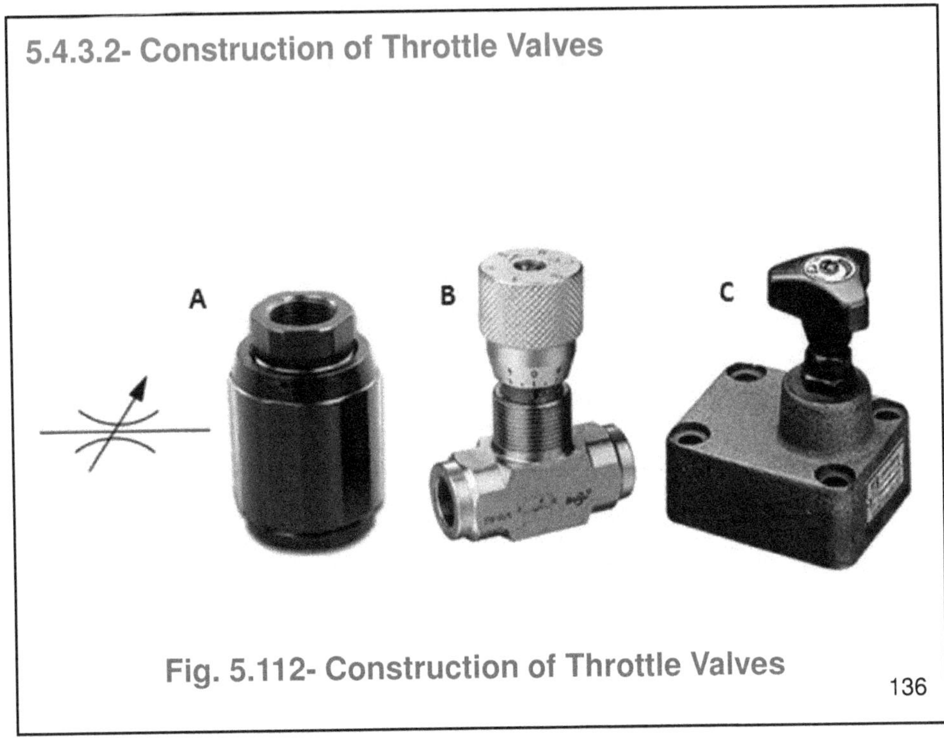

Fig. 5.112- Construction of Throttle Valves

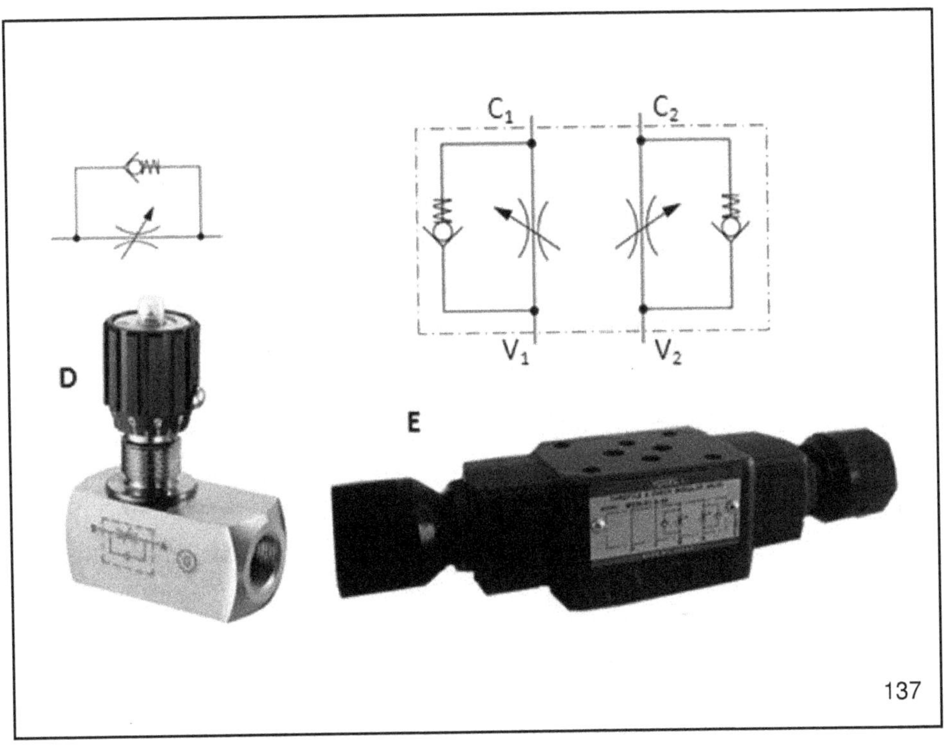

5.4.3.3- Characteristic Curves of Throttle Valves

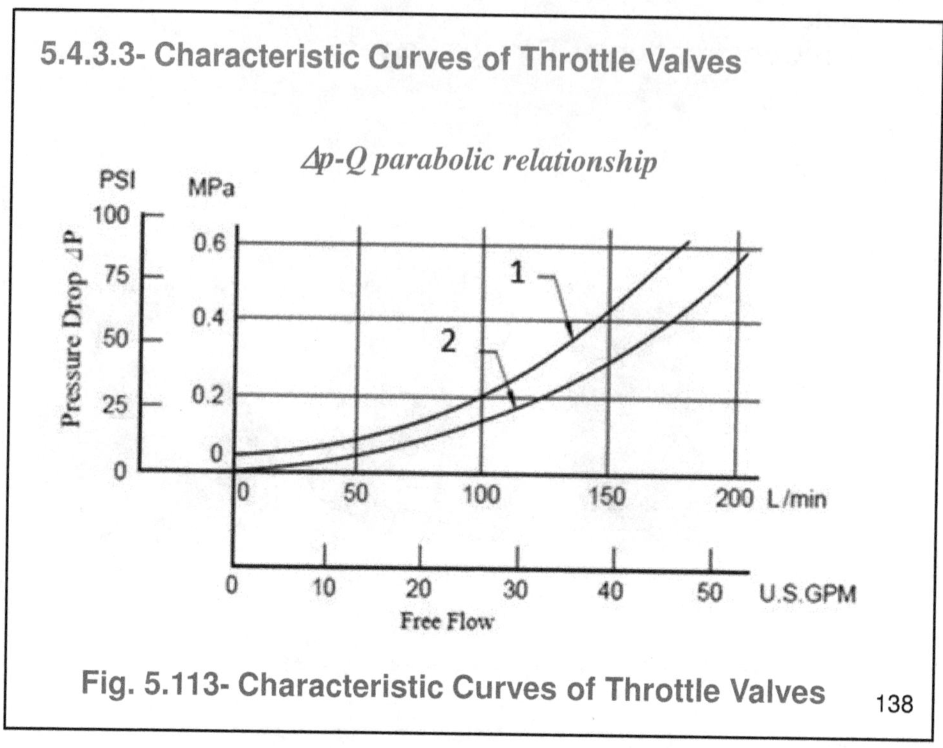

Fig. 5.113- Characteristic Curves of Throttle Valves

138

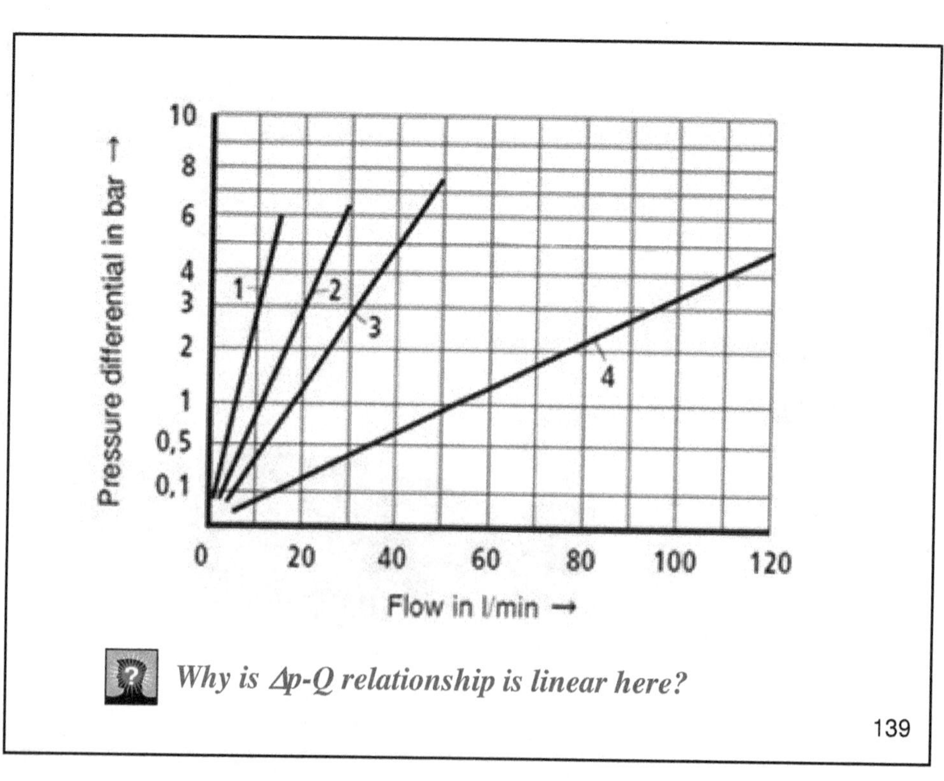

Why is Δp-Q relationship is linear here?

139

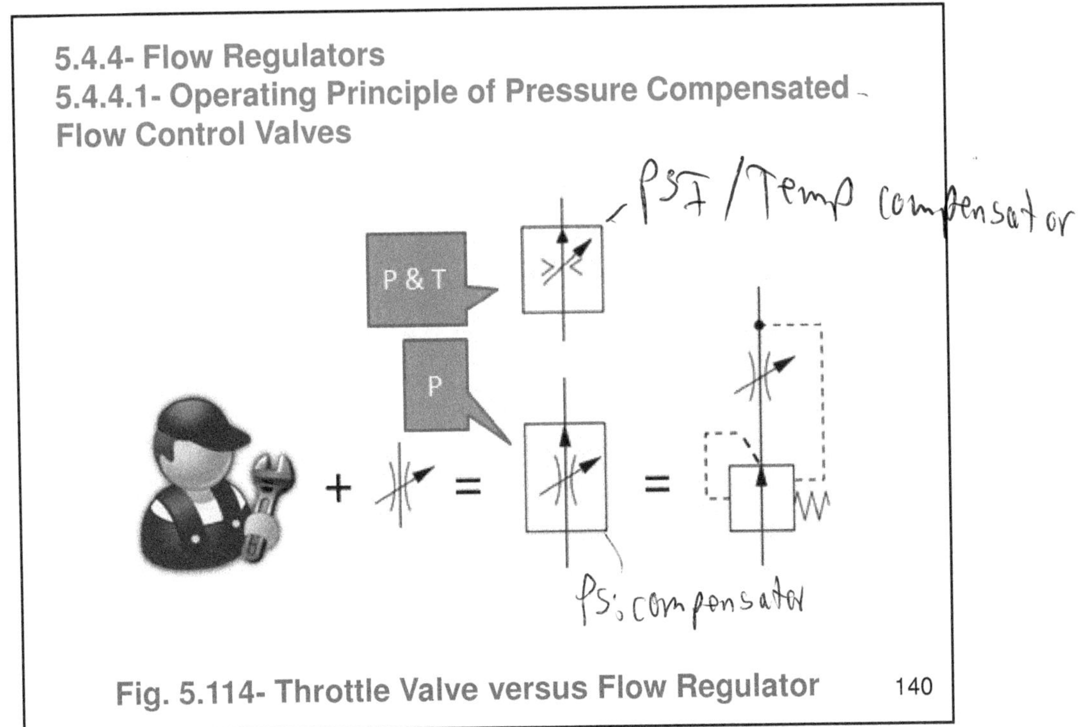

Fig. 5.114- Throttle Valve versus Flow Regulator

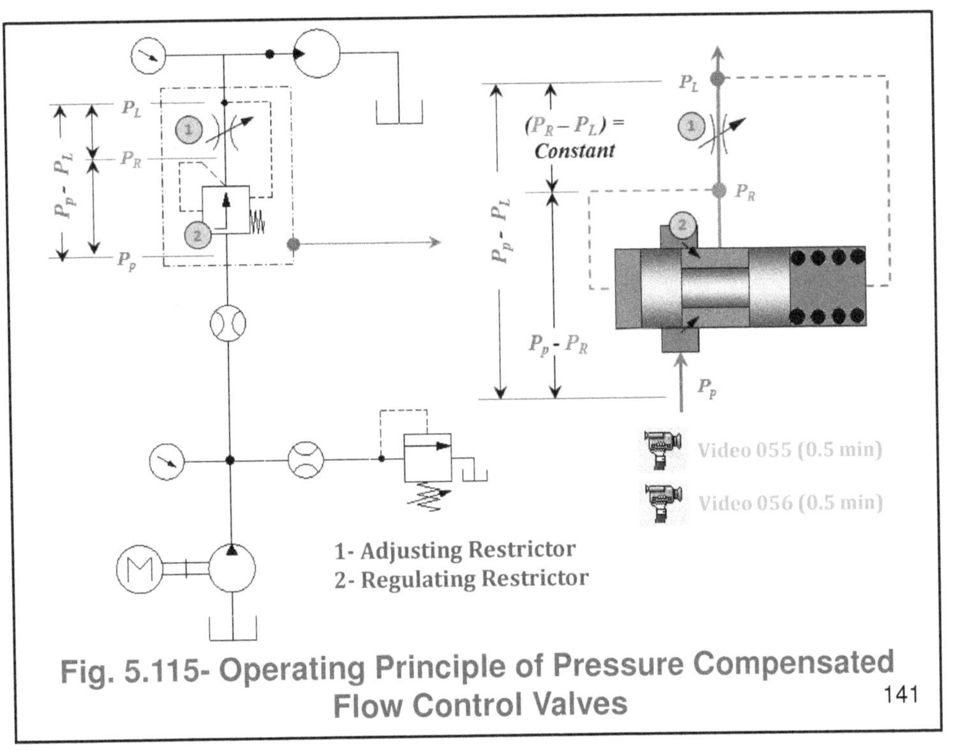

Fig. 5.115- Operating Principle of Pressure Compensated Flow Control Valves

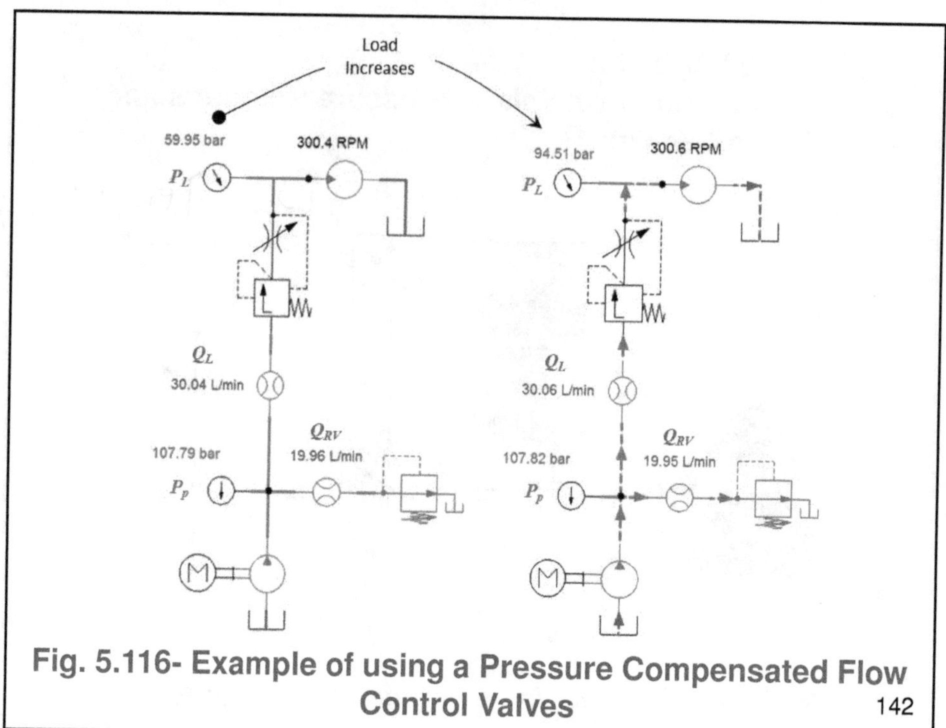

Fig. 5.116- Example of using a Pressure Compensated Flow Control Valves

142

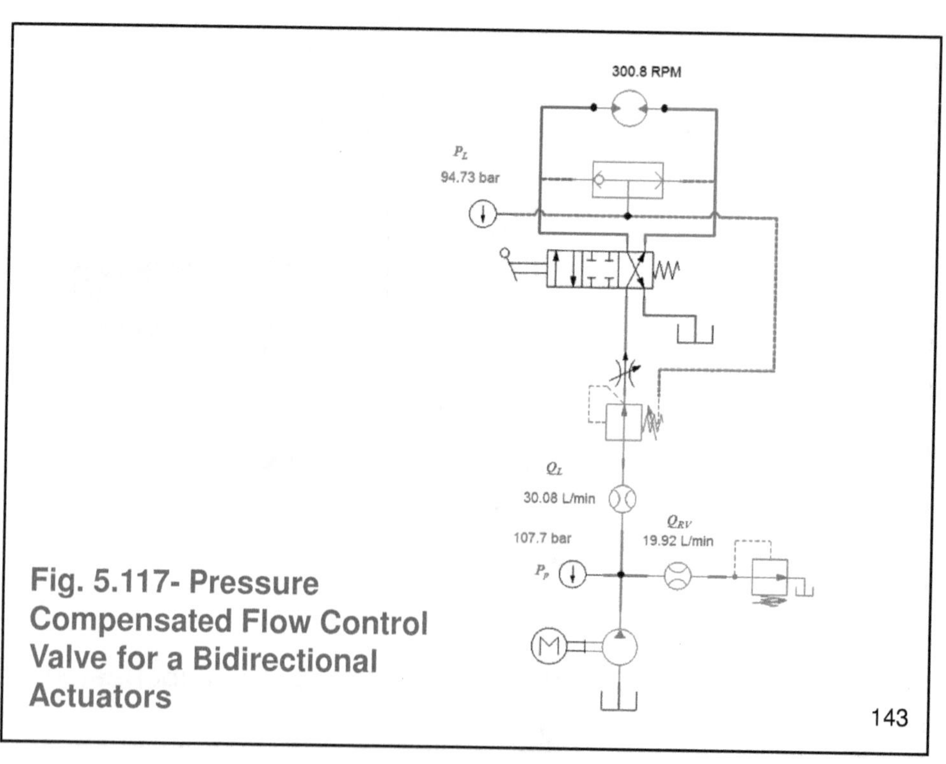

Fig. 5.117- Pressure Compensated Flow Control Valve for a Bidirectional Actuators

143

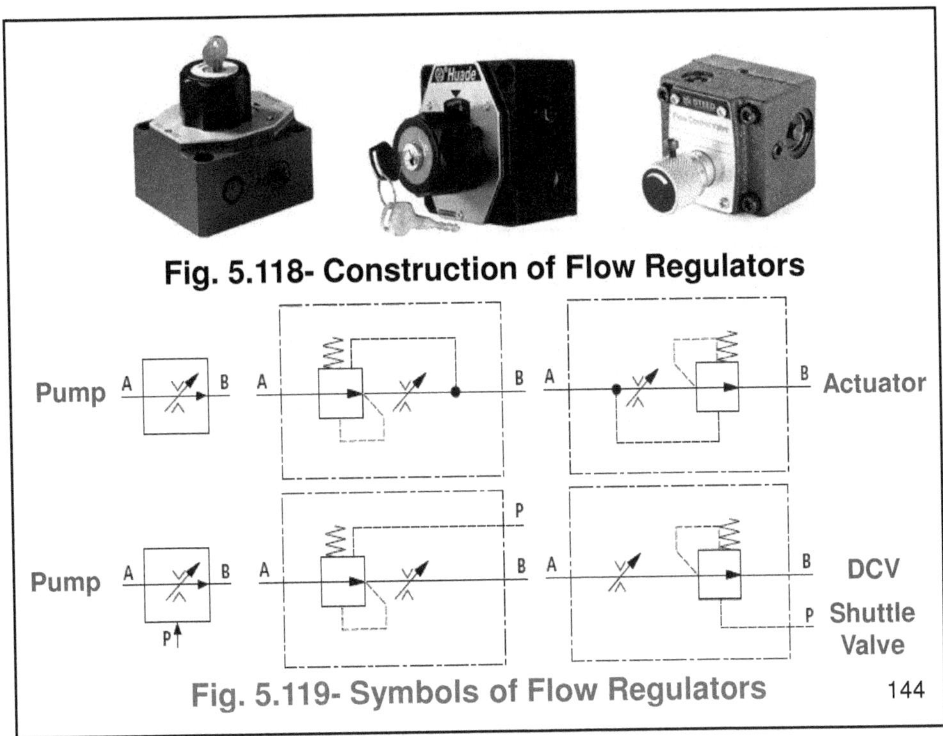

Fig. 5.118- Construction of Flow Regulators

Fig. 5.119- Symbols of Flow Regulators 144

5.4.4.3- Characteristic Curves of Flow Regulators

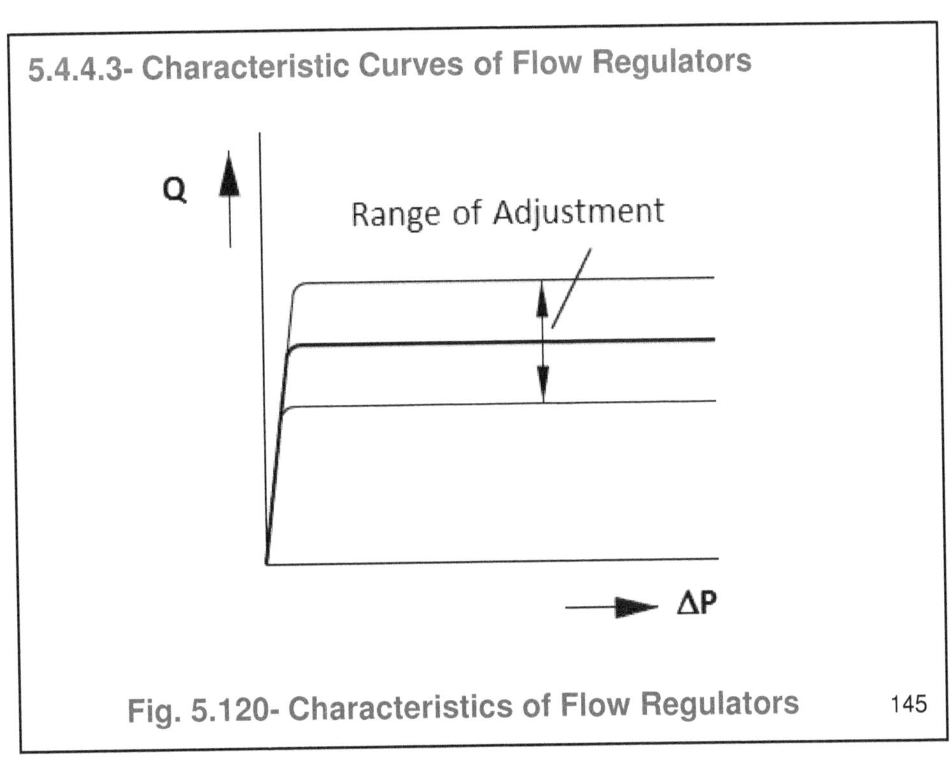

Fig. 5.120- Characteristics of Flow Regulators 145

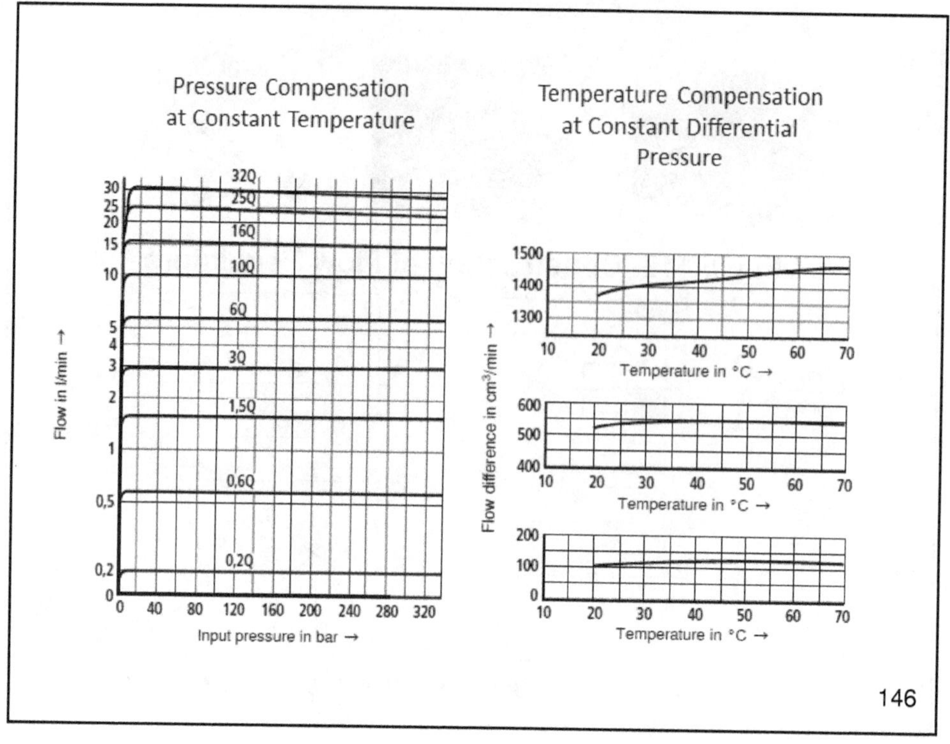

5.4.5- Flow Dividers

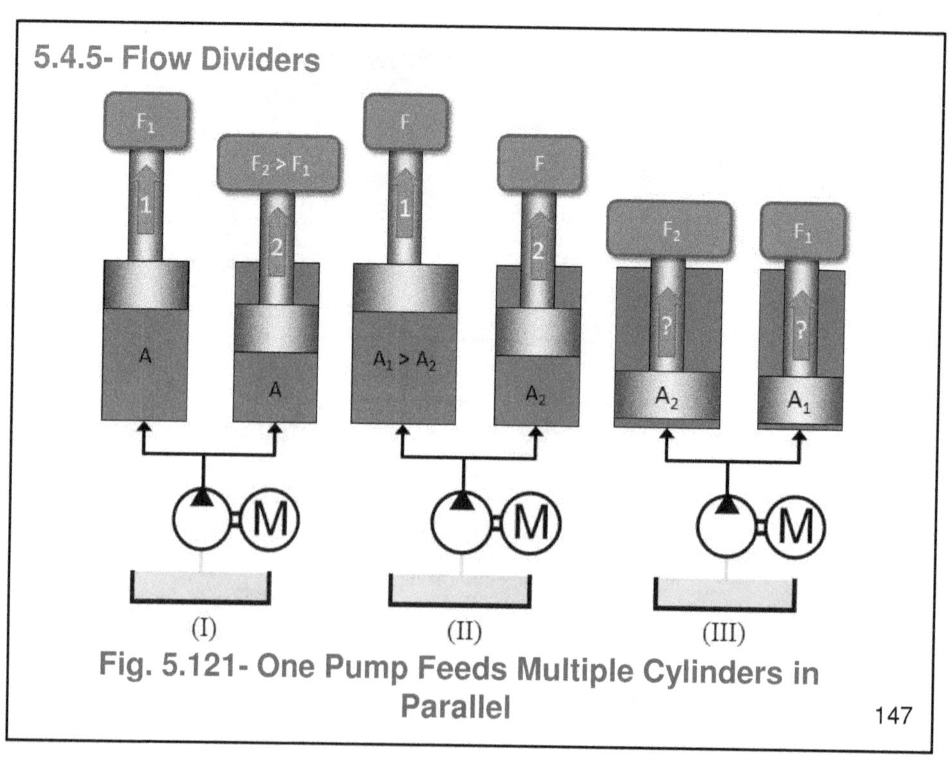

Fig. 5.121- One Pump Feeds Multiple Cylinders in Parallel

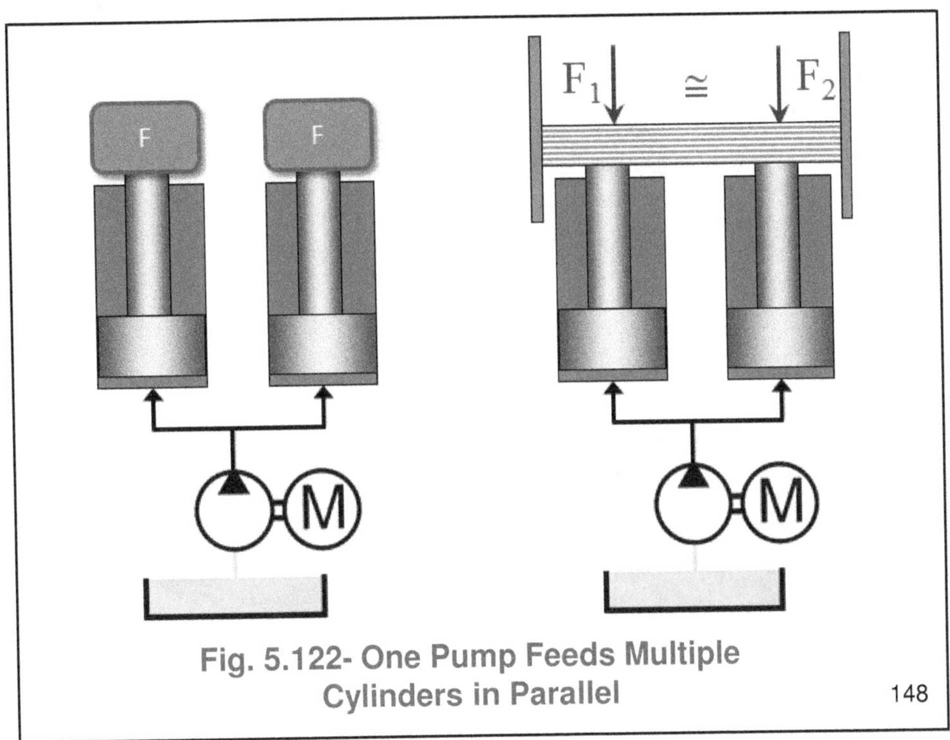

Fig. 5.122- One Pump Feeds Multiple Cylinders in Parallel

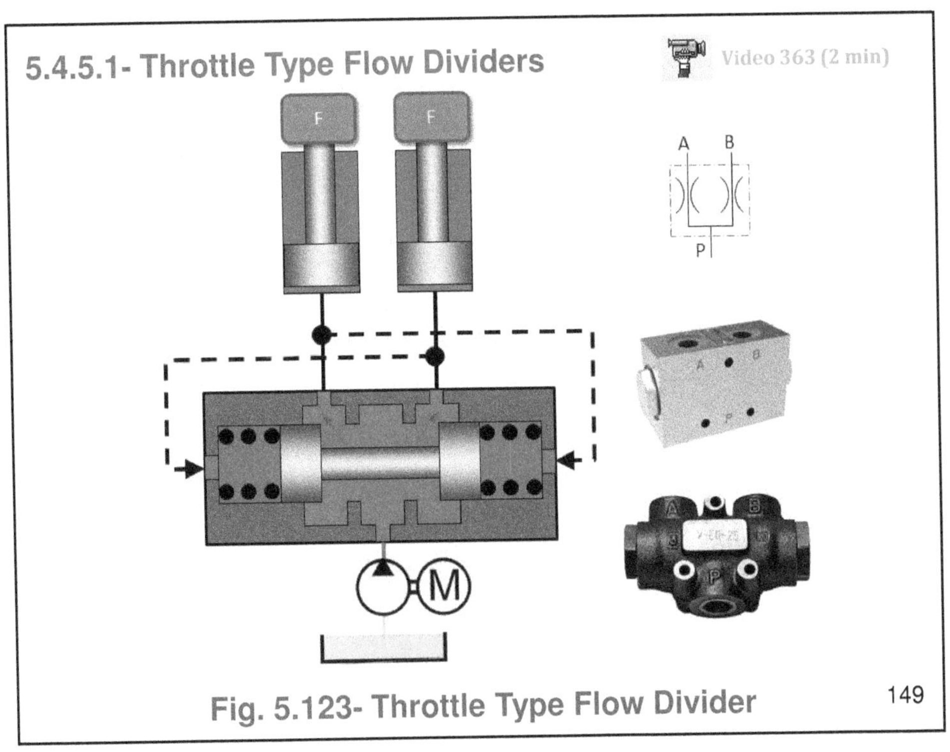

Fig. 5.123- Throttle Type Flow Divider

5.4.5.2- Motor Type Flow Dividers

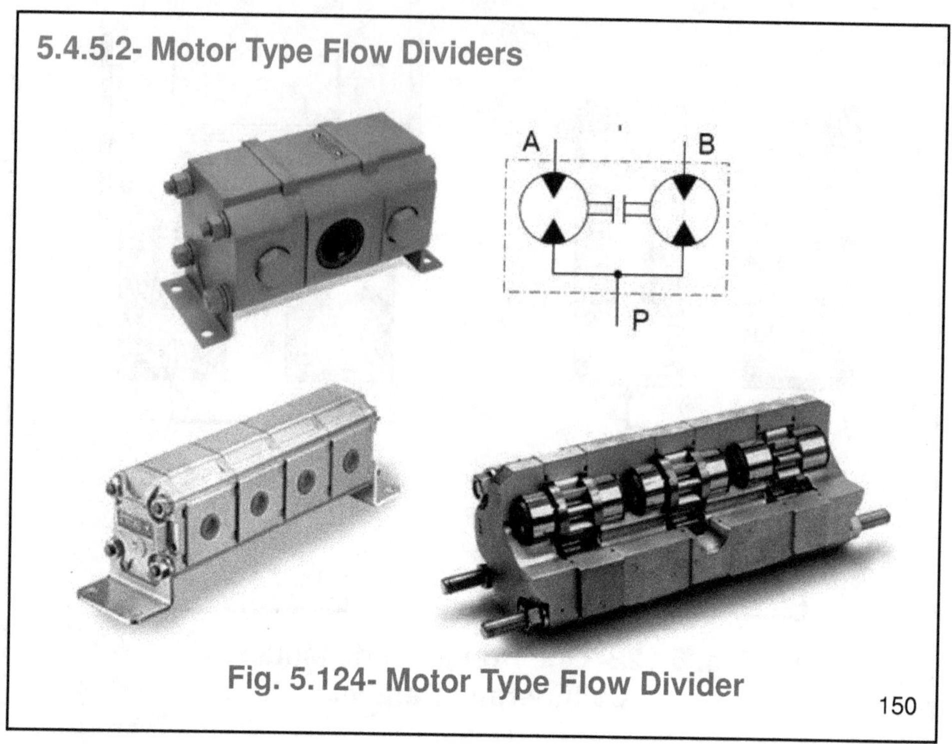

Fig. 5.124- Motor Type Flow Divider

150

5.4.6- Priority Valve

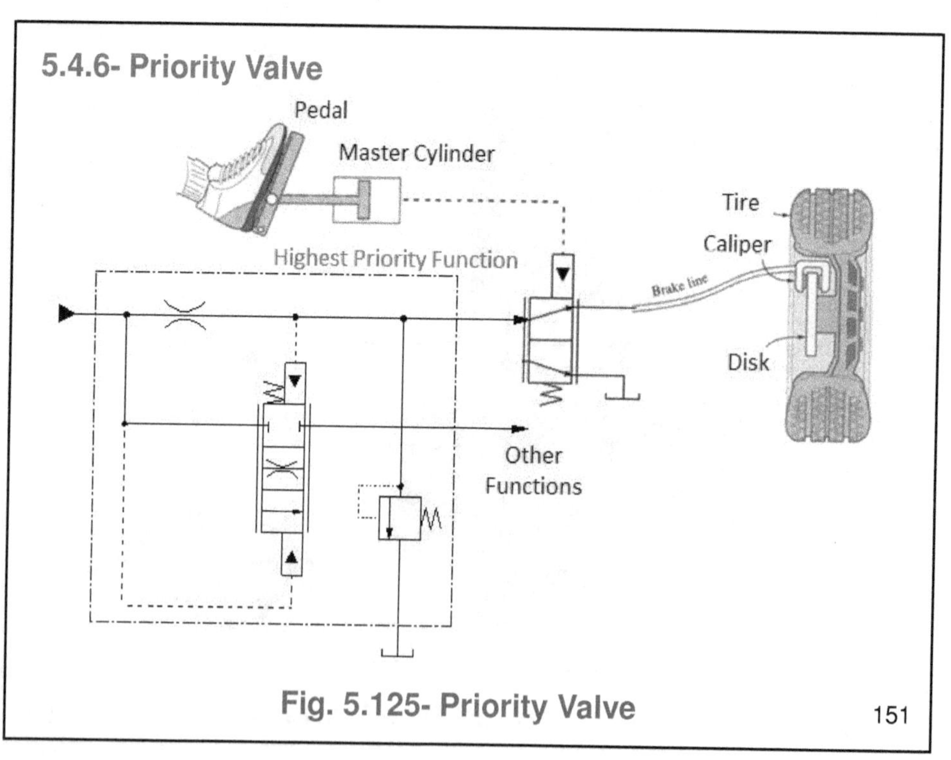

Fig. 5.125- Priority Valve

151

Chapter 5 Reviews

1. Hydraulic valves are classified into which of the **following 3 basic functions?**
   A. Pressure control, directional control and flow control.
   B. Pressure control, directional control and speed control.
   C. Force control, directional control and flow control.
   D. Pressure control, torque control and flow control.

2. Which valve design has **leakage at null** conditions?
   A. Poppet type valve.
   B. Spool type valve.
   C. Counter balance valve.
   D. Over-center valve.

3. **Direct-operated** valves are the valves that?
   A. Require a hydraulic signal for the valve to work. This signal source is internal from the same main line on which the valve is mounted.
   B. Have a pilot stage that operates the main stage.
   C. Consist of only one stage.
   D. Use a pilot signal that is supplied externally from a separate circuit other than the main line on which the valve is mounted.

4. **Pilot-operated** valves are the valves that?
   A. Require a hydraulic signal for the valve to work. This signal source is internal from the same main line on which the valve is mounted.
   B. Have a pilot stage that operates the main stage work.
   C. Consist of only one stage.
   D. Use a pilot signal that is supplied externally from a separate circuit other than the main line on which the valve is mounted.

5. **Direct-controlled** valves are the valves that?
   A. Require a hydraulic signal for the valve to work. This signal source is internal from the same main line on which the valve is mounted.
   B. Have a pilot stage that operates the main stage.
   C. Consist of only one stage.
   D. Use a pilot signal that is supplied externally from a separate circuit other than the main line on which the valve is mounted.

6. **Pilot-controlled** valves are the valves that?
   A. Require a hydraulic signal for the valve to work. This signal source is internal from the same main line on which the valve is mounted.
   B. Have a pilot stage that operates the main stage.
   C. Consist of only one stage.
   D. Use a pilot signal that is supplied externally from a separate circuit other than the main line on which the valve is mounted.

7. If **precise motion control** is required, which of the following should be used in your application?
   A. Pressure relief valve.
   B. Switching directional control valve.
   C. Proportional or servo valve.
   D. Throttle valve.

8. For the **safety of a hydraulic system**, which of the following valves is mandatory to be used with a fixed displacement pump?
   A. Counterbalance valve.
   B. Reducing valve.
   C. Check Valve.
   D. Pressure relief valve.

9. **Pressure Relief Valves** are used in a hydraulic system to?
   A. Limit the system pressure from exceeding certain maximum value.
   B. Protect the main circuit or part of the circuit.
   C. Regulate the system exerted effort (force/torque).
   D. All of the above.

10. Which of the following is the commonly known as the **set point of the PRV**?
    A. (1-5)% above the load.
    B. (10-20)% above the load.
    C. (15-20)% below the load.
    D. (30-40)% above the load.

11. A pump that discharges 10 gpm is protected by a pressure relief valve that is adjusted to open at 1714 psi. If the pump outlet is blocked, what is the **power loss value through the PRV?**
    A. 10 kW.
    B. 10 L/min.
    C. 10 psi.
    D. 10 HP.

12. A **2-stage valve** is used because of?
    A. Expected large flowrate to pass through the valve.
    B. Expected high pressure to act on the valve.
    C. Expected low leakage through the valve.
    D. Expected fast response of the valve.

13. In the symbol shown, if the solenoid is energized, this **2-stage valve** will work as?
    A. Counterbalance valve.
    B. Directional control valve.
    C. Speed control valve.
    D. Unloading valve.

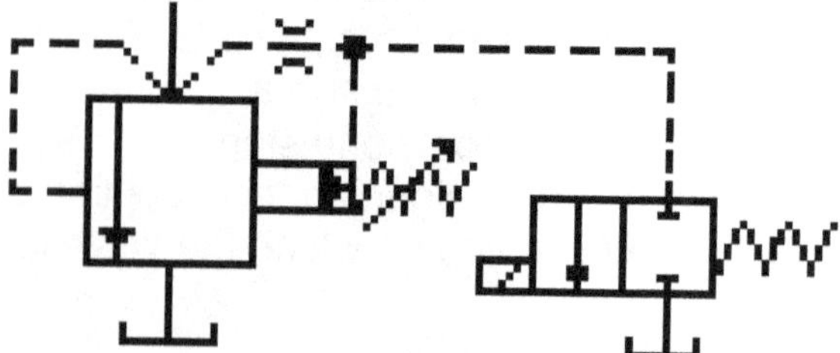

14. The part circled number 2 in the sectional view shown below is?
    A. The unloading stage.
    B. The pilot stage.
    C. The main stage.
    D. None of the above.

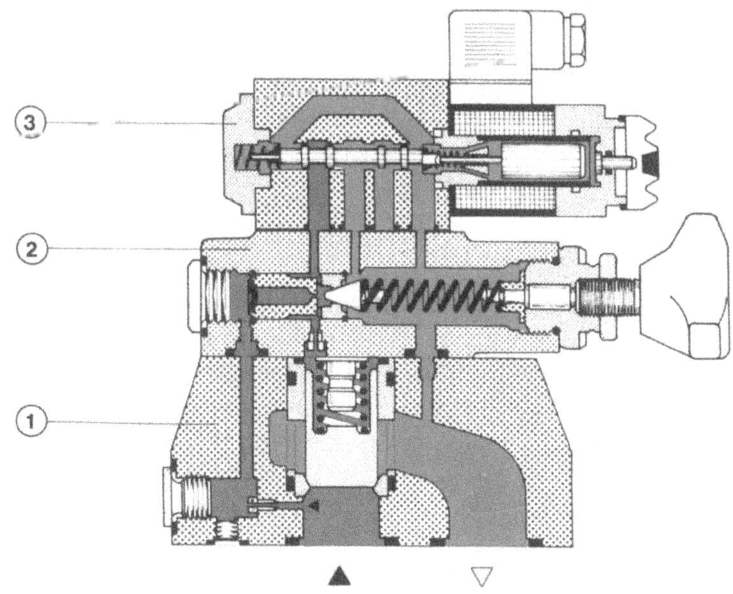

15. In the application shown below, to hold the load in the neutral position of the directional control valve, which valve should be selected for the missing component?
    A. Over-center valve.
    B. Directional control valve.
    C. Counterbalance valve.
    D. Unloading valve.

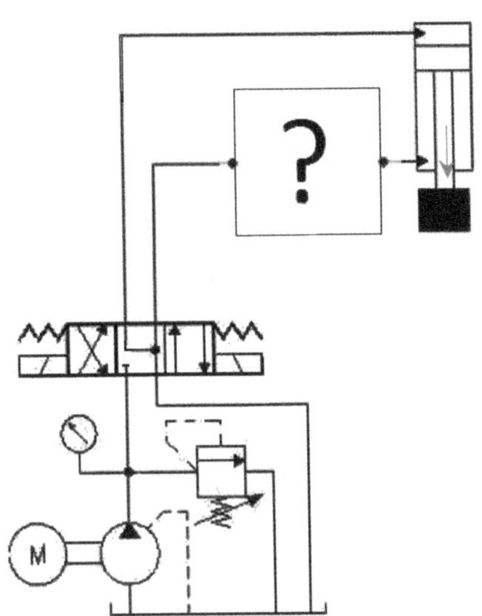

16. Which valve would be selected in an application where you want to assure that operation #1 is done before operation #2?
    A. Over-center valve.
    B. Reducing valve.
    C. Counterbalance valve.
    D. Sequence valve.

17. In a **direct-controlled 2-stage directional control valve**, pilot pressure is rated as 30 bar. Main pressure is 150 bar. Which sandwich type valve is required between the main stage and the pilot stage?
    A. Over-center valve.
    B. Reducing valve.
    C. Counterbalance valve.
    D. Sequence valve.

18. What is the difference between a pressure relief valve and a reducing valve?
    A. Relief valve is normally closed but reducing valve is normally open.
    B. Relief valve controls the upstream pressure but reducing valve controls downstream pressure.
    C. Relief valve vents its spring chamber internally but reducing valve vents it externally.
    D. All of the above is correct.

19. An **unloading valve** is adjusted to open at a 1000 psi. If the main line's pressure is 2000 psi and pilot pressure is 900 psi, can you tell what the status of the unloading valve is?
    A. Unloading valve is opened?
    B. Unloading valve is closed?
    C. Unloading valve is not affected by the pilot pressure?
    D. All of the above is incorrect.

20. In the application shown below, to hold the load in the neutral position of the directional control valve, which valve should be selected for the missing component?
    A. Double Over-center valve.
    B. Double Reducing valve.
    C. Double Counterbalance valve.
    D. Double Sequence valve.

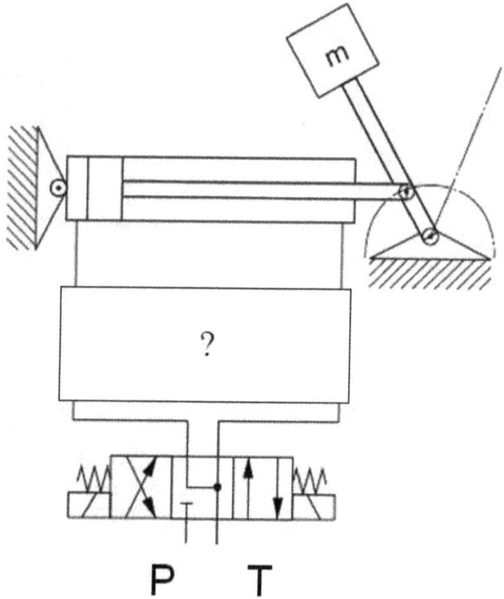

21. In a **4/3 directional control valve**?
    A. 4 ports are connected by 3 different positions of the valve.
    B. 3 ports are connected by 4 different positions of the valve.
    C. 4 ports are connected by 4 different positions of the valve.
    D. 3 ports are connected by 3 different positions of the valve.

22. **Center position spool configuration of the pilot stage** of a pilot-operated directional control valve should be of which type?
    A. Tandem center type.
    B. Open center type.
    C. Closed center type.
    D. Float center type.

23. **Center position spool configuration of the main stage** of the valve shown below is of which type?
    A. Tandem center type.
    B. Open center type.
    C. Closed center type.
    D. Float center type.

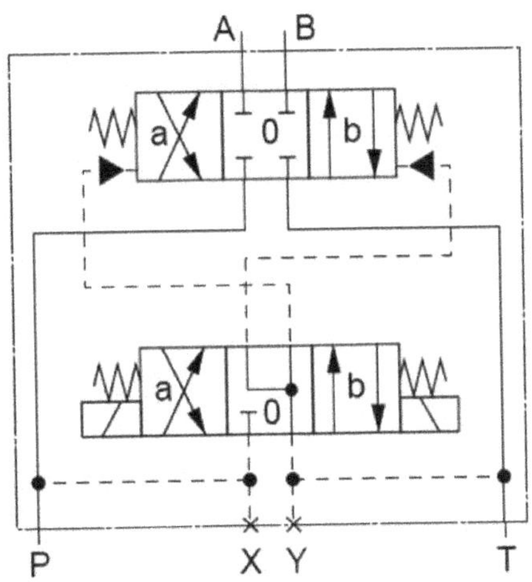

24. **Pressure compensated flow control valves** are used to?
    A. Achieve synchronized motion of identical actuators regardless the load distribution.
    B. Achieve cylinder's constant speed regardless of the supply flowrate.
    C. Achieve constant speed of an actuator's motion regardless of load variations.
    D. Achieve motor's constant rpm regardless the motor size.

25. **Flow dividers of 1:1 ratio** are used to?
    A. Achieve synchronized speed of identical actuators regardless the load distribution.
    B. Equalize the pressure of identical actuators regardless the load distribution.
    C. Distribute the input flowrate based on the size of the actuators.
    D. Distribute the input flowrate based on the motor's torque.

Chapter 5 Assignment

Student Name: -- Student ID: ------------------

Date: -- Score: ------------------------

A: Calculate the valve coefficient of a directional valve that has 10 bar differential pressure when 60 lit/min of hydraulic fluid passes through it. The hydraulic fluid has a specific gravity of 0.8.

B: Draw the power distribution chart for the speed control system shown below.

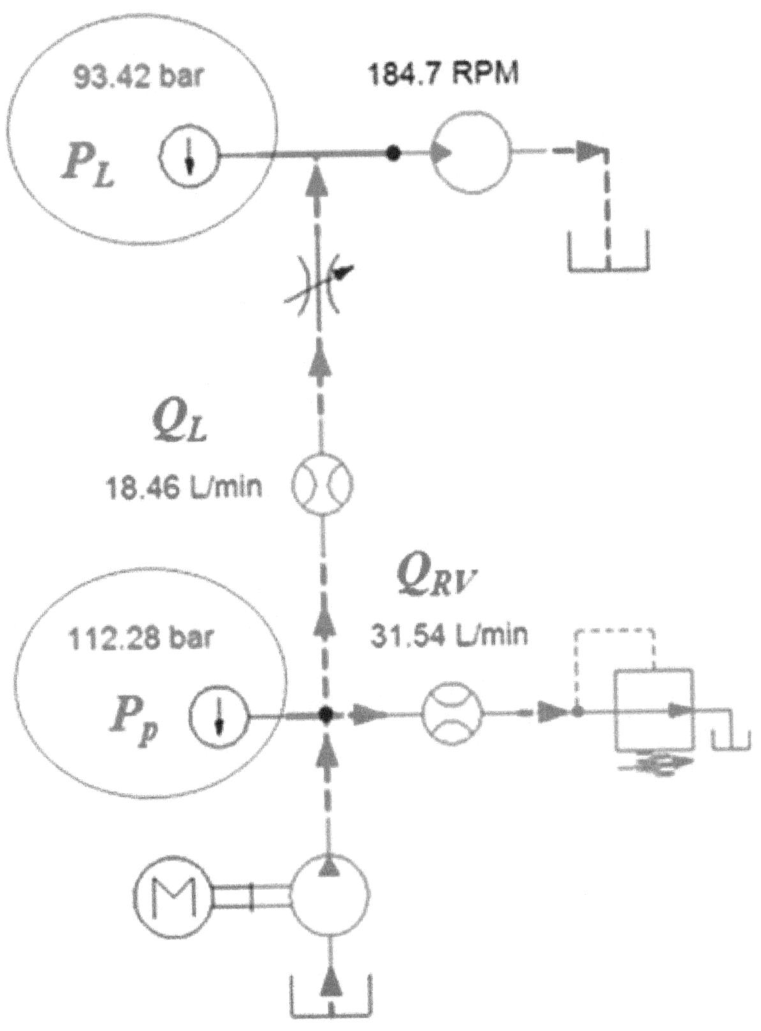

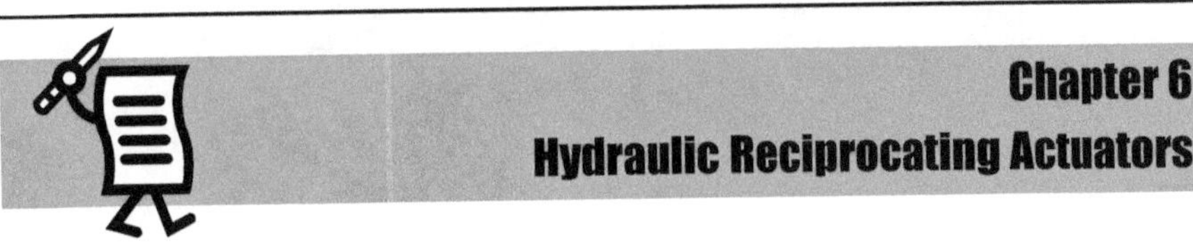

Objectives:

This chapter covers two types of hydraulic reciprocating actuators; namely cylinders and rotary actuators. The chapter presents the various types of hydraulic cylinders, the effect of using a differential cylinder on the pressure and flow distribution in the system, and the calculations for sizing a hydraulic cylinder. The chapter also presents the various types of rotary actuators and the reasons for using them. Symbols for reciprocating actuators are presented.

0

Brief Contents:

6.1- Reciprocating Hydraulic Actuators
6.2- Hydraulic Cylinders
6.3- Rotary Actuators

1

6.1 - Reciprocating Hydraulic Actuators

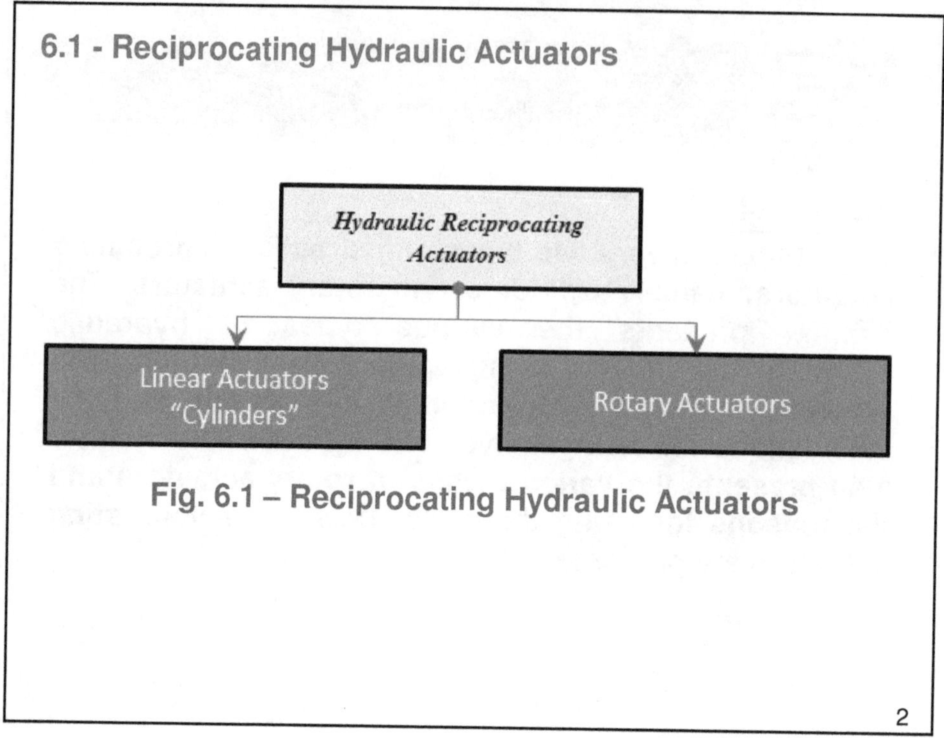

Fig. 6.1 – Reciprocating Hydraulic Actuators

2

	Cylinders	*Rotary Actuators*
Motion	Linear	Angular
Supply of Hydraulic Power	Single/Double acting	Double acting
Pressure Requirements	Function of load (force) and size of the cylinder	Function of load (torque) and size of the rotary actuator
Flow Requirements	Function of cylinder size and speed	Function of rotary actuator size and angular speed

Table 6.1 – Cylinders versus Rotary Actuators

3

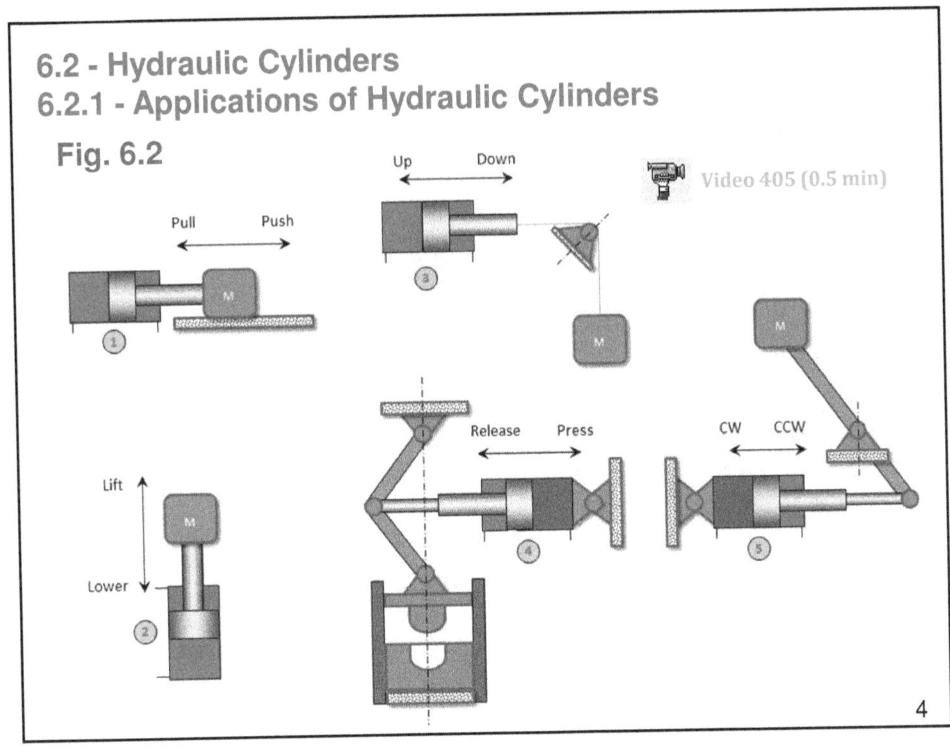
6.2 - Hydraulic Cylinders
6.2.1 - Applications of Hydraulic Cylinders
Fig. 6.2
Video 405 (0.5 min)
Pull
Push
M
1
Up
Down
3
M
Lift
Lower
M
2
Release
Press
4
CW
CCW
M
5
4

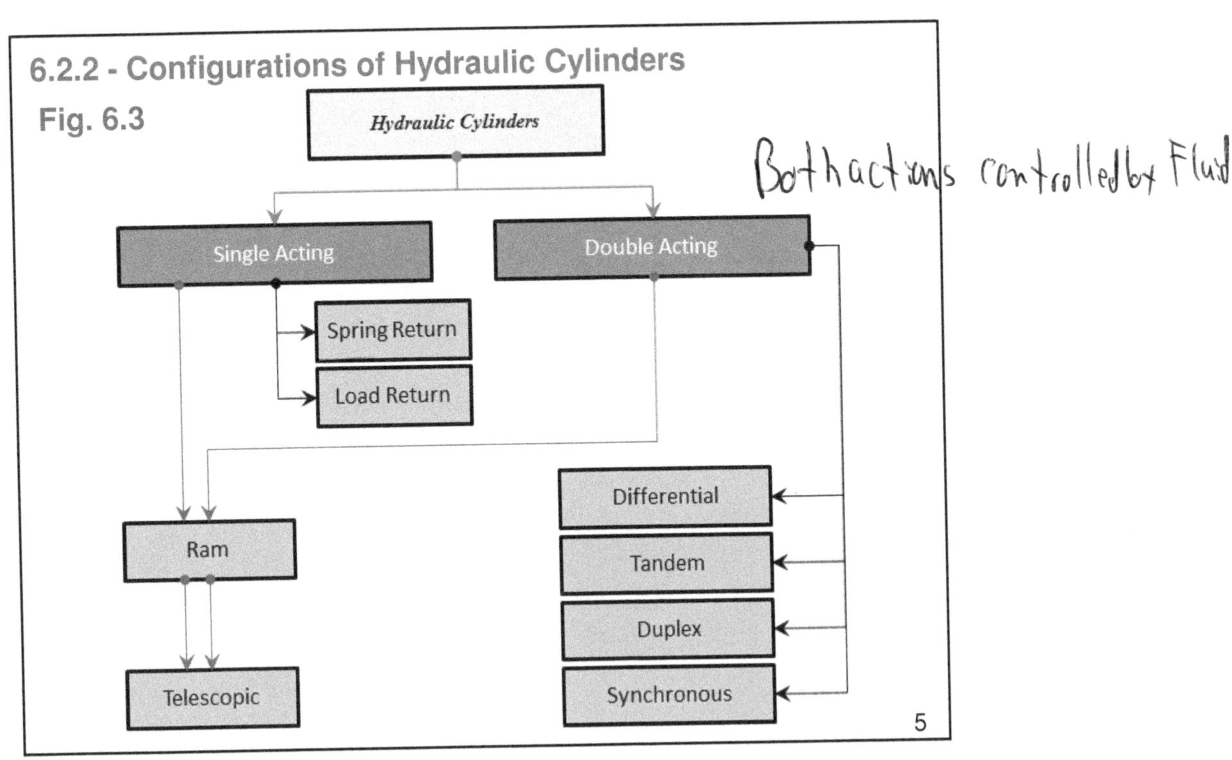
6.2.2 - Configurations of Hydraulic Cylinders
Fig. 6.3
Hydraulic Cylinders
Both actions controlled by Fluid
Single Acting
Double Acting
Spring Return
Load Return
Differential
Tandem
Duplex
Synchronous
Ram
Telescopic
5

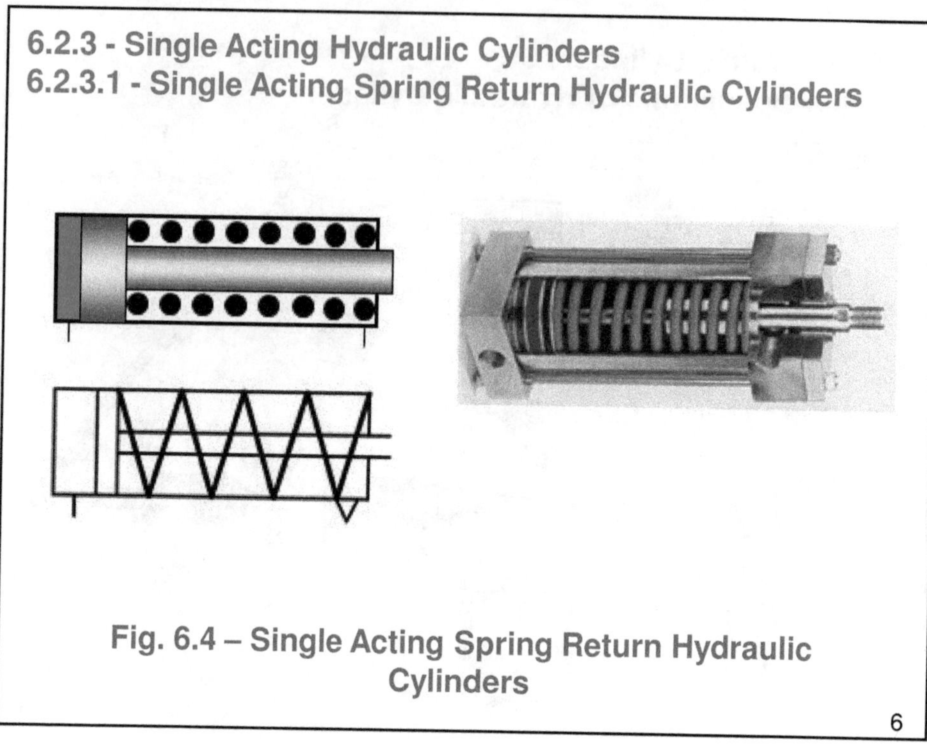

Fig. 6.4 – Single Acting Spring Return Hydraulic Cylinders

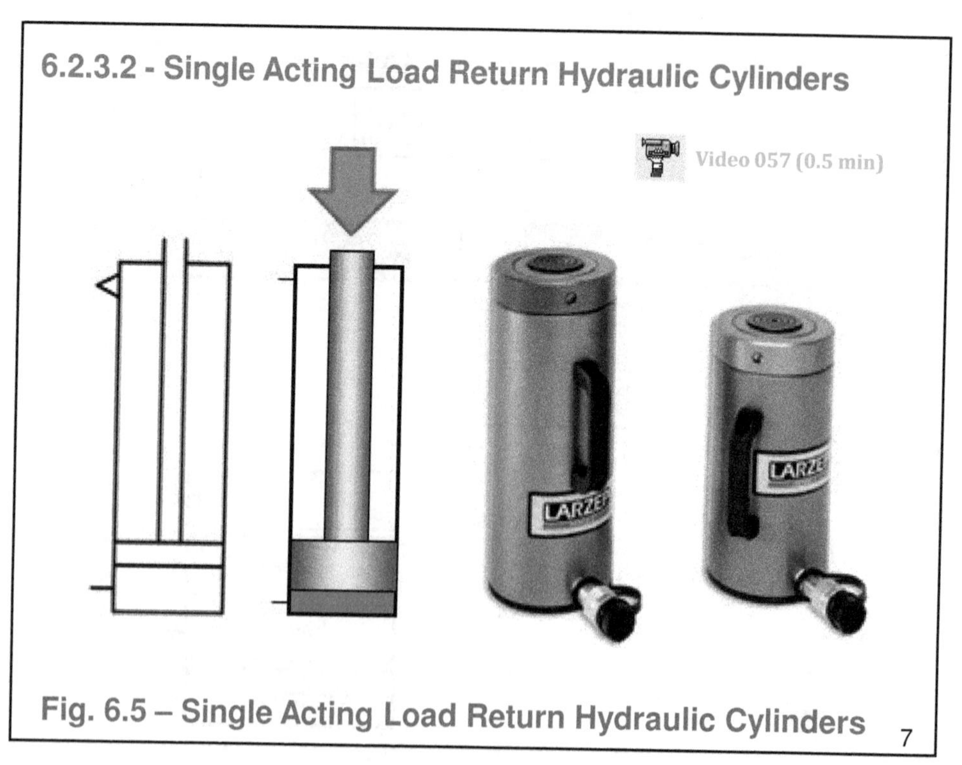

Fig. 6.5 – Single Acting Load Return Hydraulic Cylinders

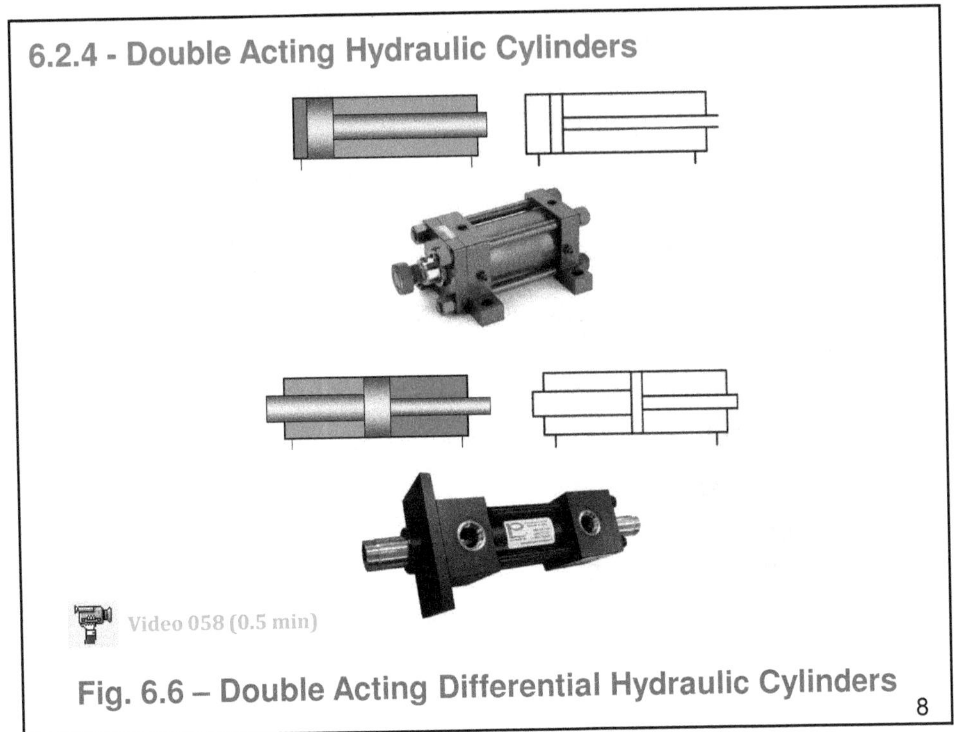

Fig. 6.6 – Double Acting Differential Hydraulic Cylinders

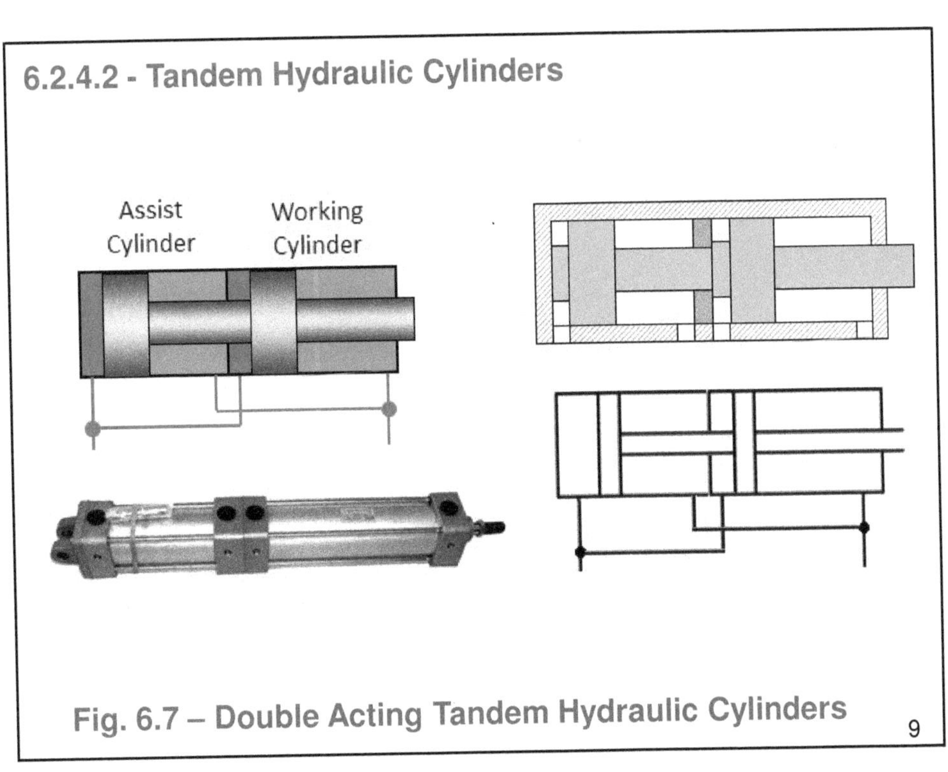

Fig. 6.7 – Double Acting Tandem Hydraulic Cylinders

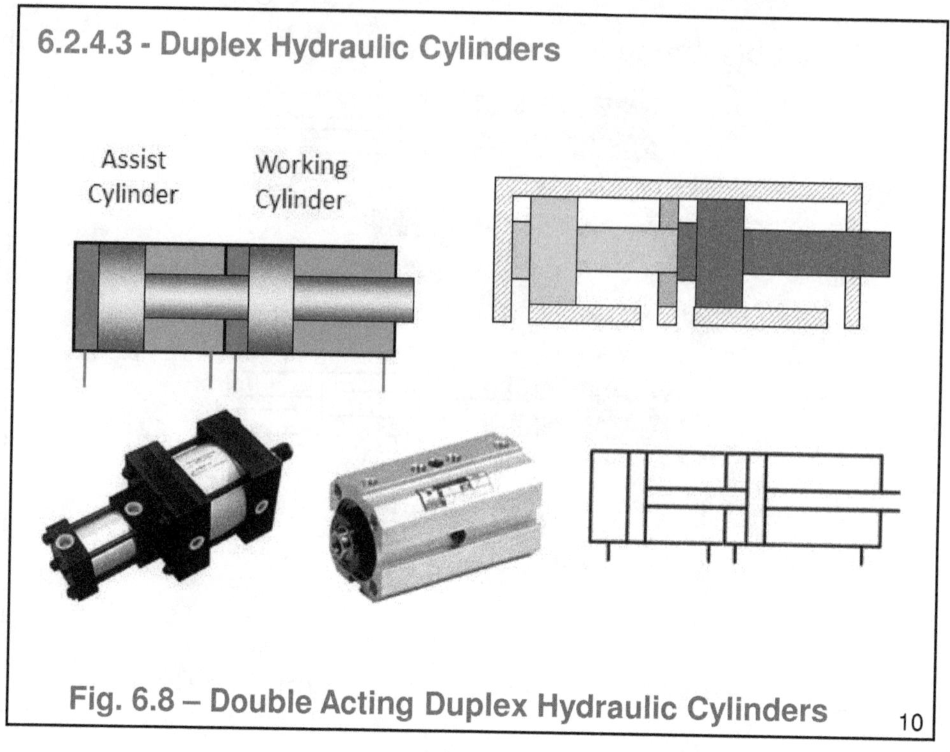

Fig. 6.8 – Double Acting Duplex Hydraulic Cylinders

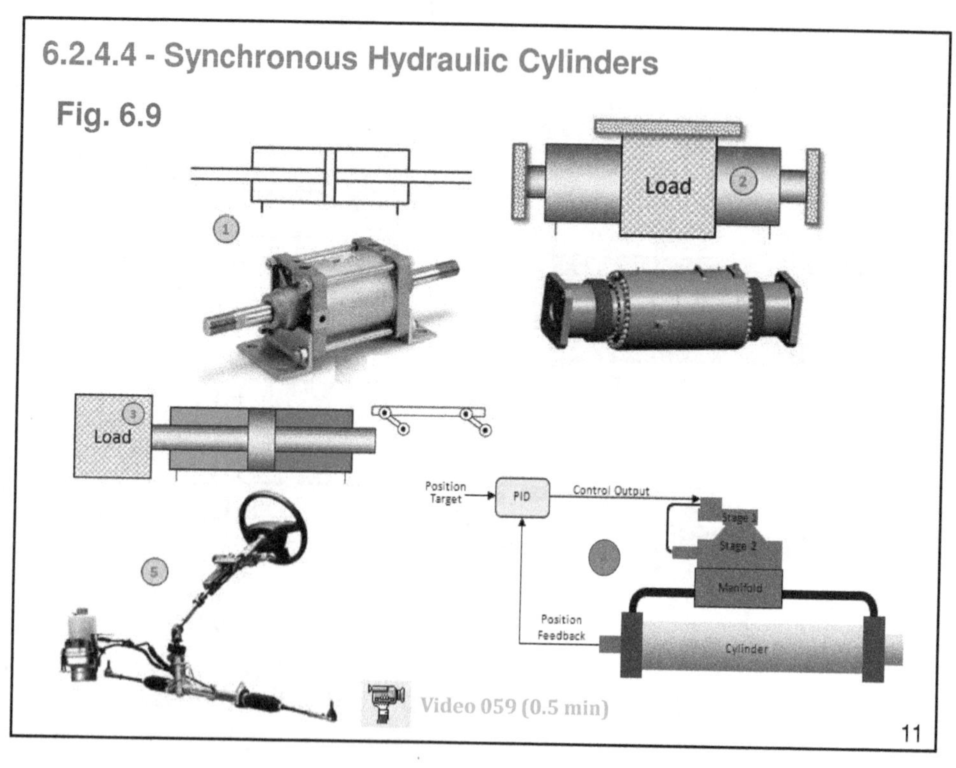

Fig. 6.9

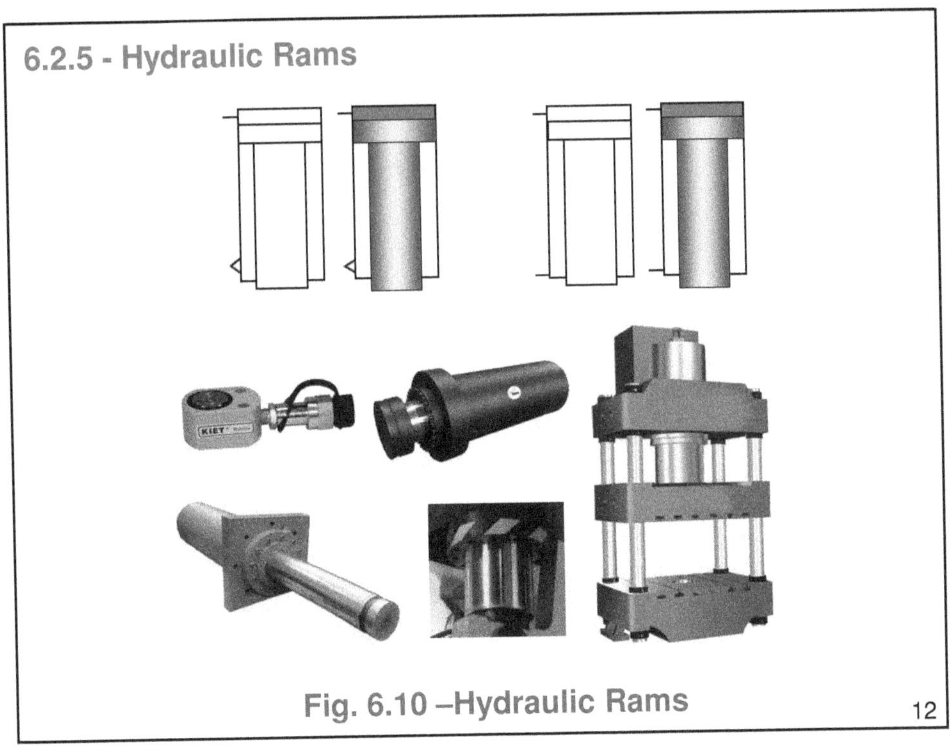

Fig. 6.10 –Hydraulic Rams

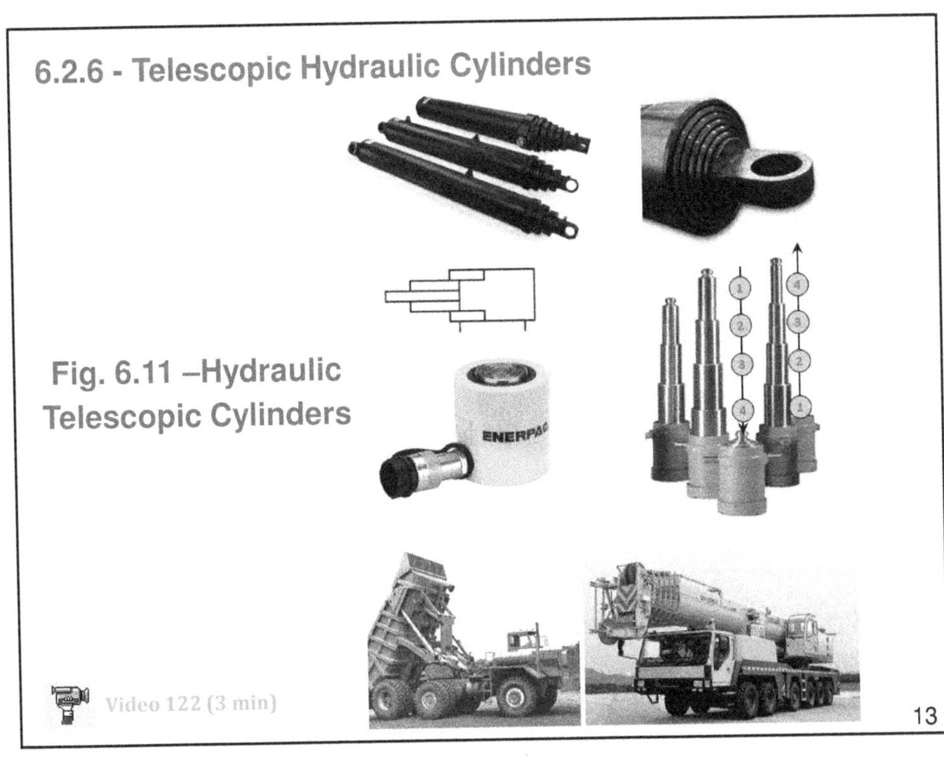

Fig. 6.11 –Hydraulic Telescopic Cylinders

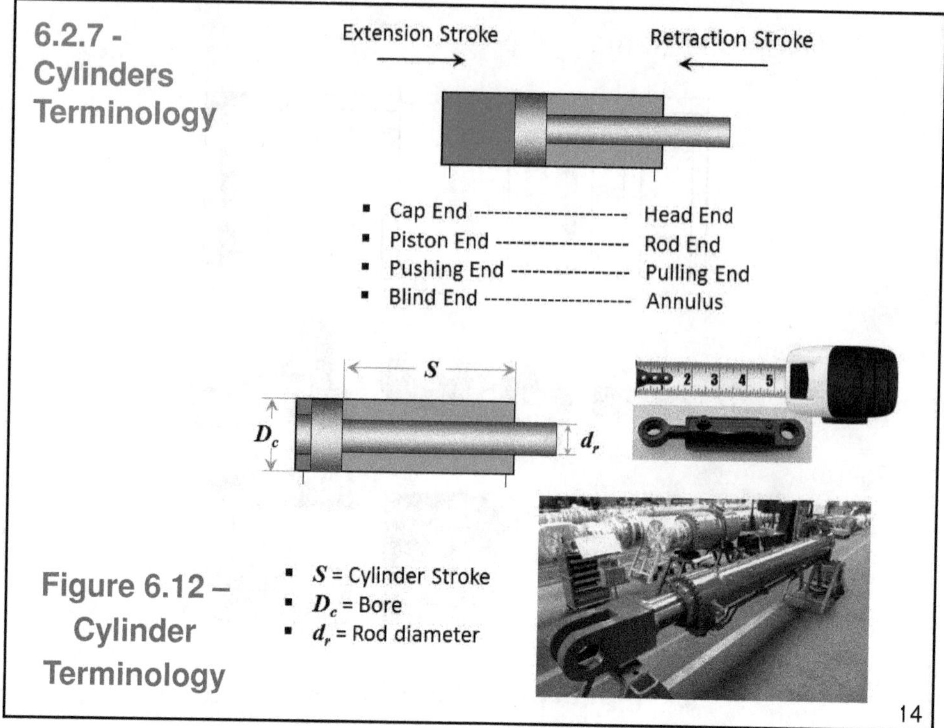
6.2.7 -
Cylinders
Terminology
Extension Stroke
Retraction Stroke
Cap End ---------------------- Head End
Piston End ------------------- Rod End
Pushing End ----------------- Pulling End
Blind End -------------------- Annulus
S
D_c
d_r
Figure 6.12 – Cylinder Terminology
S = Cylinder Stroke
D_c = Bore
d_r = Rod diameter
14

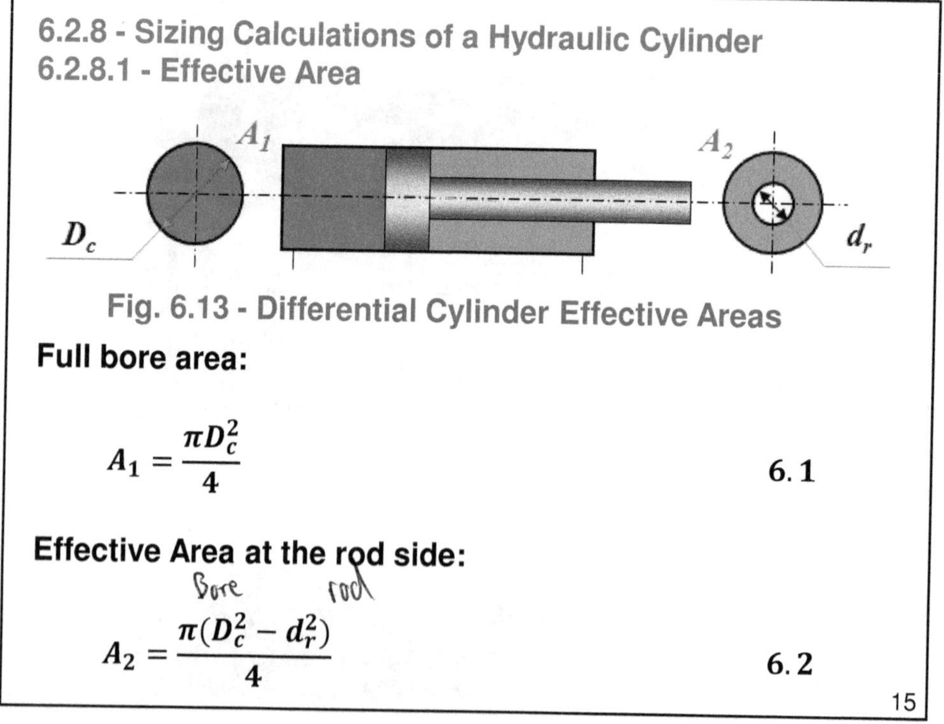
6.2.8 - Sizing Calculations of a Hydraulic Cylinder
6.2.8.1 - Effective Area
A_1
A_2
D_c
d_r
Fig. 6.13 - Differential Cylinder Effective Areas
Full bore area:
$$A_1 = \frac{\pi D_c^2}{4} \quad 6.1$$
Effective Area at the rod side:
Bore
rod
$$A_2 = \frac{\pi(D_c^2 - d_r^2)}{4} \quad 6.2$$
15

6.2.8.2 - Pressure Requirements of a Hydraulic Cylinder

In Metric System of Units:

$$Working\ Pressure = p\ (bar)$$

$$= \frac{F\ (N) \times 10^{-5}}{A\ (m^2)} = \frac{F\ (kg) \times 9.81}{A\ (cm^2) \times 10} \qquad 6.3.A$$

In English System of Units:

$$Working\ Pressure = p\ (psi)$$

$$= \frac{F\ (lb)}{A\ (in^2)} \qquad 6.3.B$$

16

6.2.8.3 - Flow Requirements of a Hydraulic Cylinder

In Metric System of Units:

$$Q\left[\frac{l}{min}\right] = \frac{v\ [cm/s] \times A[cm^2] \times 60}{1000} \qquad 6.4.A$$

In English System of Units:

$$Q[gpm] = \frac{v\ [fps] \times A[in^2]}{0.321} \qquad 6.4.B$$

Which area to be used in the equations?

17

6.2.8.4 - Effect of Area Ratio on Working Pressure

Assuming:

- **The cylinder extends at constant speed against no external load.**
- **The return oil from the rod side is restricted.**
- **Neglect the pressure drop across the valve and hydraulic lines.**

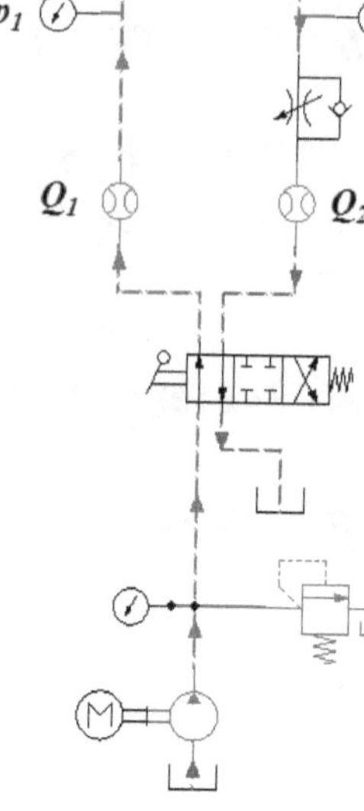

Fig. 6.14 - Effect of Area Ratio on Working Pressure

Which pressure is higher, p_1 or p_2?

18

Pressure Intensification:

$$p_1 \times A_1 = p_2 \times A_2$$

$$\Rightarrow p_2 = p_1 \frac{A_1}{A_2} \qquad 6.5.A$$

In case the cylinder experiences an external force F:

$$p_1 \times A_1 = p_2 \times A_2 + F$$

$$\Rightarrow p_2 = \frac{(p_1 \times A_1) - F}{A_2} \qquad 6.5.B$$

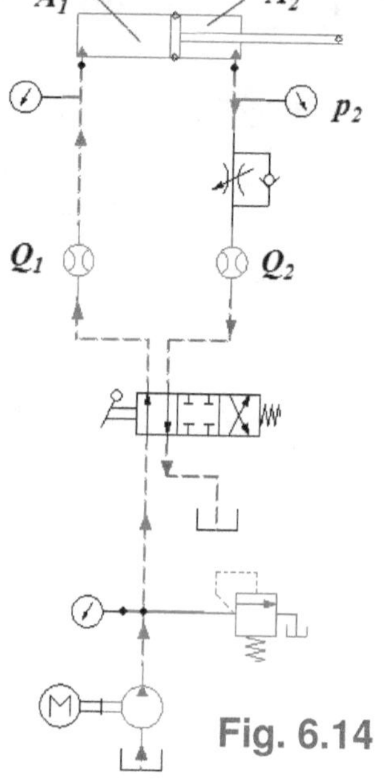

Fig. 6.14

Video 386 (3 min)

19

6.2.8.5 - Effect of Area Ratio on Working Flow

$$\frac{Q_1}{A_1} = \frac{Q_2}{A_2} \Rightarrow Q_2 = Q_1 \frac{A_2}{A_1} \qquad 6.6$$

If pump discharges 40 lit/min, what is the best selection for the size of the directional valve?

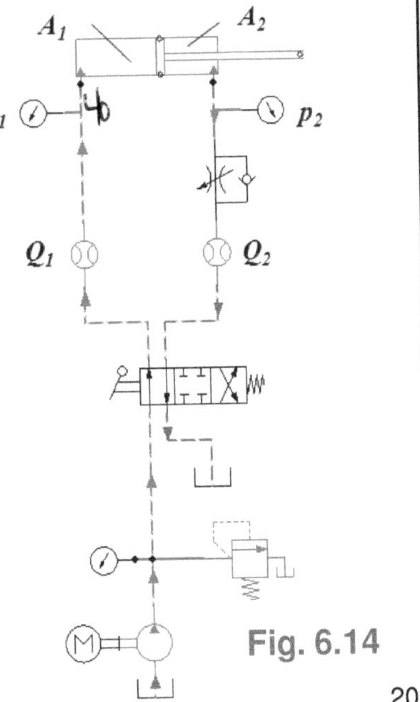

Fig. 6.14

- **Valve size "A" that handles flow up to 30 lit/min?**
- **Valve size "B" that handles flow from 30-60 lit/min?**

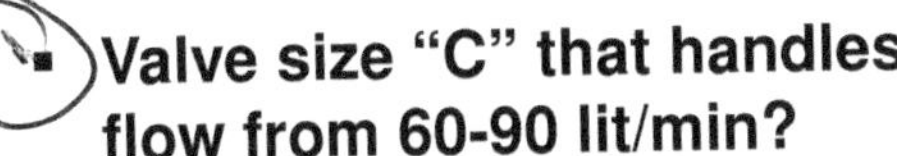

- **Valve size "C" that handles flow from 60-90 lit/min?**

What about Motors & Synchronous Cylinders?

20

Differential cylinders have more flow when retracted

6.2.8.6 - Case Study of Sizing a Hydraulic Cylinder

Given a differential cylinder of:

- **Area Ratio = 2:1; Bore: D_c = 100 mm.**
- **Cylinder Stroke: S = 1000 mm;**
- **Rod Side Pressure: p_2 = 20 bar.**
- **External Force: F = 10 Tons.**
- **Seal Friction: k_{vf} = 10 N/(cm/s); Q_1 = 48 lit/min.**

Solution:

- **Area Ratio = 2:1 → d_r = 70.71 mm**
- **Equation 6.1 → A_1 = 78.5 cm^2**
- **Equation 6.2 → A_2 = 39.25 cm^2**
- **Equation 6.4.A → v = 10.19 cm/s. So, the cylinder takes 9.81 seconds to complete the stoke.**
- **Equation 6.5.B → p_1 = 135.1 bar. Note that seal friction force is added.**
- **Equation 6.6 → Q_2 = 24 lit/min.**

21

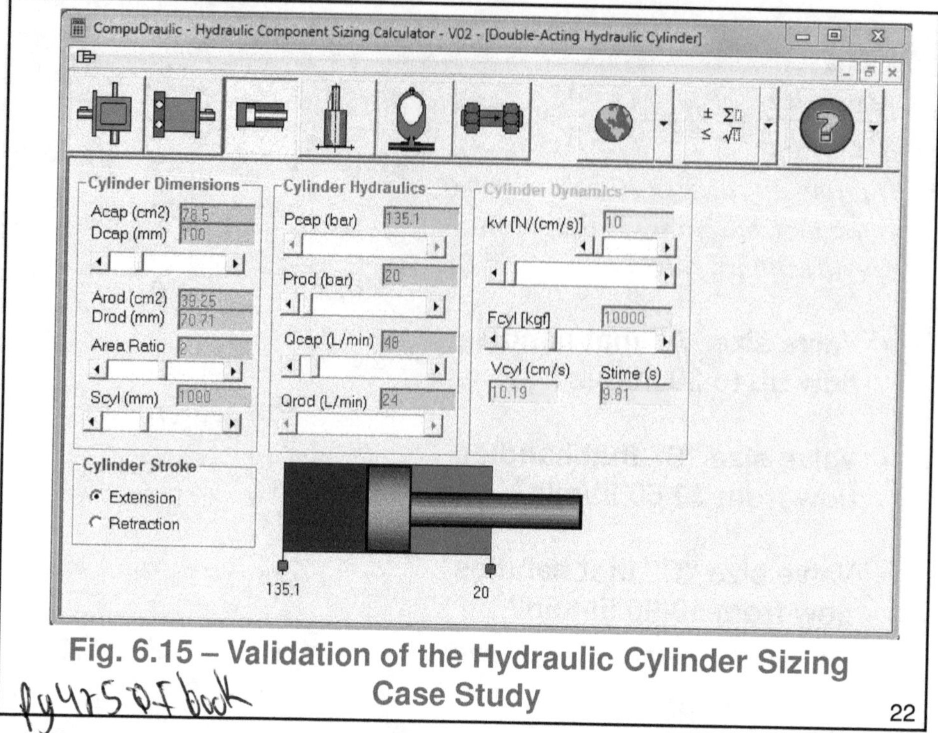

Fig. 6.15 – Validation of the Hydraulic Cylinder Sizing Case Study

6.2.9- Body Styles of Hydraulic Cylinders

6.2.9.1 - Tie-Rod Cylinders

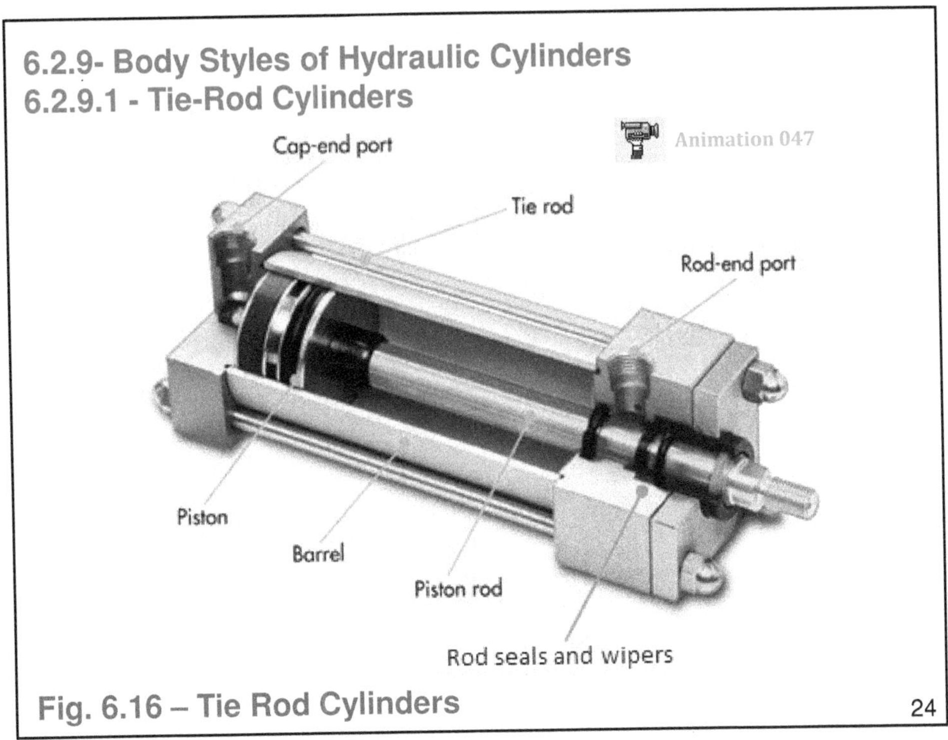

Fig. 6.16 – Tie Rod Cylinders

24

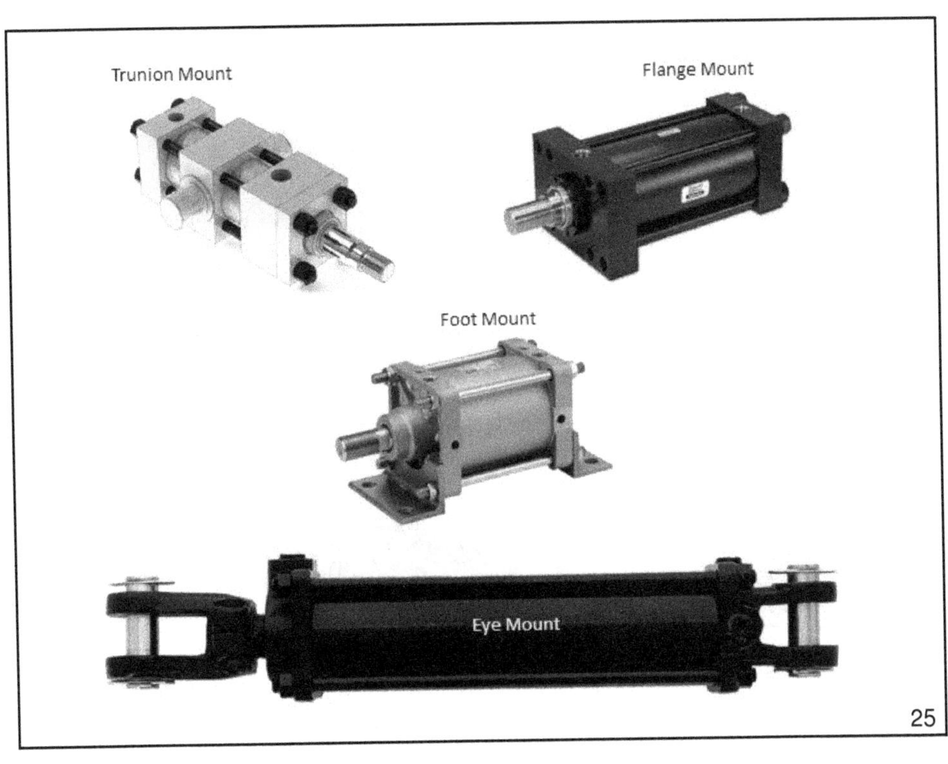

25

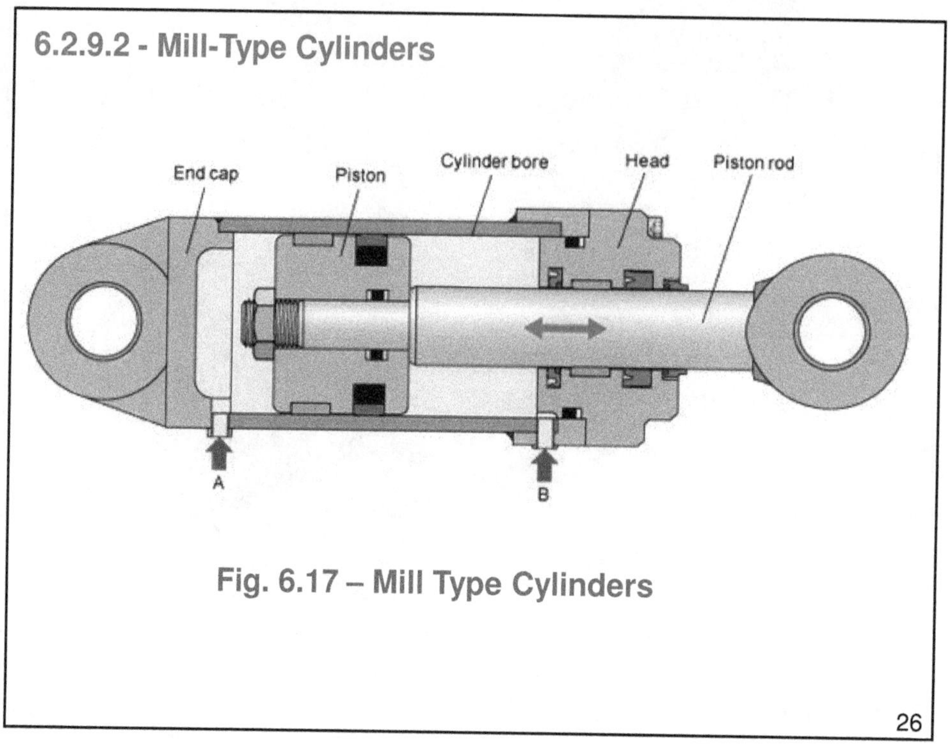
6.2.9.2 - Mill-Type Cylinders
End cap
Piston
Cylinder bore
Head
Piston rod
A
B
Fig. 6.17 – Mill Type Cylinders
26

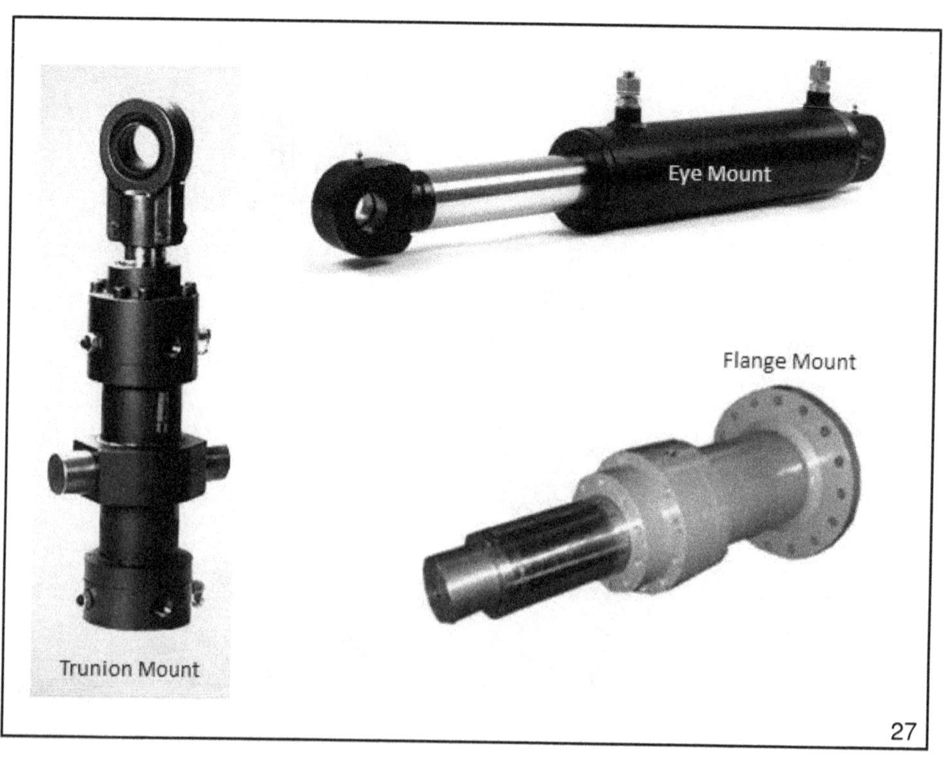
Eye Mount
Flange Mount
Trunion Mount
27

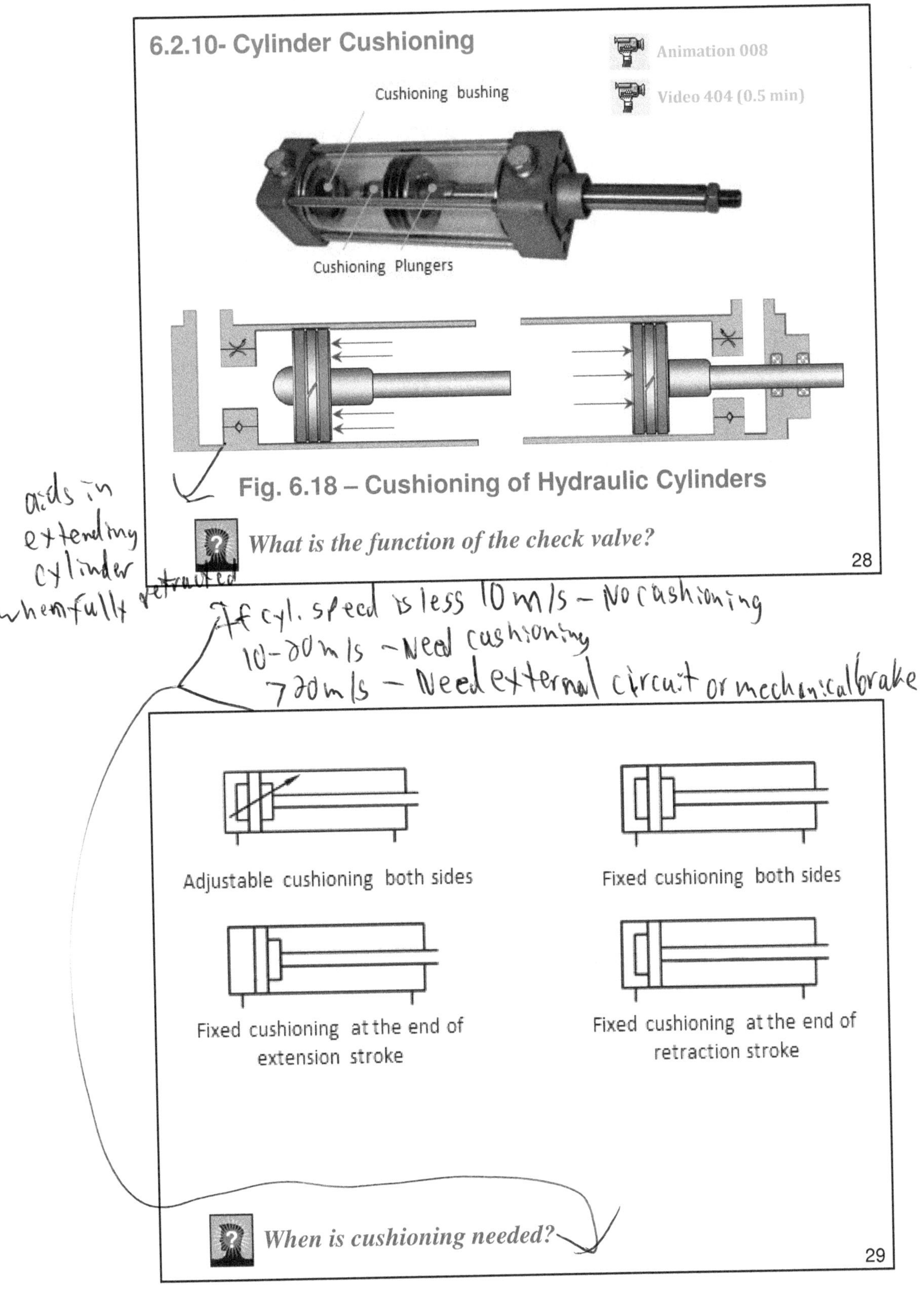
6.2.10- Cylinder Cushioning
Animation 008
Video 404 (0.5 min)
Cushioning bushing
Cushioning Plungers
Fig. 6.18 – Cushioning of Hydraulic Cylinders
What is the function of the check valve?
28
aids in extending cylinder when fully retracted
If cyl. speed is less 10 m/s – No cushioning
10-20 m/s – Need cushioning
>20 m/s – Need external circuit or mechanical brake
Adjustable cushioning both sides
Fixed cushioning both sides
Fixed cushioning at the end of extension stroke
Fixed cushioning at the end of retraction stroke
When is cushioning needed?
29

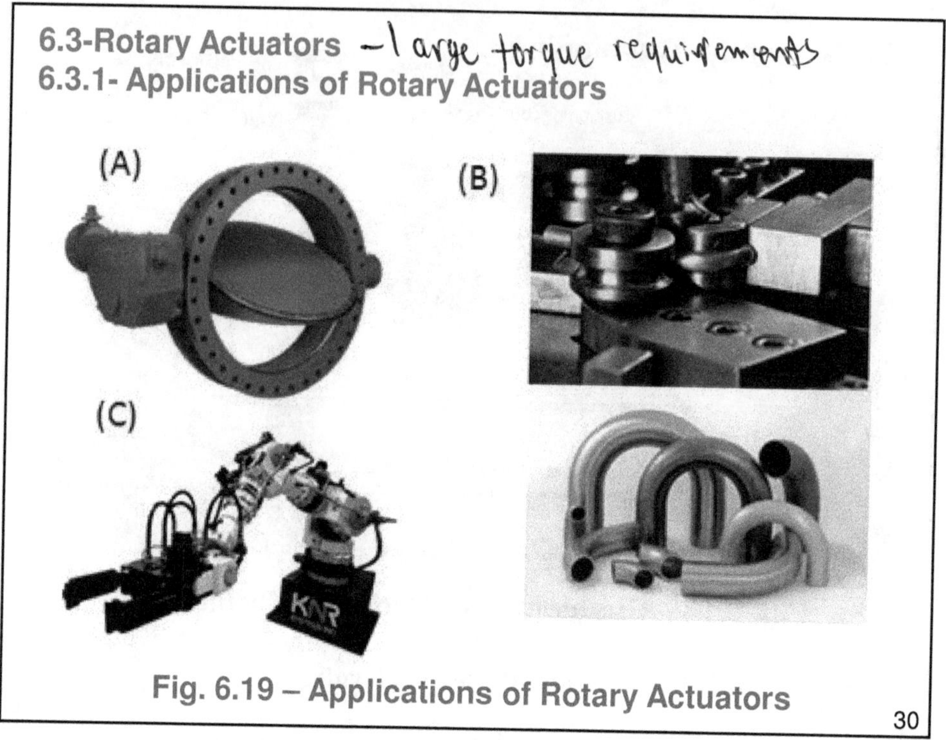

Fig. 6.19 – Applications of Rotary Actuators

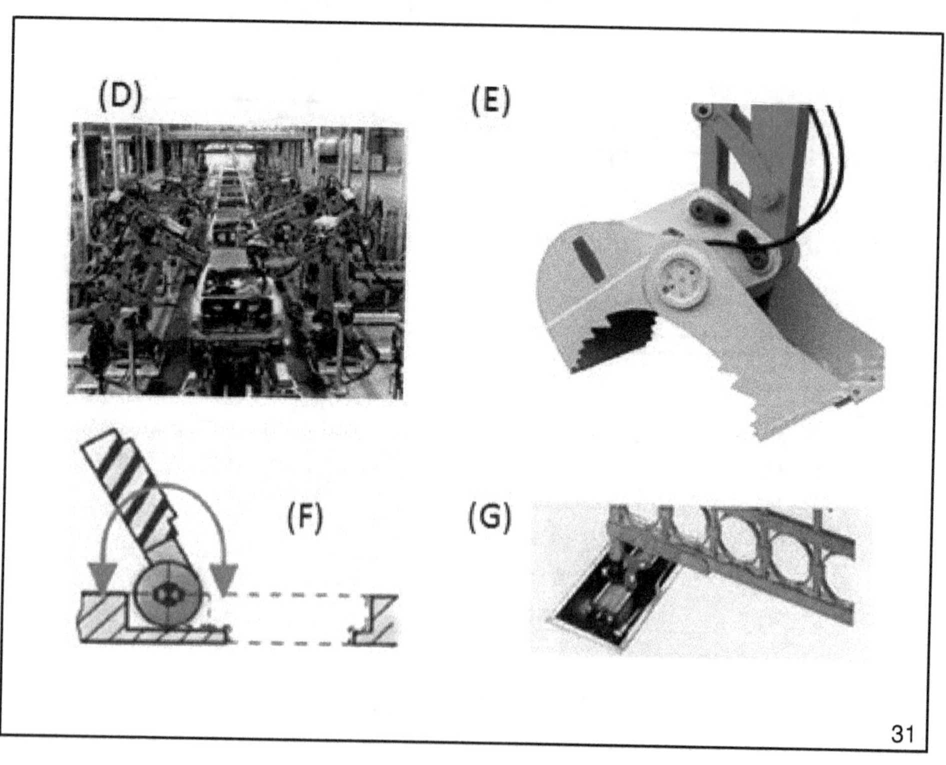

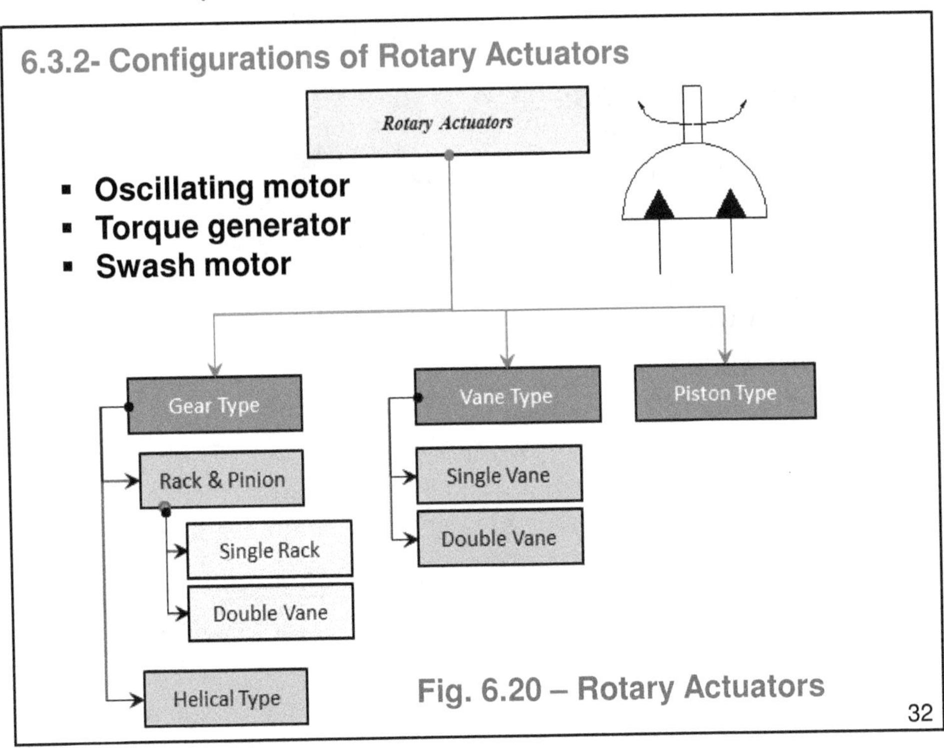

Fig. 6.20 – Rotary Actuators

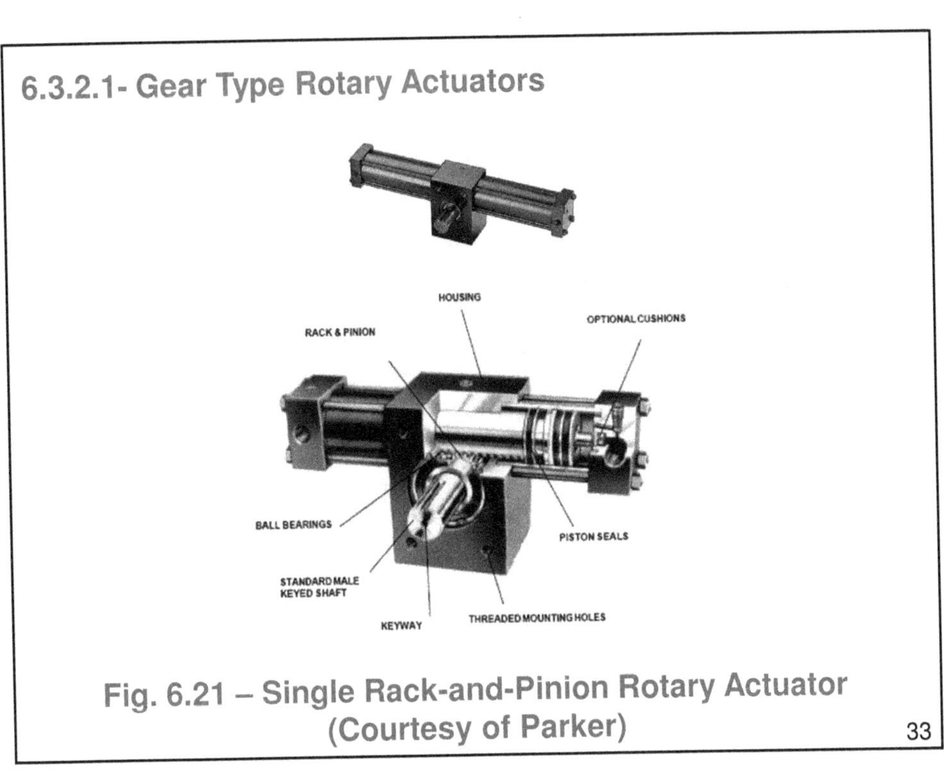

Fig. 6.21 – Single Rack-and-Pinion Rotary Actuator (Courtesy of Parker)

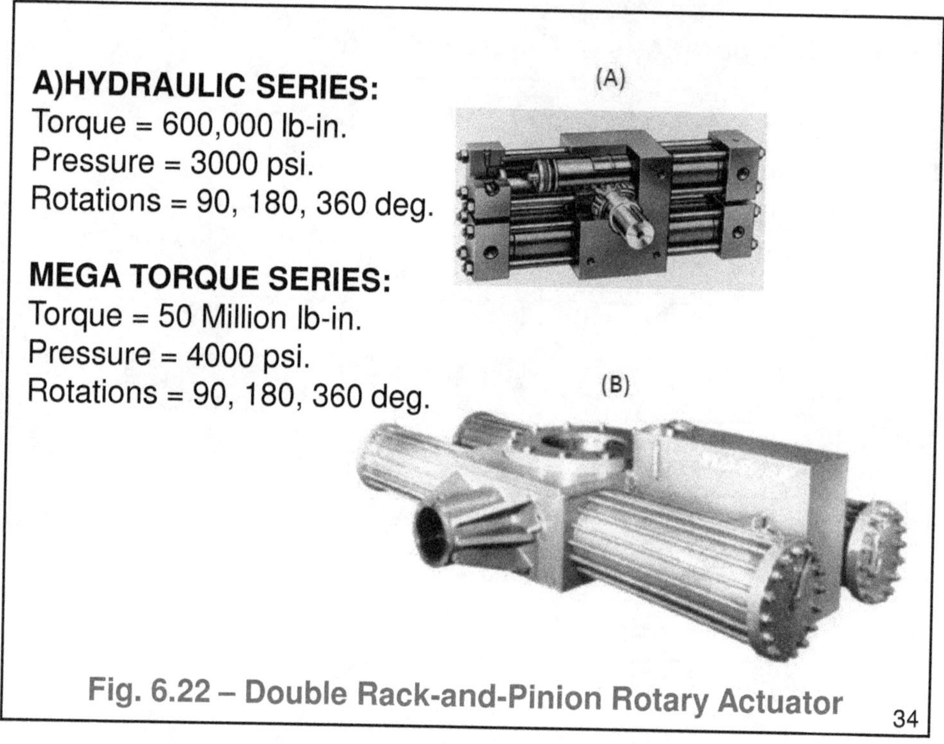

Fig. 6.22 – Double Rack-and-Pinion Rotary Actuator

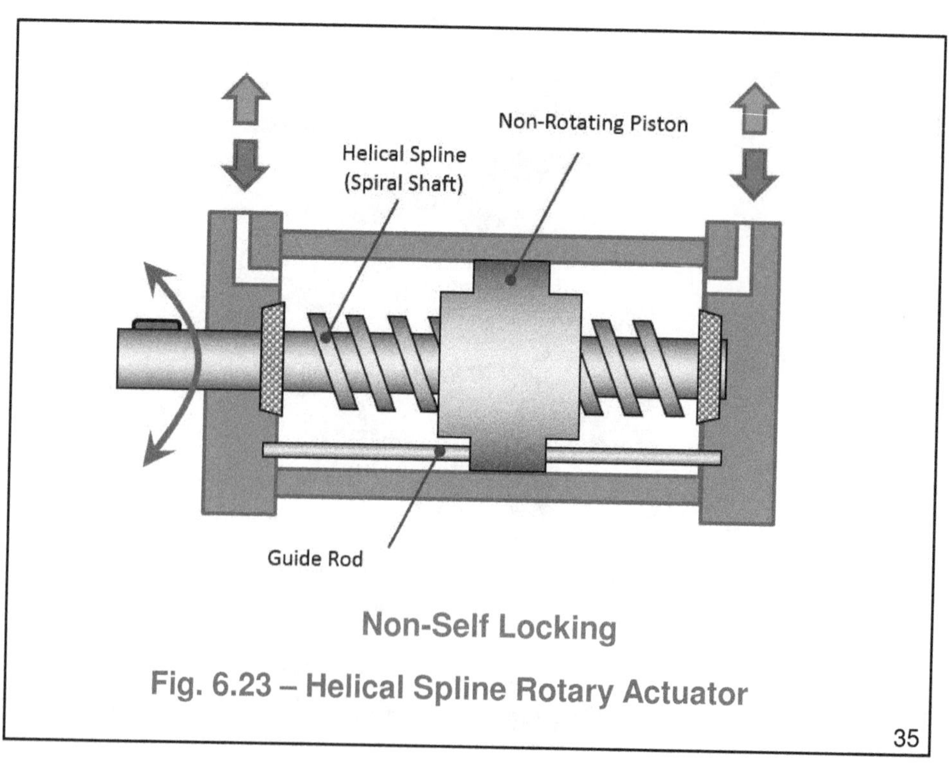

Fig. 6.23 – Helical Spline Rotary Actuator

Product Advantages

Powerful

- High Torque
- High Bearing

Drift Free Positioning

- Zero Internal Leakage
- Smooth Operation
- No External Brake Required

Backdrives in Overload Conditions

- Hydraulic Fuse
- Prevents Mechanical Damage

Ultra-Compact

- High Power Density
- Fits Tight Spaces

Streamline Design

- Eliminates Linkages and Bearings
- Reduced Bill of Materials

Durable

- Suitable for Harsh Environments
- Moving Parts Enclosed

HELAC

Self-Locking

36

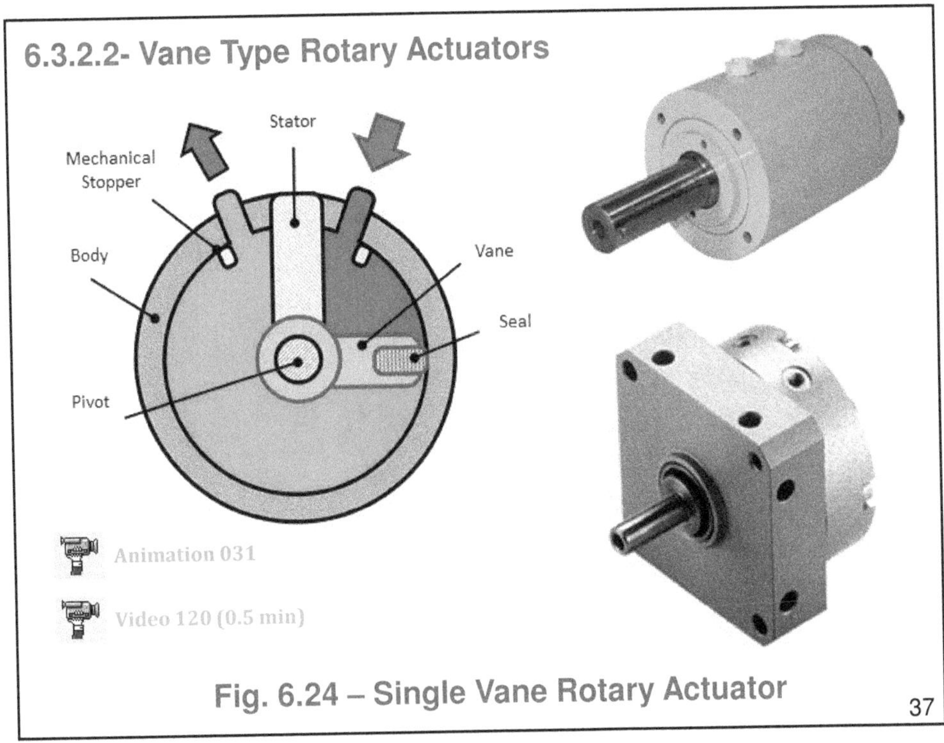

Fig. 6.24 – Single Vane Rotary Actuator

37

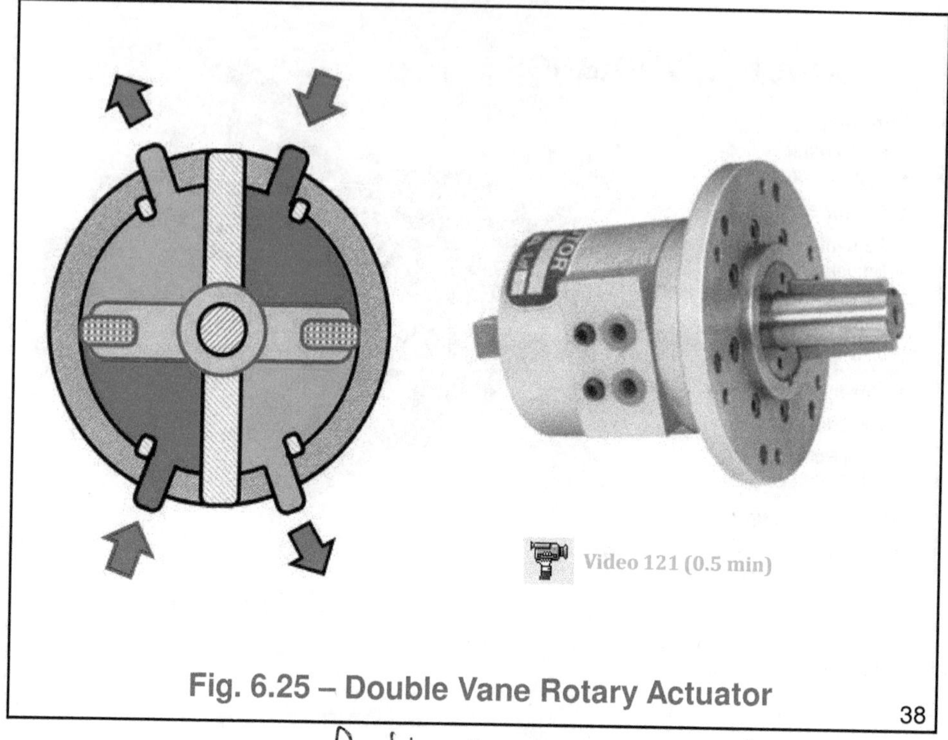

Fig. 6.25 – Double Vane Rotary Actuator

38

Doubles Torque over single vane

6.3.2.3- Piston Type Rotary Actuators

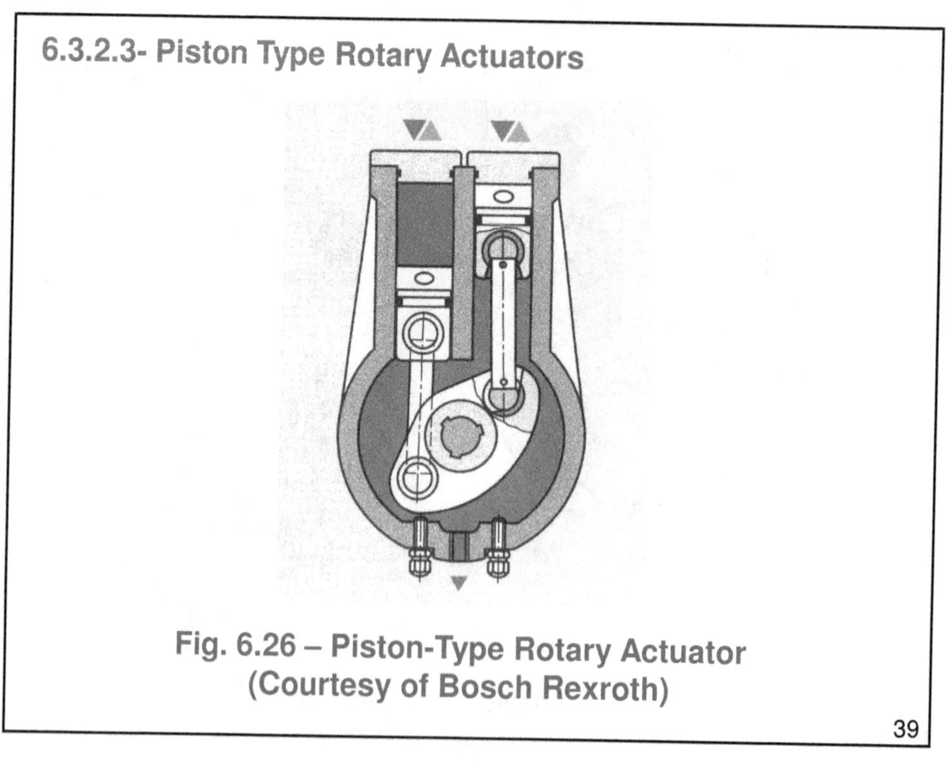

Fig. 6.26 – Piston-Type Rotary Actuator
(Courtesy of Bosch Rexroth)

39

6.3.3- Rotary Actuator Sizing Calculations

6.3.3.1- Pressure Requirements of a Rotary Actuator

In Metric System of Units:

$$\Delta p_m\,[bar] = \frac{20\pi\ x\ T(N.m)}{D_{RA}[cc]} \qquad 6.7.A$$

In English System of Units:

$$\Delta p_m\,[psi] = \frac{2\pi\ x\ T(lb.in)}{D_{RA}[in^3]} \qquad 6.7.B$$

40

6.3.3.2 - Flow Requirements of a Rotary Actuator

In Metric System of Units:

Flow rate   actuator

$$Q_{RA}(Lit/min) = \frac{D_{RA}(cc) \times 60}{t \times 1000} \qquad 6.8.A$$

In English System of Units:

$$Q_{RA}(gpm) = \frac{D_{RA}(in^3) \times 60}{t \times 231} \qquad 6.8.A$$

41

Chapter 6 Reviews

1. Which application most commonly uses **synchronous cylinders**?
   A. Hydraulic presses.
   B. Injection molding machines.
   C. Steering system.
   D. Hydraulic elevators.

2. Which equation represents the **force balance** on the shown **differential cylinder?**
   A. $P_1A_1 = P_2A_2 - F +$ (mass x acceleration).
   B. $P_1A_1 = P_2A_2 + F -$ (mass x acceleration)
   C. $P_1A_1 = P_2A_2 + F +$ (mass x acceleration).
   D. None of the above.

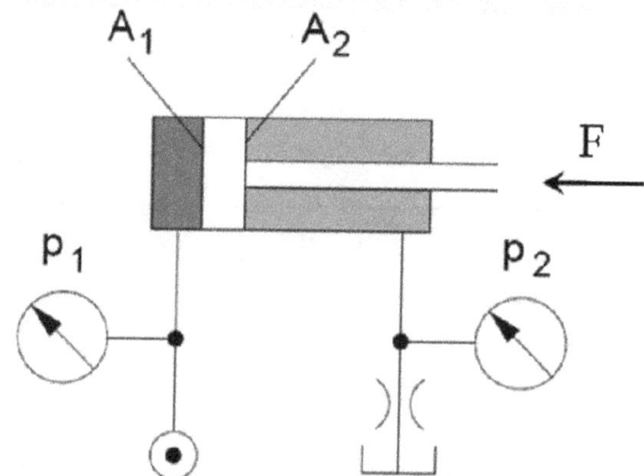

3. In a differential cylinder with an **area ratio 2:1**, if the piston side is receiving 10 L/min, how much flow will exit the rod side of the cylinder?
   A. 10 L/min.
   B. 20 L/min.
   C. 5 L/min.
   D. 15 L/min.

4. In an application that is using a **differential cylinder with an area ratio 2:1**, if the pump flowrate is 10 L/min, what size return line filter should be used?
   A. Filter with rated flow of maximum 5 L/min.
   B. Filter with rated flow of maximum 15 L/min.
   C. Filter with rated flow of maximum 30 L/min.
   D. None of the above

5. What is the maximum velocity above which the cylinder designer must consider adding **cushioning** on both sides of cylinder piston?
   A. 10 cm/s.
   B. 10 in/s.
   C. 20 cm/s.
   D. 30 cm/s.

6. Based on cylinder symbol shown below, the **cushioning arrangement** is?
   A. Adjustable cushioning for the end of the extension stroke.
   B. Adjustable cushioning for the end of the retraction stroke.
   C. Adjustable cushioning for the end of the both strokes.
   D. Fixed cushioning for the end of the both strokes.

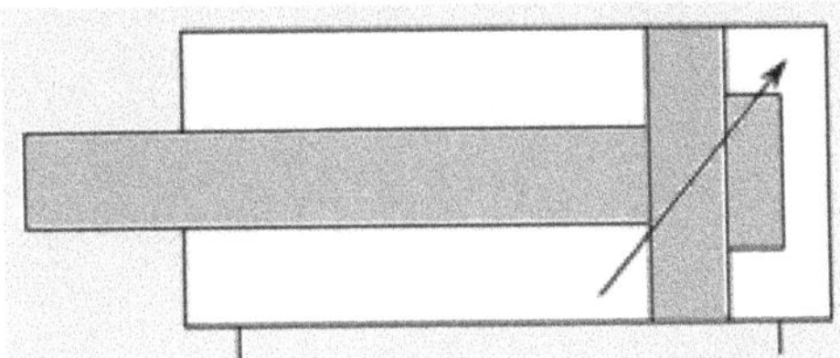

7. Which actuator is the best for an application requiring torque output for a limited angular displacement?
   A. Differential cylinder.
   B. Hydraulic motor.
   C. Rotary actuator.
   D. Synchronous cylinder.

8. For a **telescopic cylinder**, which of the following statements are True?
   A. Larger stage extends first.
   B. Smaller stage extends faster.
   C. Small stage retracts first.
   D. All above sentences are true.

9. A rotary actuator could be of a?
   A. Swash plate type.
   B. Variable displacement type.
   C. Vane type.
   D. Differential type.

10. A helical spline rotary actuator has the following advantage over the rack and pinion type?
    A. Larger angular displacement.
    B. Self-locking capability.
    C. Larger torque.
    D. Higher working pressure.

Chapter 6 Assignment

Student Name: -- Student ID: ------------------

Date: -- Score: -----------------------

In the circuit below, if the area ratio is 2:1, find the missing pressure and flow readings. Also find the maximum flow in the circuit?

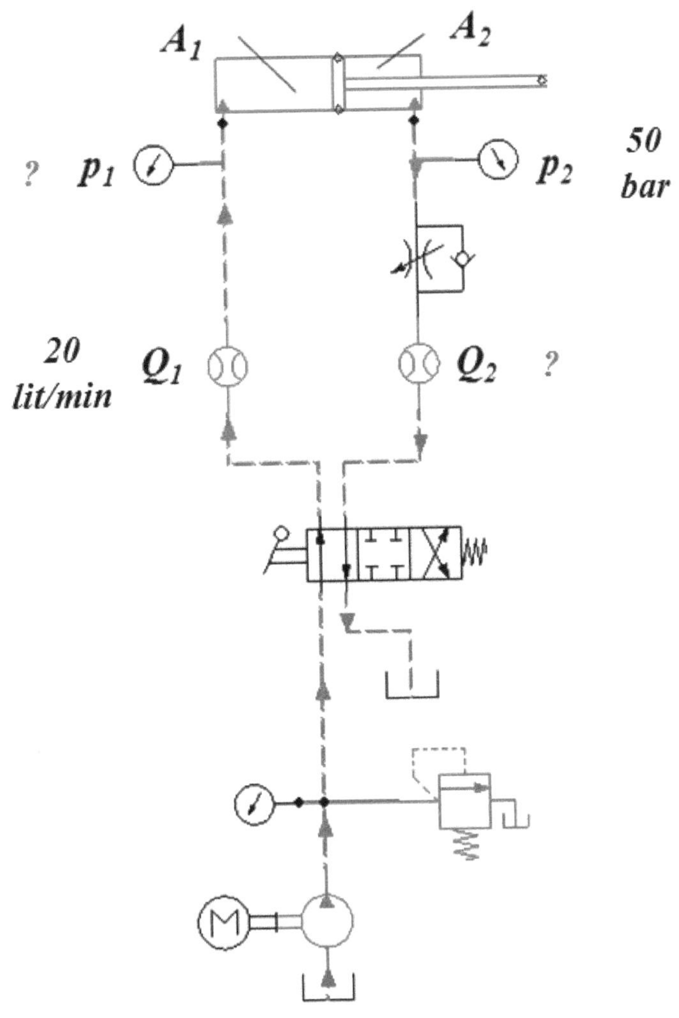

Objectives:

This chapter covers the construction, operating principles and symbols of the three types of accumulators: piston, bladder and diaphragm. The chapter discusses the applications of energy storage and shock absorption using accumulators.

Brief Contents:

7.1- Operating Principles of Hydraulic Accumulators.
7.2- Hydraulic Accumulators Sizing Calculations.
7.3- Construction of Hydraulic Accumulators.

0

0

7.1- Operating Principles of Hydraulic Accumulators

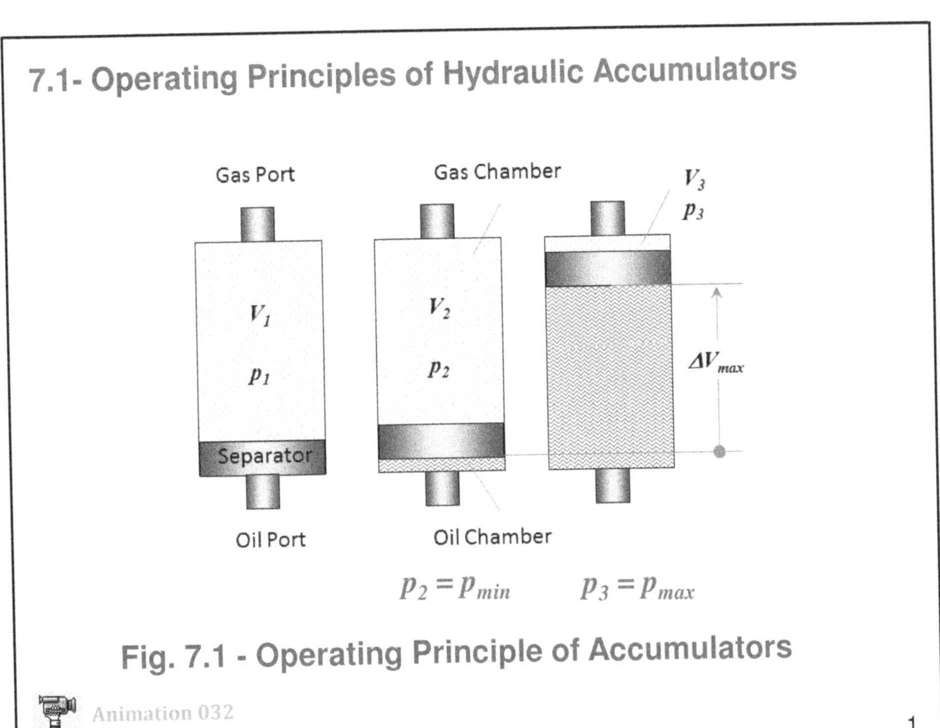

Fig. 7.1 - Operating Principle of Accumulators

Animation 032

1

1

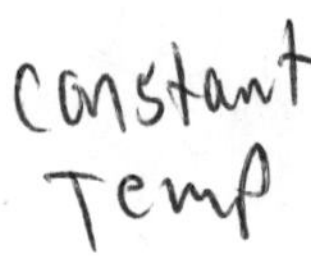

7.2- Hydraulic Accumulators Sizing Calculations

Isothermal (Slow) Process: $p \times V = Constant$

$$p_1 \times V_1 = p_2 \times V_2 = p_3 \times V_3 \qquad 7.1$$

Adiabatic (Fast) Process:

$$\frac{p \times V}{T} = Constant \Rightarrow \frac{p_1 \times V_1}{T_1} = \frac{p_2 \times V_2}{T_2} = \frac{p_3 \times V_3}{T_3} \qquad 7.2.A$$

$$p \times V^{\gamma} = Constant \Rightarrow p_1 \times V_1^{\gamma} = p_3 \times V_3^{\gamma} = p_1 \times V_3^{\gamma} \qquad 7.2.B$$

2

2

Absolute Pressure & Temperature:

In Metric System of Units:

$$^{o}K = {}^{o}C + 273 \qquad 7.3.A$$

In English System of Units:

$$^{o}R = {}^{o}F + 459.7 \qquad 7.3.B$$

3

3

Average Absolute Pressure $(p_2 + p_3)/2$		*Adiabatic Exponent*				
		Operating Temperature				
psi	bar	**24 °C**	**38 °C**	**60 °C**	**77 °C**	**93 °C**
		75 °F	**100 °F**	**140 °F**	**170 °F**	**200 °F**
100	7	1.4	1.4	1.4	1.4	1.4
150	10	1.4	1.4	1.4	1.4	1.4
200	14	1.4	1.4	1.4	1.4	1.4
250	17	1.5	1.4	1.4	1.4	1.4
300	21	1.5	1.5	1.5	1.5	1.5
350	24	1.5	1.5	1.5	1.5	1.5
400	28	1.5	1.5	1.5	1.5	1.5
500	34	1.5	1.5	1.5	1.5	1.5

Table 7.1 - Adiabatic Exponent versus Operating Pressures and Temperature 4

4

Average Absolute Pressure $(p_2 + p_3)/2$		*Adiabatic Exponent*				
		Operating Temperature				
psi	bar	**24 °C**	**38 °C**	**60 °C**	**77 °C**	**93 °C**
		75 °F	**100 °F**	**140 °F**	**170 °F**	**200 °F**
600	41	1.5	1.5	1.5	1.5	1.5
700	48	1.5	1.5	1.5	1.5	1.5
800	55	1.6	1.5	1.5	1.5	1.5
900	62	1.6	1.6	1.6	1.5	1.5
1000	69	1.6	1.6	1.6	1.6	1.6
1250	86	1.6	1.6	1.6	1.6	1.6
1500	103	1.7	1.7	1.6	1.6	1.6
2000	138	1.8	1.7	1.7	1.7	1.7
2500	172	1.9	1.8	1.8	1.7	1.7
3000	207	1.9	1.9	1.8	1.8	1.7

5

5

Stored Energy:

In Metric System of Units:

$$Energy\,[kJ) = \frac{p\,(bar) \times 10^5 \times V\,(\frac{lit}{1000})}{1000}$$

$$= \frac{p\,(bar) \times \Delta V\,(lit)}{10} \qquad 7.4.A$$

In English System of Units:

$$Energy\,[lb.in) = p\,(psi) \times \Delta V\,(in^3) \qquad 7.4.B$$

6

6

The following is a case study of sizing an accumulator:

Given values for a piston accumulator:

- p_2 = 100 bar and p_3 = 200 bar.
- Isothermal gas process.
- Accumulator nominal volume, V_1 = 1 liter.
- Accumulator pre-charge pressure, p_1 = 93 bar.

Solution:

- Equation 7.1 → V_2 = 0.93 liter.
- Equation 7.1 → V_3 = 0.46 liter.
- Equation 7.4.A → *Energy* = 9.4 kJ.

7

7

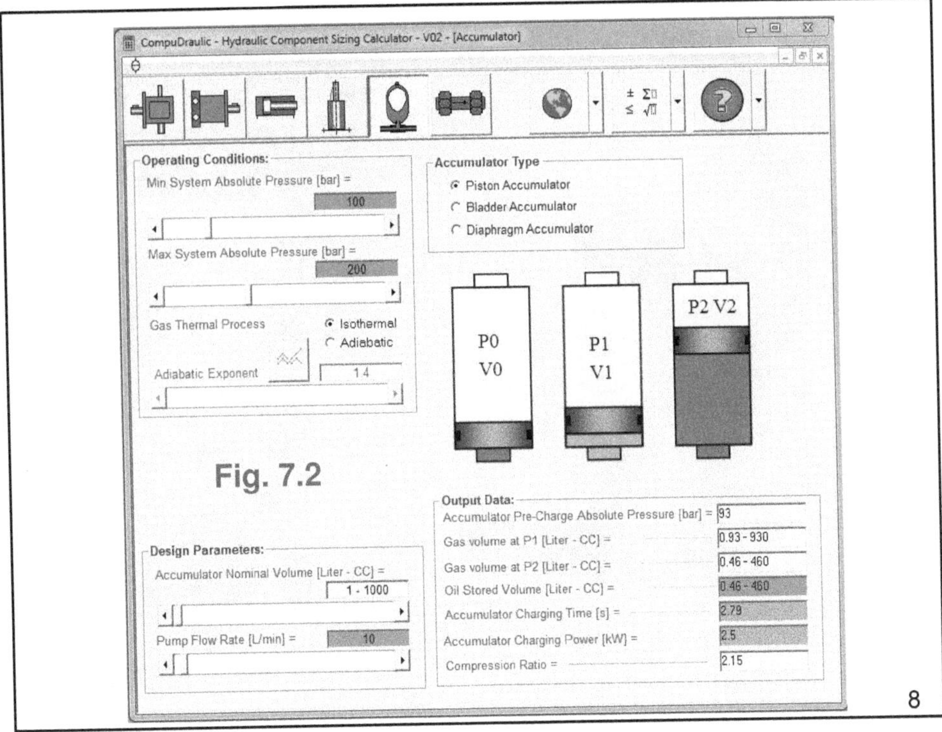

Fig. 7.2

8

8

7.3- Construction of Hydraulic Accumulators

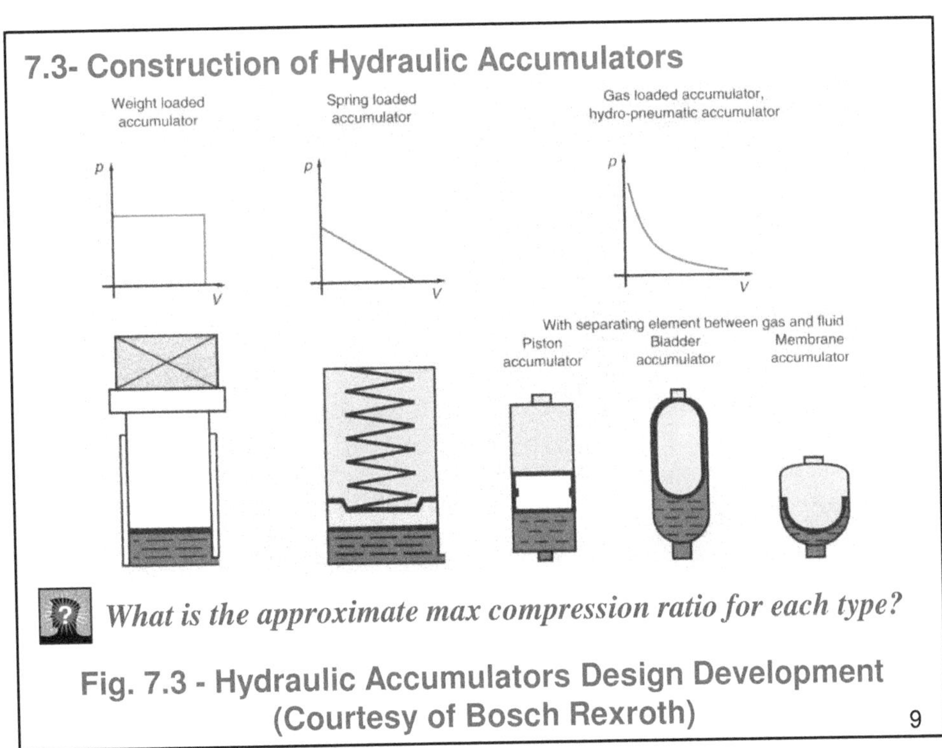

What is the approximate max compression ratio for each type?

Fig. 7.3 - Hydraulic Accumulators Design Development (Courtesy of Bosch Rexroth)

9

9

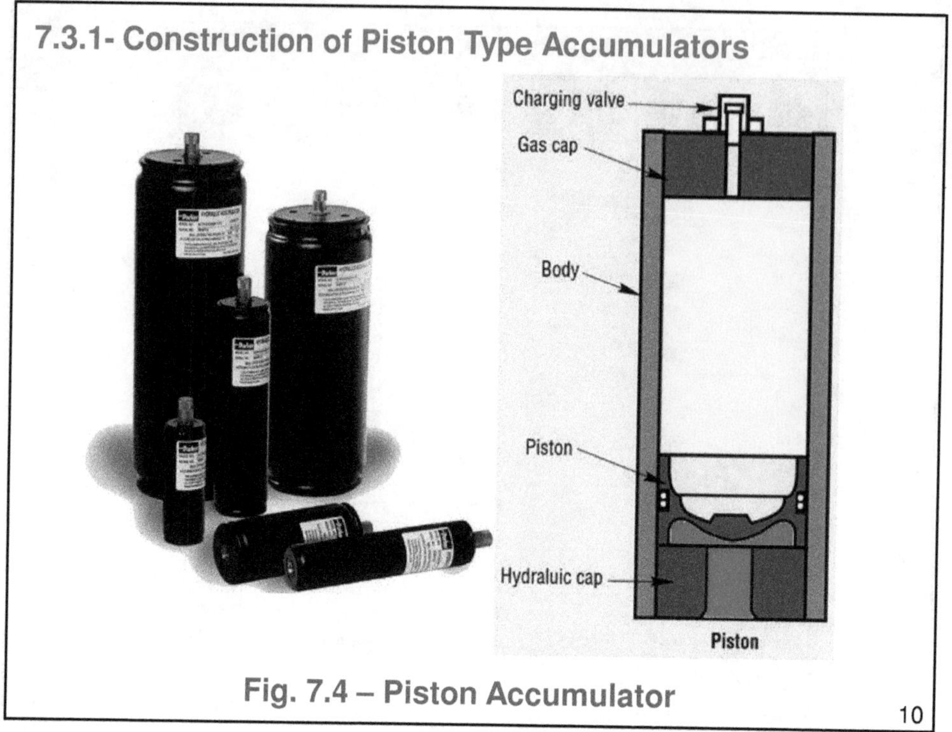

Fig. 7.4 – Piston Accumulator

10

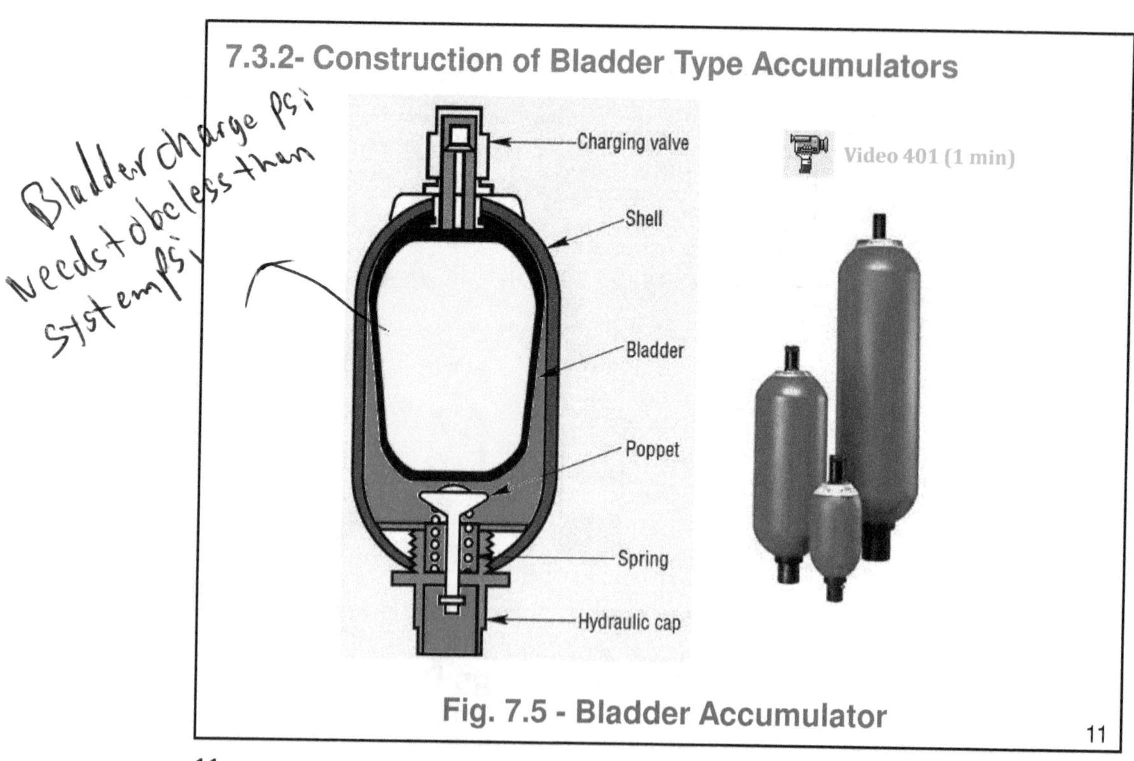

Fig. 7.5 - Bladder Accumulator

11

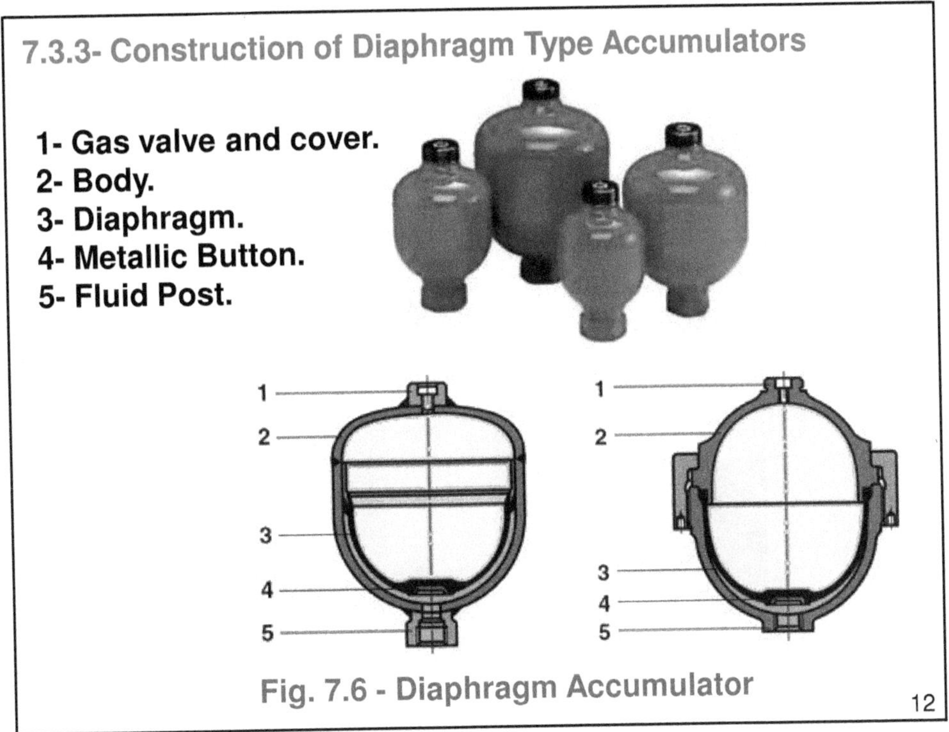
7.3.3- Construction of Diaphragm Type Accumulators
1- Gas valve and cover.
2- Body.
3- Diaphragm.
4- Metallic Button.
5- Fluid Post.
1
2
3
4
5
1
2
3
4
5
Fig. 7.6 - Diaphragm Accumulator
12

12

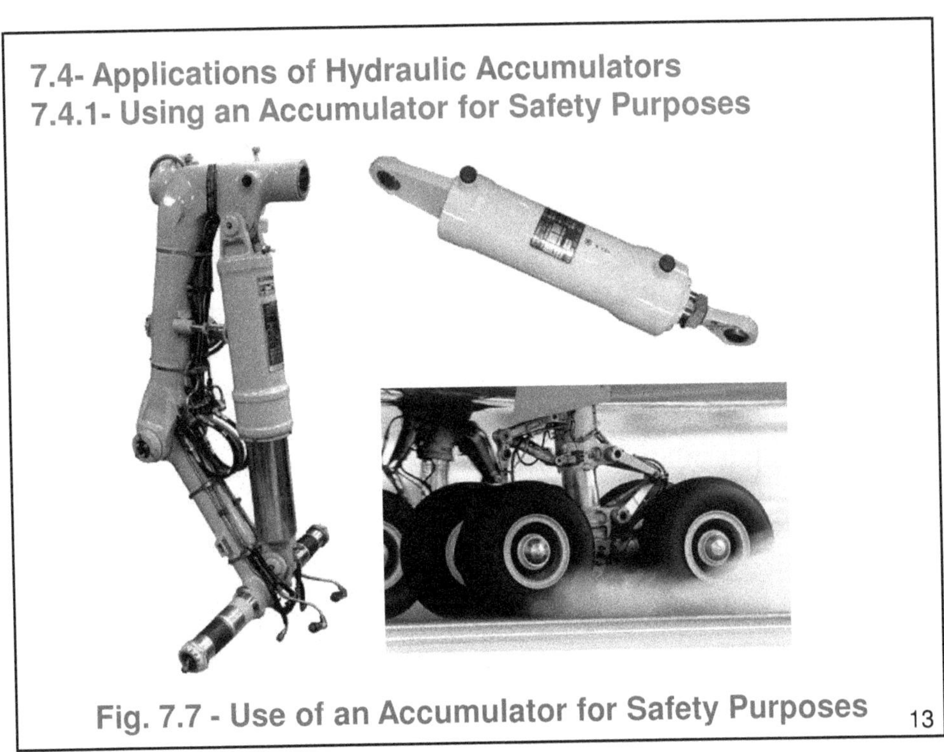
7.4- Applications of Hydraulic Accumulators
7.4.1- Using an Accumulator for Safety Purposes
Fig. 7.7 - Use of an Accumulator for Safety Purposes
13

13

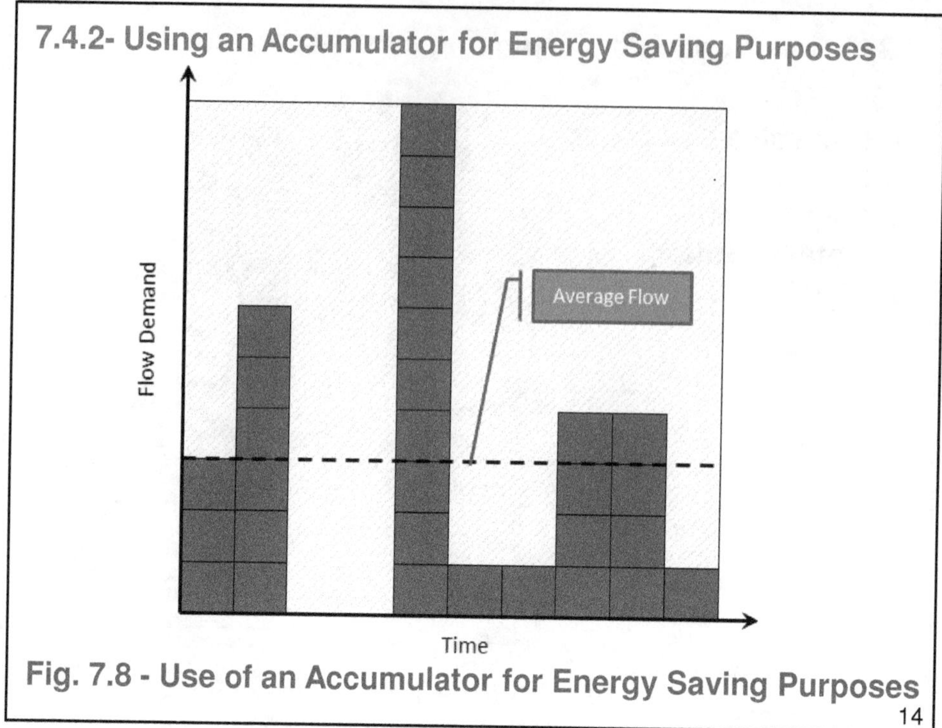

Fig. 7.8 - Use of an Accumulator for Energy Saving Purposes

14

14

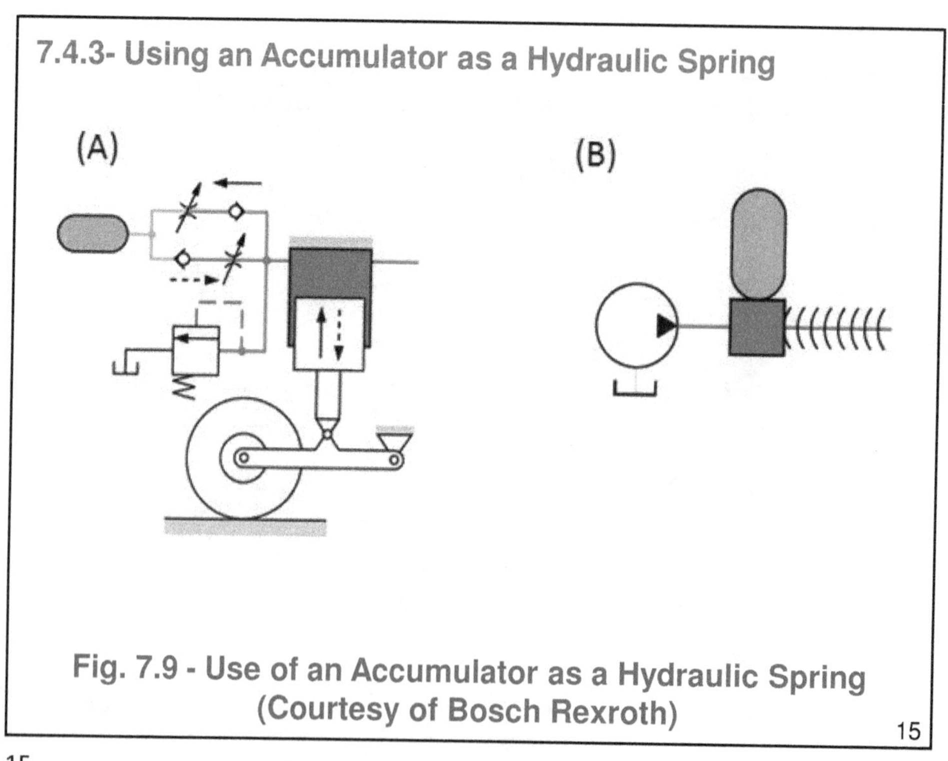

Fig. 7.9 - Use of an Accumulator as a Hydraulic Spring (Courtesy of Bosch Rexroth)

15

15

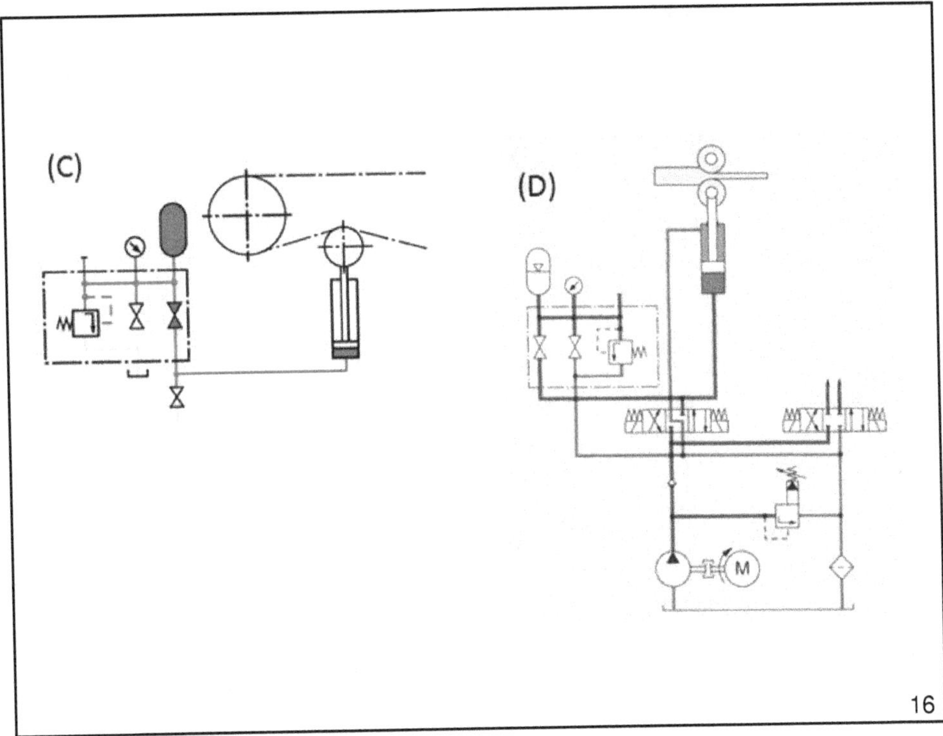

16

Chapter 7 Reviews

1. Which hydraulic component is designed to **store hydraulic energy**?
   A. Accumulator.
   B. Intensifier.
   C. Tank.
   D. Telescopic cylinder.

2. Which accumulator has the **highest compression ratio**?
   A. Diaphragm accumulator.
   B. Bladder accumulator.
   C. Piston accumulator.
   D. None of the above.

3. The **poppet valve** in a bladder accumulator is used to?
   A. Prevent over charging of the accumulator.
   B. Regulate the discharge rate of the accumulator.
   C. Prevent bladder extrusion after the accumulator is fully discharged.
   D. None of the above.

4. The **Isothermal process** means that?
   A. The accumulator charging and discharging process is slow.
   B. The accumulator charging and discharging process is fast.
   C. The accumulator charging and discharging process is accompanied with variable temperature.
   D. None of the above.

5. The **normal compression ratio** of a bladder accumulator is?
   A. 1.5 -2.
   B. 4-5.
   C. 5-9.
   D. None of the above.

Chapter 7 Assignment

Student Name: -- Student ID: -------------------

Date: --- Score: ------------------------

Find the maximum volume of oil stored in an accumulator that has a 1 liter size, initial gas pressure of 50 bar gauge pressure, and maximum pressure is 60 bar gauge pressure. Assume the process is isothermal.

Chapter 8
Hydraulic Circuits for Basic Applications

Objectives:

This chapter covers the basic safety and energy saving requirements of a hydraulic system. The chapter also presents ideas for motion control of a single hydraulic actuator against resistive and overrunning loads, multiple hydraulic actuators in parallel and series, speed control of a hydraulic actuator, boosting speed of a hydraulic actuator, sequence of operation, and hydrostatic transmission.

0

Brief Contents:

1

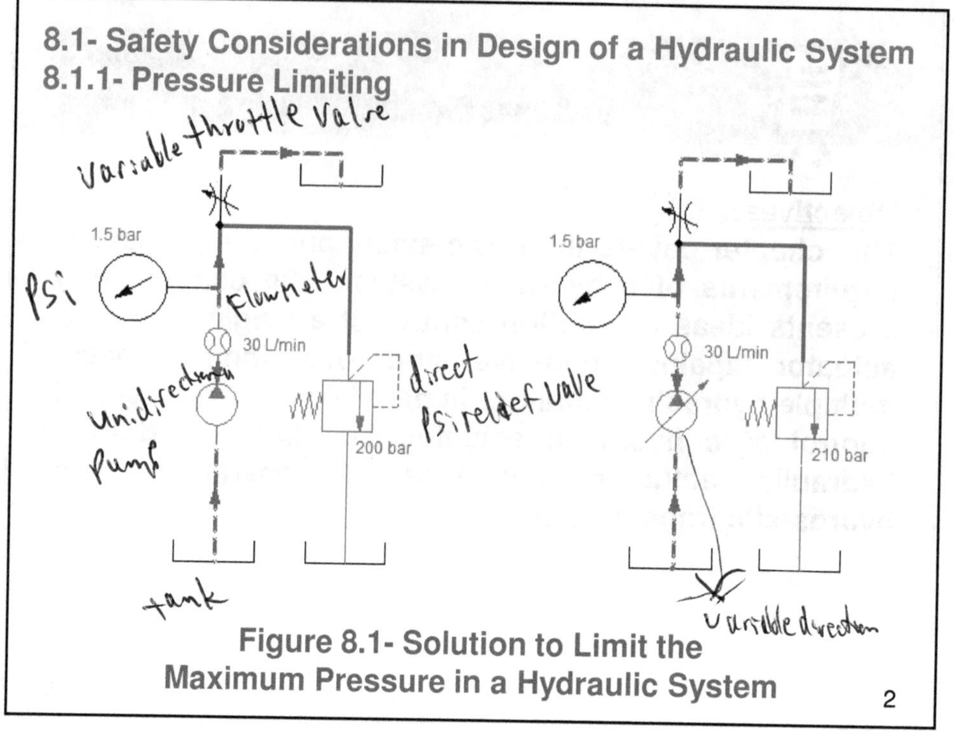
8.1- Safety Considerations in Design of a Hydraulic System
8.1.1- Pressure Limiting
variable throttle valve
1.5 bar
psi
flowmeter
30 L/min
unidirection pump
direct psi relief valve
200 bar
tank
1.5 bar
30 L/min
210 bar
variable direction
Figure 8.1- Solution to Limit the Maximum Pressure in a Hydraulic System
2

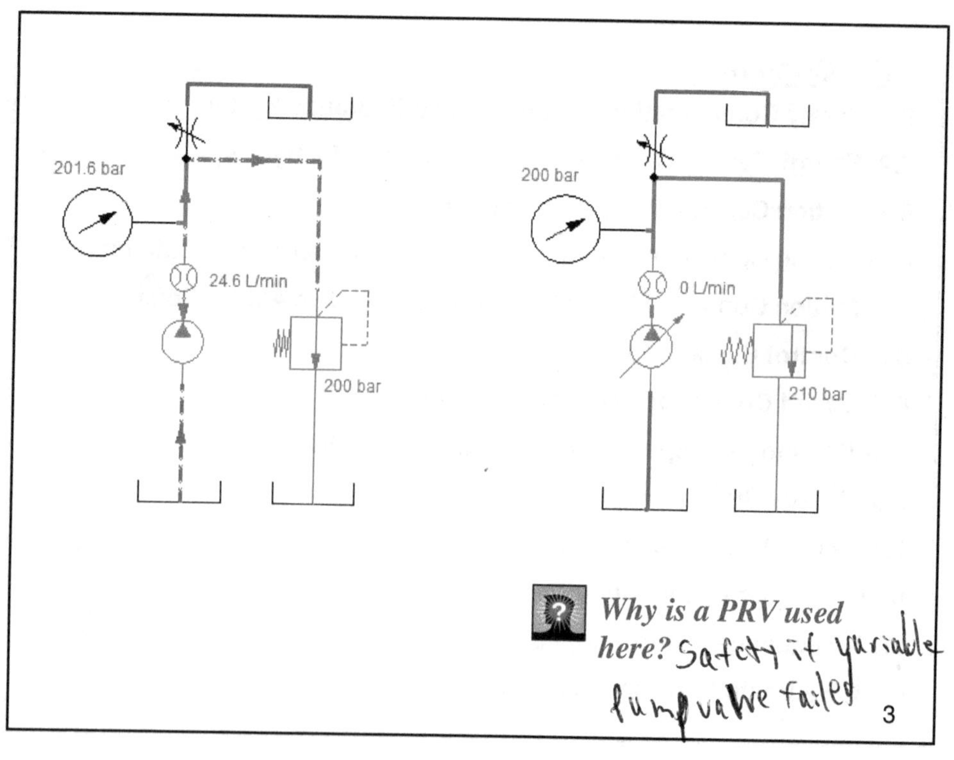
201.6 bar
24.6 L/min
200 bar
200 bar
0 L/min
210 bar
Why is a PRV used here? Safety if variable pump valve failed
3

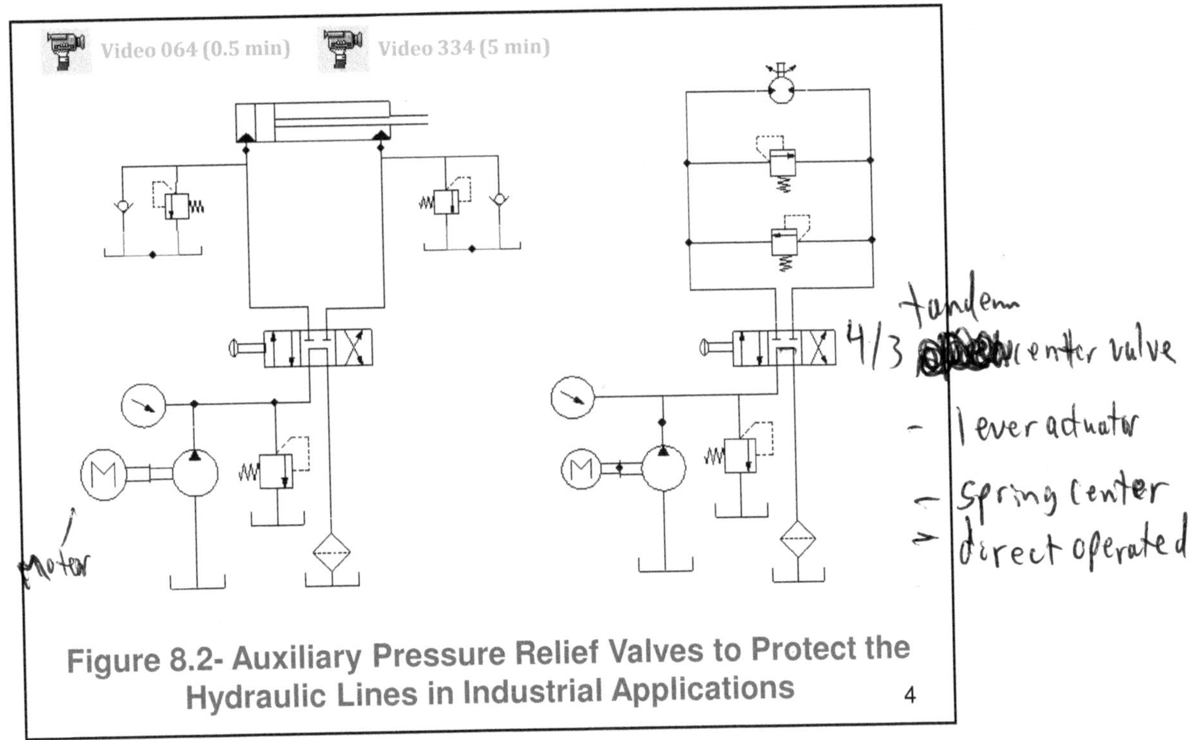

Figure 8.2- Auxiliary Pressure Relief Valves to Protect the Hydraulic Lines in Industrial Applications 4

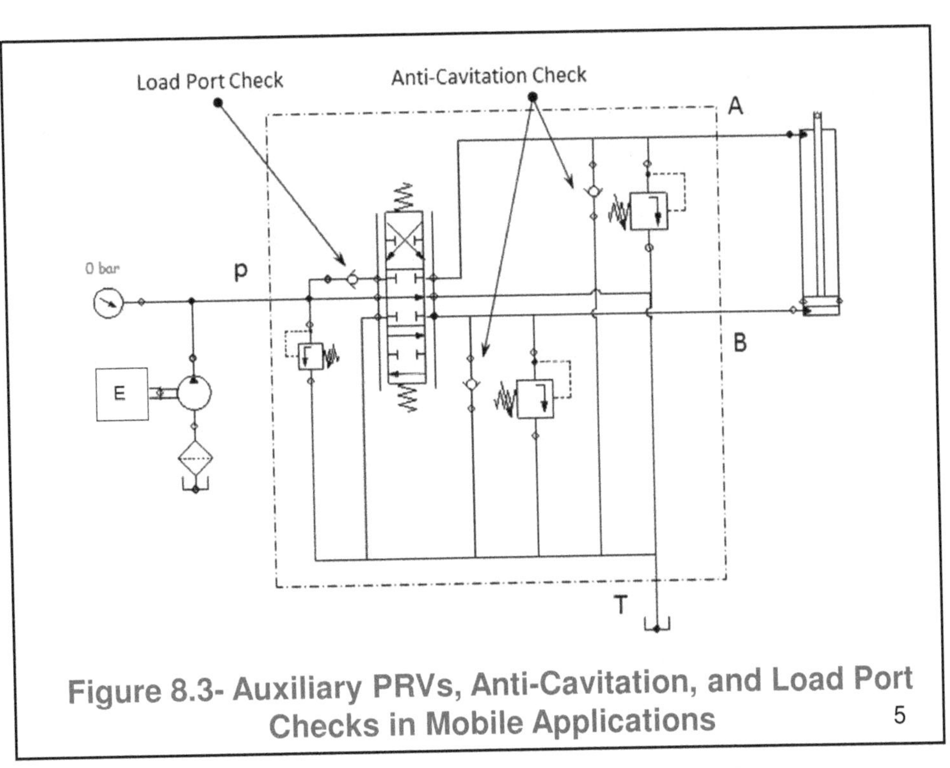

Figure 8.3- Auxiliary PRVs, Anti-Cavitation, and Load Port Checks in Mobile Applications 5

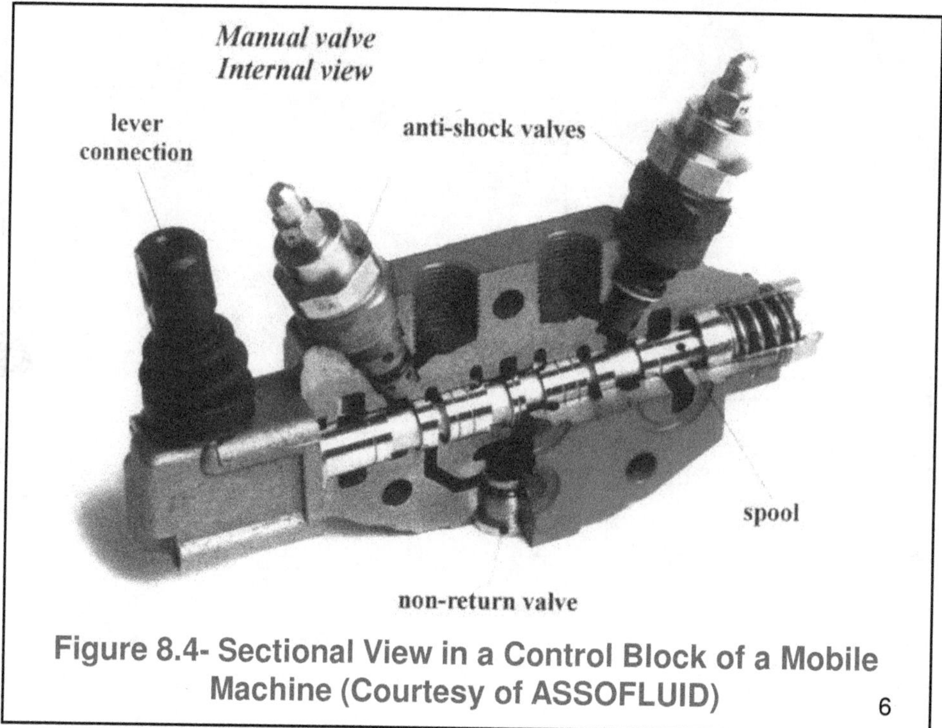

Figure 8.4- Sectional View in a Control Block of a Mobile Machine (Courtesy of ASSOFLUID)

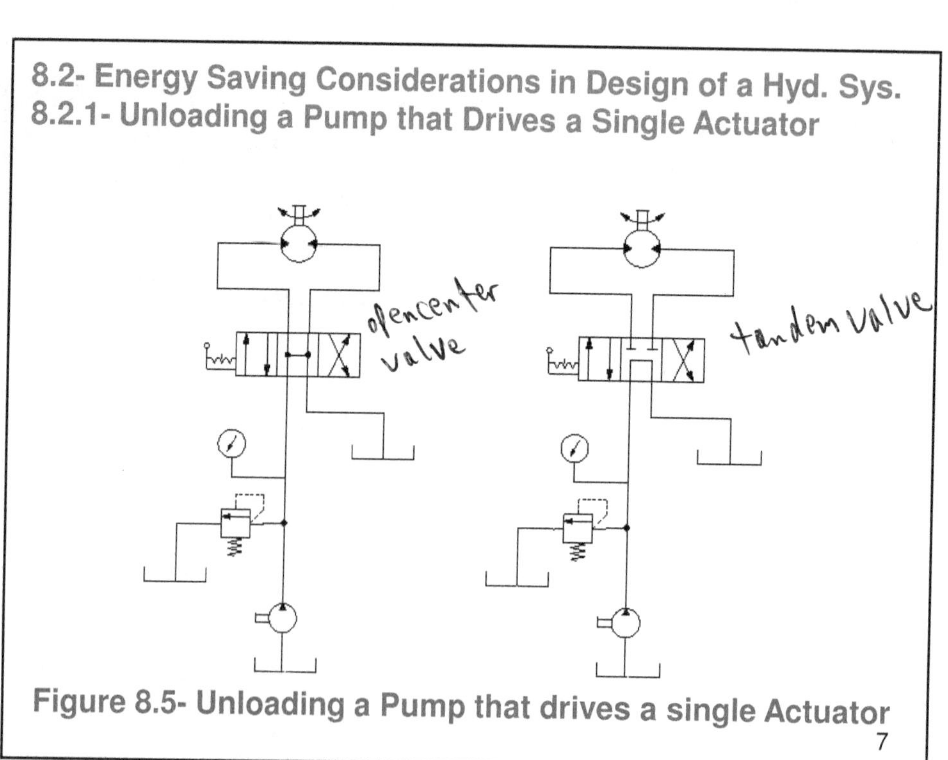

Figure 8.5- Unloading a Pump that drives a single Actuator

8.2.2- Unloading a Pump that Drives Multiple Actuators

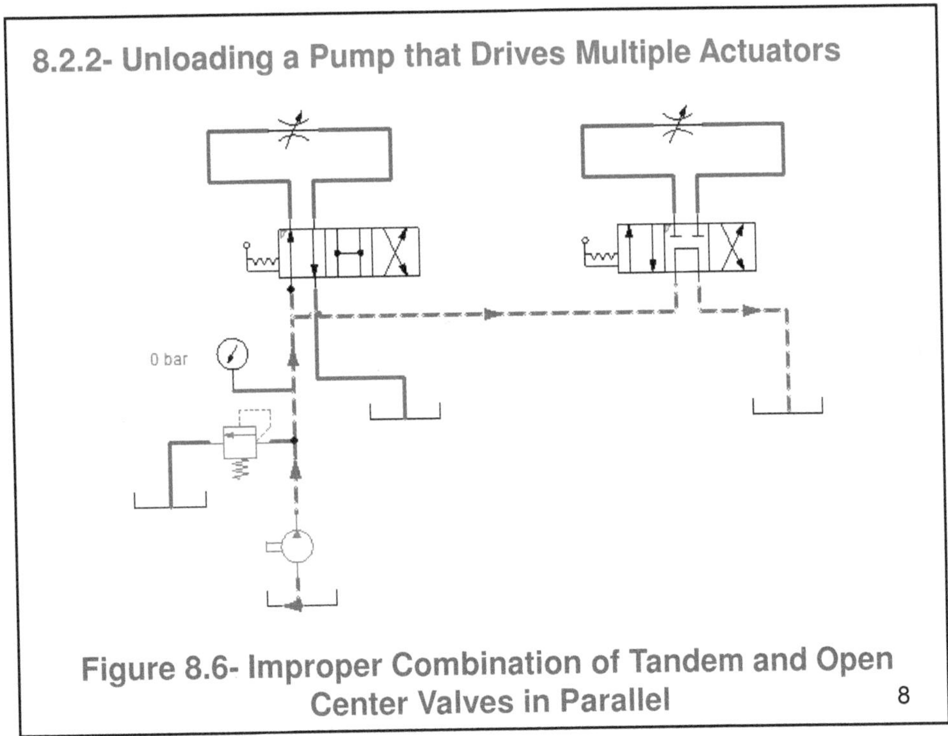

Figure 8.6- Improper Combination of Tandem and Open Center Valves in Parallel 8

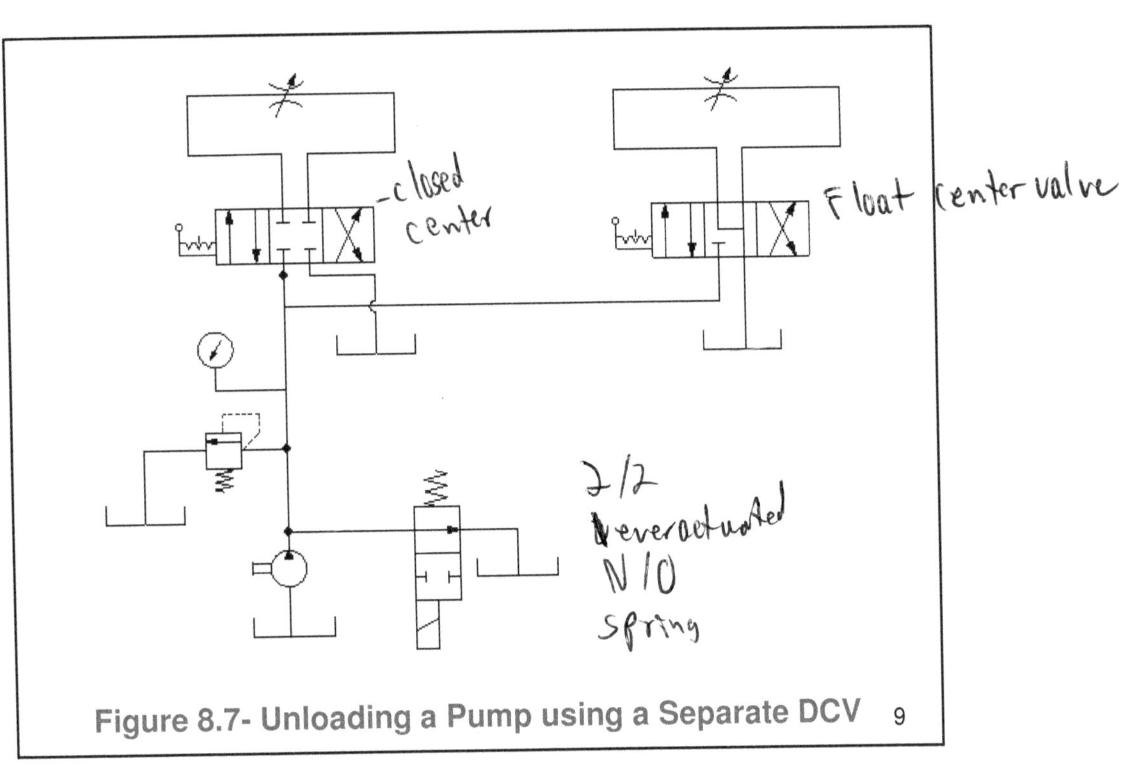

Figure 8.7- Unloading a Pump using a Separate DCV 9

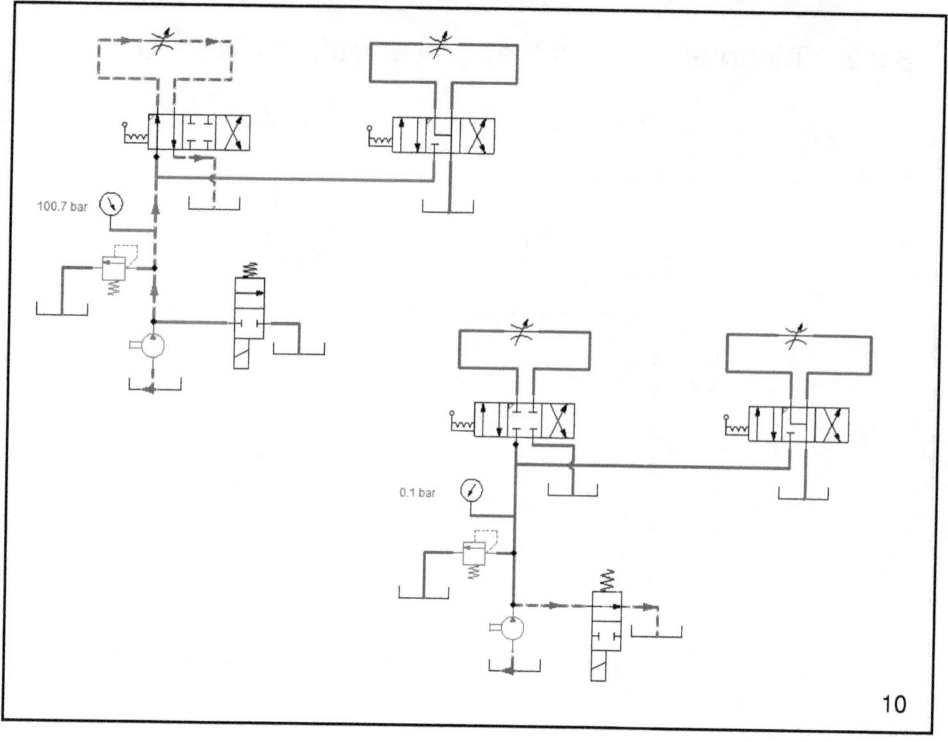

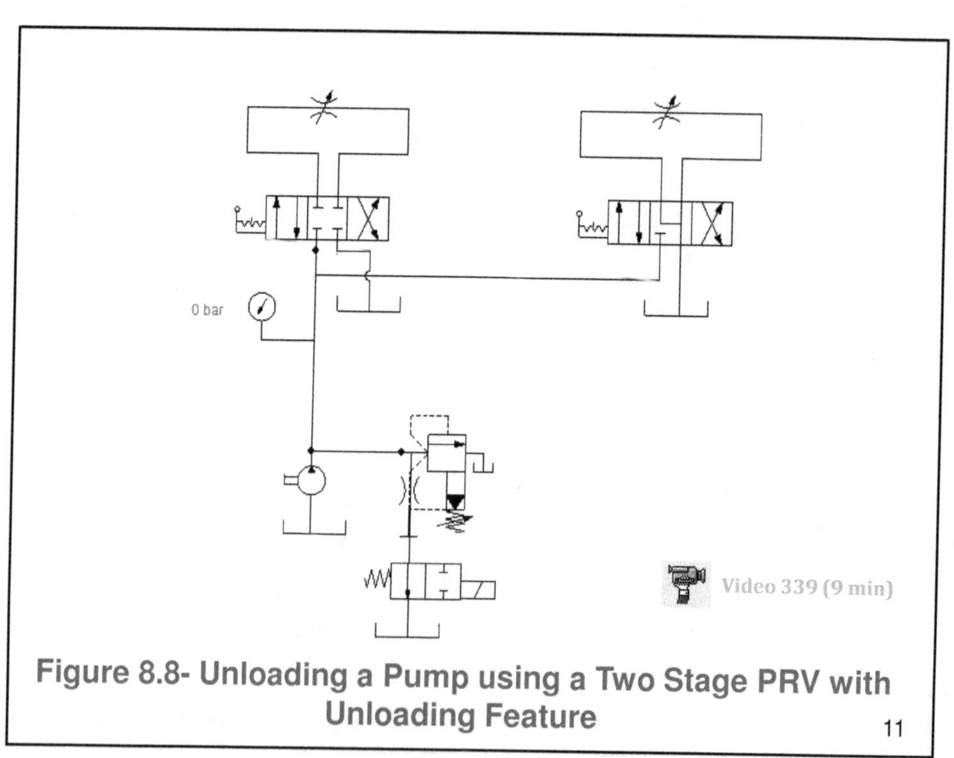

Figure 8.8- Unloading a Pump using a Two Stage PRV with Unloading Feature

8.2.4- Unloading a Pump by a Pressure Unloading Valve

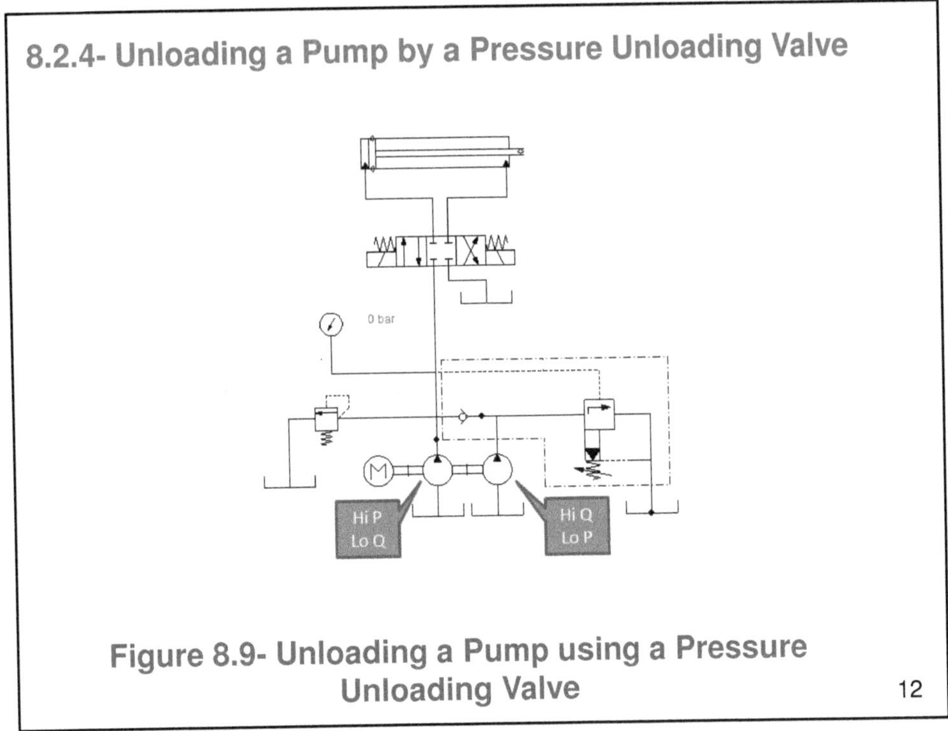

Figure 8.9- Unloading a Pump using a Pressure Unloading Valve

12

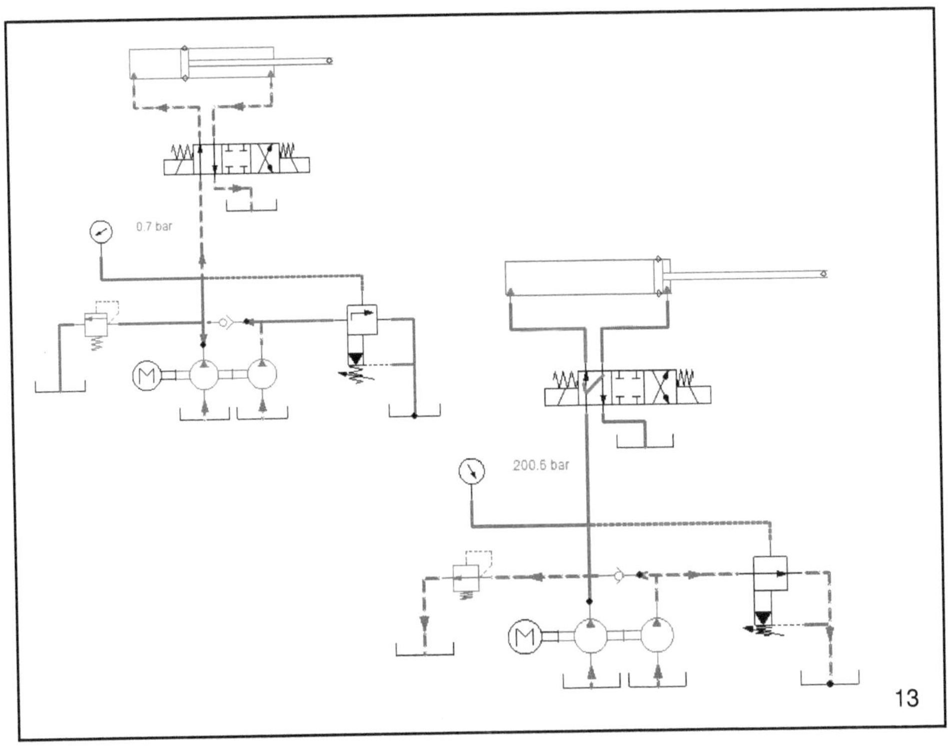

13

8.2.5- Unloading a Pump in Mobile Applications

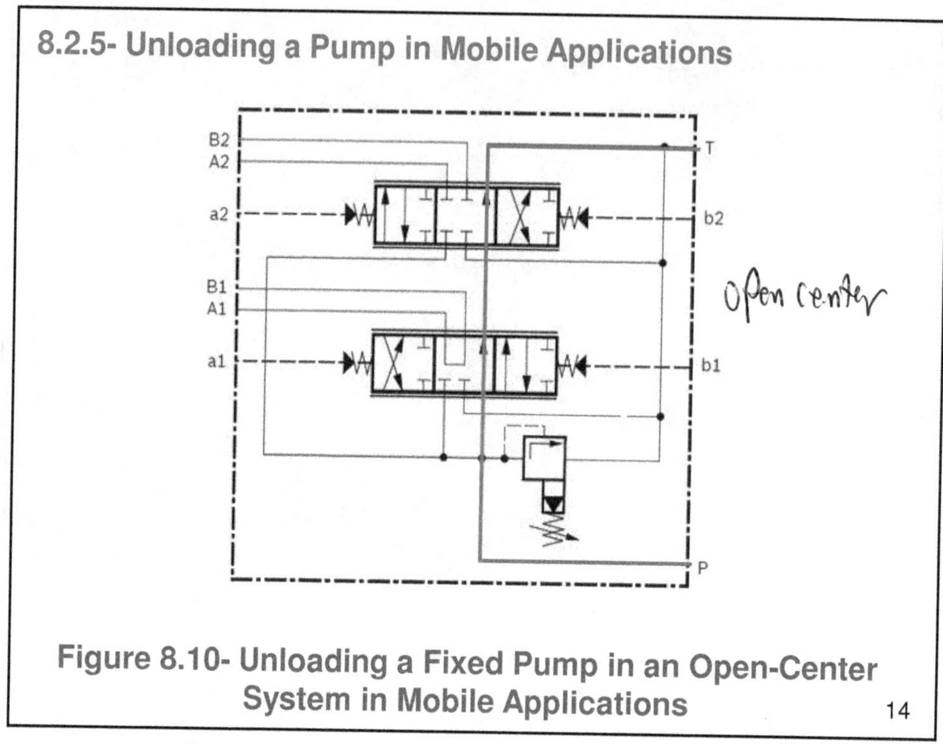

Figure 8.10- Unloading a Fixed Pump in an Open-Center System in Mobile Applications

14

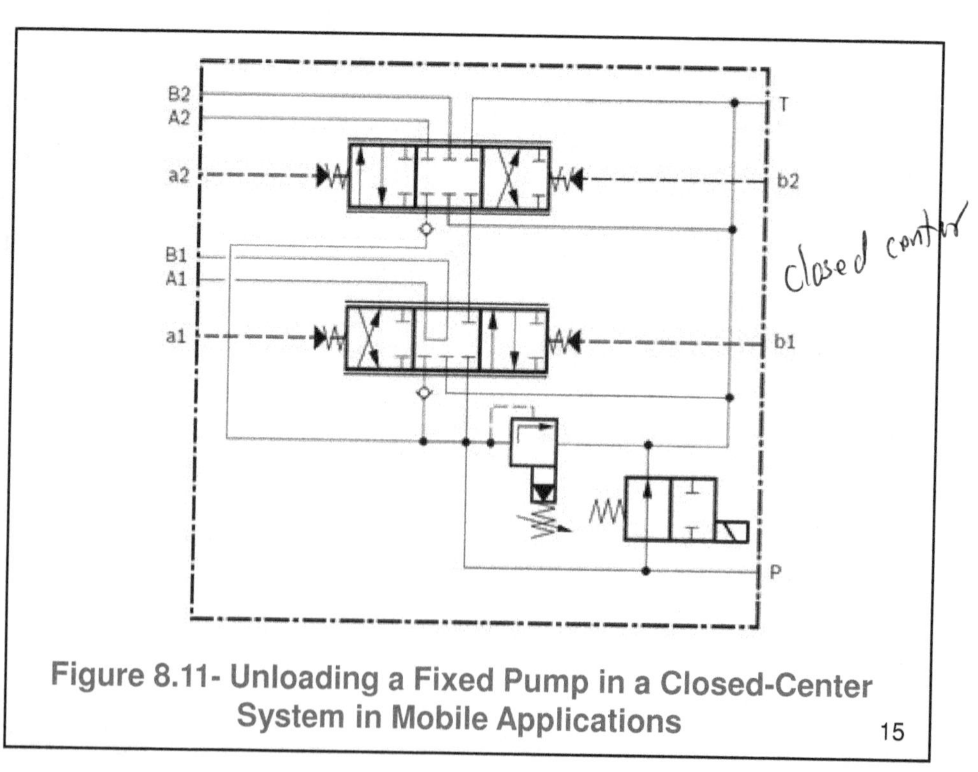

Figure 8.11- Unloading a Fixed Pump in a Closed-Center System in Mobile Applications

15

Features	*Open versus Closed Center System*
Design Complexity:	▪ An open-center system is simple in design and does not require a special arrangement to unload the pump. ▪ A closed-center system requires either a variable pump or an additional built-in module to unload the pump if it has a fixed displacement.
Cost:	▪ If both systems have the same number of sections, obviously the open-center one will be less expensive.
Energy Saving:	▪ In idle conditions of an open-center system, the pump wastes energy continuously through the consecutive sections. ▪ In idle conditions of a closed-center system, the pump is unloaded.
Pressure Readiness & Response:	▪ In an open-center system, by shifting one of the spools, the pump supplies the power immediately to the actuator. ▪ In a closed-center system, the pump takes time to be loaded.

Table 8.1- Open-Center System versus Closed-Center System in Mobile Applications

16

8.3- Motion Control of a Single Actuator

Positive (Resistive) Load:

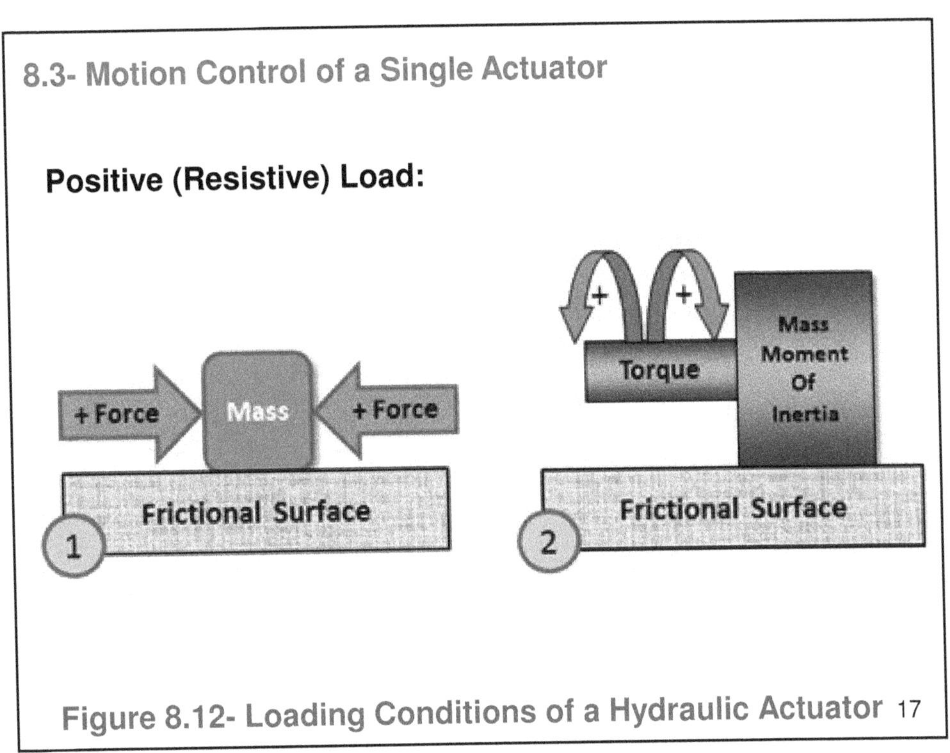

Figure 8.12- Loading Conditions of a Hydraulic Actuator 17

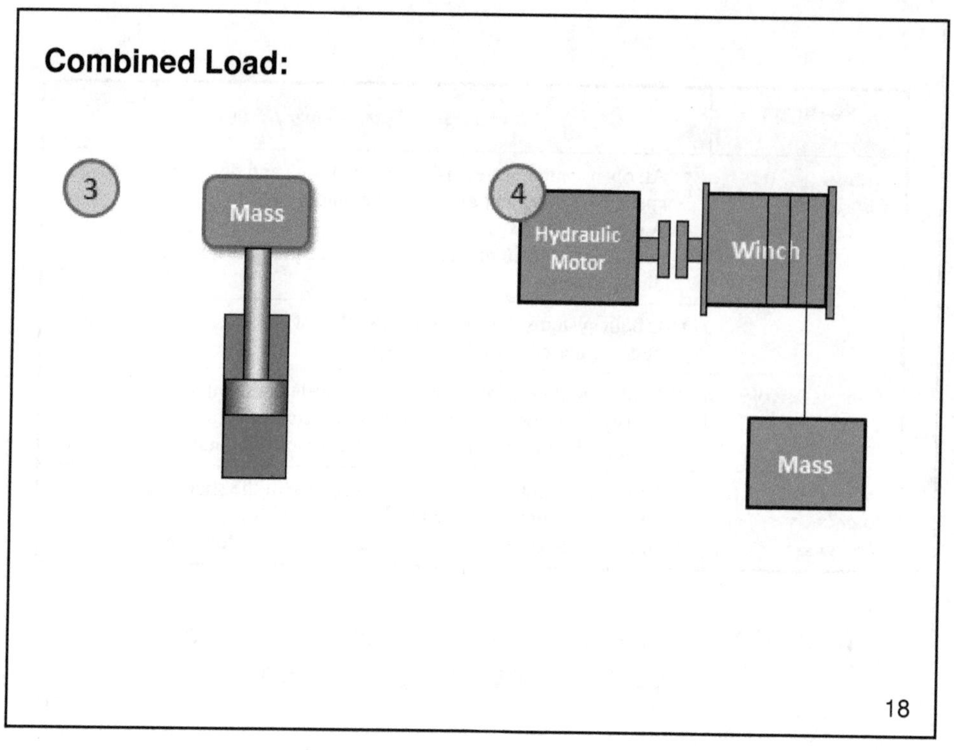
Combined Load:
3
Mass
4
Hydraulic Motor
Winch
Mass
18

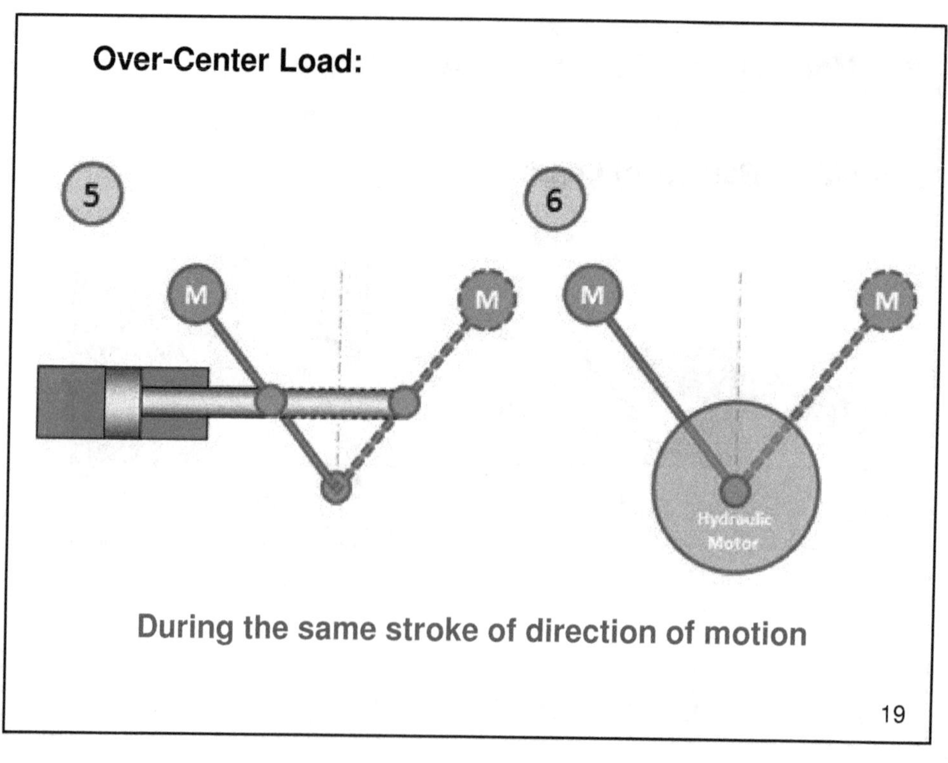
Over-Center Load:
5
M
M
6
M
M
Hydraulic Motor
During the same stroke of direction of motion
19

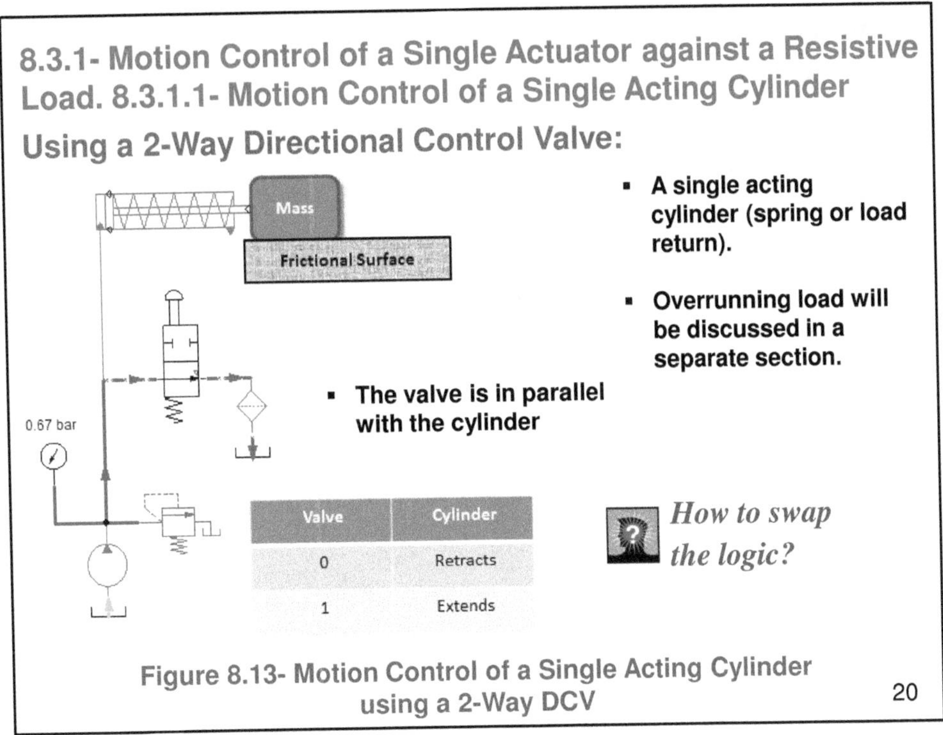

Figure 8.13- Motion Control of a Single Acting Cylinder using a 2-Way DCV

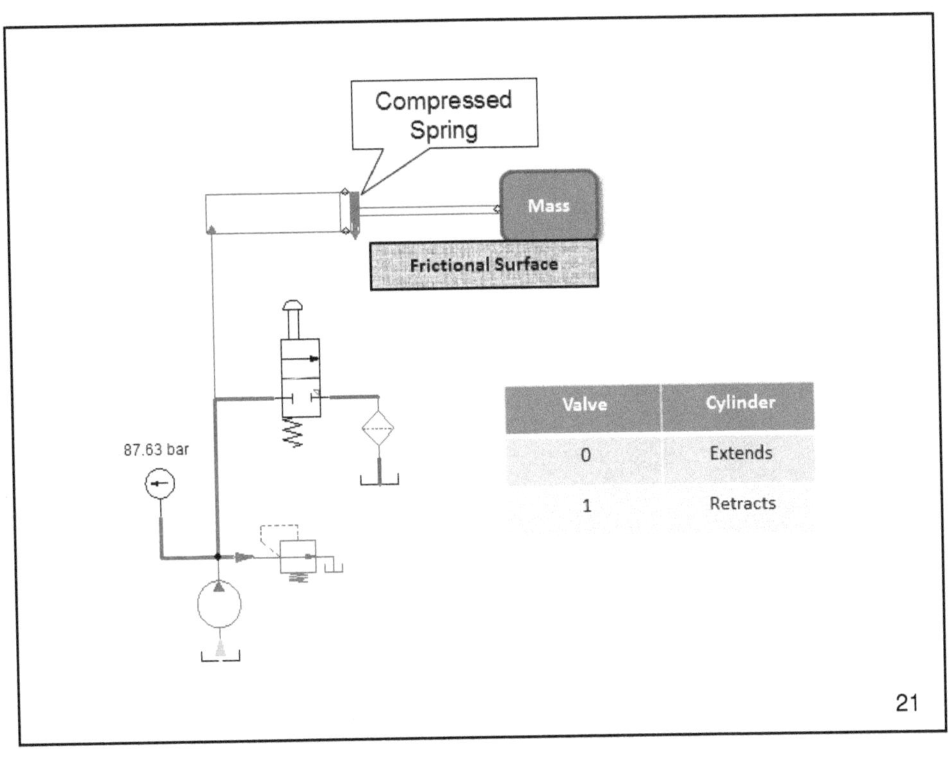

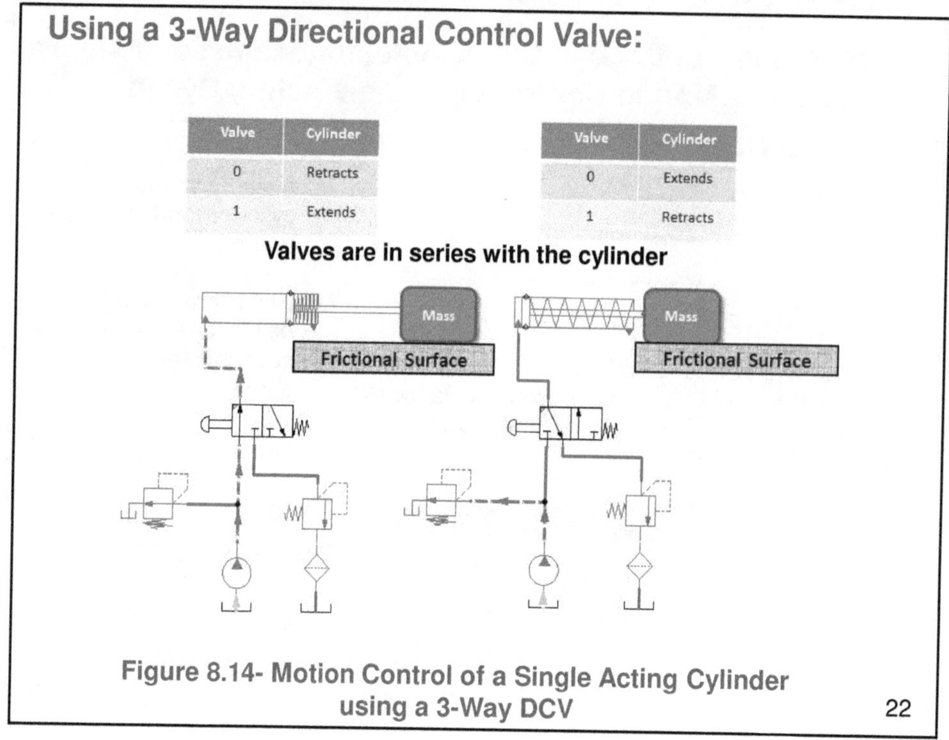

Figure 8.14- Motion Control of a Single Acting Cylinder using a 3-Way DCV

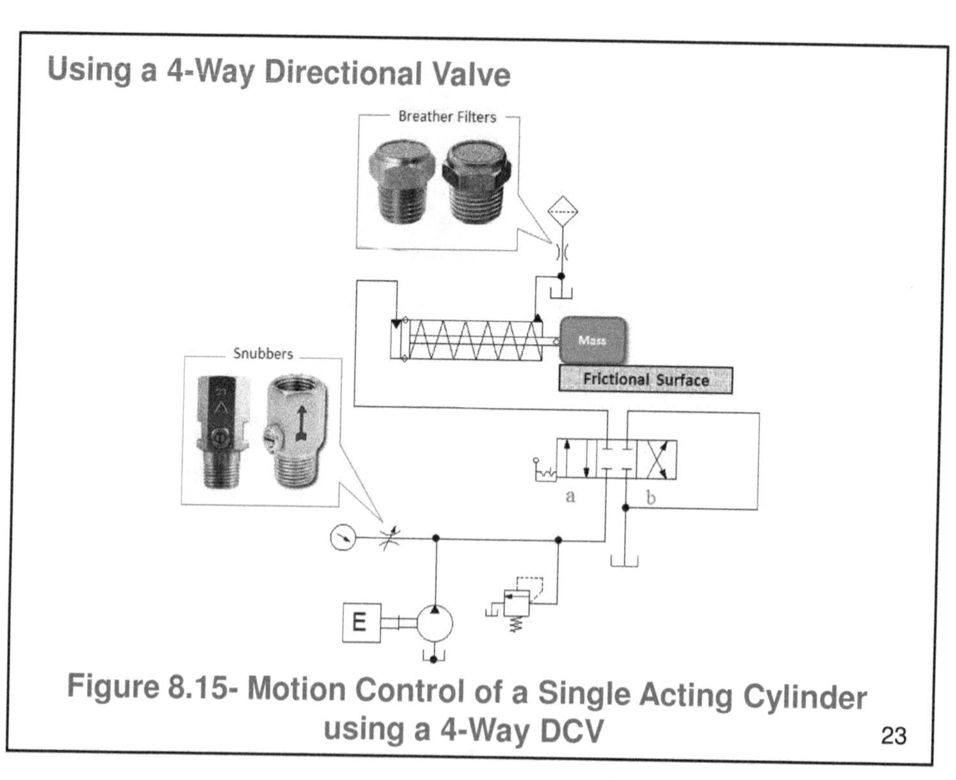

Figure 8.15- Motion Control of a Single Acting Cylinder using a 4-Way DCV

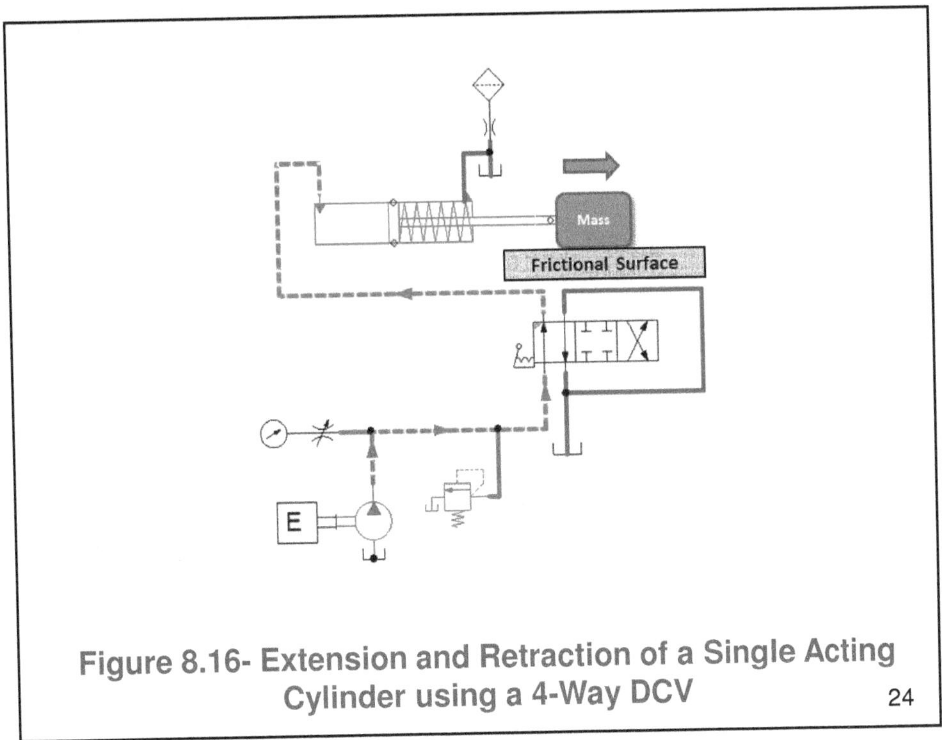

Figure 8.16- Extension and Retraction of a Single Acting Cylinder using a 4-Way DCV

24

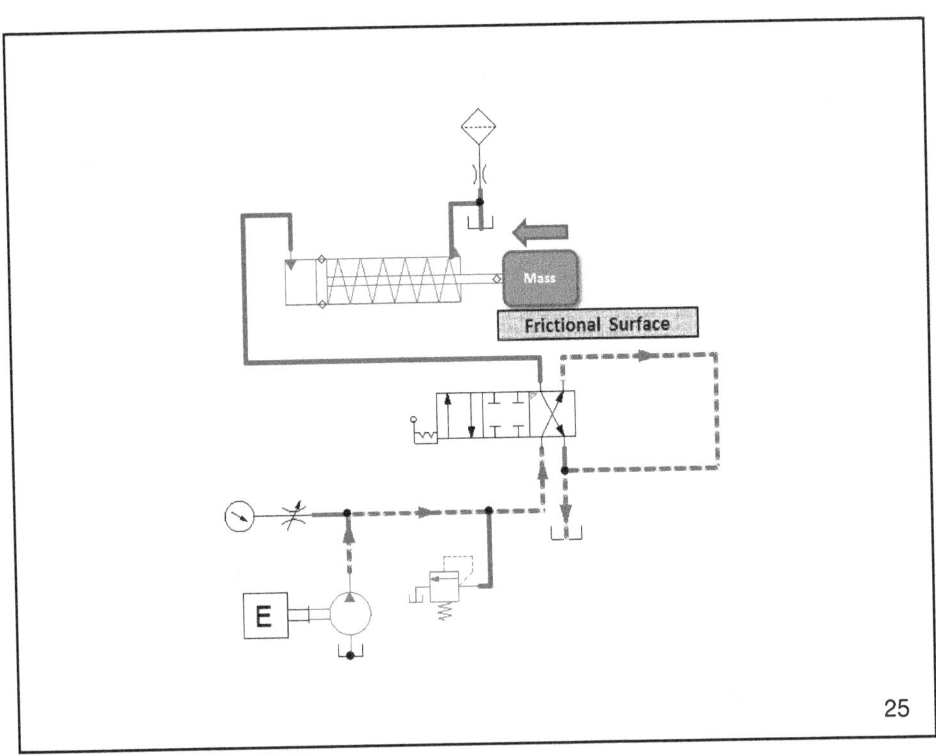

25

8.3.1.2- Motion Control of a Unidirectional Motor

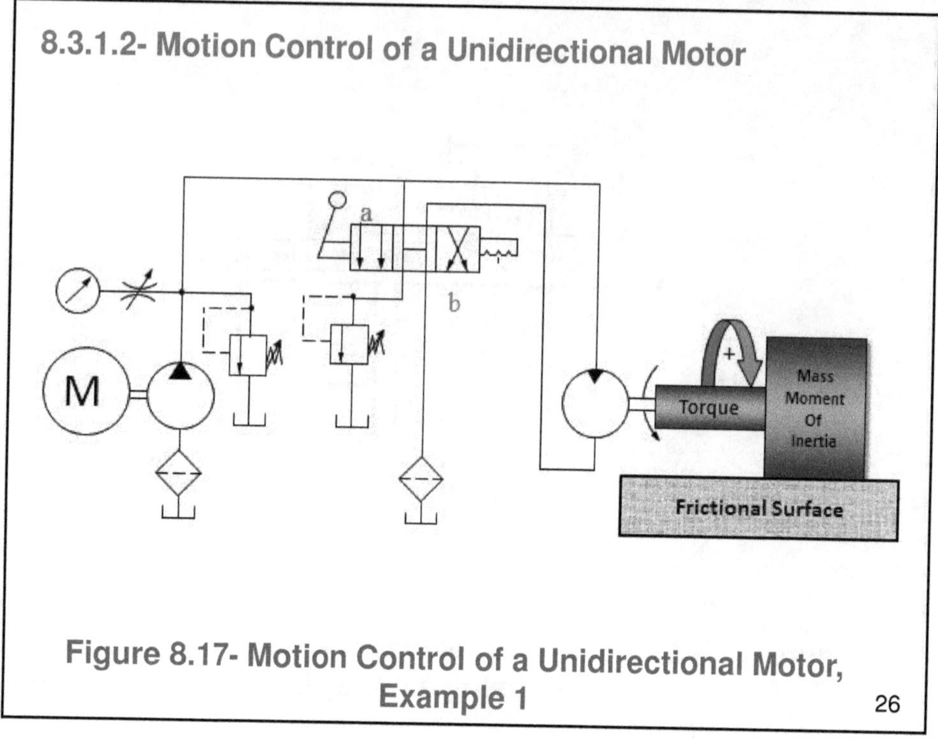

Figure 8.17- Motion Control of a Unidirectional Motor, Example 1

26

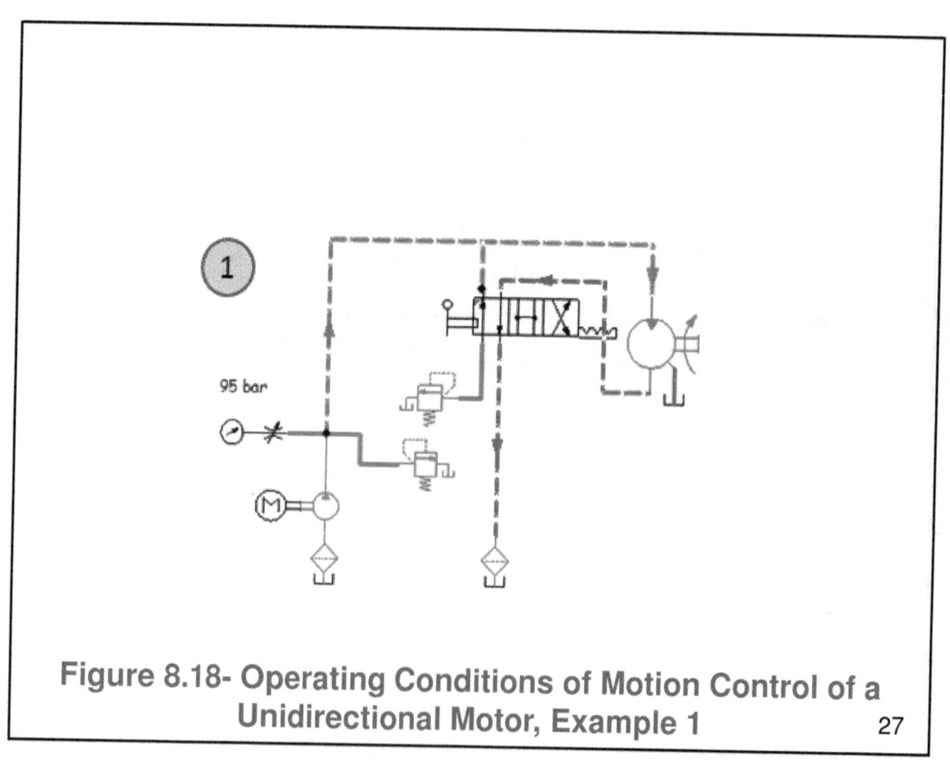

Figure 8.18- Operating Conditions of Motion Control of a Unidirectional Motor, Example 1

27

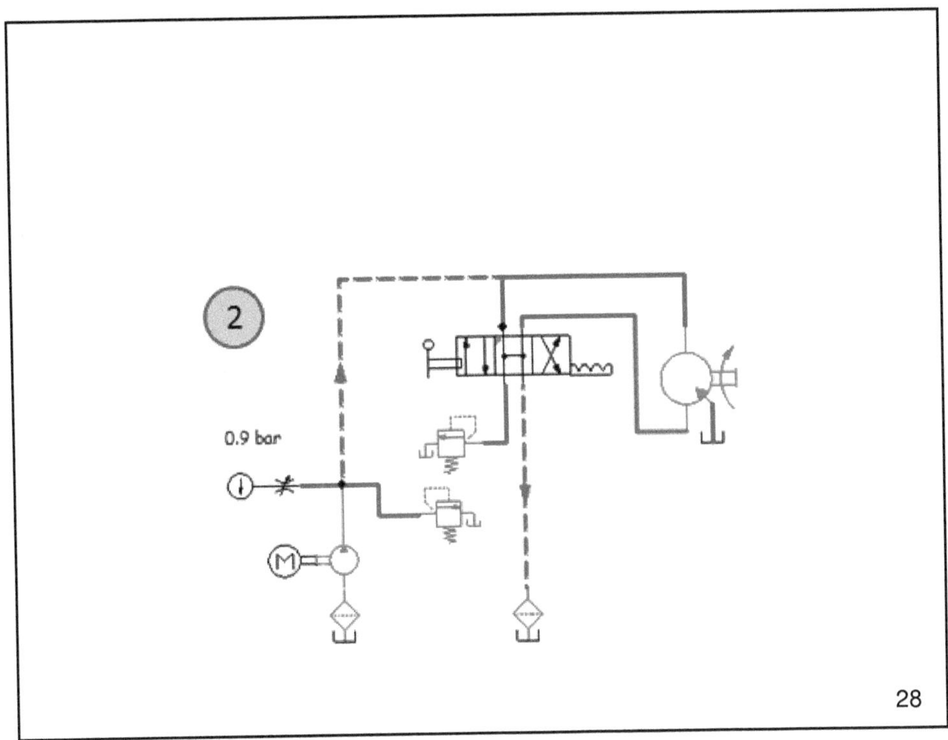
2
0.9 bar
M
28

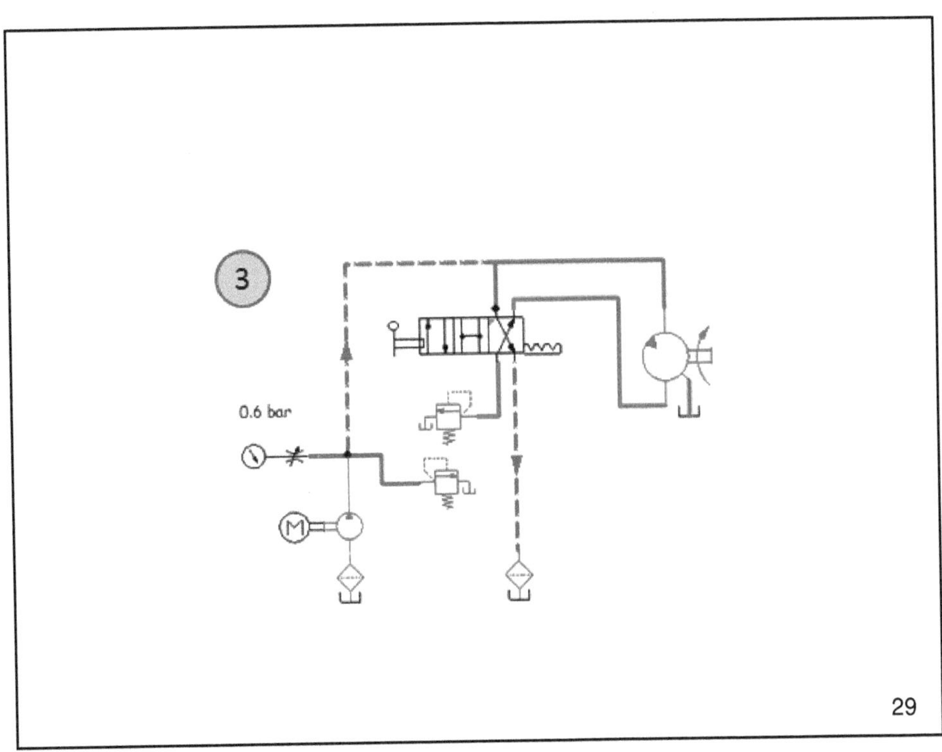
3
0.6 bar
M
29

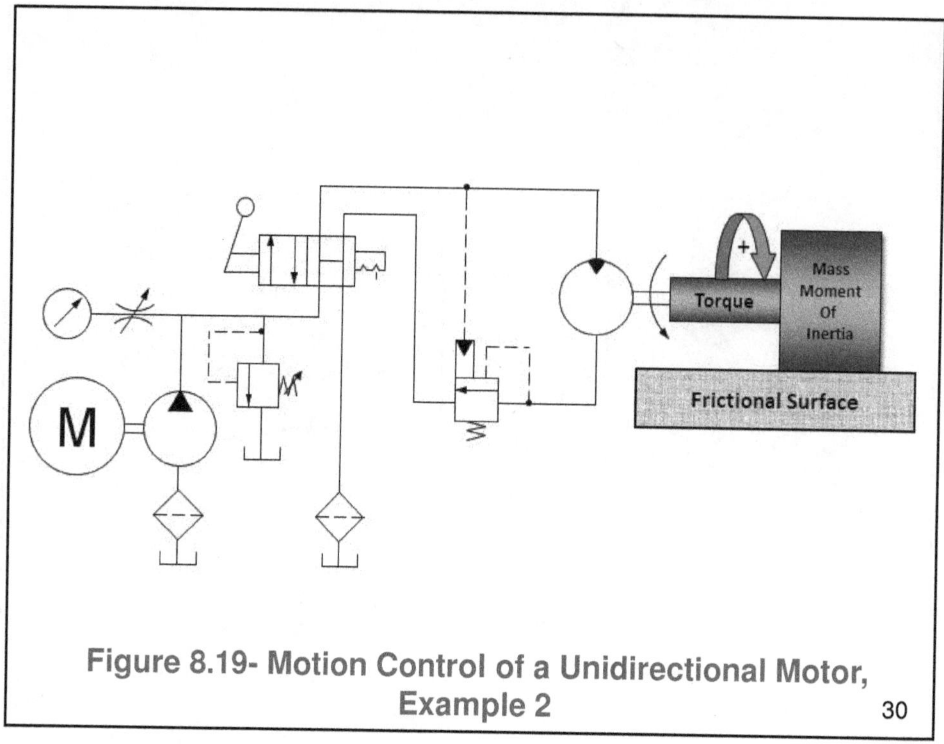

Figure 8.19- Motion Control of a Unidirectional Motor, Example 2

30

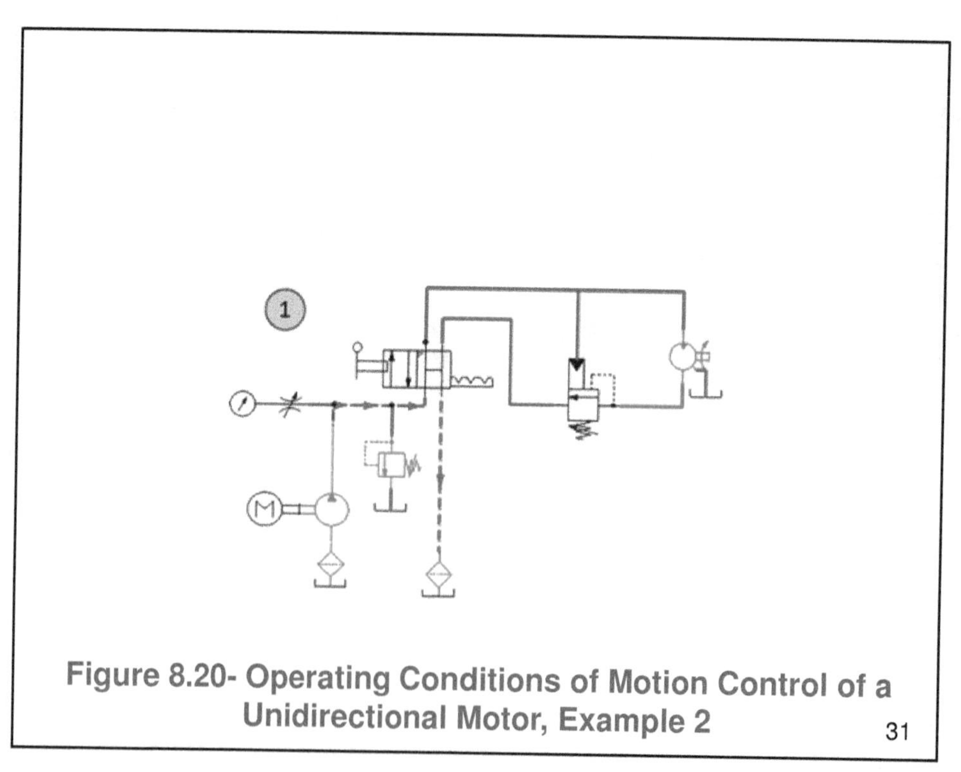

Figure 8.20- Operating Conditions of Motion Control of a Unidirectional Motor, Example 2

31

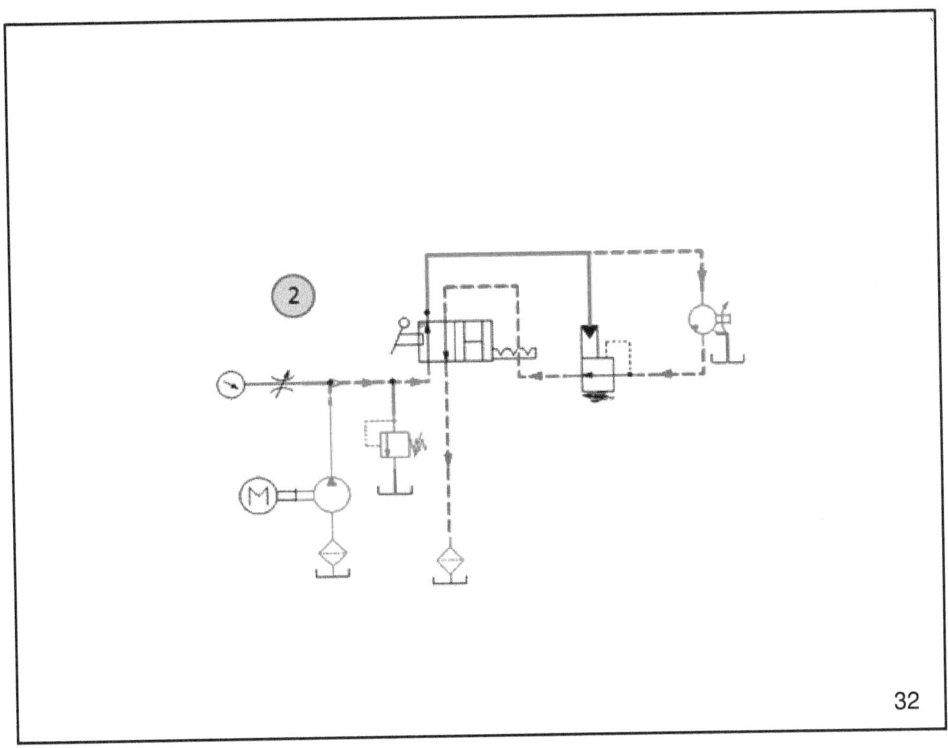
2
32

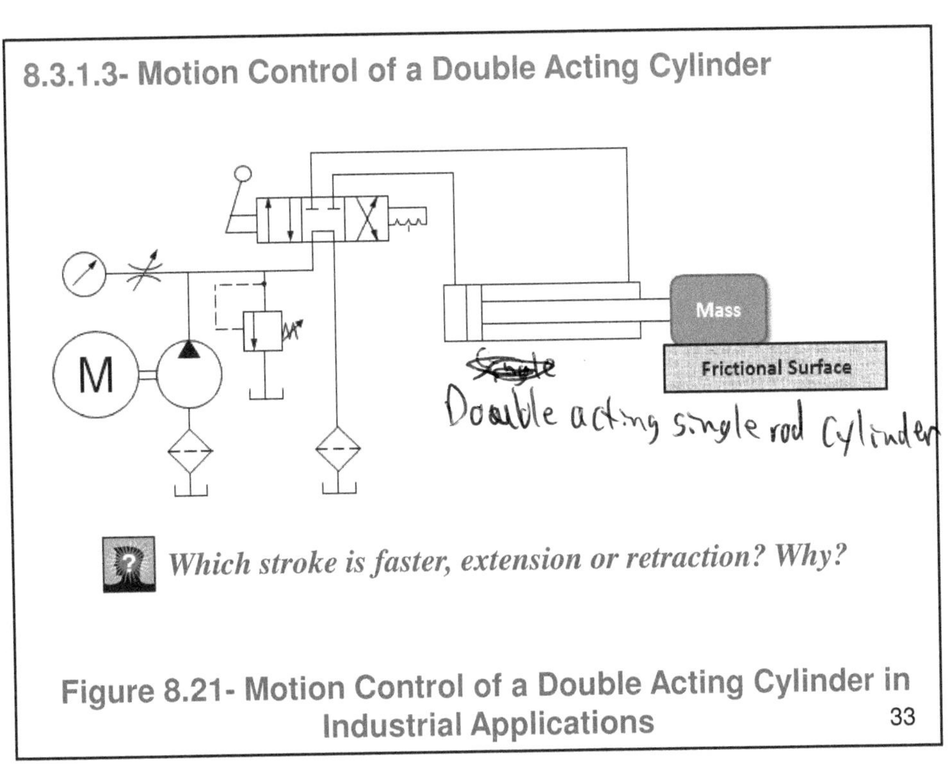
8.3.1.3- Motion Control of a Double Acting Cylinder
M
Mass
Frictional Surface
Double acting single rod Cylinder
Which stroke is faster, extension or retraction? Why?
Figure 8.21- Motion Control of a Double Acting Cylinder in Industrial Applications
33

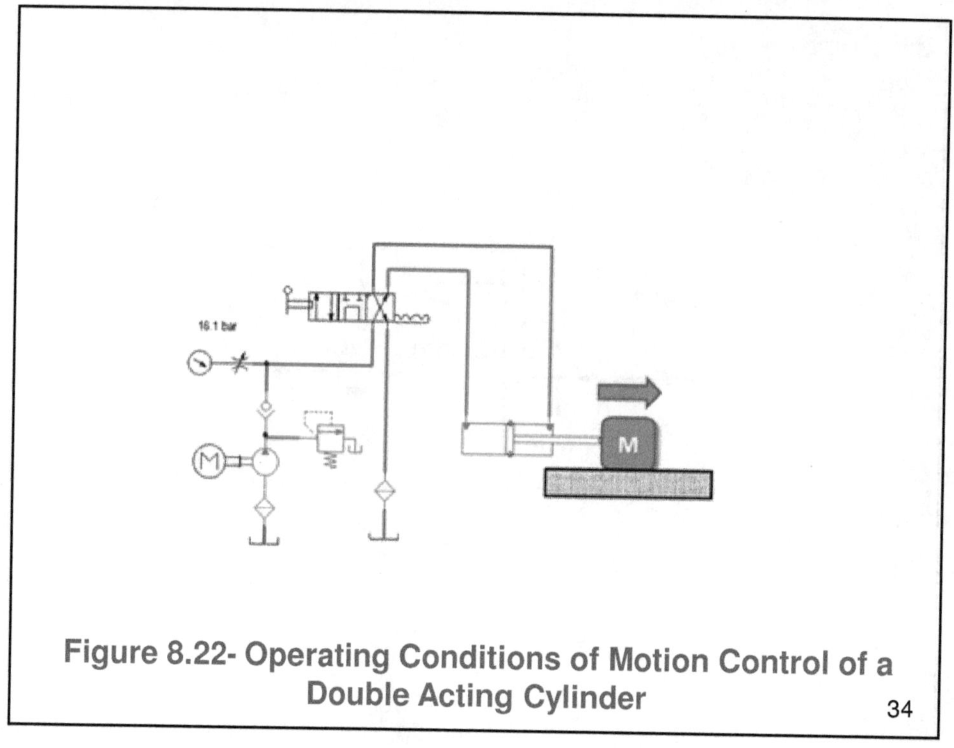

Figure 8.22- Operating Conditions of Motion Control of a Double Acting Cylinder

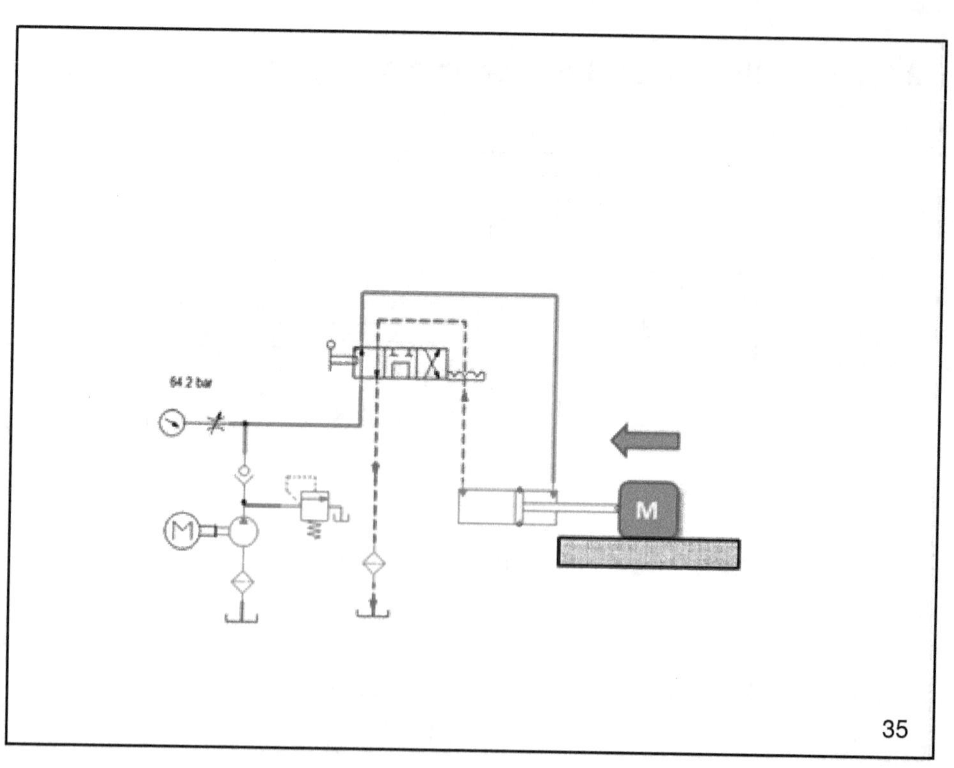

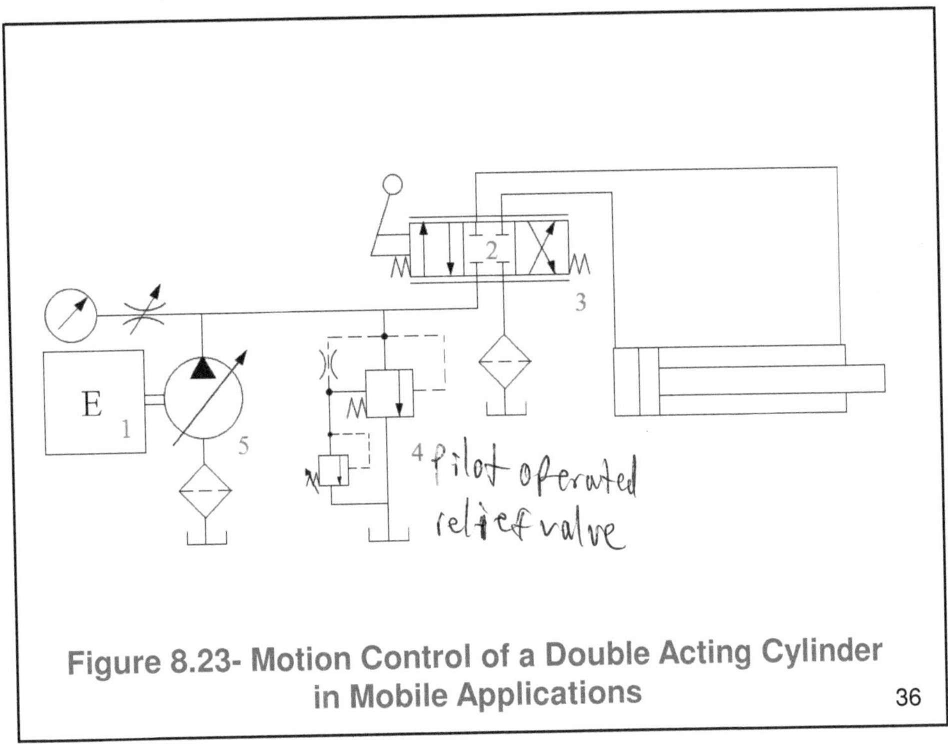

Figure 8.23- Motion Control of a Double Acting Cylinder in Mobile Applications

36

8.3.1.4- Motion Control of a Bidirectional Motor

Case 1:

Speed: **Slow,** Load: **Small,** Idle Conditions: **brake**

Tandem-Center DCV:
Recommended with a fixed pump that is unloaded through the central position of the valve.

Closed-Center DCV:
Recommended with a pressure compensated pump that de-strokes itself at the central position of the valve so that it saves energy.

37

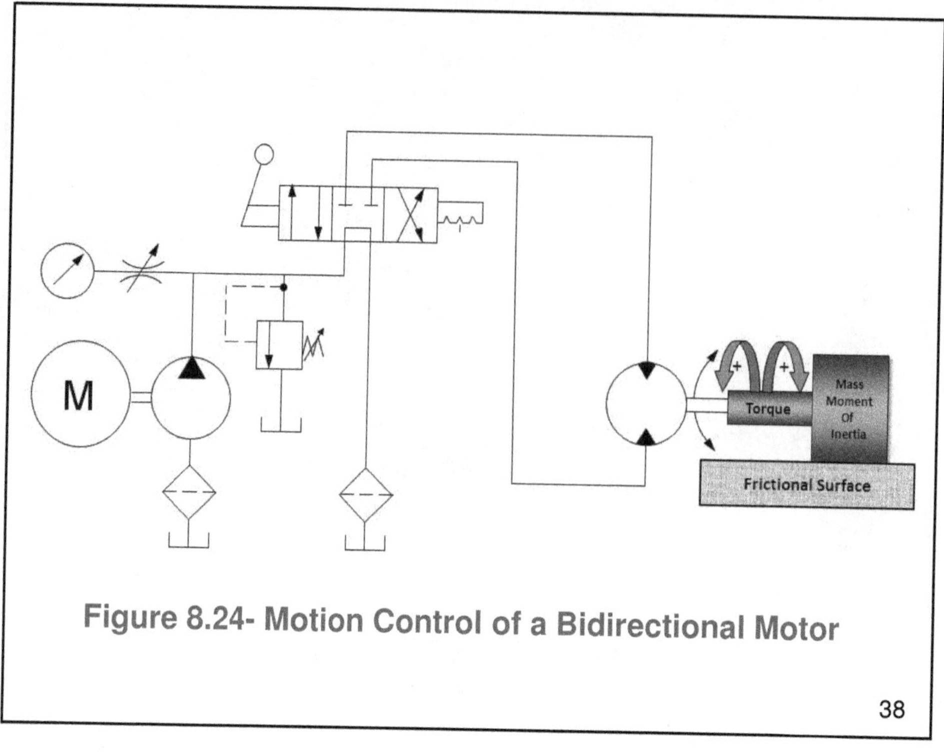

Figure 8.24- Motion Control of a Bidirectional Motor

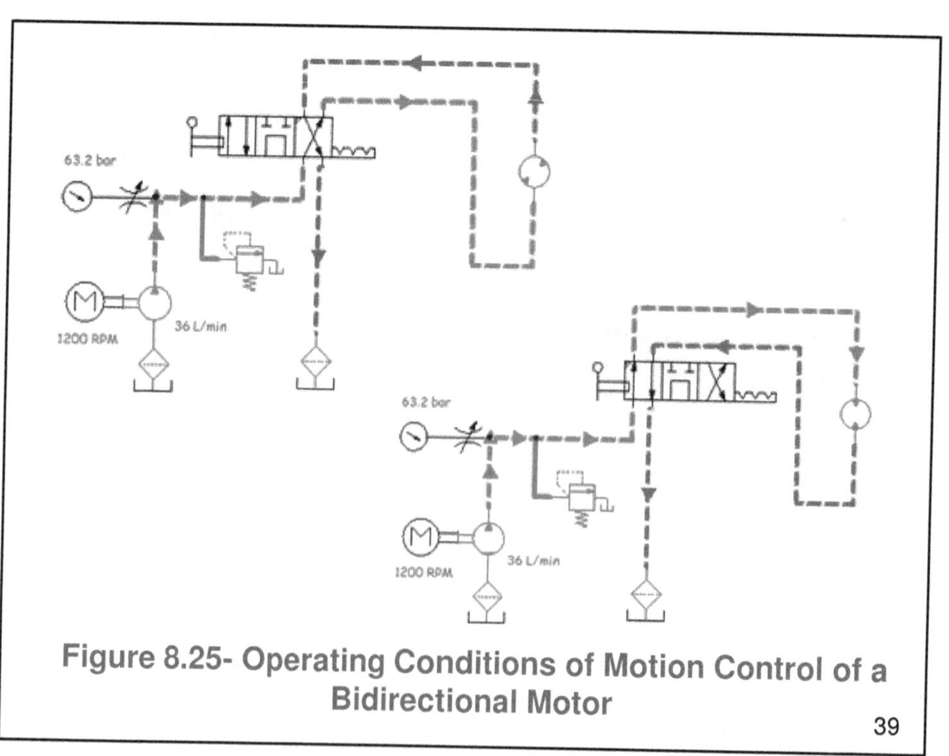

Figure 8.25- Operating Conditions of Motion Control of a Bidirectional Motor

Case 2:

Speed: **High,** Load: **Large,** Idle Conditions: **brake**

Open or Float-Center DCV with a double motor brake valve.

Case 3:
Speed: **Regardless,** Load: **Regardless,**
Idle Conditions: **float**

Open-Center DCV: **Recommended with a** fixed pump

Float-Center DCV: **Recommended with a** pressure compensated pump.

40

8.3.1.5- Motion Control of a Rotary Actuator

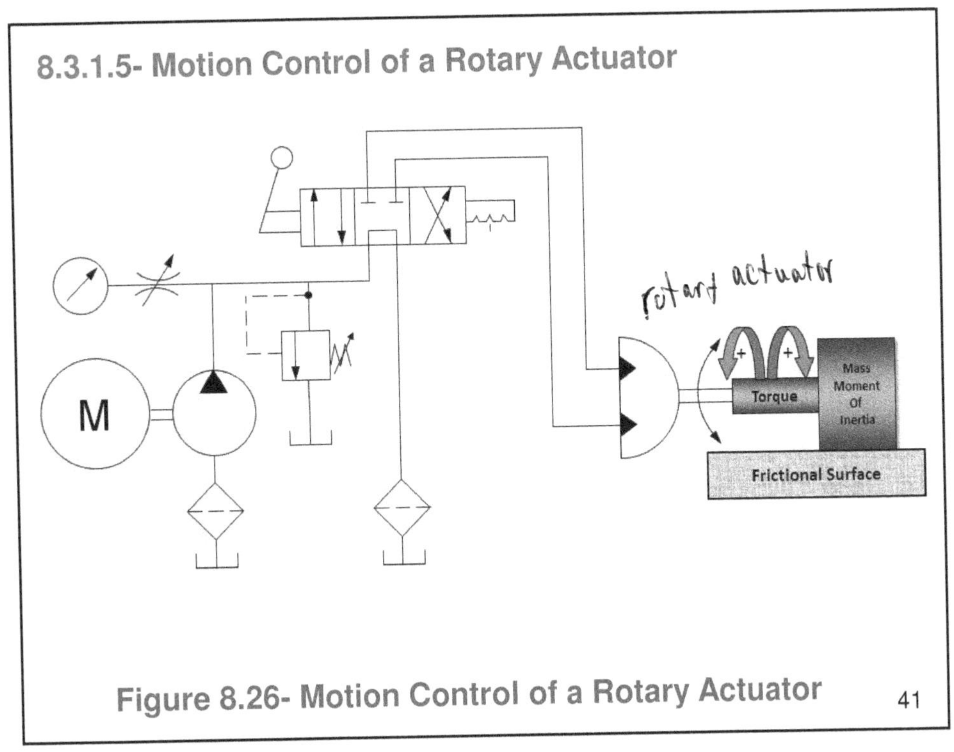

Figure 8.26- Motion Control of a Rotary Actuator

41

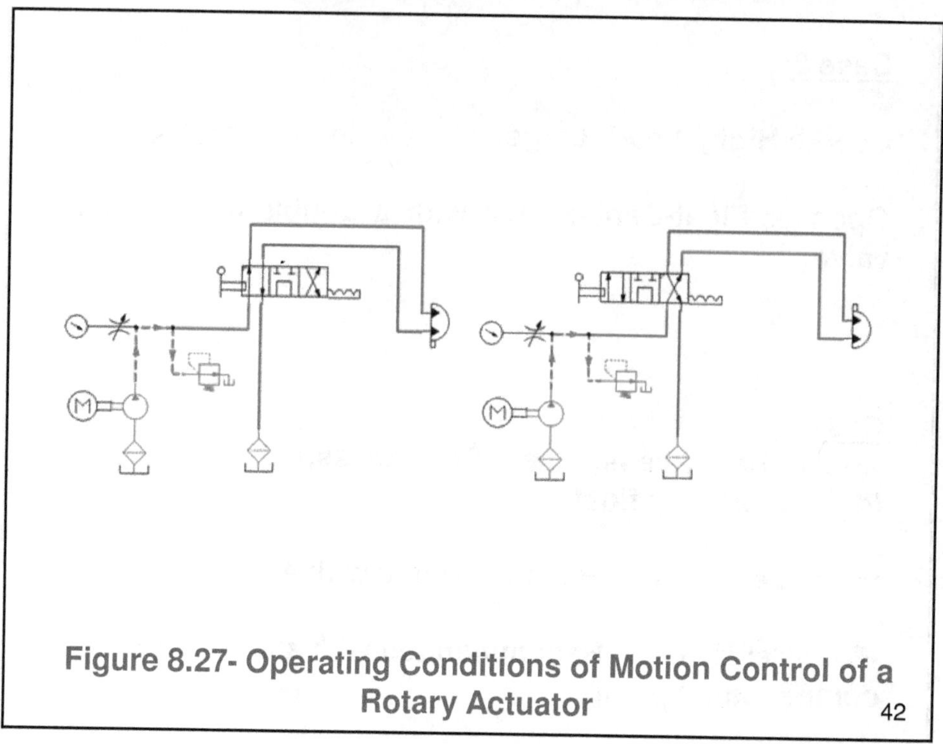

Figure 8.27- Operating Conditions of Motion Control of a Rotary Actuator

42

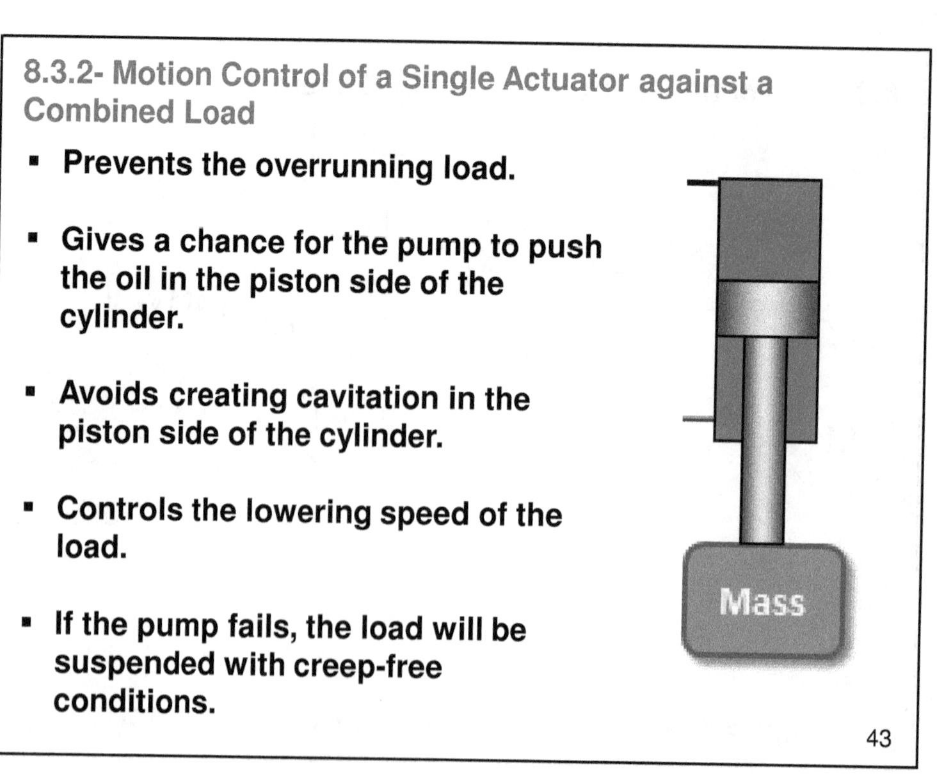

8.3.2- Motion Control of a Single Actuator against a Combined Load

- **Prevents the overrunning load.**
- **Gives a chance for the pump to push the oil in the piston side of the cylinder.**
- **Avoids creating cavitation in the piston side of the cylinder.**
- **Controls the lowering speed of the load.**
- **If the pump fails, the load will be suspended with creep-free conditions.**

43

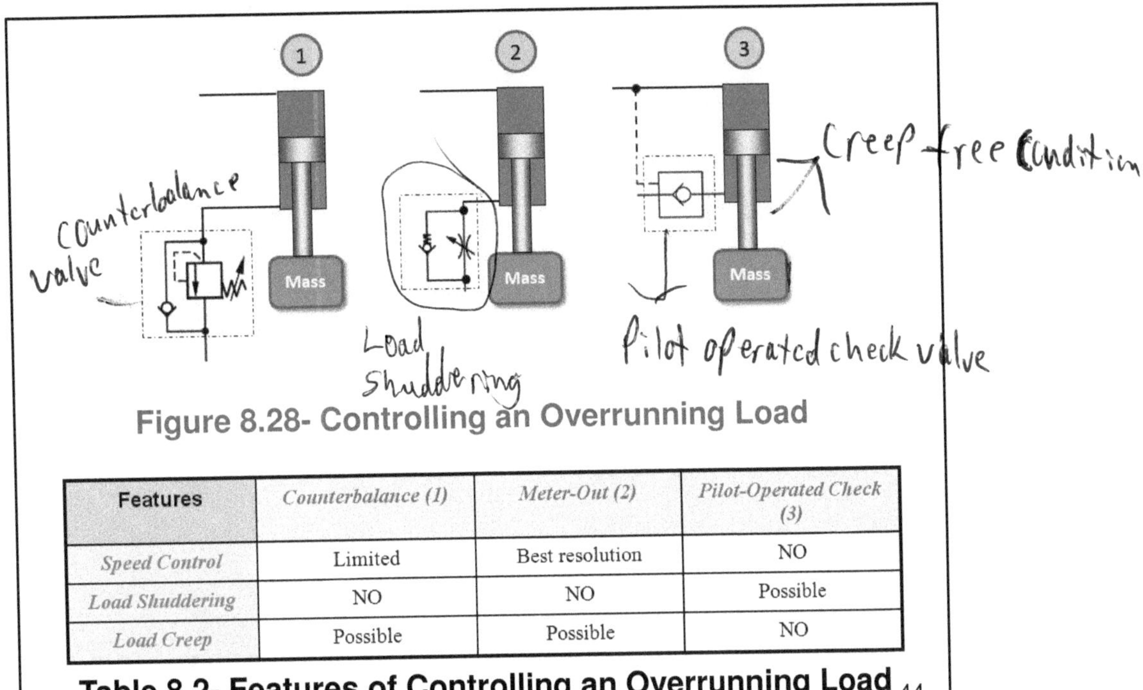

Figure 8.28- Controlling an Overrunning Load

Features	*Counterbalance (1)*	*Meter-Out (2)*	*Pilot-Operated Check (3)*
Speed Control	Limited	Best resolution	NO
Load Shuddering	NO	NO	Possible
Load Creep	Possible	Possible	NO

Table 8.2- Features of Controlling an Overrunning Load

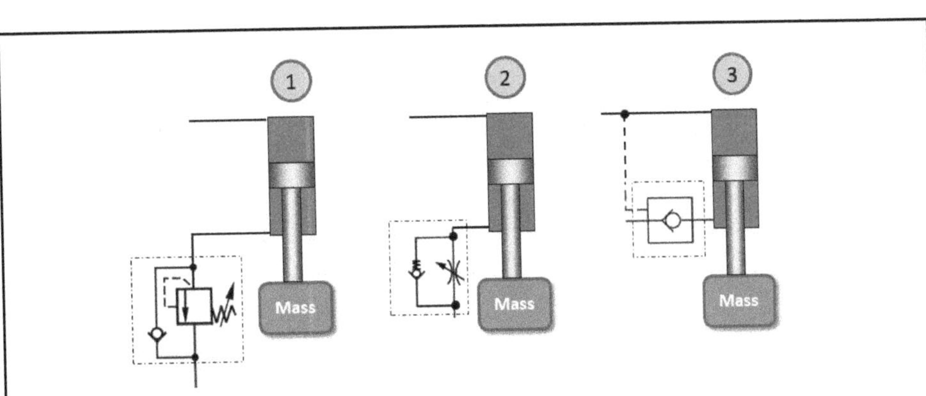

For Safety Purposes:

- **Float-Center type DCV.**
- **Valve built in with the cylinder.**
- **Pressure intensification at the rod side must be considered.**
- **If no choice but using a line between the valve and the cylinder, DO NOT use flexible hose.**

For Energy Saving Purposes:

- **Use of pressure compensated pump with the Float-Center type DCV.**

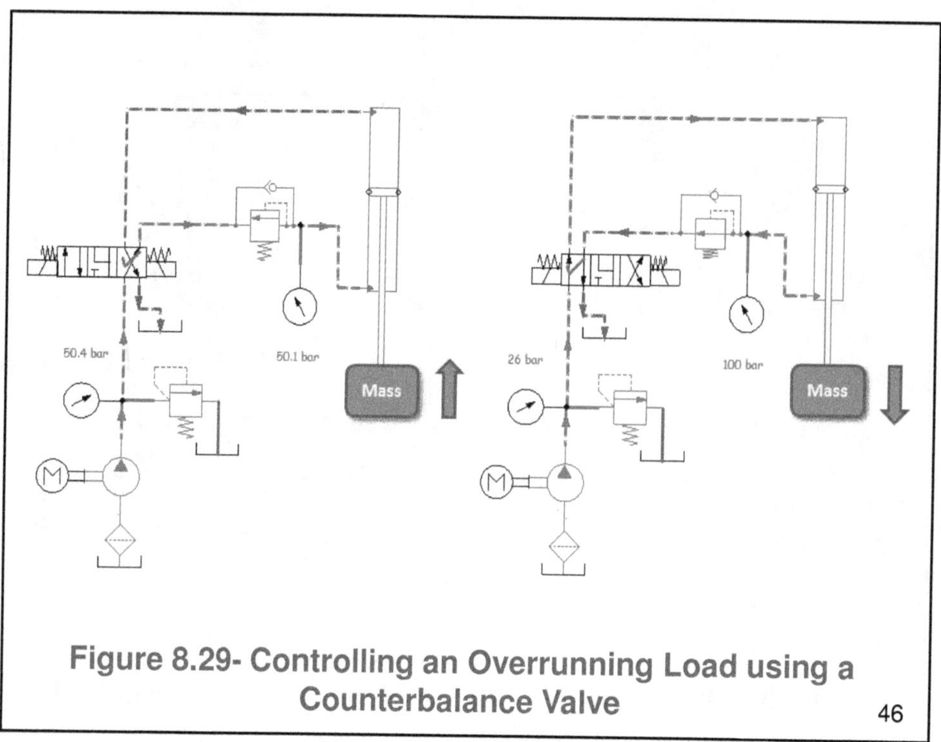

Figure 8.29- Controlling an Overrunning Load using a Counterbalance Valve

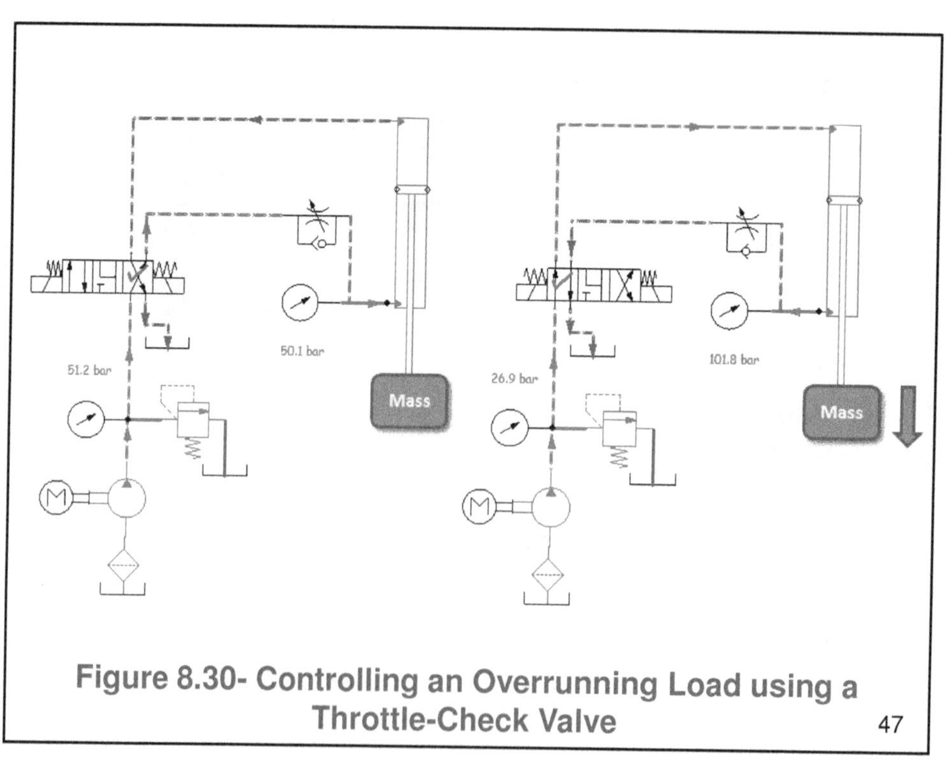

Figure 8.30- Controlling an Overrunning Load using a Throttle-Check Valve

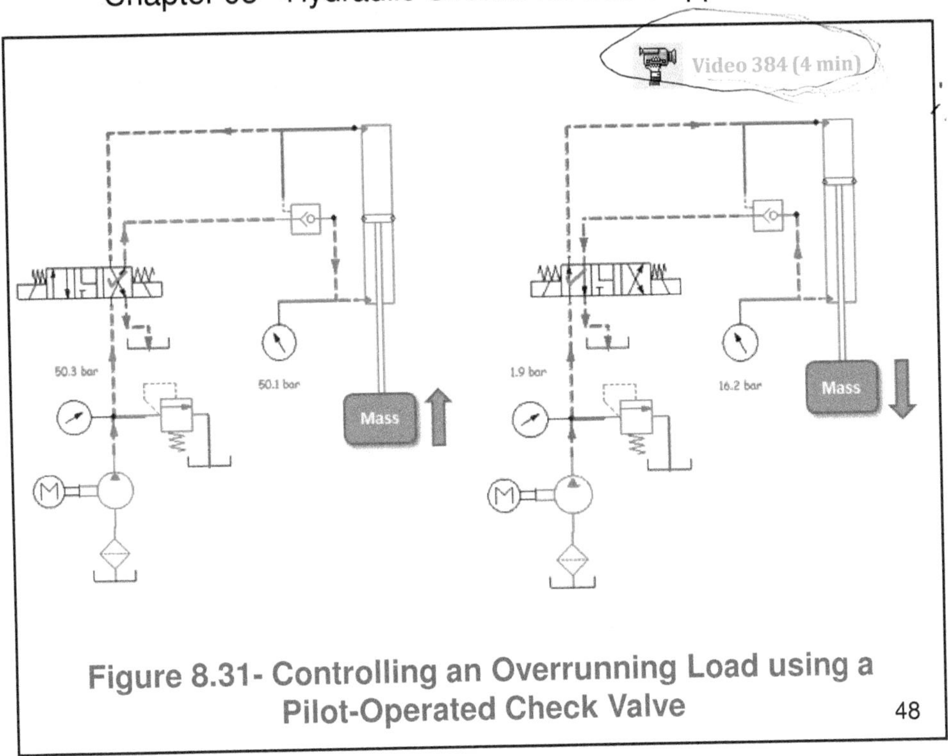

Figure 8.31- Controlling an Overrunning Load using a Pilot-Operated Check Valve

48

8.3.3- Motion Control of a Single Actuator against an Over-Center Load

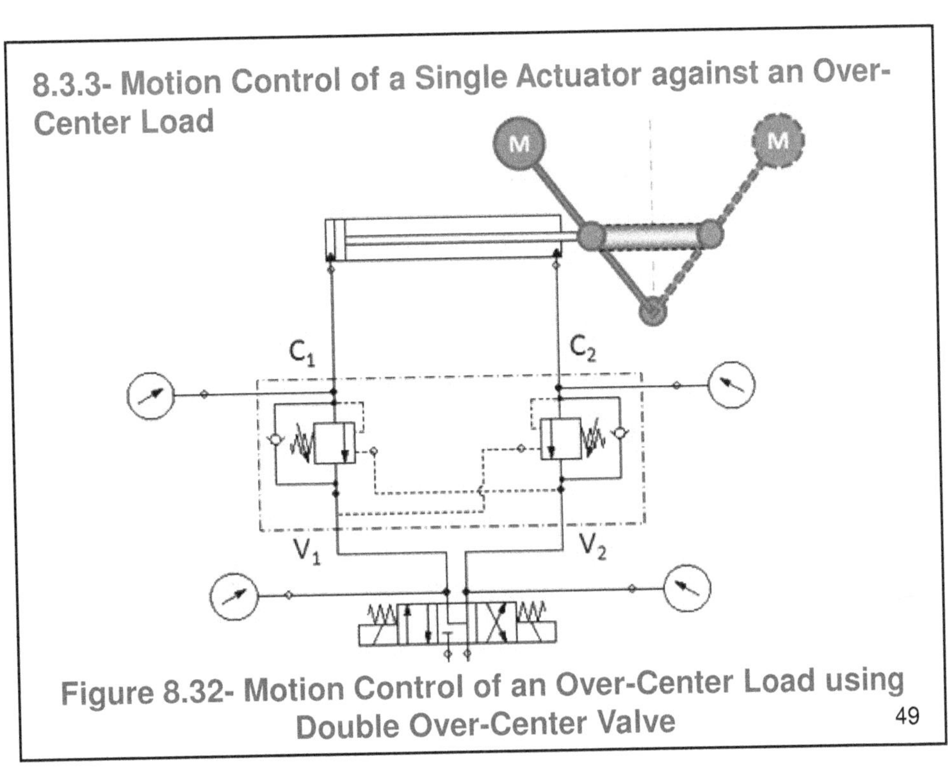

Figure 8.32- Motion Control of an Over-Center Load using Double Over-Center Valve

49

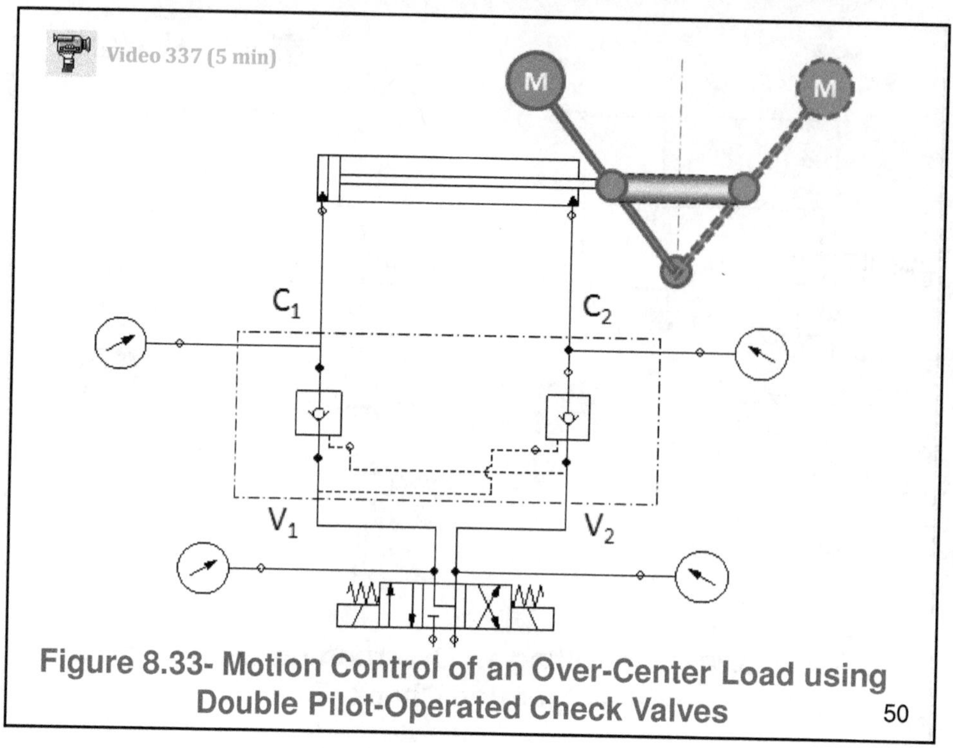

Figure 8.33- Motion Control of an Over-Center Load using Double Pilot-Operated Check Valves

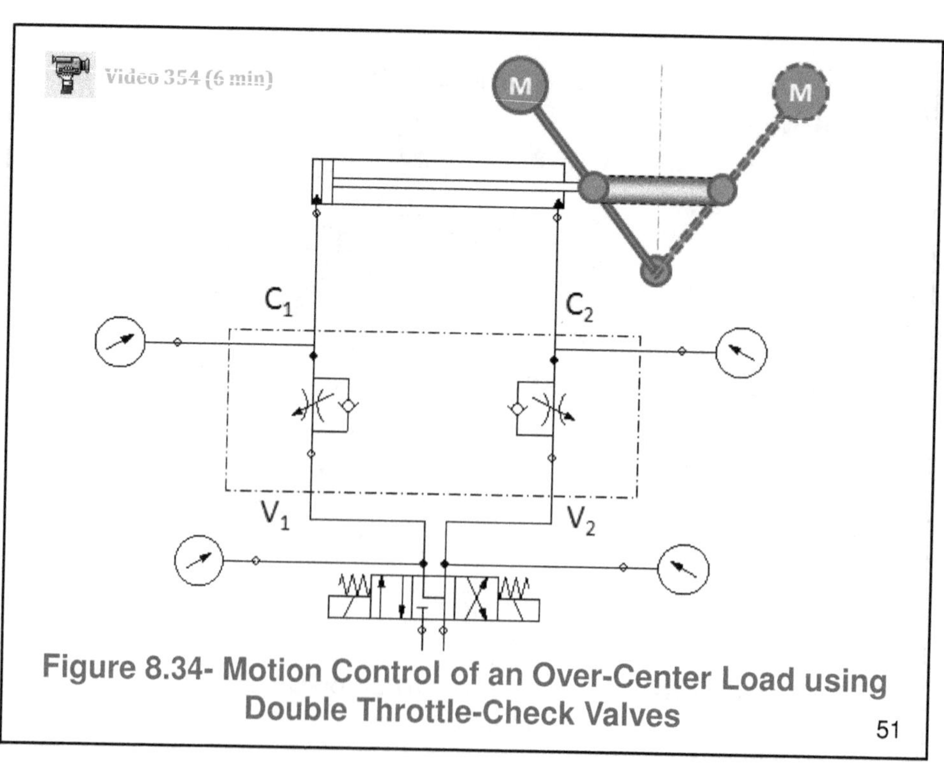

Figure 8.34- Motion Control of an Over-Center Load using Double Throttle-Check Valves

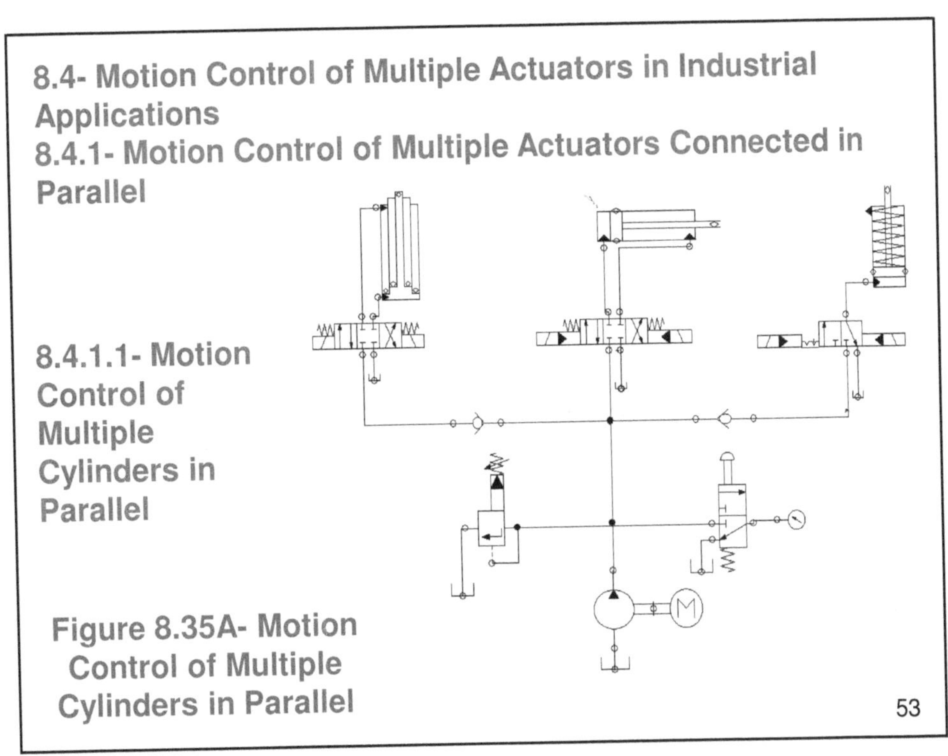

Figure 8.35A- Motion Control of Multiple Cylinders in Parallel

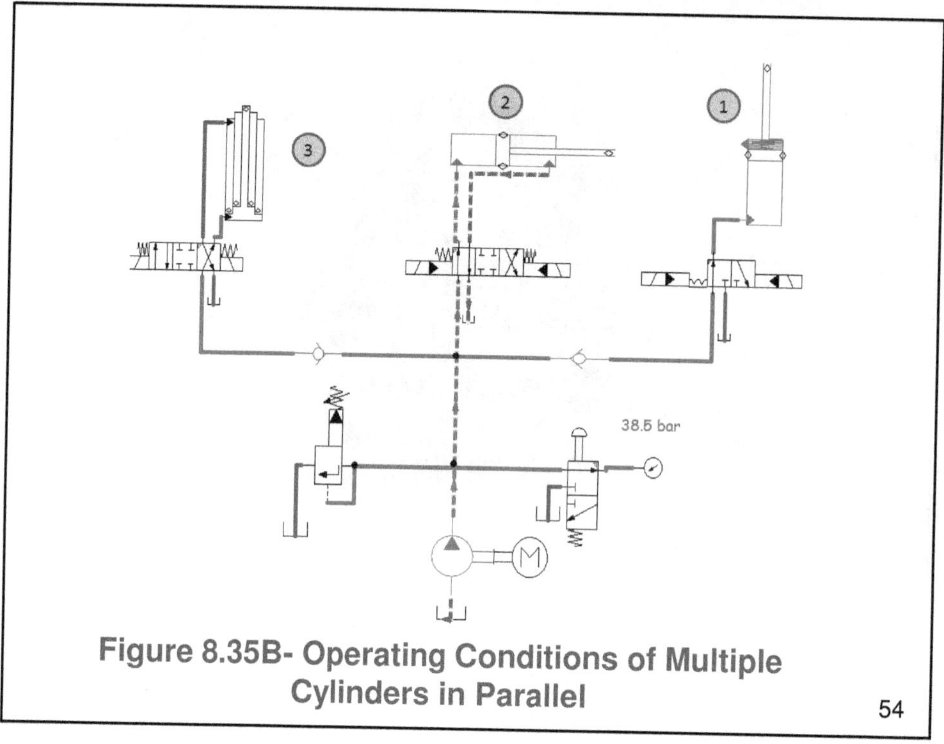

Figure 8.35B- Operating Conditions of Multiple Cylinders in Parallel

54

8.4.1.2- Motion Control of Multiple Motors Connected in Parallel

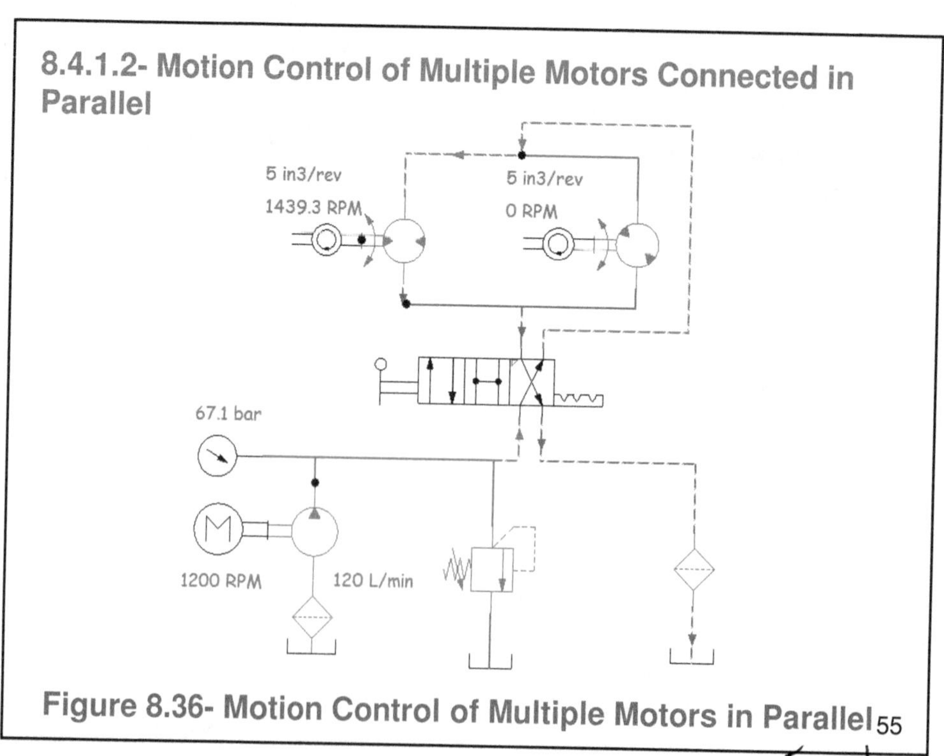

Figure 8.36- Motion Control of Multiple Motors in Parallel 55

Cylinders in Parallel Move in series

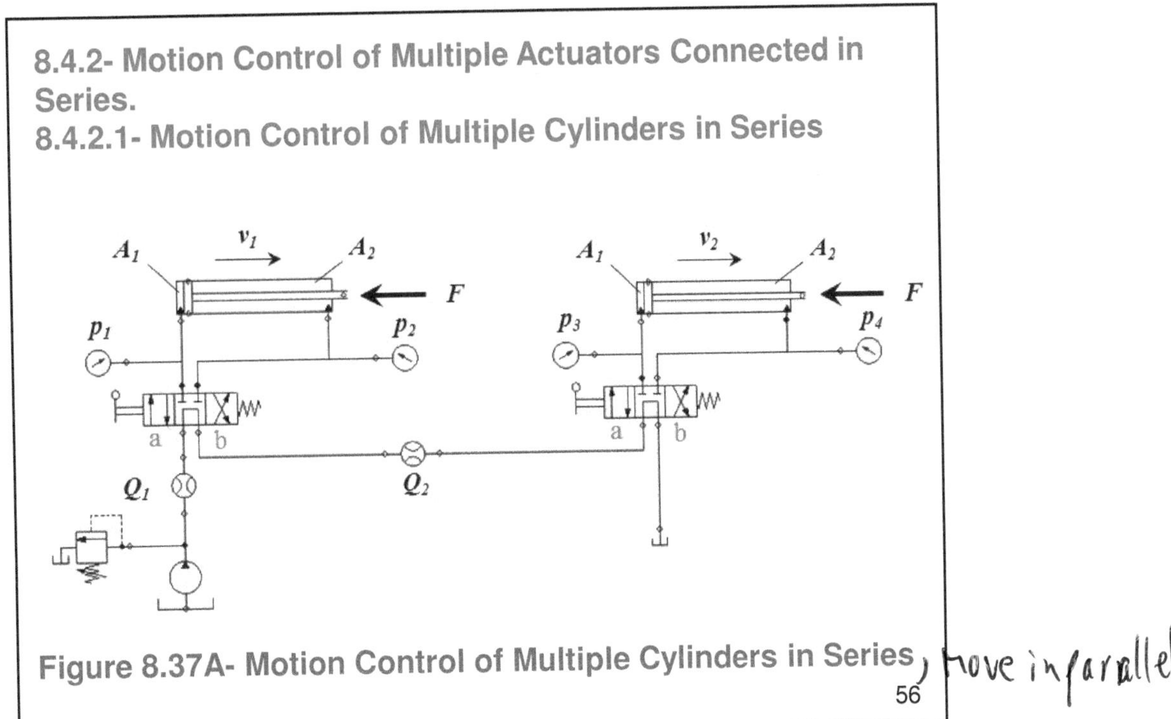
8.4.2- Motion Control of Multiple Actuators Connected in Series.
8.4.2.1- Motion Control of Multiple Cylinders in Series
A1
v1
A2
F
p1
p2
a
b
Q1
Q2
A1
v2
A2
F
p3
p4
a
b
Figure 8.37A- Motion Control of Multiple Cylinders in Series
56

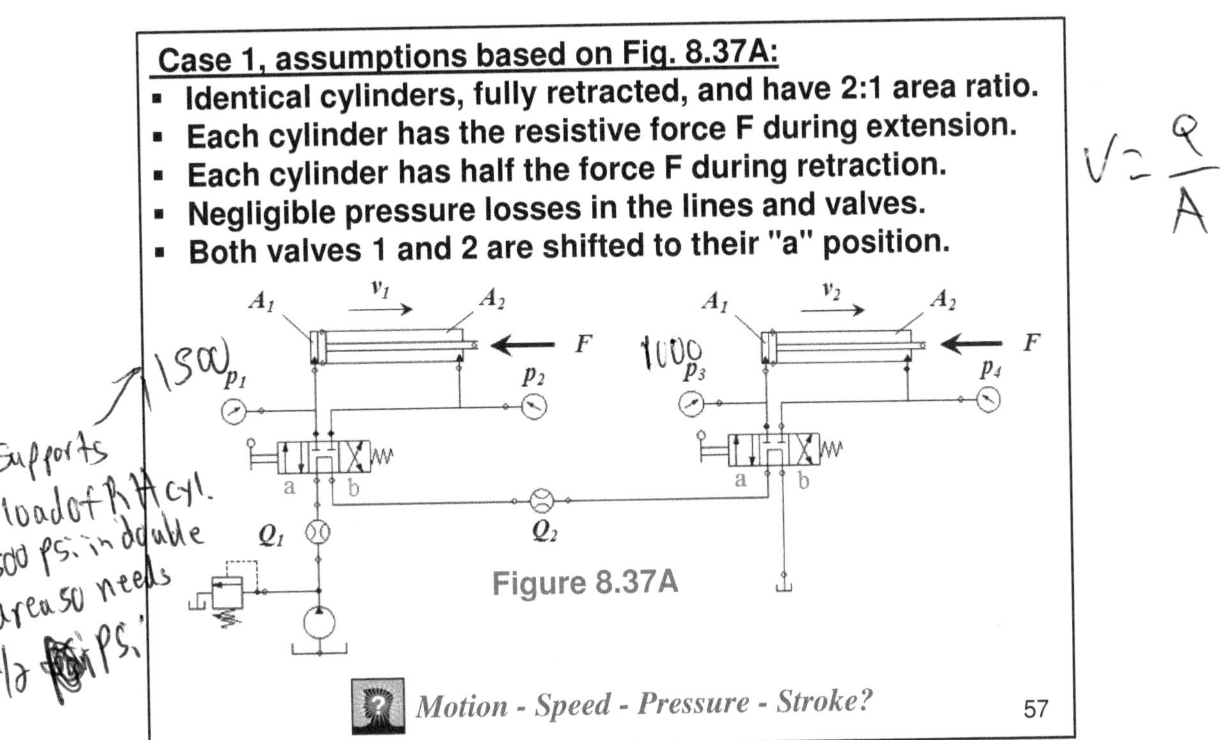
Case 1, assumptions based on Fig. 8.37A:
▪ Identical cylinders, fully retracted, and have 2:1 area ratio.
▪ Each cylinder has the resistive force F during extension.
▪ Each cylinder has half the force F during retraction.
▪ Negligible pressure losses in the lines and valves.
▪ Both valves 1 and 2 are shifted to their "a" position.
A1
v1
A2
F
p1
p2
a
b
Q1
Q2
A1
v2
A2
F
p3
p4
a
b
Figure 8.37A
Motion - Speed - Pressure - Stroke?
57

$$Speed: \; v_1 = \frac{Q_1}{A_1} = \frac{Q_2}{A_2} \Rightarrow Q_2 = \frac{A_2}{A_1} \times Q_1 = 0.5\, Q_1 \;;$$

$$v_2 = \frac{Q_2}{A_1} = \frac{0.5\, Q_1}{A_1} \Rightarrow v_2 = 0.5\, v_1 \qquad 8.1$$

MI-2 6.01 cm/s
MI-3 2.48 cm/s
p_1 150.8 bar
p_2 101 bar
p_3 100.8 bar
p_4 0.1 bar
Q_1 23.61 L/min
Q_2 11.72 L/min

Figure 8.37B- Motion, Speed and Pressure Distribution of Multiple Cylinders in Series, Case 1

58

$$Pressure: \; p_1A_1 = F + p_2A_2 \Rightarrow p_1 = \frac{F}{A_1} + \frac{p_2A_2}{A_1} \;;$$

$$p_3 = \frac{F}{A_1} = ; \; p_2 = p_3 \Rightarrow p_1 = p_3 + 0.5p_3 = 1.5\, p_3 \qquad 8.2$$

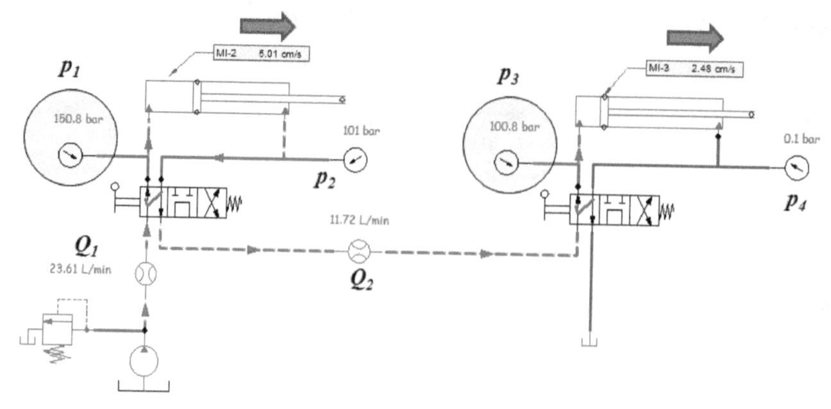

Figure 8.37B

59

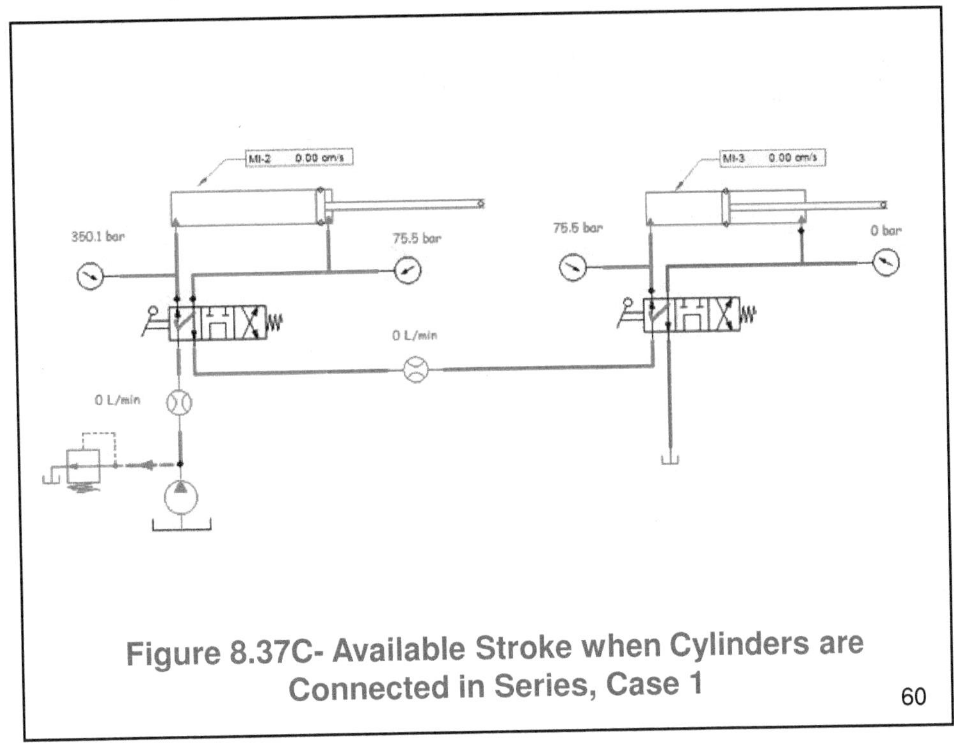

Figure 8.37C- Available Stroke when Cylinders are Connected in Series, Case 1

60

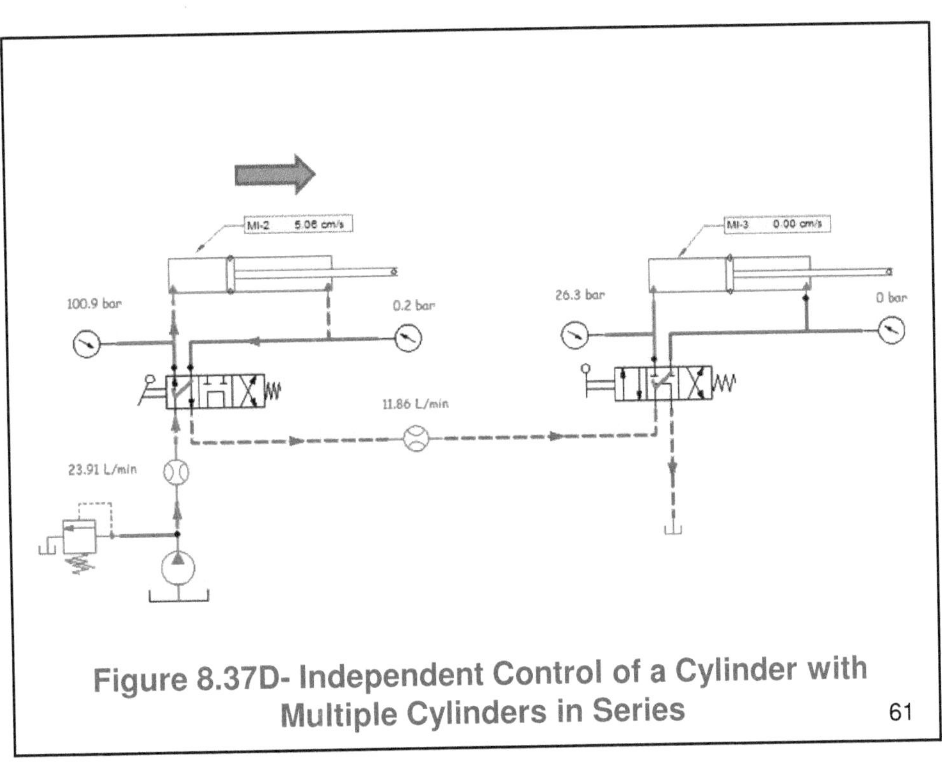

Figure 8.37D- Independent Control of a Cylinder with Multiple Cylinders in Series

61

Case 2, assumptions based on Fig. 8.37A:

- **Identical cylinders and have 2:1 area ratio.**
- **Cylinder 1 is fully extended.**
- **Cylinder 2 is fully retracted.**
- **Valve 1 is shifted to "b" and valve 2 is shifted to "a".**
- **Same loading conditions as in case 1.**
- **Negligible pressure losses in the lines and valves.**

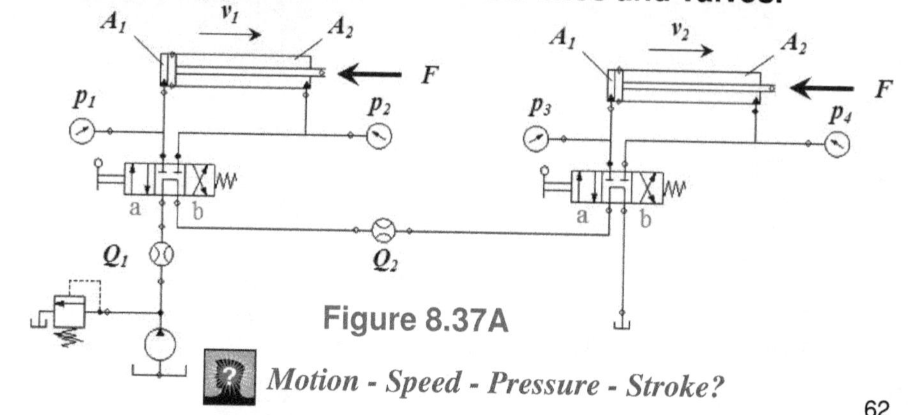

Figure 8.37A

Motion - Speed - Pressure - Stroke?

62

$$Speed: v_1 = \frac{Q_1}{A_2} = \frac{Q_2}{A_1} \Rightarrow Q_2 = \frac{A_1}{A_2} \times Q_1 = 2\, Q_1 \;;$$

$$v_2 = \frac{Q_2}{A_1} = \frac{2\, Q_1}{2\, A_2} \Rightarrow v_2 = v_1 \qquad 8.3$$

Figure 8.37E- Motion, Speed and Pressure Distribution of Multiple Cylinders in Series, Case 2

63

$$\textit{Pressure}: p_2 A_2 = 0.5F + p_1 A_1 \Rightarrow p_2 = \frac{0.5F}{A_2} + \frac{p_1 A_1}{A_2}$$

$$= \frac{0.5F}{0.5\,A_1} + 2\,p_1; \quad p_3 = \frac{F}{A_1} = ;\ p_1 = p_3$$

$$\Rightarrow p_2 = p_3 + 2p_3 = 3\,p_3$$

8.4

Figure 8.37E

64

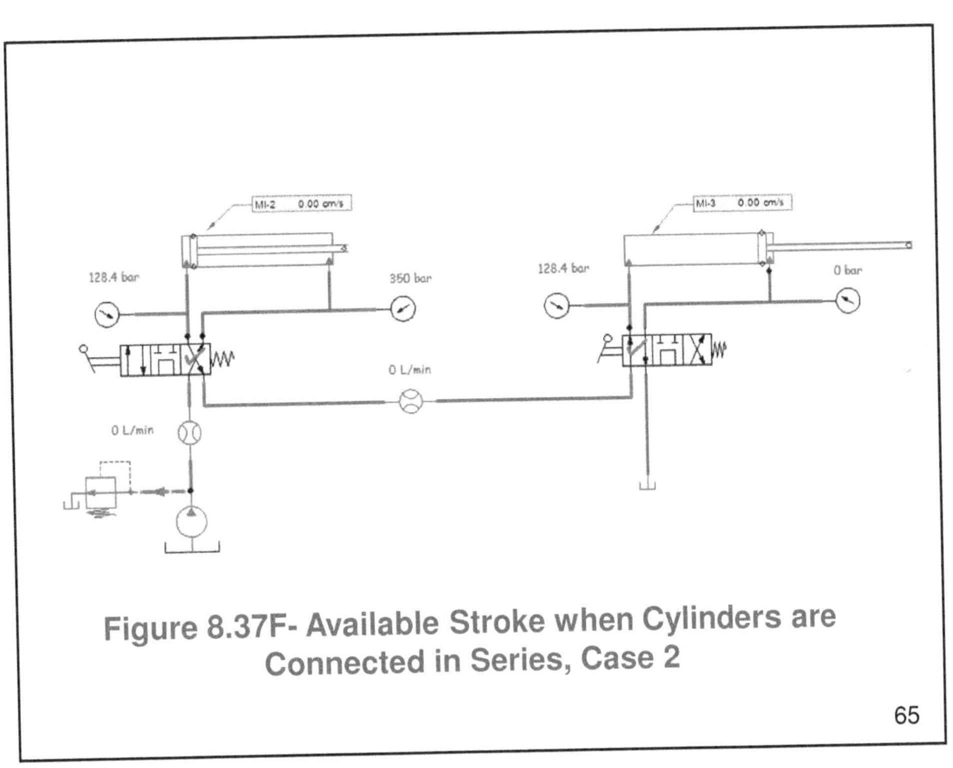

Figure 8.37F- Available Stroke when Cylinders are Connected in Series, Case 2

65

8.4.2.2- Motion Control of Multiple Motors in Series

Case 1, assumptions based on Fig. 8.38A:

- **Motors are identical in size and have 100% efficiency.**
- **Both motors support a resistive torque of an equal value for both directions of rotation.**
- **Negligible pressure losses in the lines and valves.**
- **Both valves 1 and 2 are shifted to their "a" position.**

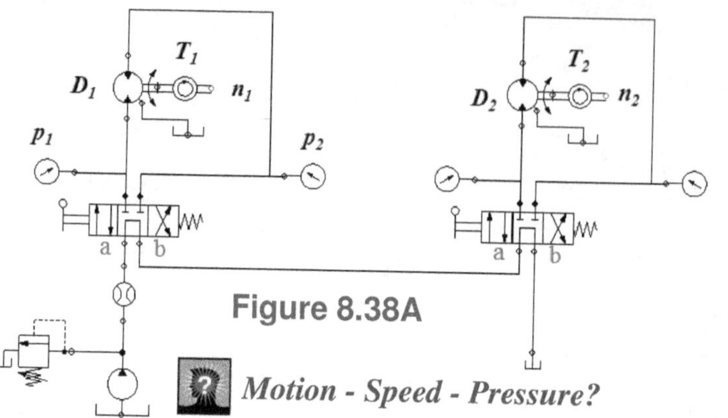

Figure 8.38A

66

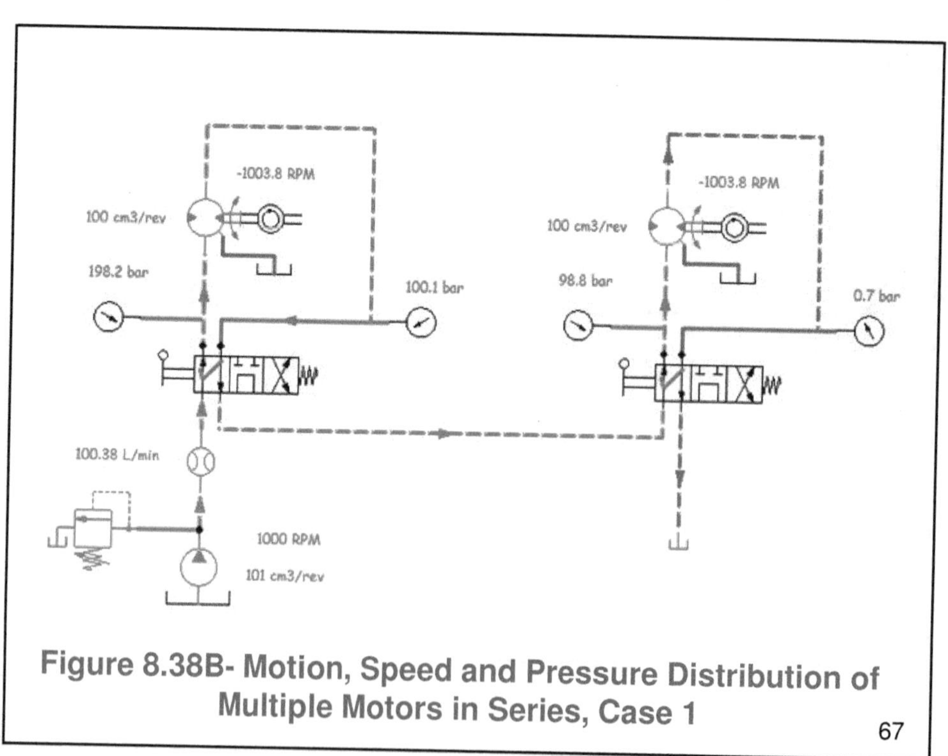

Figure 8.38B- Motion, Speed and Pressure Distribution of Multiple Motors in Series, Case 1

67

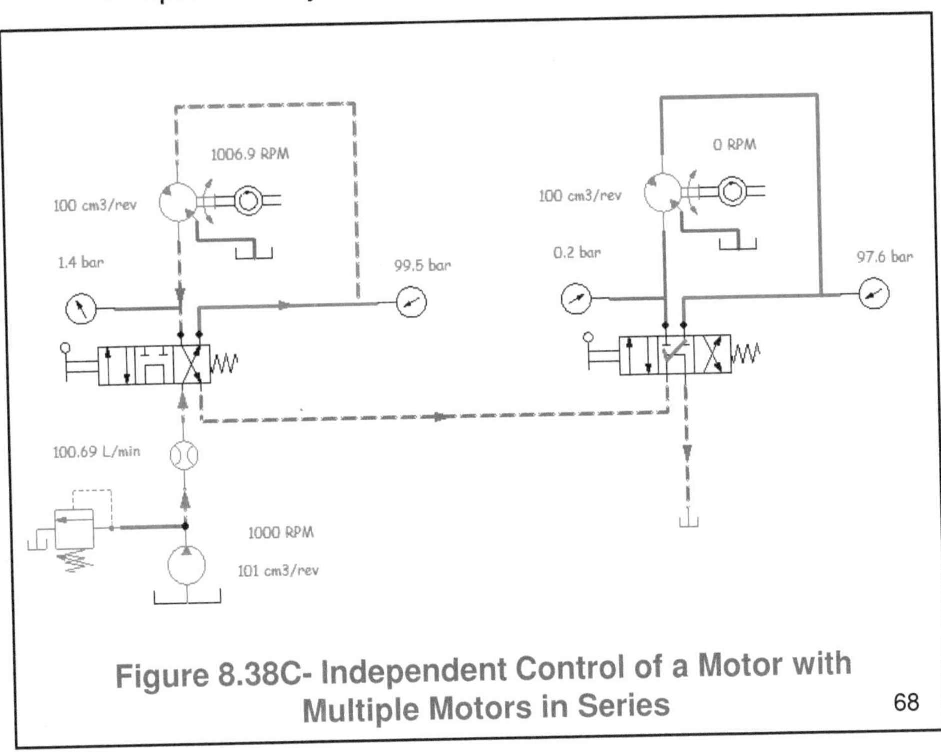

Figure 8.38C- Independent Control of a Motor with Multiple Motors in Series

68

Case 2, assumptions based on Fig. 8.38A:

- **Motor 1 has double the size of Motor 2, i.e. $D_1 = 2\, D_2$.**
- **Both valves 1 and 2 are shifted to their "a" position.**
- **Same loading conditions as in case 1.**
- **Negligible pressure losses in the lines and valves.**

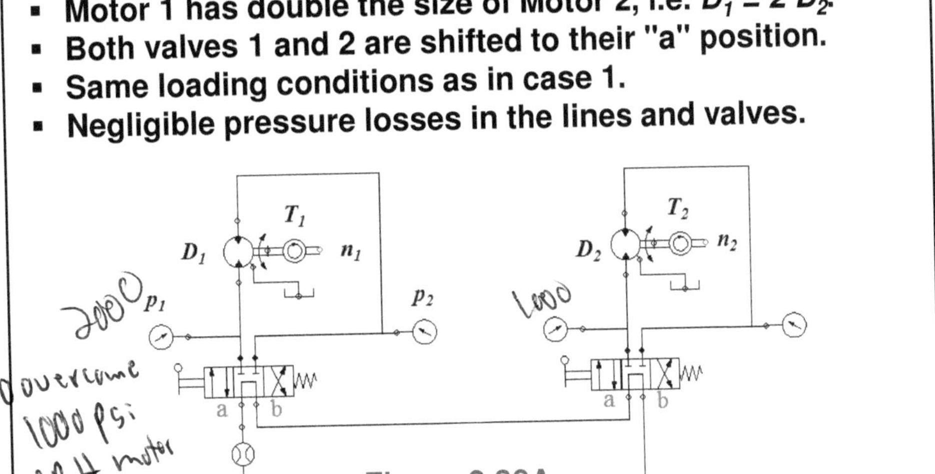

Figure 8.38A

Motion - Speed - Pressure?

69

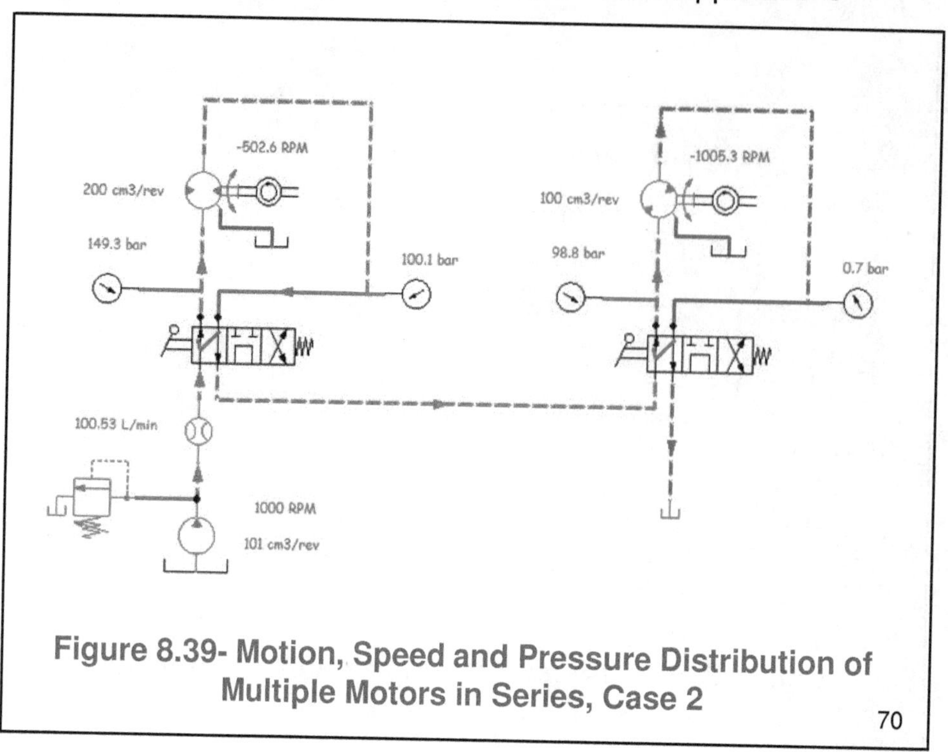

Figure 8.39- Motion, Speed and Pressure Distribution of Multiple Motors in Series, Case 2

70

8.5- Motion Control of Multiple Act. in Mobile Applications.
8.5.1- Parallel Control Blocks in Mobile Machines

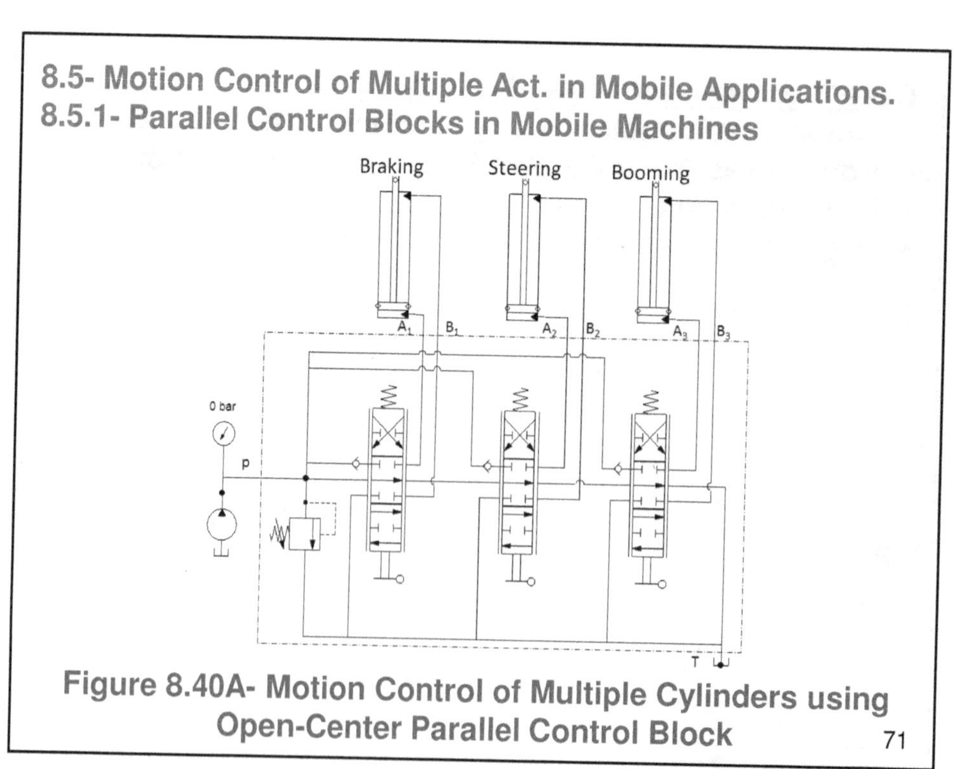

Figure 8.40A- Motion Control of Multiple Cylinders using Open-Center Parallel Control Block

71

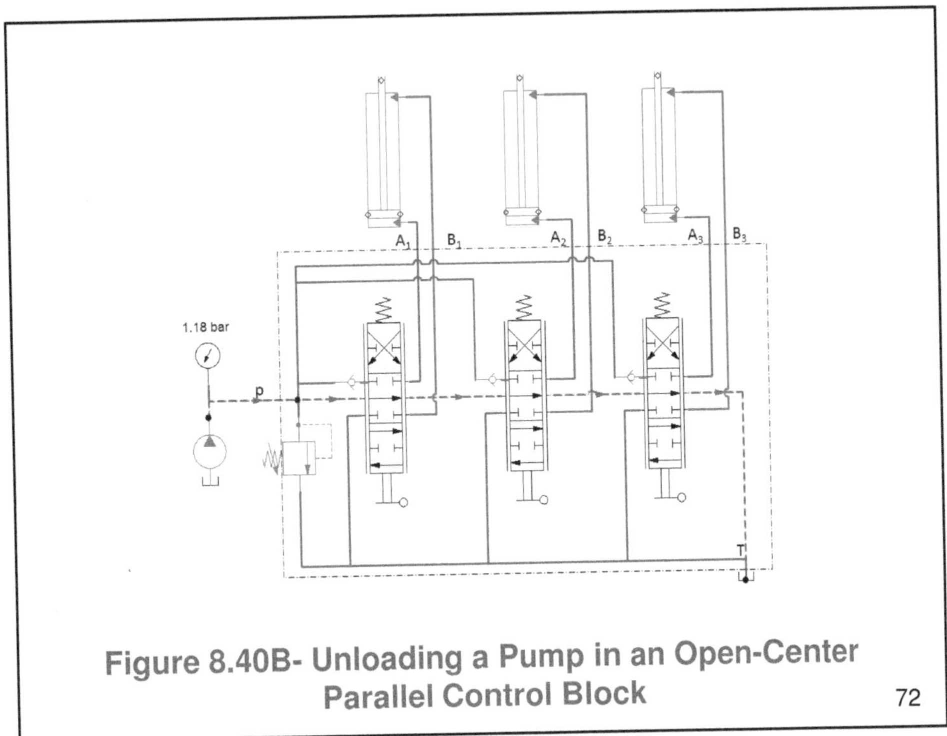

Figure 8.40B- Unloading a Pump in an Open-Center Parallel Control Block 72

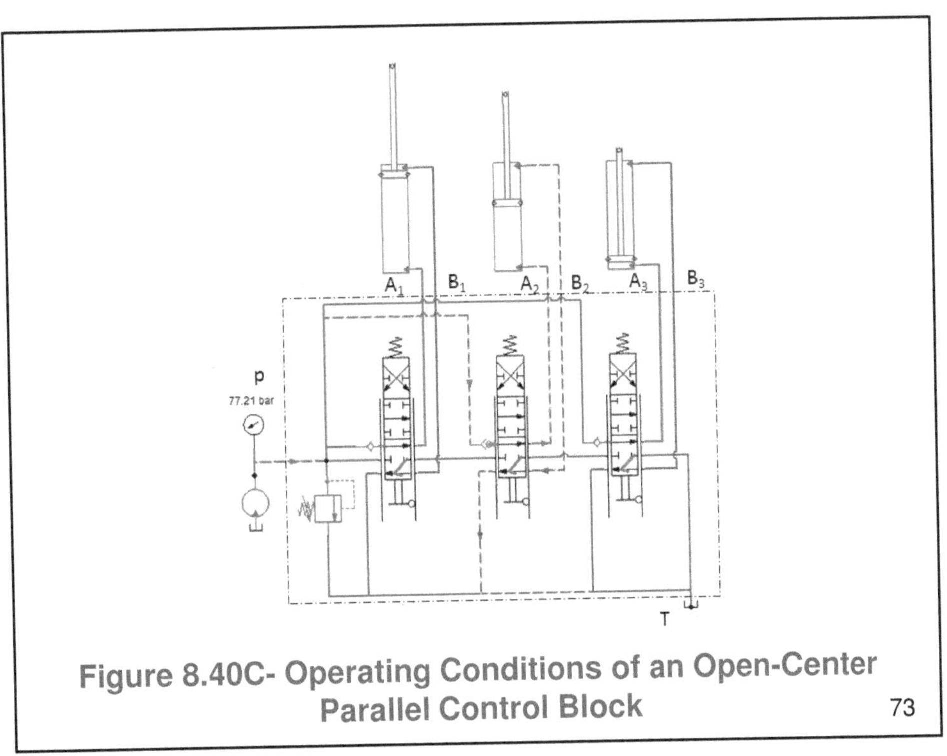

Figure 8.40C- Operating Conditions of an Open-Center Parallel Control Block 73

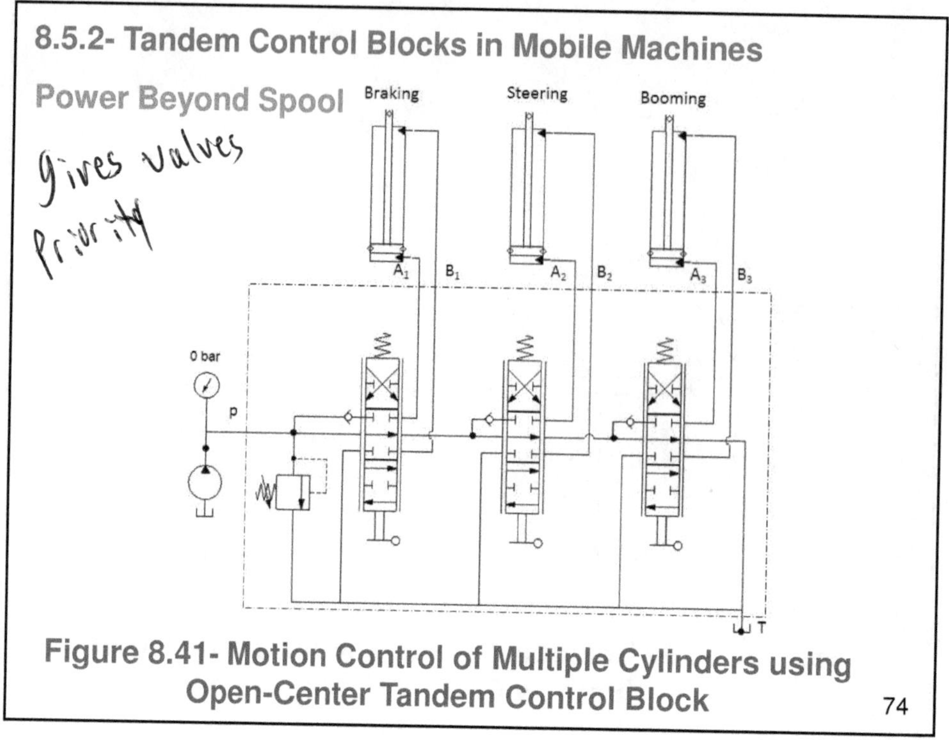

Figure 8.41- Motion Control of Multiple Cylinders using Open-Center Tandem Control Block

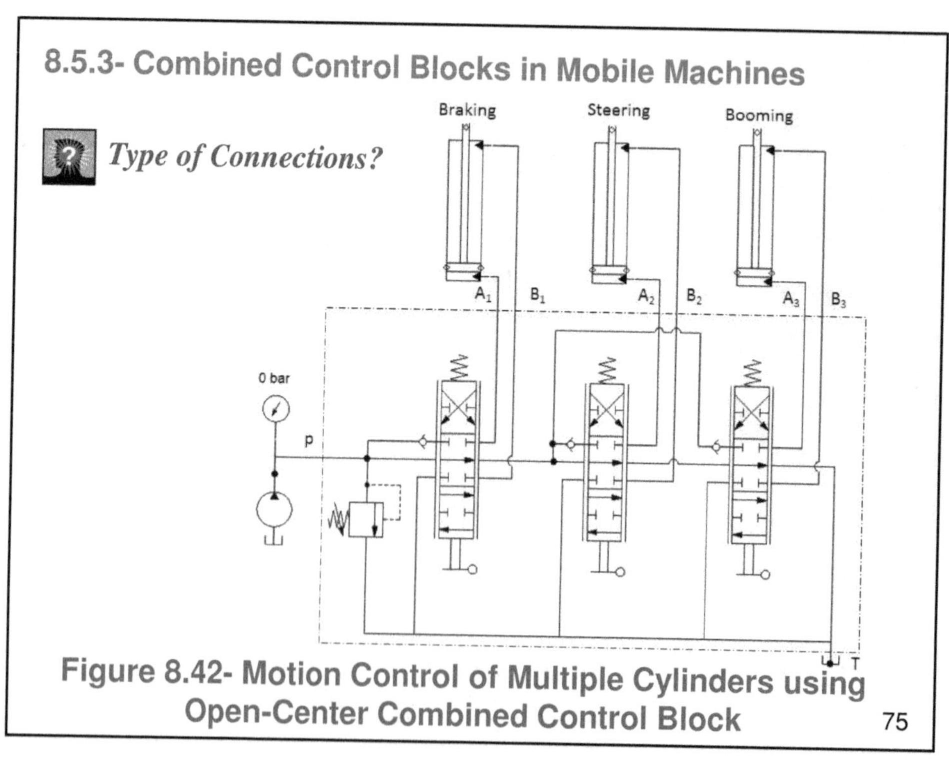

Figure 8.42- Motion Control of Multiple Cylinders using Open-Center Combined Control Block

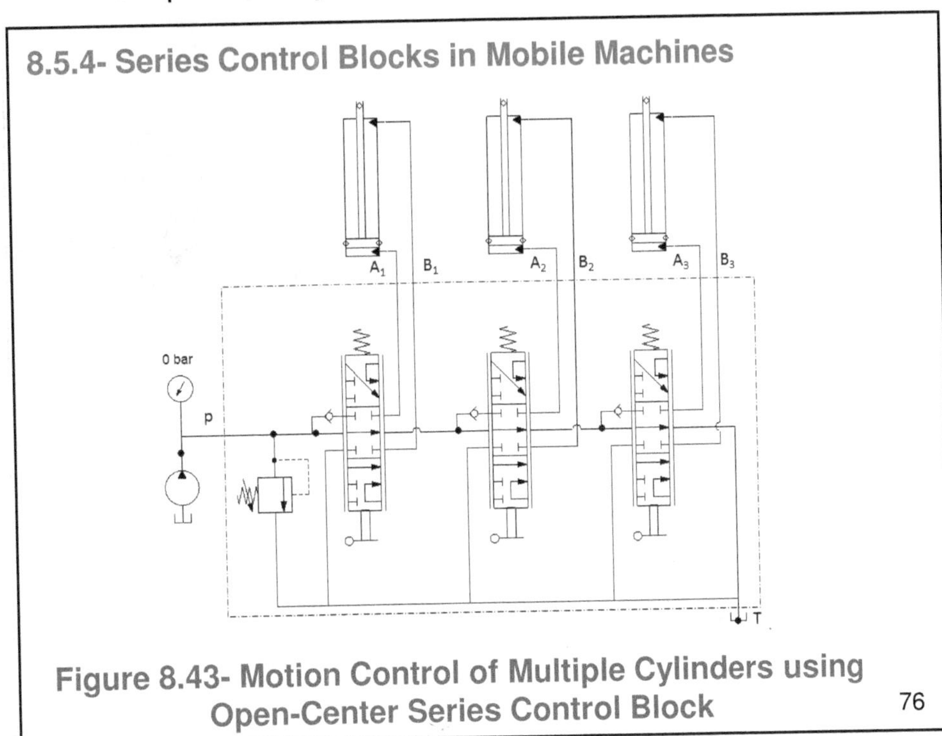
8.5.4- Series Control Blocks in Mobile Machines
A1
B1
A2
B2
A3
B3
0 bar
p
T
Figure 8.43- Motion Control of Multiple Cylinders using Open-Center Series Control Block
76

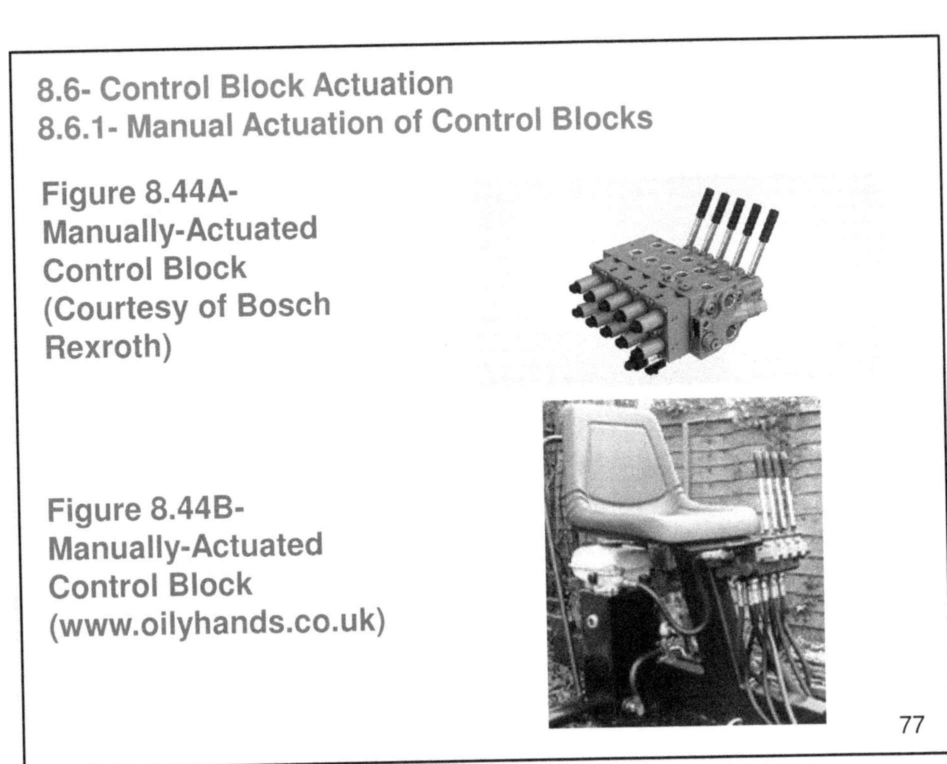
8.6- Control Block Actuation
8.6.1- Manual Actuation of Control Blocks
Figure 8.44A- Manually-Actuated Control Block (Courtesy of Bosch Rexroth)
Figure 8.44B- Manually-Actuated Control Block (www.oilyhands.co.uk)
77

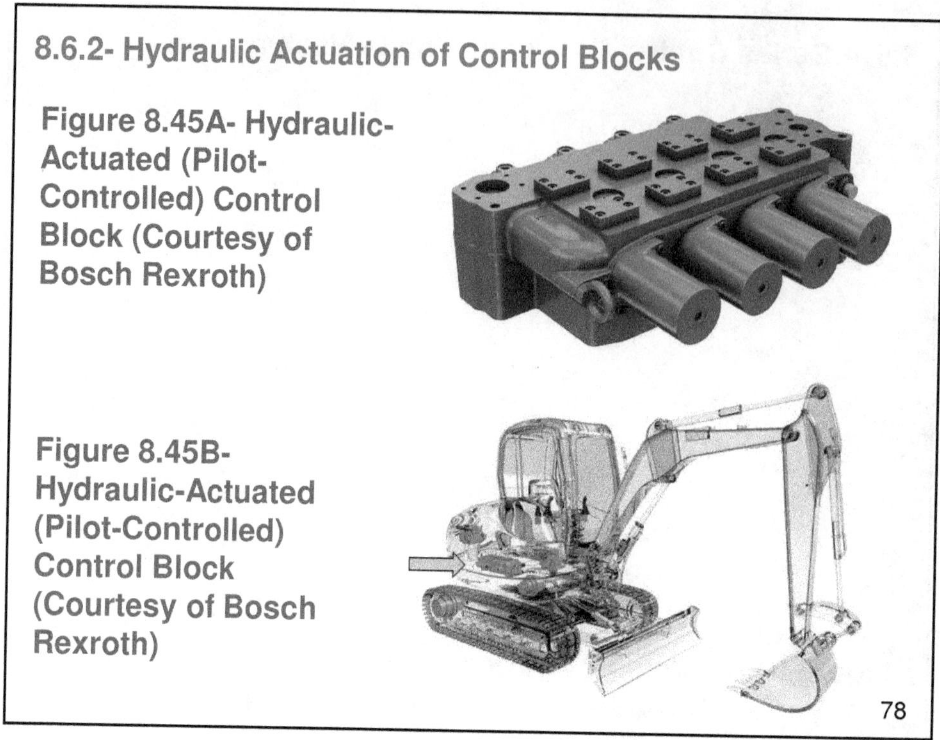

8.6.2- Hydraulic Actuation of Control Blocks

Figure 8.45A- Hydraulic-Actuated (Pilot-Controlled) Control Block (Courtesy of Bosch Rexroth)

Figure 8.45B- Hydraulic-Actuated (Pilot-Controlled) Control Block (Courtesy of Bosch Rexroth)

78

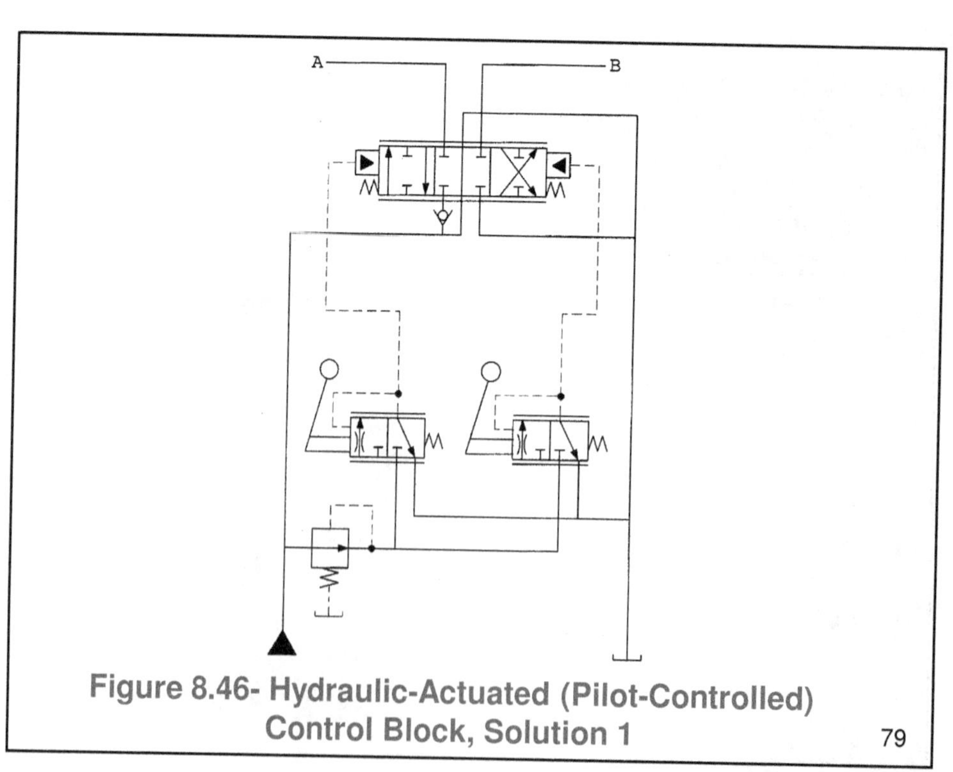

Figure 8.46- Hydraulic-Actuated (Pilot-Controlled) Control Block, Solution 1

79

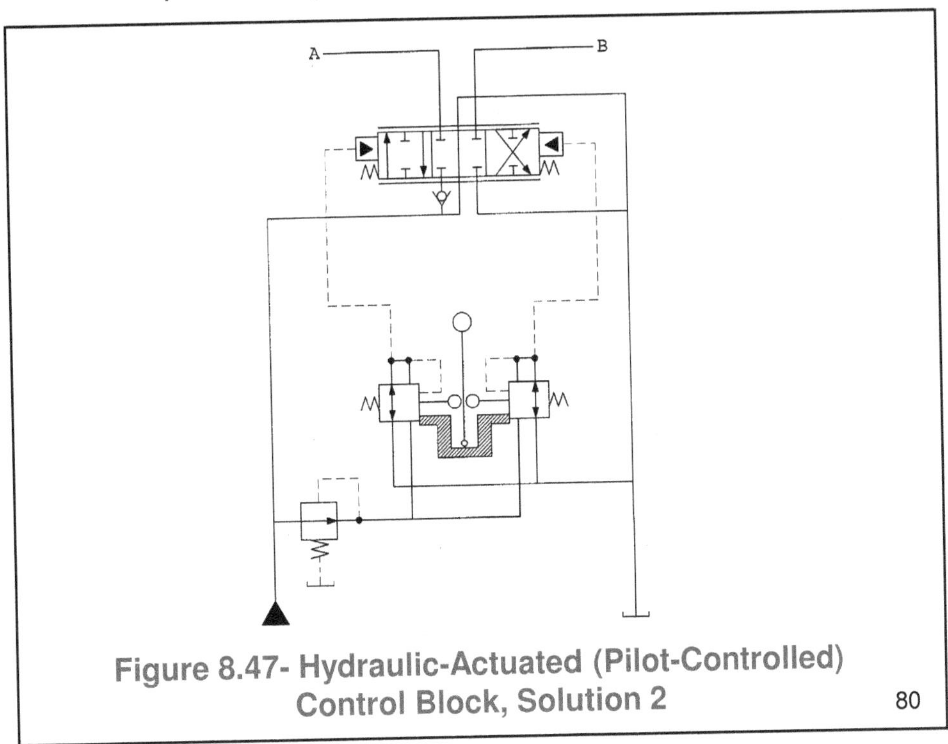

Figure 8.47- Hydraulic-Actuated (Pilot-Controlled) Control Block, Solution 2

80

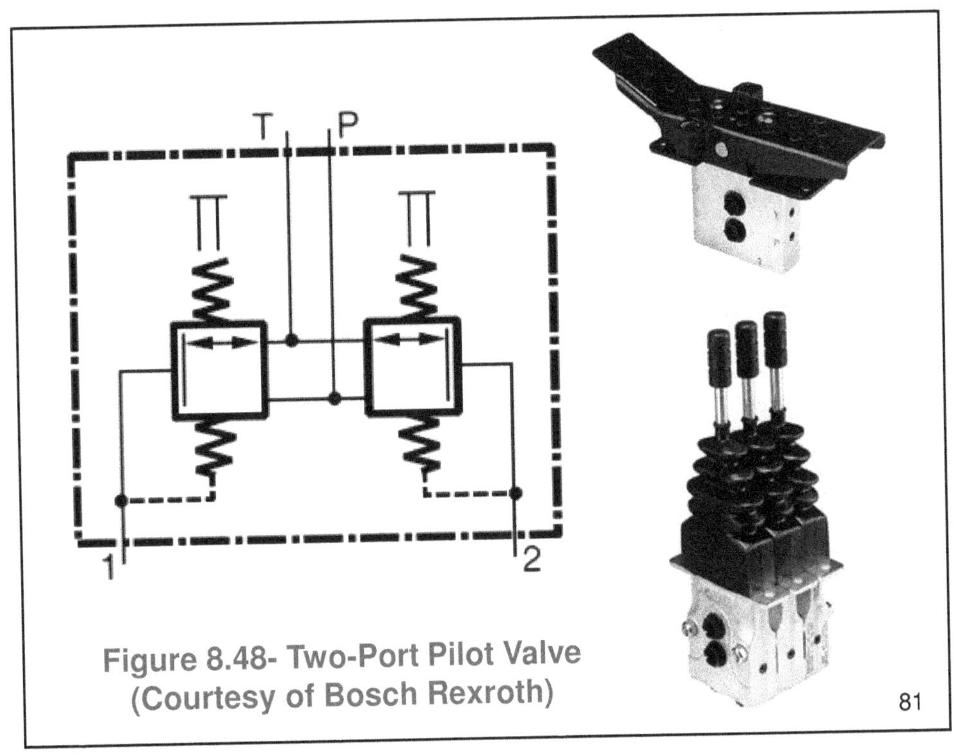

Figure 8.48- Two-Port Pilot Valve (Courtesy of Bosch Rexroth)

81

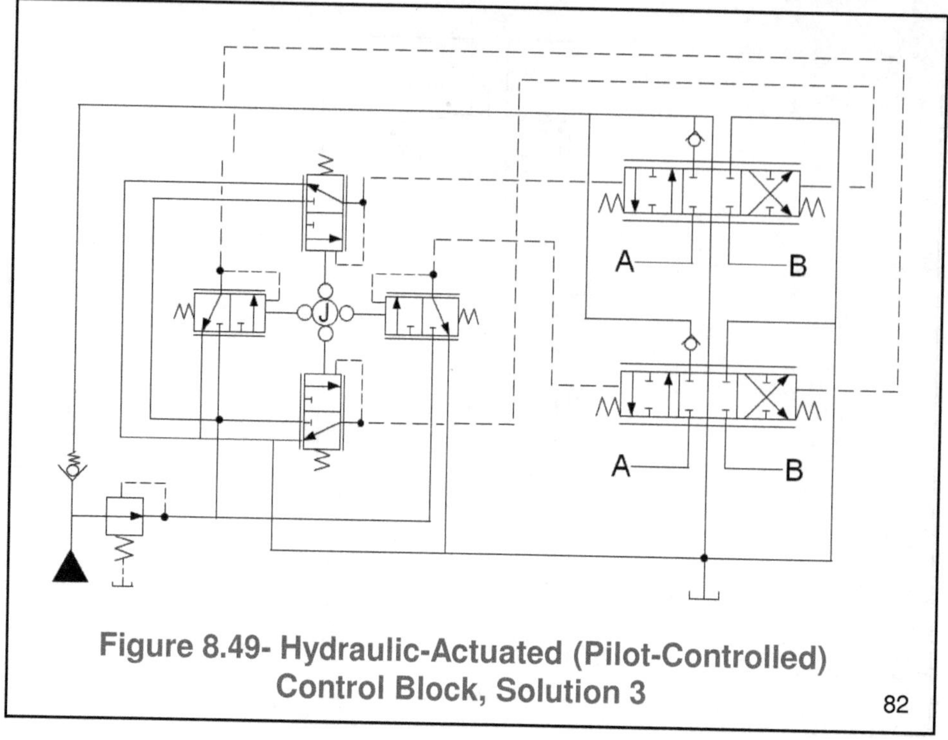

Figure 8.49- Hydraulic-Actuated (Pilot-Controlled) Control Block, Solution 3

82

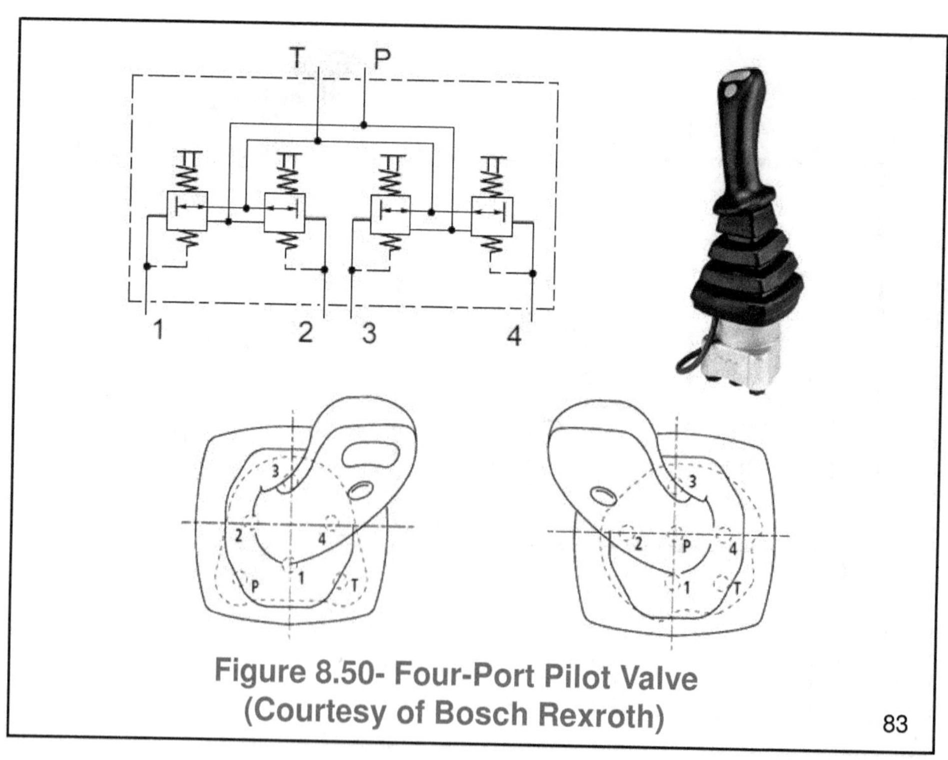

Figure 8.50- Four-Port Pilot Valve (Courtesy of Bosch Rexroth)

83

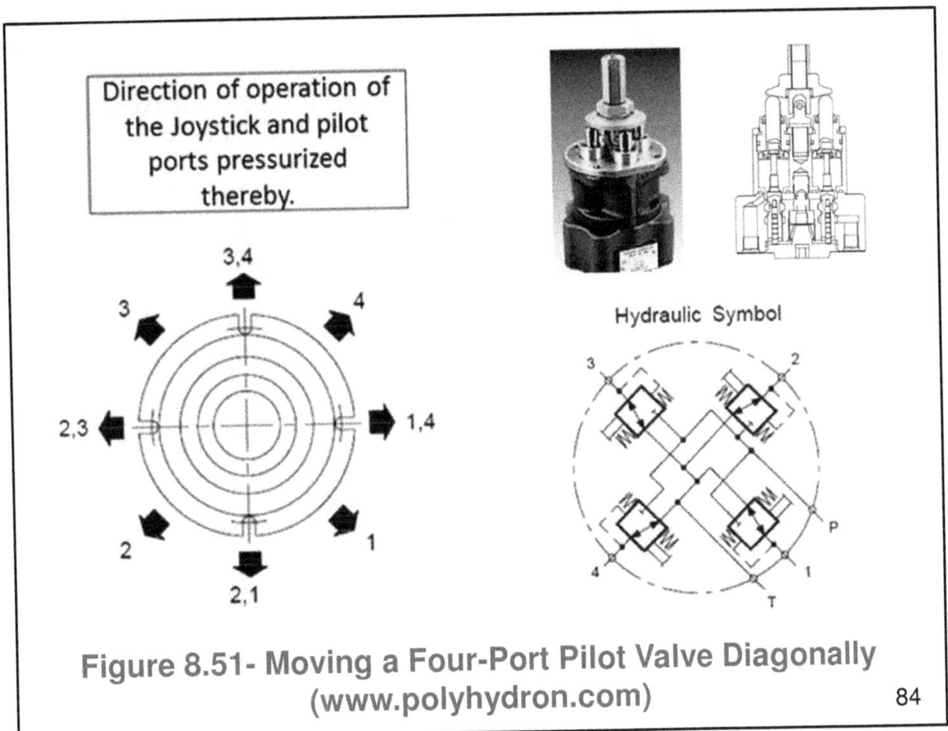

Figure 8.51- Moving a Four-Port Pilot Valve Diagonally (www.polyhydron.com)

8.6.3- Electric Actuation of Control Blocks

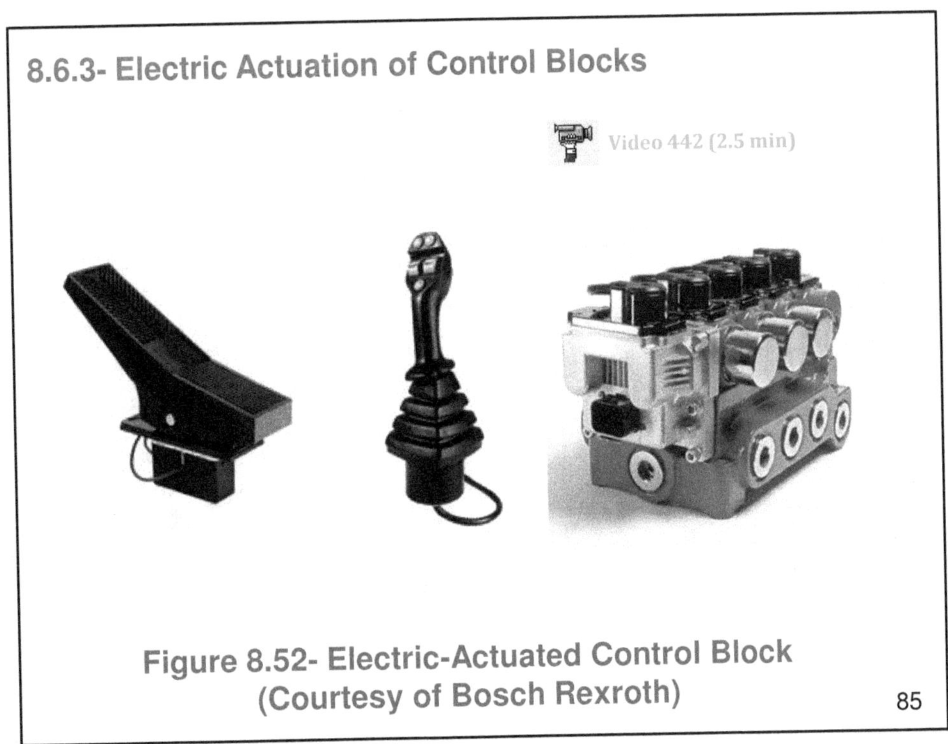

Figure 8.52- Electric-Actuated Control Block (Courtesy of Bosch Rexroth)

8.7- Speed Control of a Hydraulic Actuator

8.7.1- Valve-Controlled Actuators versus Pump-Controlled Actuators

Mobile Heavy Equipment - High Efficiency Excavator (Test Bed 1)

In a typical hydraulic excavator, flow from the main pump(s) is routed through a control valve to the various actuators that allow the machine to drive, rotate, and dig.

CONTROL VALVES

ENGINE

PUMPS

EXCAVATOR

ILLUSTRATION BY TIM MONTGOMERY/TJMONTGOMERY.COM

Figure 8.53- Valve-Controlled Actuators Concept (Courtesy of CCEFP)

86

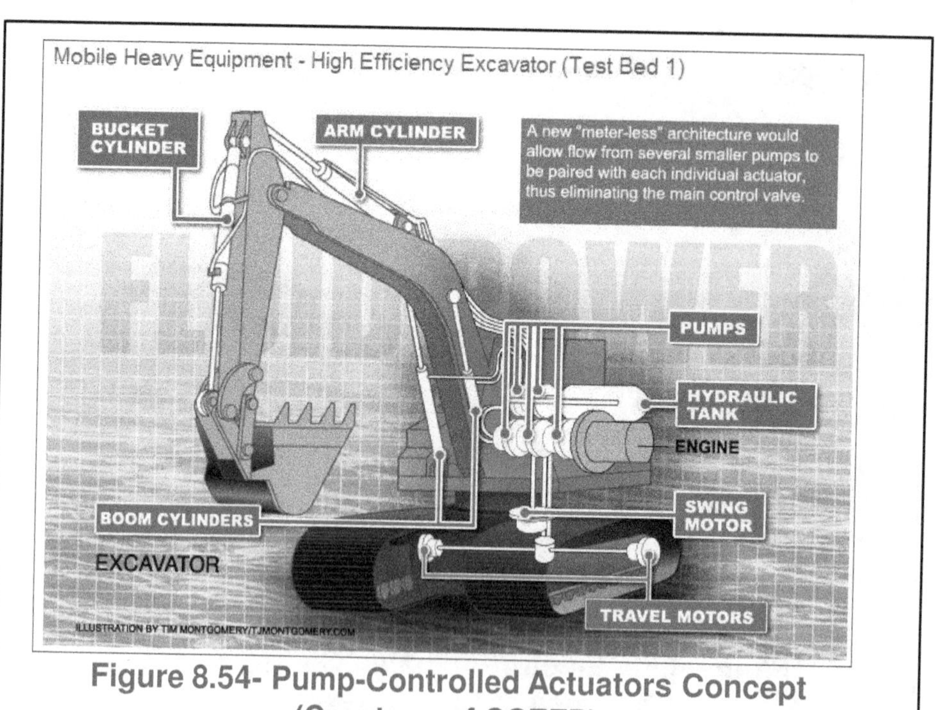

Figure 8.54- Pump-Controlled Actuators Concept (Courtesy of CCEFP)

87

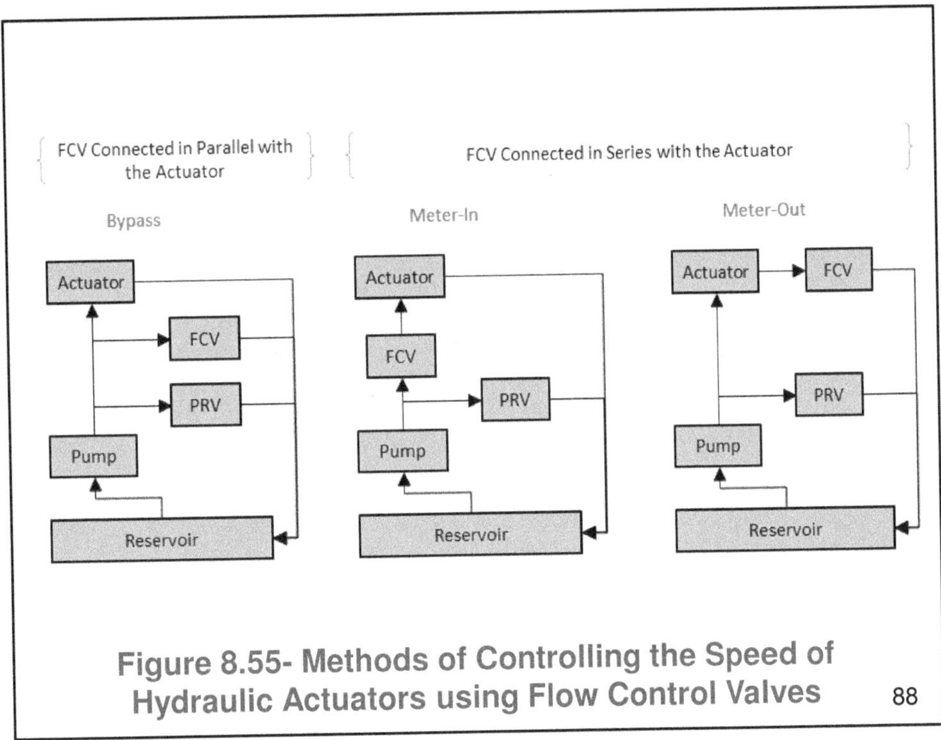

Figure 8.55- Methods of Controlling the Speed of Hydraulic Actuators using Flow Control Valves

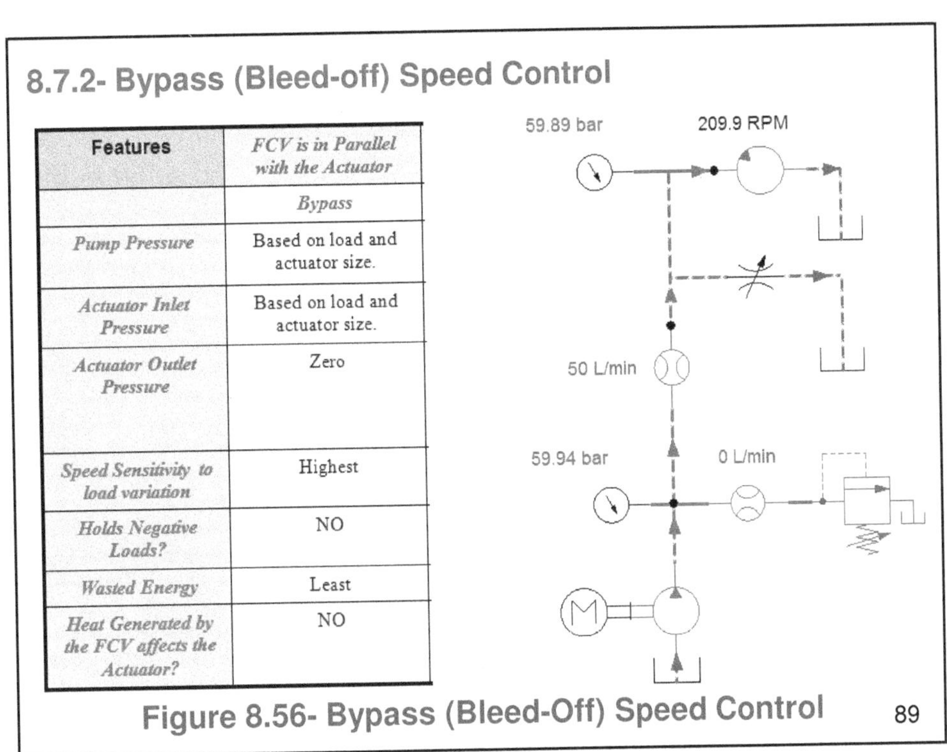

8.7.2- Bypass (Bleed-off) Speed Control

Features	*FCV is in Parallel with the Actuator*
	Bypass
Pump Pressure	Based on load and actuator size.
Actuator Inlet Pressure	Based on load and actuator size.
Actuator Outlet Pressure	Zero
Speed Sensitivity to load variation	Highest
Holds Negative Loads?	NO
Wasted Energy	Least
Heat Generated by the FCV affects the Actuator?	NO

Figure 8.56- Bypass (Bleed-Off) Speed Control

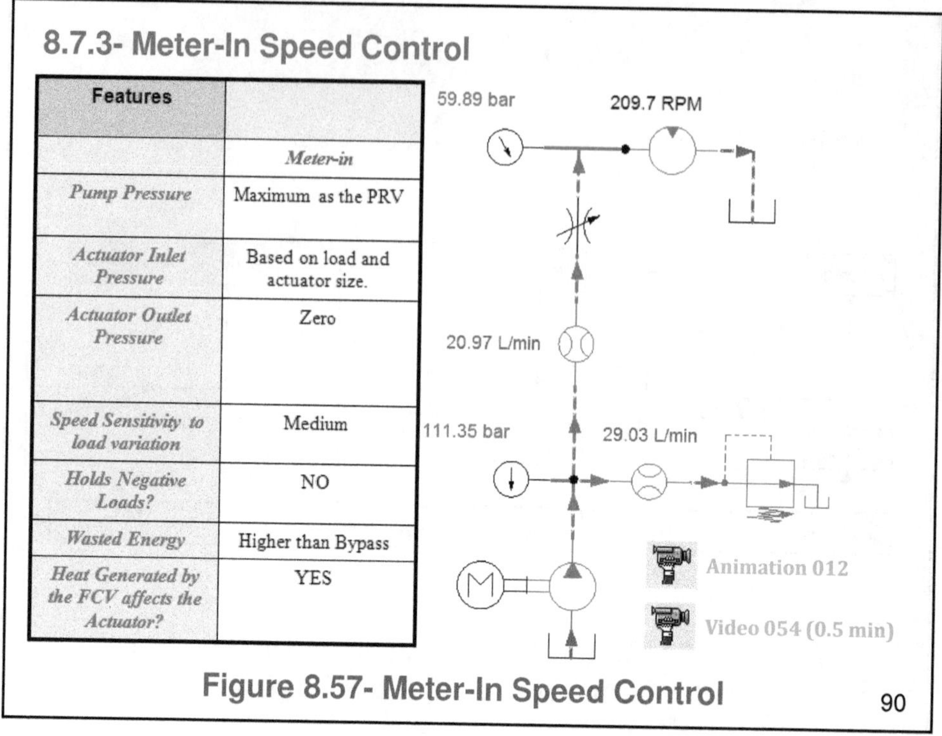

8.7.3- Meter-In Speed Control

Features	
	Meter-in
Pump Pressure	Maximum as the PRV
Actuator Inlet Pressure	Based on load and actuator size.
Actuator Outlet Pressure	Zero
Speed Sensitivity to load variation	Medium
Holds Negative Loads?	NO
Wasted Energy	Higher than Bypass
Heat Generated by the FCV affects the Actuator?	YES

Figure 8.57- Meter-In Speed Control

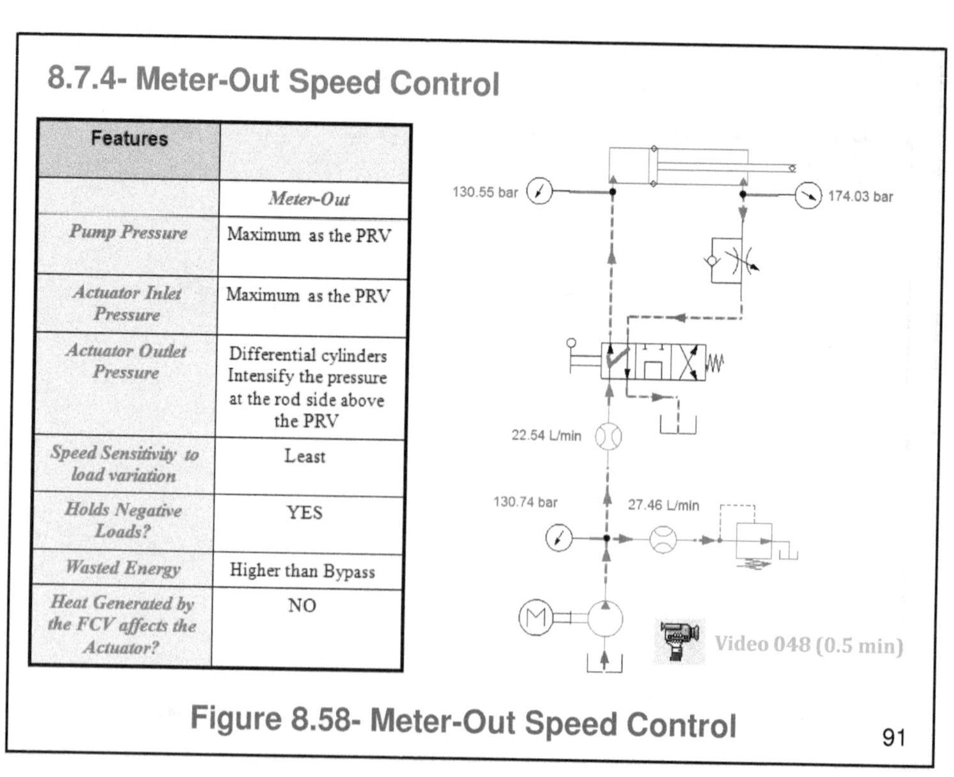

8.7.4- Meter-Out Speed Control

Features	
	Meter-Out
Pump Pressure	Maximum as the PRV
Actuator Inlet Pressure	Maximum as the PRV
Actuator Outlet Pressure	Differential cylinders Intensify the pressure at the rod side above the PRV
Speed Sensitivity to load variation	Least
Holds Negative Loads?	YES
Wasted Energy	Higher than Bypass
Heat Generated by the FCV affects the Actuator?	NO

Figure 8.58- Meter-Out Speed Control

8.7.5- Speed Control Comparative Study

Video 380 (2 min)

Features	*FCV is in Parallel with the Actuator*	*FCV is in Series with the Actuator*	
	Bypass	*Meter-in*	*Meter-Out*
Pump Pressure	Based on load and actuator size.	Maximum as the PRV	Maximum as the PRV
Actuator Inlet Pressure	Based on load and actuator size.	Based on load and actuator size.	Maximum as the PRV
Actuator Outlet Pressure	Zero	Zero	Differential cylinders Intensify the pressure at the rod side above the PRV
Speed Sensitivity to load variation	Highest	Medium	Least
Holds Negative Loads?	NO	NO	YES
Wasted Energy	Least	Higher than Bypass	Higher than Bypass
Heat Generated by the FCV affects the Actuator?	NO	YES	NO

Table 8.3- Speed Control Comparative Study

92

8.7.6- Placement of the Flow Control Valve versus the Directional Control Valve

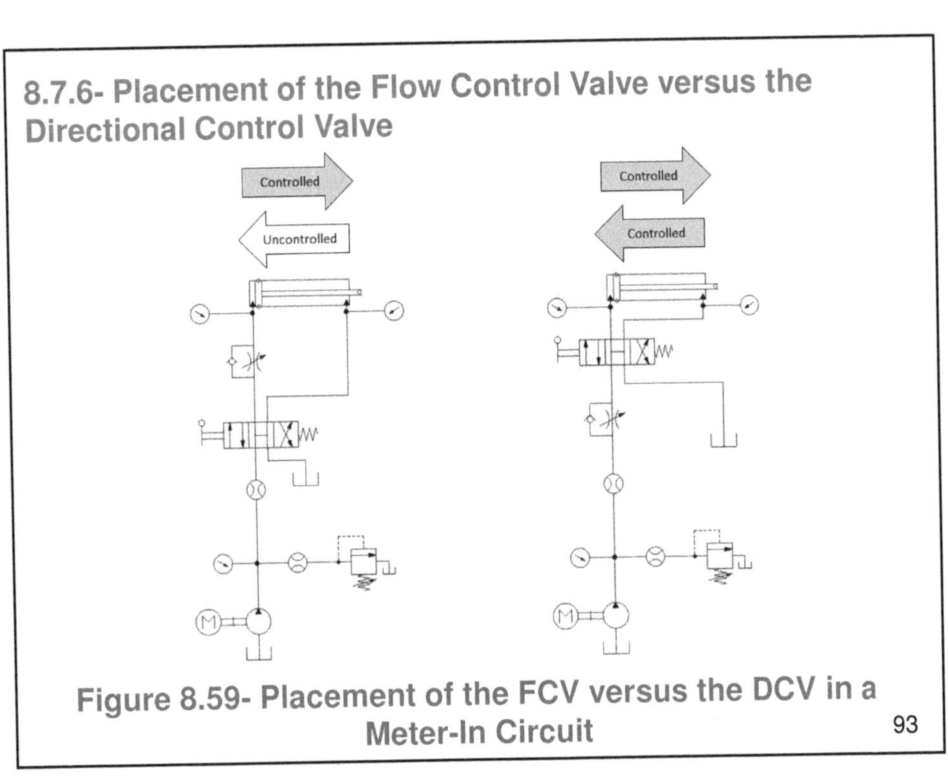

Figure 8.59- Placement of the FCV versus the DCV in a Meter-In Circuit

93

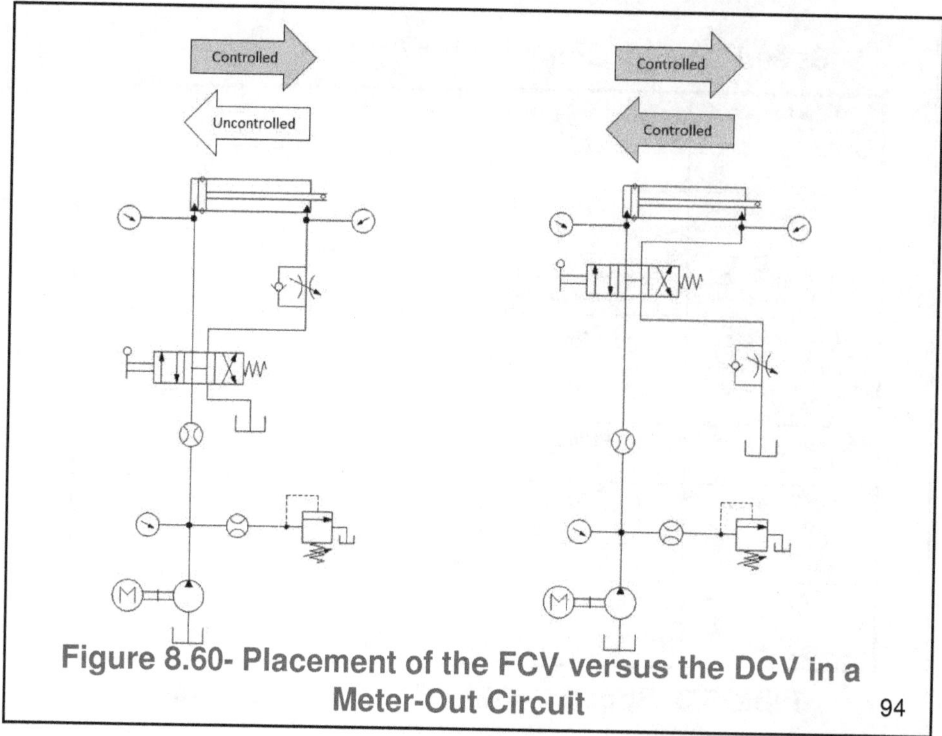

Figure 8.60- Placement of the FCV versus the DCV in a Meter-Out Circuit

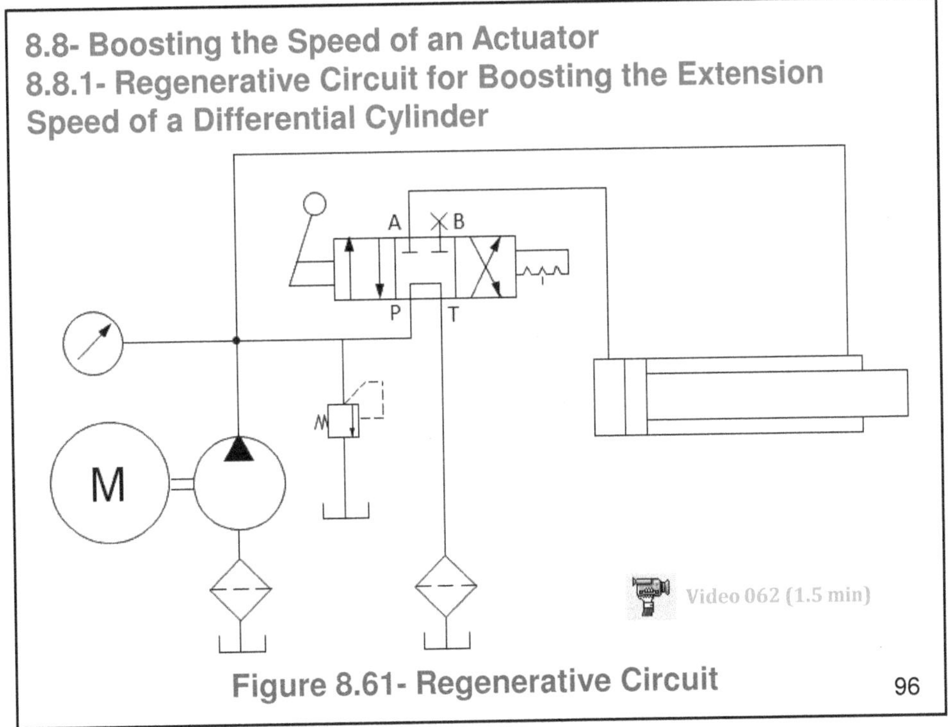

Figure 8.61- Regenerative Circuit

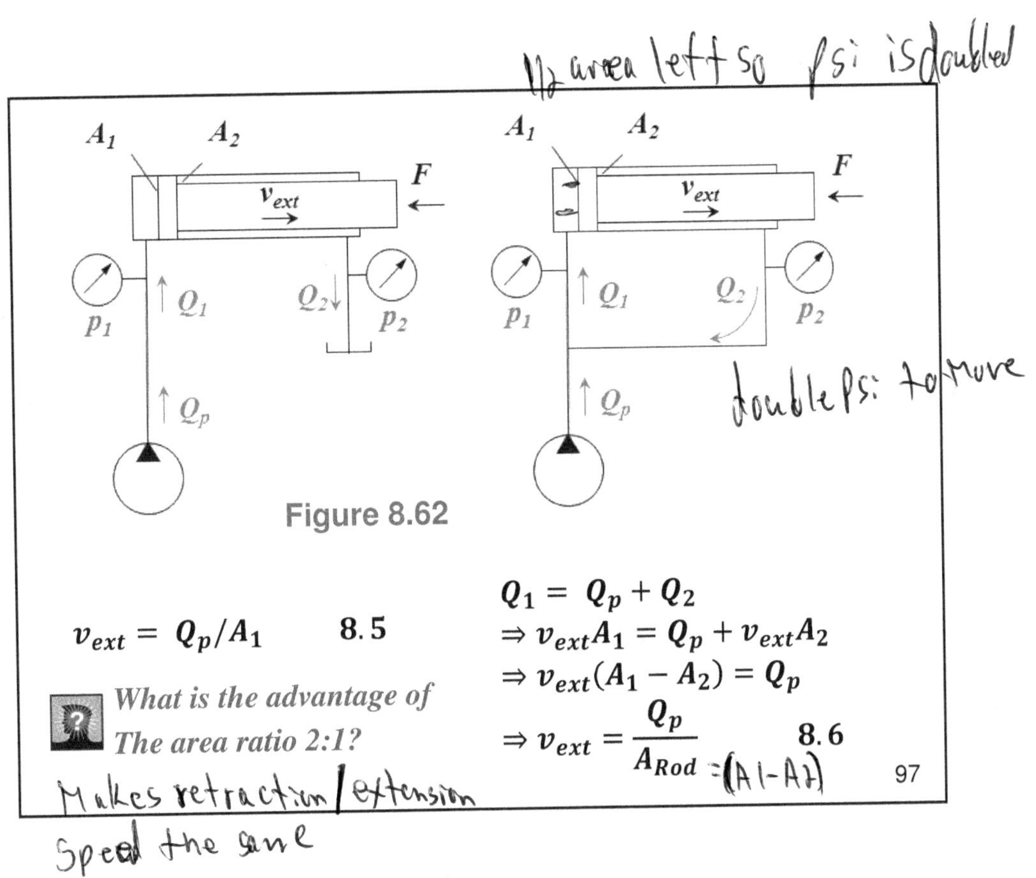

Figure 8.62

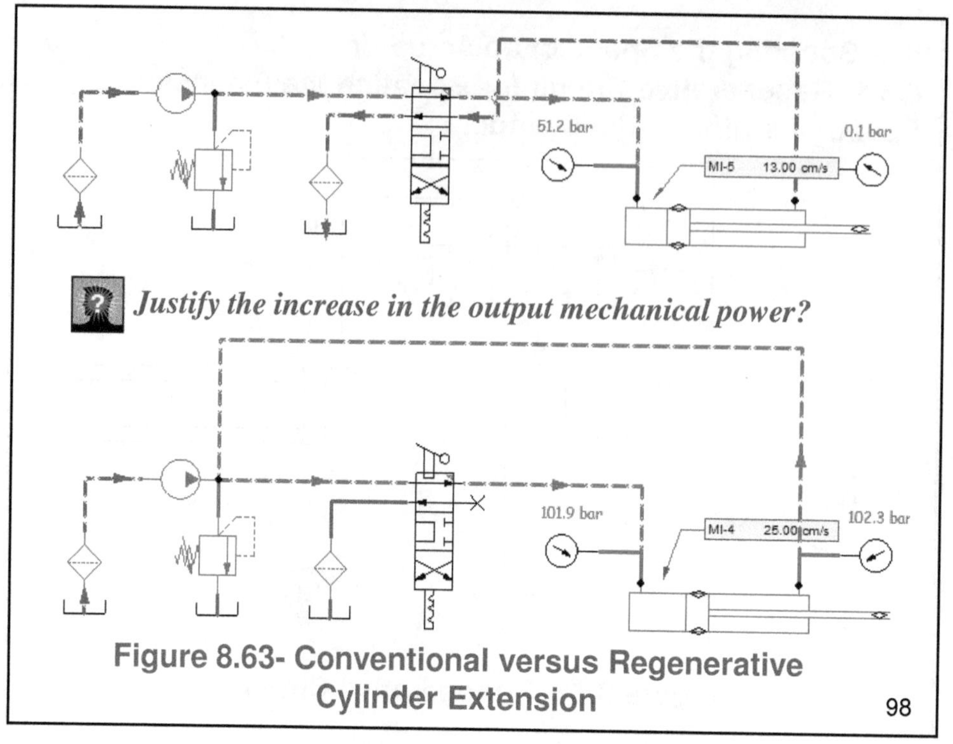

Figure 8.63- Conventional versus Regenerative Cylinder Extension

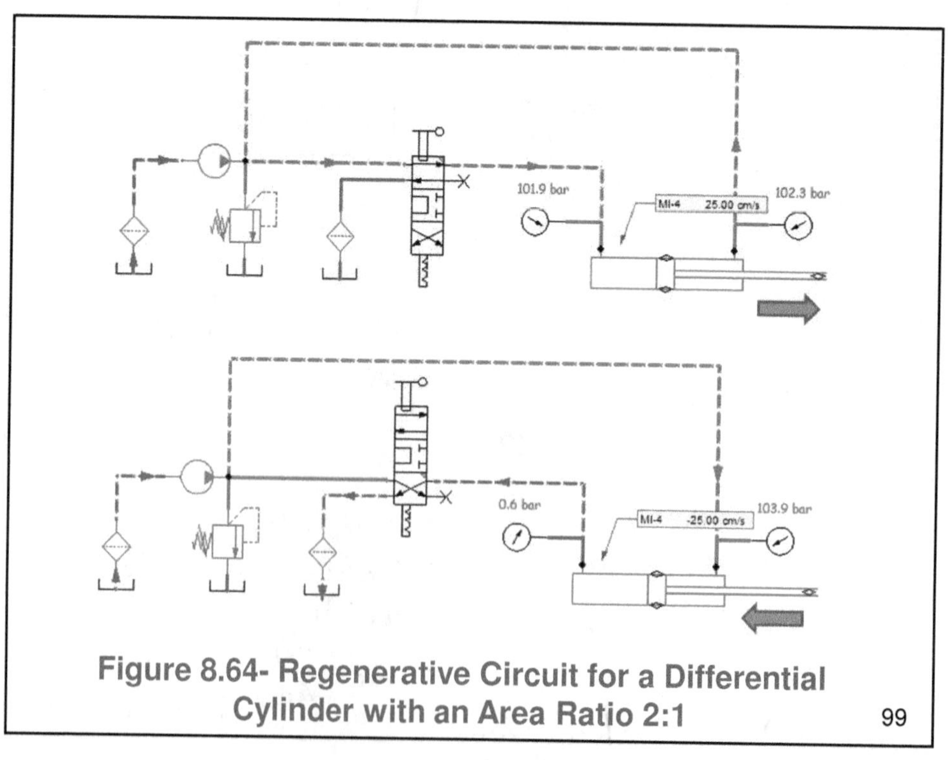

Figure 8.64- Regenerative Circuit for a Differential Cylinder with an Area Ratio 2:1

8.8.1.2- Effect of the Regenerative Connection on the Operating Pressure

$p_1A_1 = F + p_2A_2$; $p_1 = p_2$ ***(assuming no line or valve losses)***

$$\Rightarrow p_1(A_1 - A_2) = F \Rightarrow p_1 = \frac{F}{A_{Rod}} \qquad 8.7$$

100

8.8.2- Boosting Actuator Speed by using an Accumulator

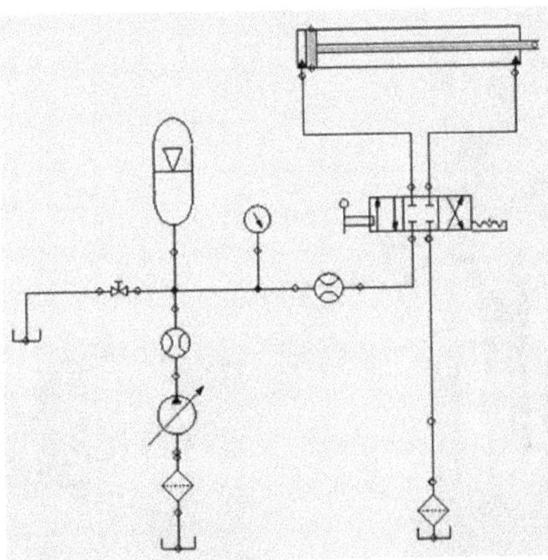

How to control the discharge rate from the accumulator?

Figure 8.65- Boosting an Actuator Speed by an Accumulator

101

Figure 8.66- Boosting the Speed of a Cylinder in Die Casting Machines

102

8.9- Energy Storage Application

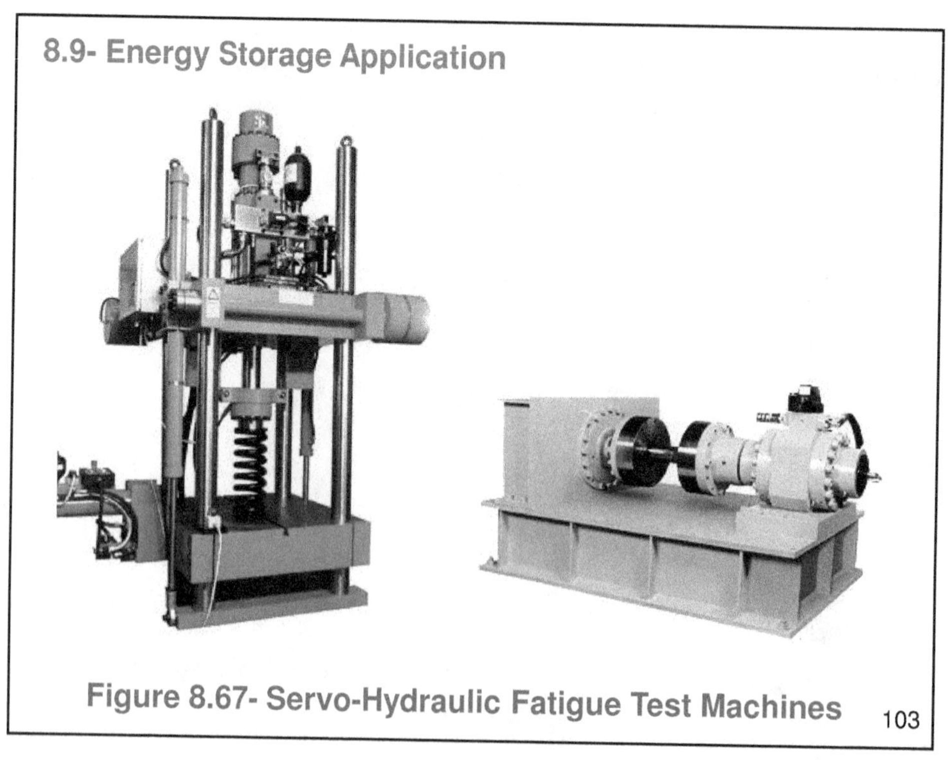

Figure 8.67- Servo-Hydraulic Fatigue Test Machines

103

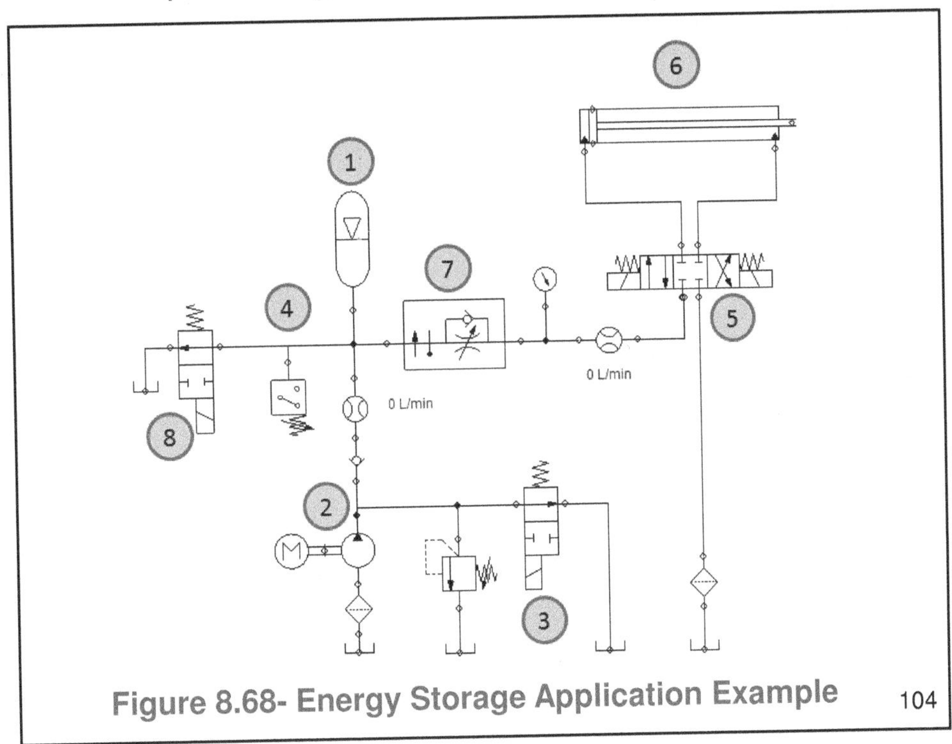

Figure 8.68- Energy Storage Application Example

104

Ex.7

105

8.10- Actuators Synchronization Application

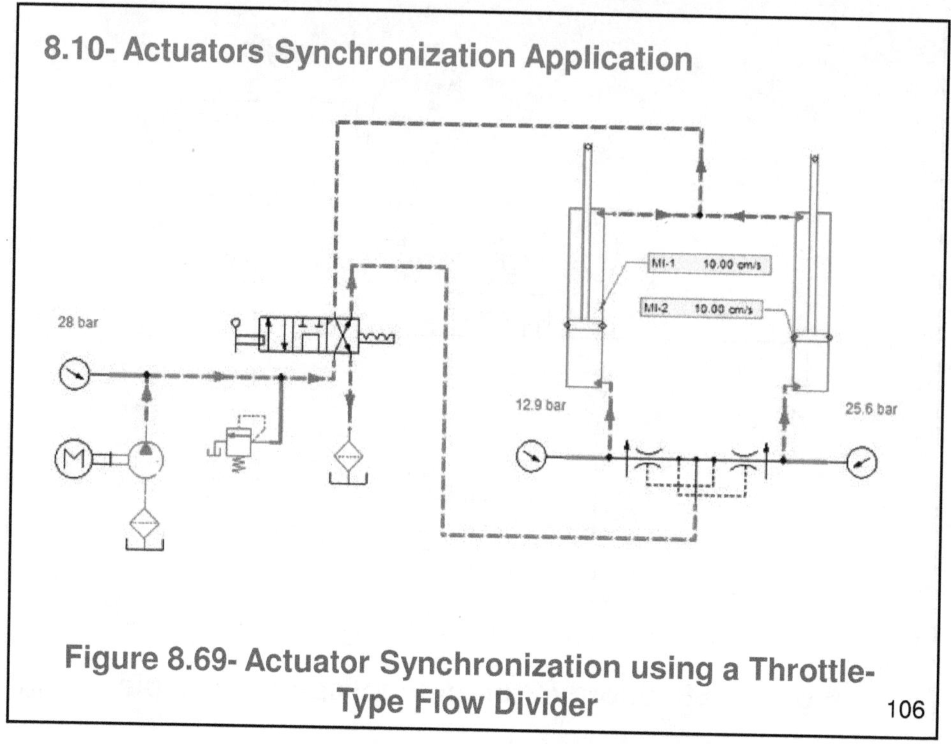

Figure 8.69- Actuator Synchronization using a Throttle-Type Flow Divider 106

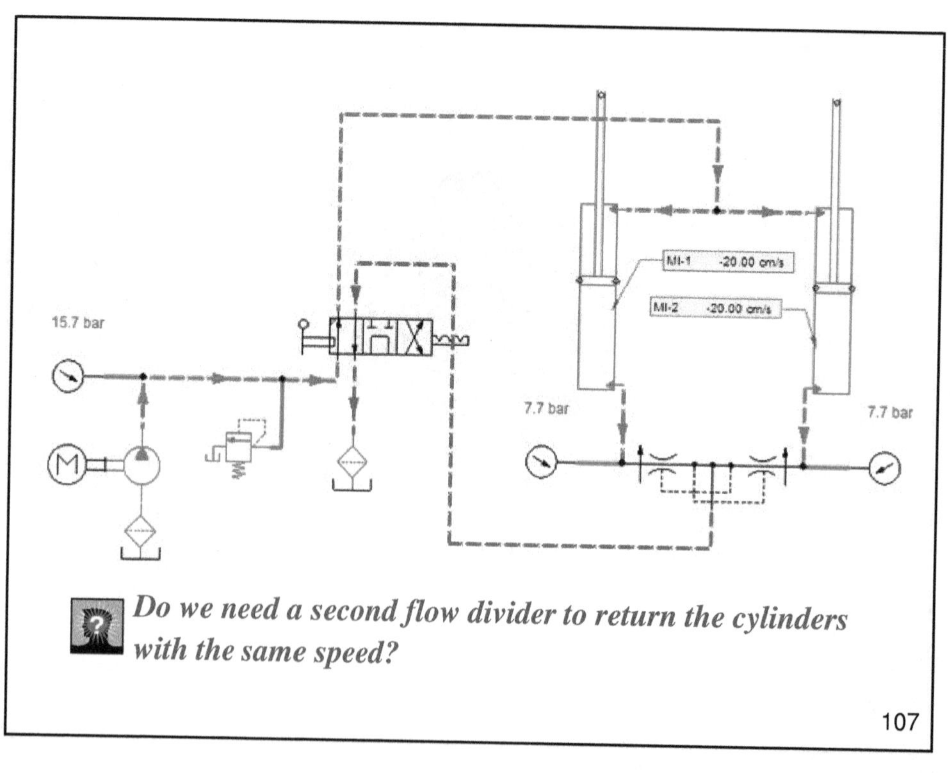

Do we need a second flow divider to return the cylinders with the same speed?

107

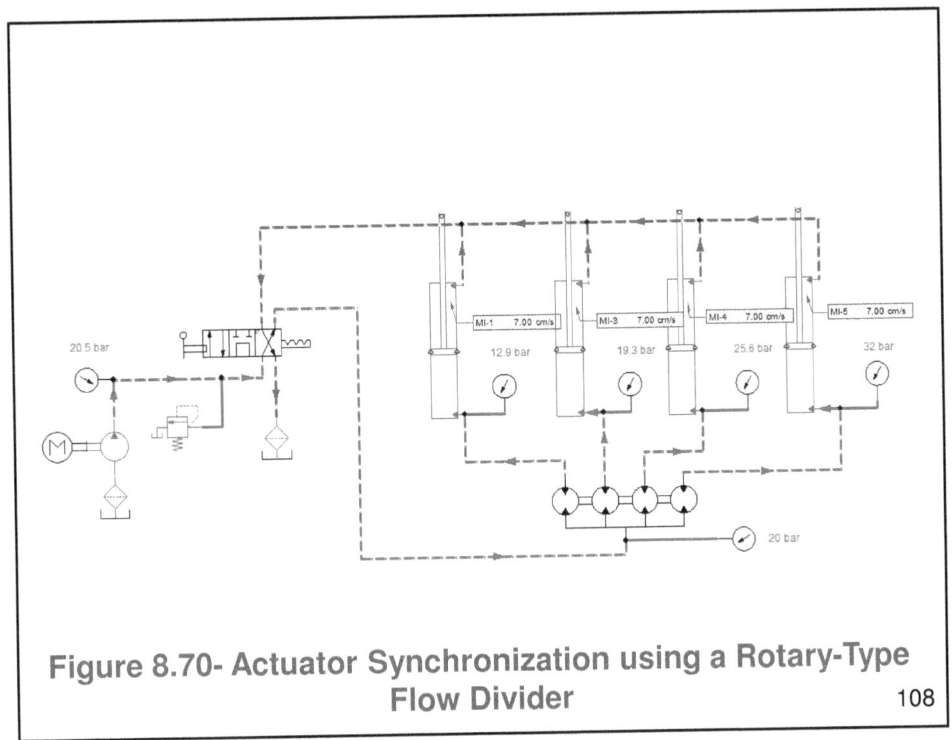

Figure 8.70- Actuator Synchronization using a Rotary-Type Flow Divider

108

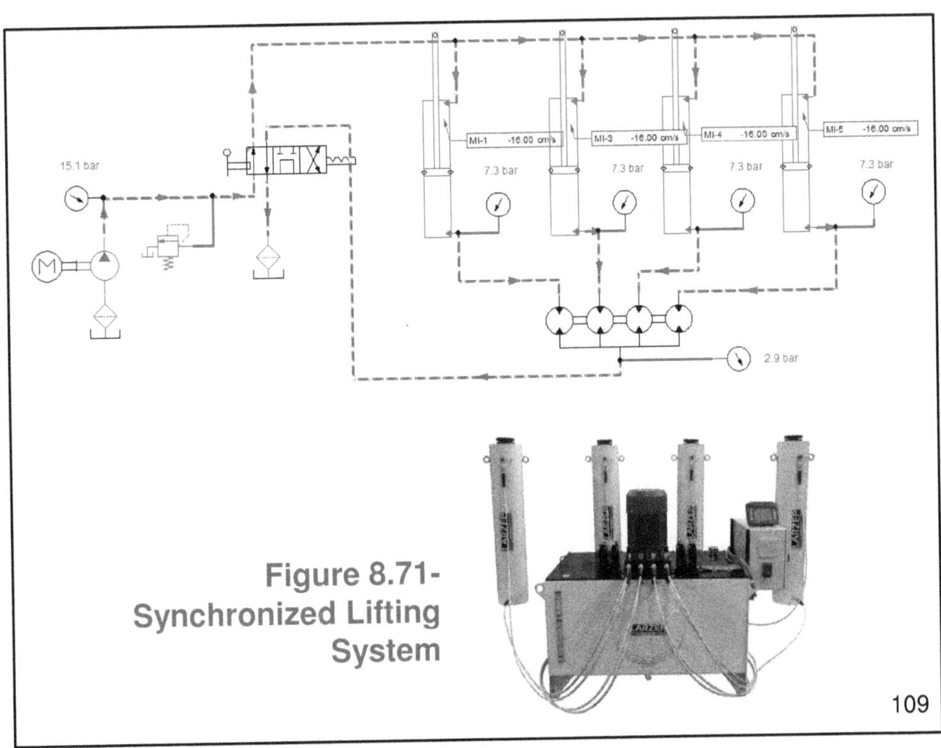

Figure 8.71- Synchronized Lifting System

109

8.11- Cylinder Deceleration Application

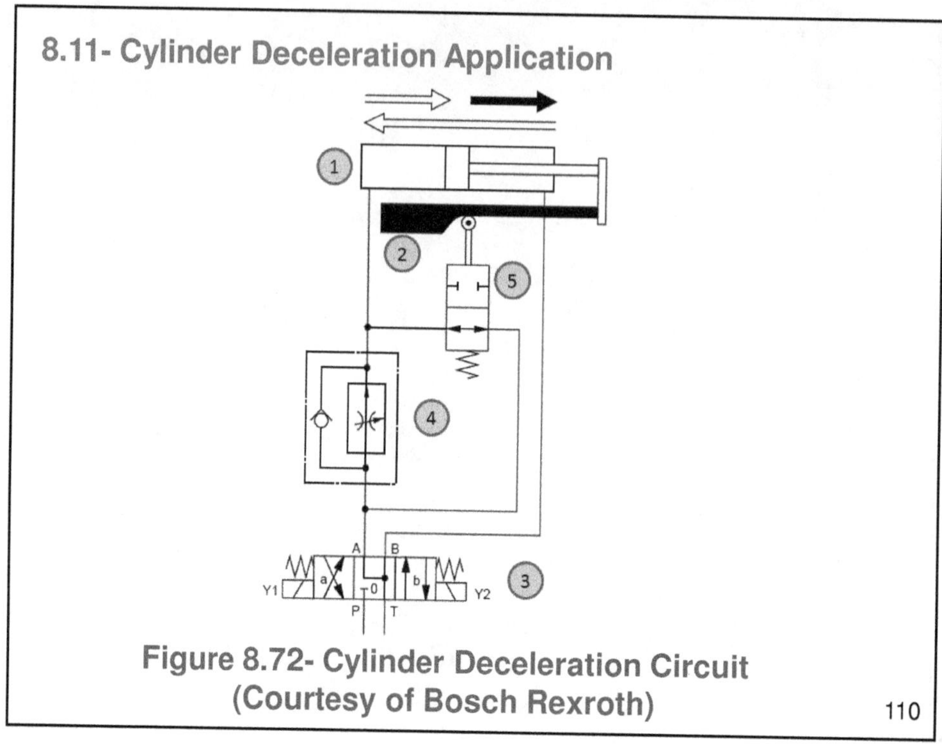

Figure 8.72- Cylinder Deceleration Circuit
(Courtesy of Bosch Rexroth)

110

8.12- Sequence Control Application

How to manage the forces?

Video 063 (0.5 min)

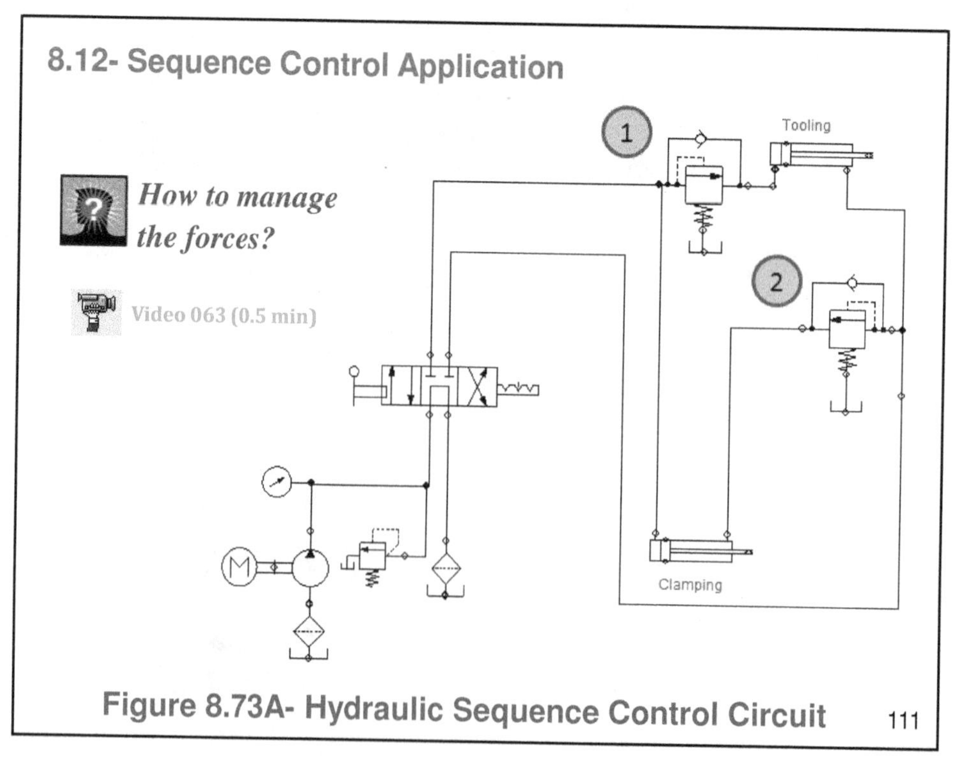

Figure 8.73A- Hydraulic Sequence Control Circuit

111

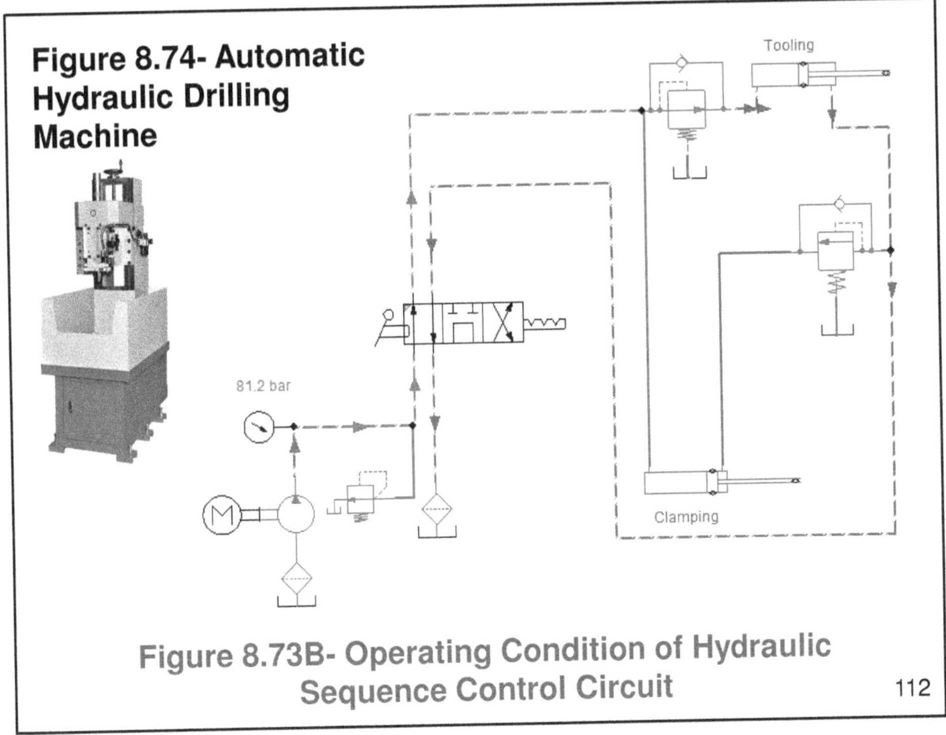

Figure 8.73B- Operating Condition of Hydraulic Sequence Control Circuit

8.13- Hydrostatic Transmission

Video 029 (0.5 min)

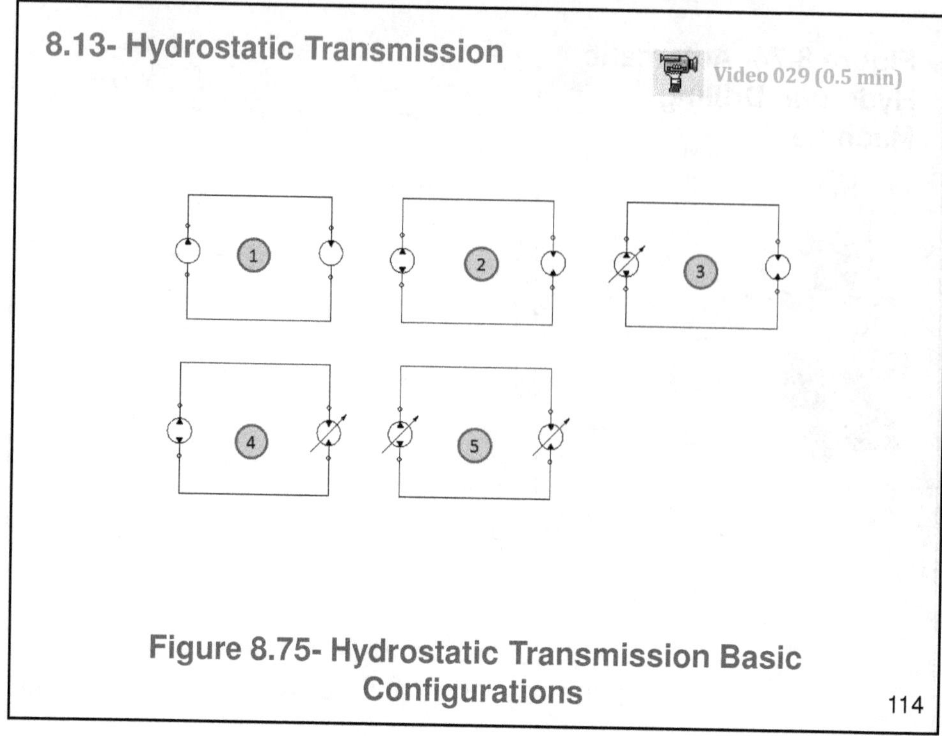

Figure 8.75- Hydrostatic Transmission Basic Configurations

114

Features	Displacement			Motor Output	
	Pump	Motor	Direction	Speed	Torque
1	Fixed	Fixed	1	Fixed	Fixed
2	Fixed	Fixed	2	Fixed	Fixed
3	Variable	Fixed	2	Variable	Fixed
4	Fixed	Variable	2	Variable	Variable
5	Variable	Variable	2	Variable	Variable

Table 8.4- Hydrostatic Transmission Characteristics

115

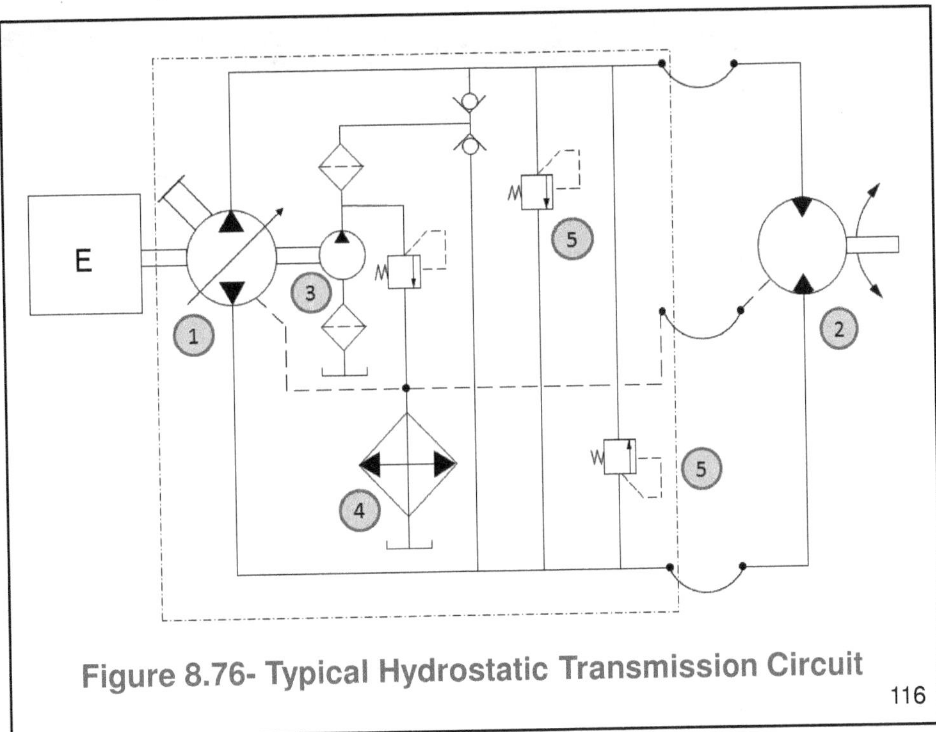

Figure 8.76- Typical Hydrostatic Transmission Circuit

116

Hydrostatic transmission has the following benefits as compared with the conventional mechanical power transmission:

- Rapid response and easy machine operation that makes it more productive.
- Stepless speed change rather than gear shifting.
- Separate feed and control of each motor.
- Applicable techniques to stabilize the motor speed against load variation.
- Considerable installation flexibility.
- Adaptable to integration with electronic control to add more operation intelligence.
- Cost effective with large power applications.

117

Chapter 8 Reviews

1. Find the correct match between the following names of the components and the numbers of the **symbols** shown below.

   A. (1=I, 2=II, 3=III, 4=IV).
   B. (16=I, 18=II, 19=III, 4=IV).
   C. (1=I, 18=II, 19=III, 17=IV).
   D. (16=I, 18=II, 19=III, 17=IV).

   I. Direct operated pressure relief valve.
   II. Pilot operated pressure relief valve.
   III. Telescopic cylinder.
   IV. Pilot operated valve with unloading valve.

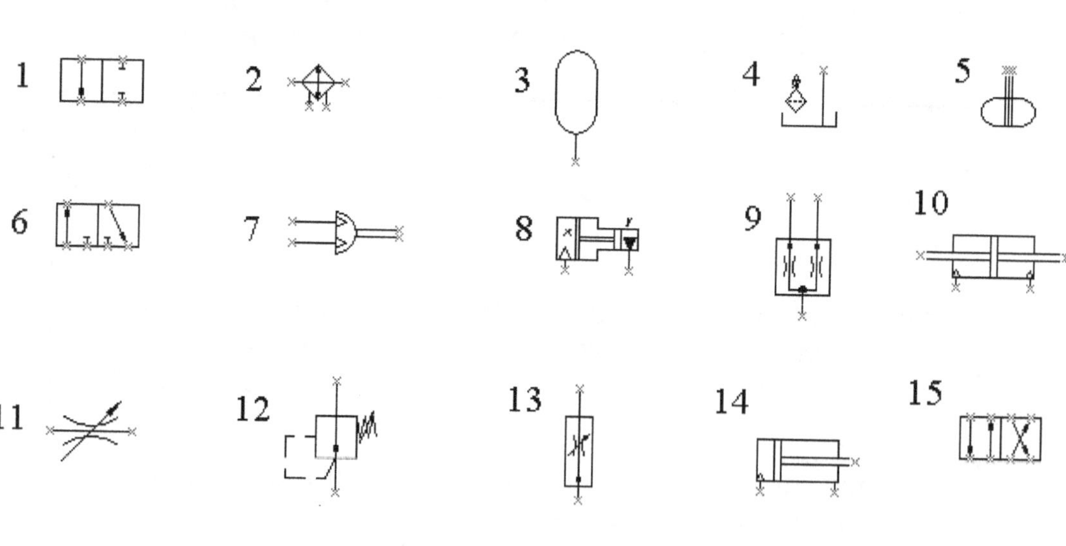

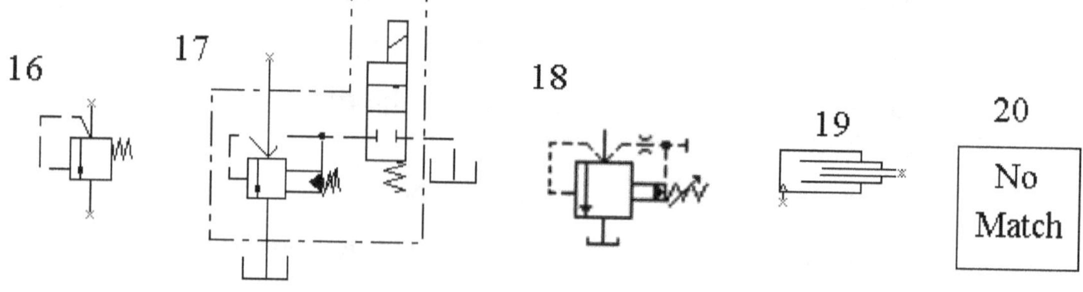

2. Find the correct match between the following names of the components and the numbers of the **symbols** shown before.

   A. (9=I, 13=II, 6=III, 7=IV).
   B. (9=I, 1= II, 6= III, 7= IV).
   C. (9=I, 1= II, 6= III, 5= IV).
   D. (1=I, 13= II, 6= III, 7= IV).

   I. Flow divider.
   II. Compensated flow control valve.
   III. Directional control valve – 3/2.
   IV. Rotary Actuator.

   ----

3. Find the correct match between the following names of the components and the numbers of the **symbols** shown before.

   A. (3=I, 12=II, 2=III, 5=IV)
   B. (3= I, 12= II, 2= III, 6= IV)
   C. (3= I, 12= II, 2= III, 20= IV)
   D. (3= I, 12= II, 2= III, 7= IV)

   I. Accumulator
   II. Reducing valve
   III. Oil Cooler
   IV. Pump-motor unit

   ----

4. Which **symbols** represents the sectional view of this valve?

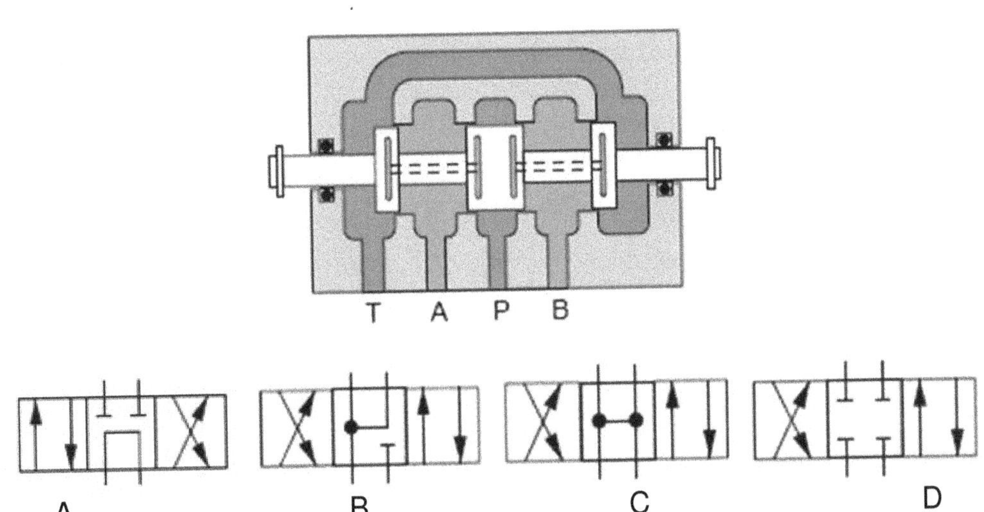

5. In an application where a backup pressure relief valve is installed with a **pressure compensated pump,** which statement is correct?
   A. Pressure compensator of the pump should be set higher than the backup PRV.
   B. Pressure compensator of the pump should be set lower than the backup PRV.
   C. Pressure compensator of the pump should be set at the same pressure as the backup PRV.
   D. None of the above.

6. The **mobile machine circuit** shown below is classified as an?
   A. Open circuit.
   B. Open center valve.
   C. Open loop.
   D. Open center system.

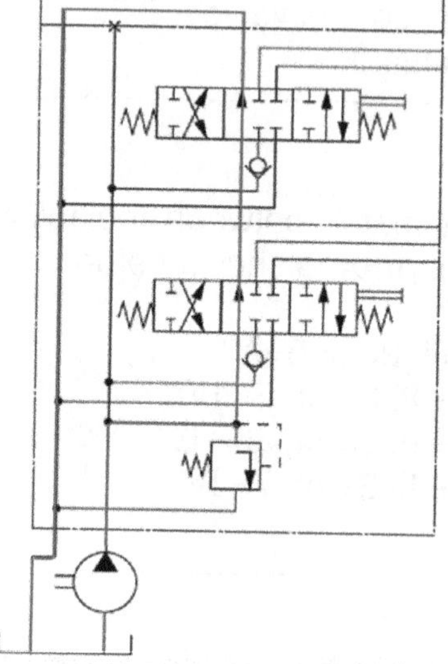

7. In the **cylinders circuit** shown below, the cylinders are identical and each has an area of ratio 2:1. If both valves are shifted simultaneously to the position A, what is your expectation about the motion of the cylinders?
   A. Cylinders I and II will extend simultaneously. Cylinder I will move slower.
   B. Cylinders I and II will extend simultaneously. Cylinder II will move slower.
   C. Cylinder I will be extend until it is deadheaded, then cylinder II will start to extend.
   D. Cylinders I and II will retract simultaneously. Cylinder II will be slower.

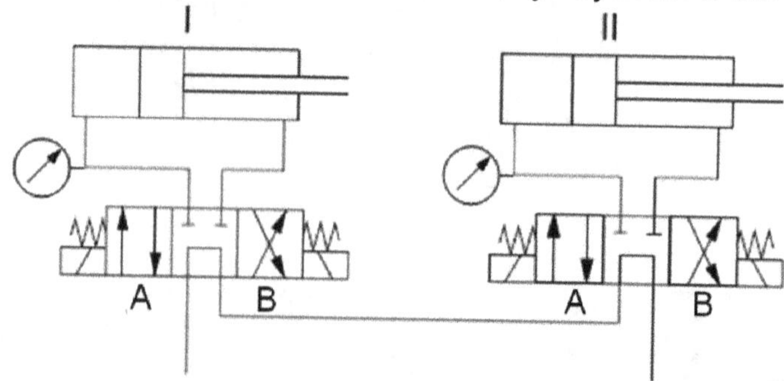

8. In the **motors circuit** shown below, both motors are identical in size and have the same external torque. If the valve is shifted so that motor I receives flow before

motor II and the reading of pressure gauge p_I is 2000 psi, what should be the reading of the pressure gauge p_{II}?

A. 500 psi.
B. 1000 psi.
C. 1500 psi.
D. 2000 psi.

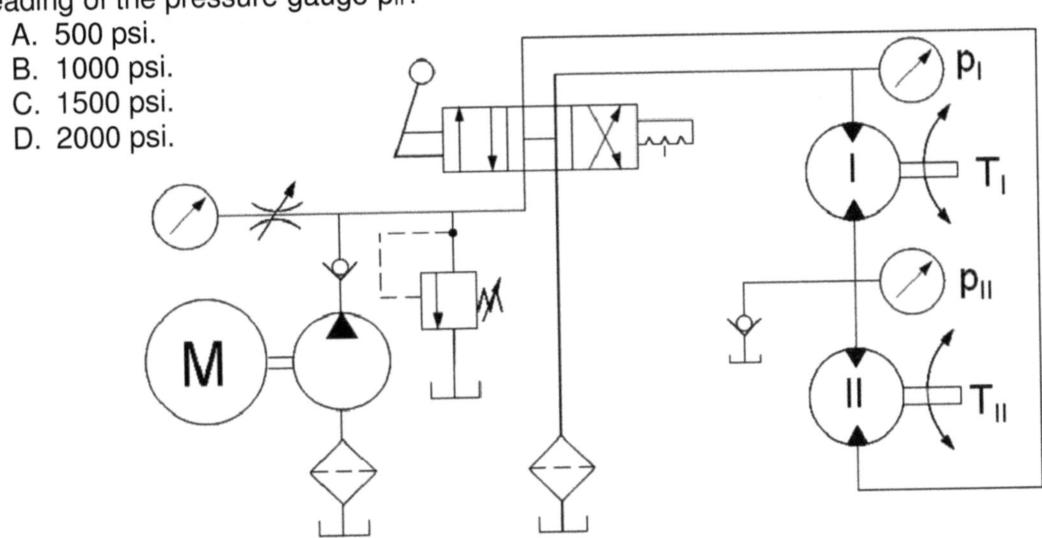

9. In the circuit shown below, assuming that the spools are fully shifted, which statement is correct?
   A. Braking and steering functions can be energized simultaneously.
   B. Steering and booming functions can be energized simultaneously.
   C. Any of the functions can be energized simultaneously.
   D. If any of the functions are performed, the other downstream functions will stop.

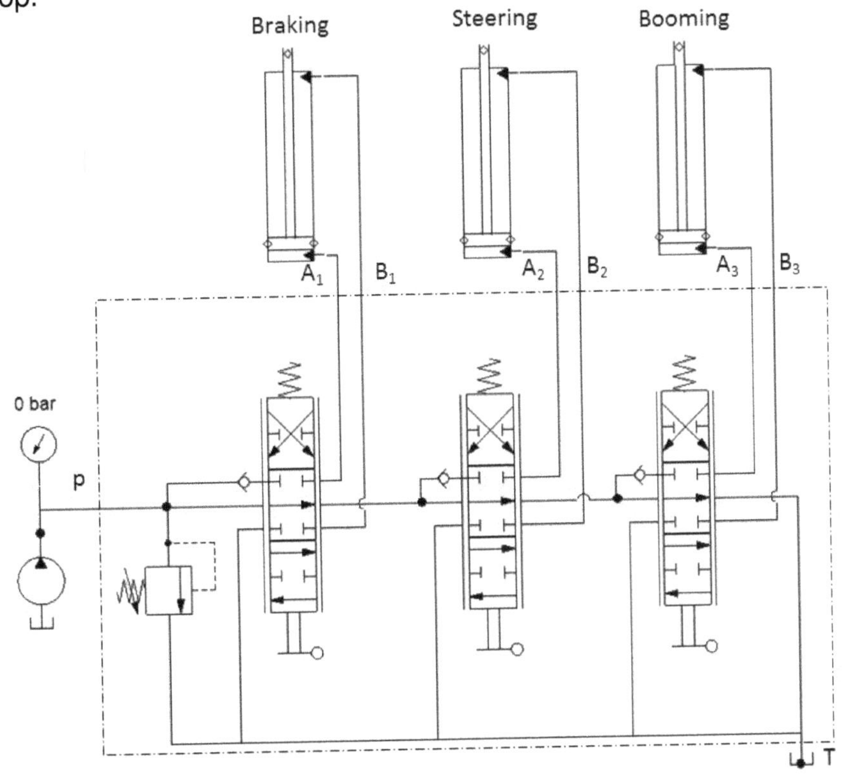

10. In the circuit shown below, how are the three sections of the **mobile control block** connected?
    A. Parallel connection.
    B. Series connection.
    C. Tandem connection.
    D. Combined tandem and parallel connection.

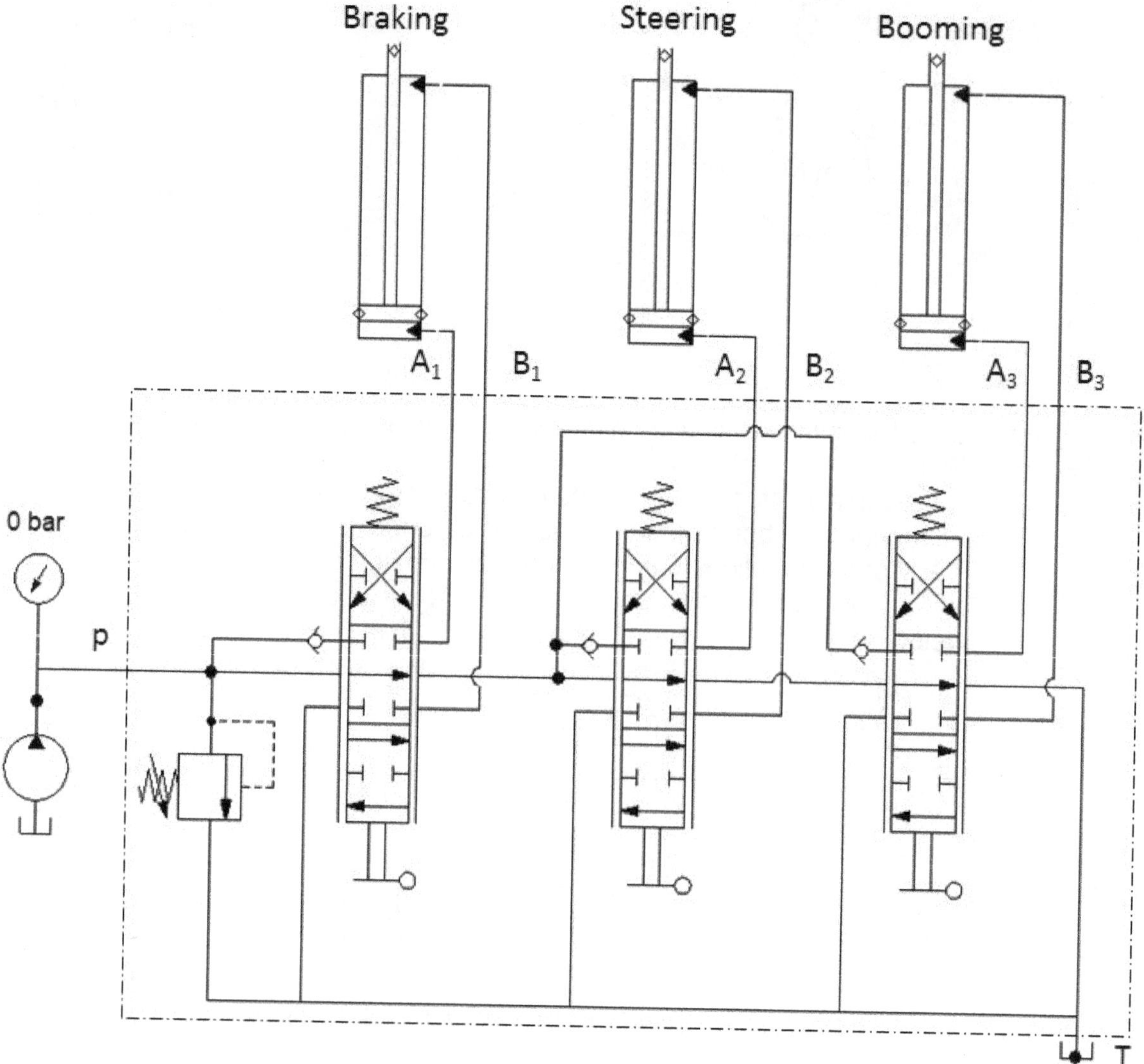

11. Which circuit is used for **controlling speed of an overrunning load**?
    A. Meter in circuit.
    B. Meter out circuit.
    C. By pass circuit.
    D. None of the above.

12. What is the **purpose** of the **hydraulic circuit** shown below**?**
    A. Meter in circuit to control the cylinder's extension speed.
    B. Meter in circuit to control the cylinder's retraction speed.
    C. Meter out circuit to control the cylinder's extension speed.
    D. Meter out circuit to control the cylinder's retraction speed.

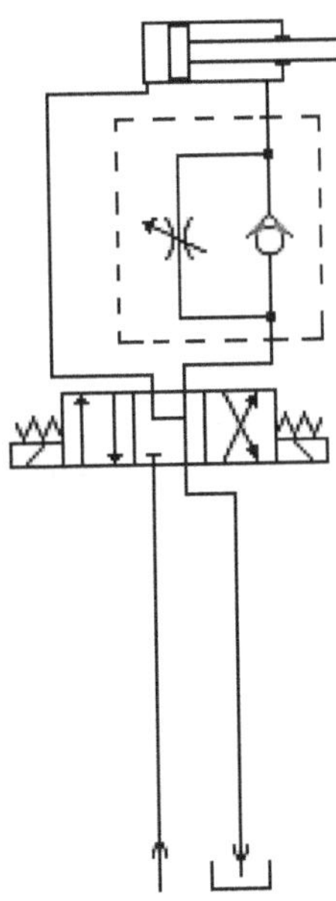

13. Which circuit is used to reciprocate a differential cylinder with an **area ratio 2:1** in which the cylinder extension speed and retraction speeds are the same?
    A. Hi-Lo circuit.
    B. Closed circuit.
    C. Open circuit
    D. Regenerative circuit.

14. In a regenerative cylinder circuit, since the pressure is equal on both sides of the cylinder, what is your expectation about the **cylinder extension speed?**
    A. The cylinder will extend with boosted speed.
    B. The cylinder will retract.
    C. The cylinder will be held in its current position.
    D. None of the above.

15. What is the **name of the missing valve?**
    A. Pressure relief valve.
    B. Motor brake valve.
    C. Unloading valve.
    D. Counterbalance valve.

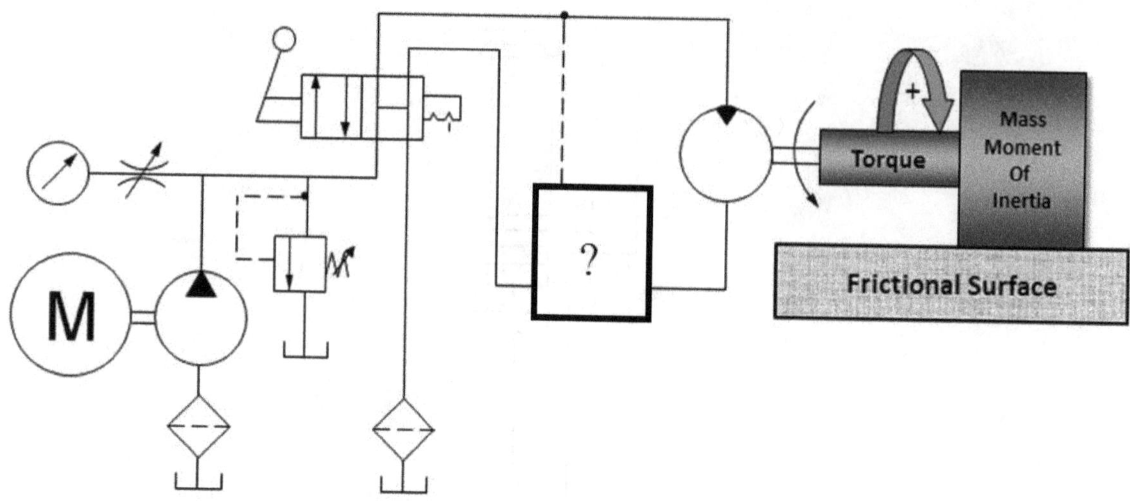

16. In the circuit shown below, the cylinder has an area ratio of 2:1. If the rod side is open to the tank without any restriction, it has been found that the pressure required to extend the cylinder against the resistive load F is p_1 = 50 bar. The rod side of the cylinder has been restricted to build back pressure of p_2 = 50 bar during the cylinder extension. Due to the concept of the **pressure intensification**, what is your expectation for the pressure reading p_1 if the cylinder is resisting the same force F**?**
    A. 25 bar.
    B. 50 bar.
    C. 75 bar.
    D. 100 bar.

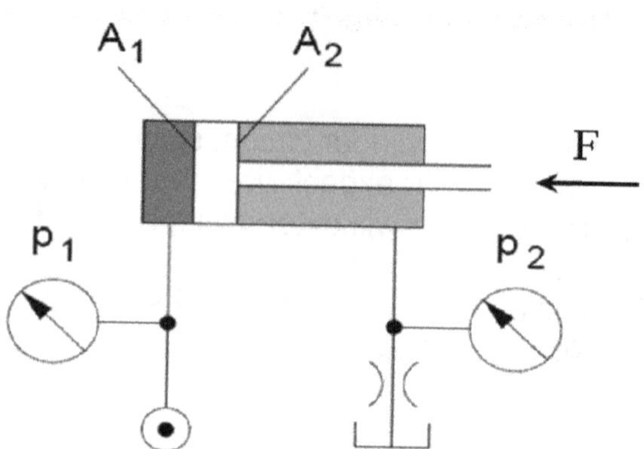

17. In the **sequence circuit** shown below, which of the following cylinders will be retracted first?
    A. Cylinder 1.
    B. Cylinder 2.
    C. Both are retracted at the same time.
    D. None of the above.

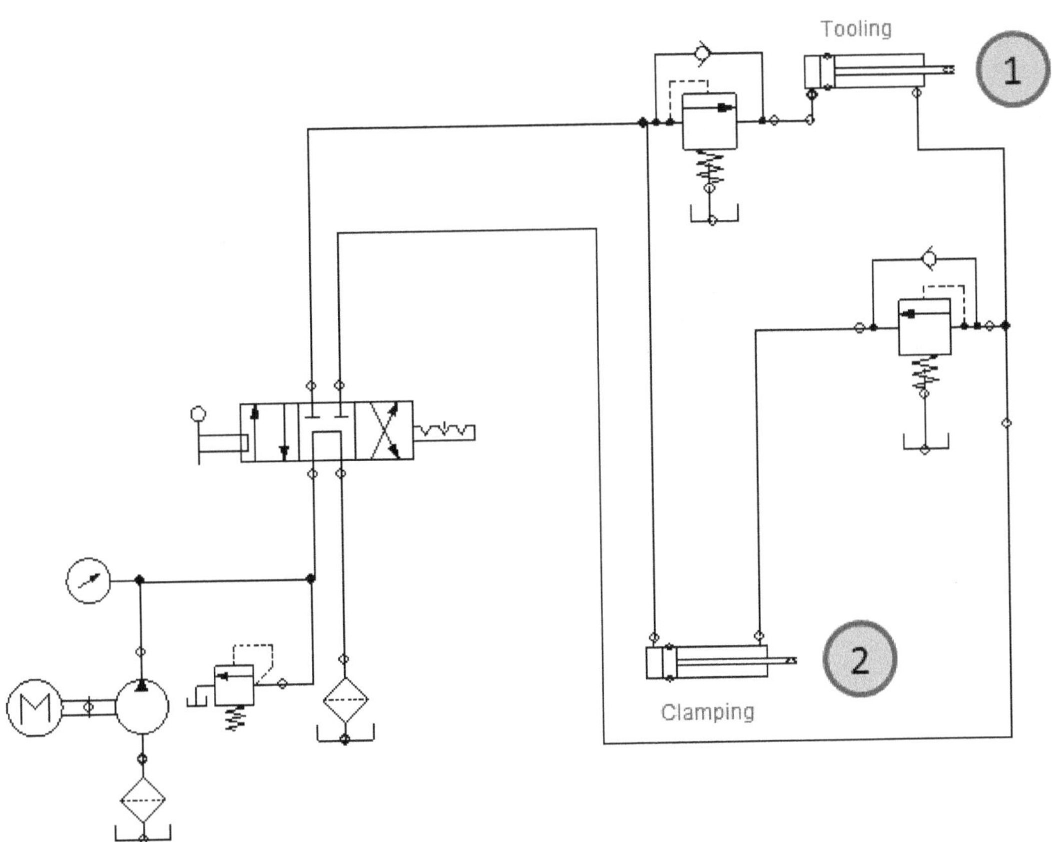

18. In a **hydrostatic transmission**, what is the most commonly used pump-motor configuration?
    A. Piston pump – vane motor.
    B. Swash plate axial piston pump – bent axis axial piston motor.
    C. Gear pump – vane motor.
    D. None of the above.

19. In the **hydrostatic transmission circuit** shown below, the type of the main pump should be?
    A. Variable displacement bi-directional pump.
    B. Fixed displacement over-center pump.
    C. Variable displacement over-center pump.
    D. Fixed displacement over-center pump.

20. In the hydrostatic transmission circuit shown below, the main purpose for using **boost pump** is to?
    A. Make up for leakage in the pump and motor.
    B. Prevent cavitation the main pump.
    C. Maintain a positive pressure above atmospheric on the low side of the system.
    D. All of the above.

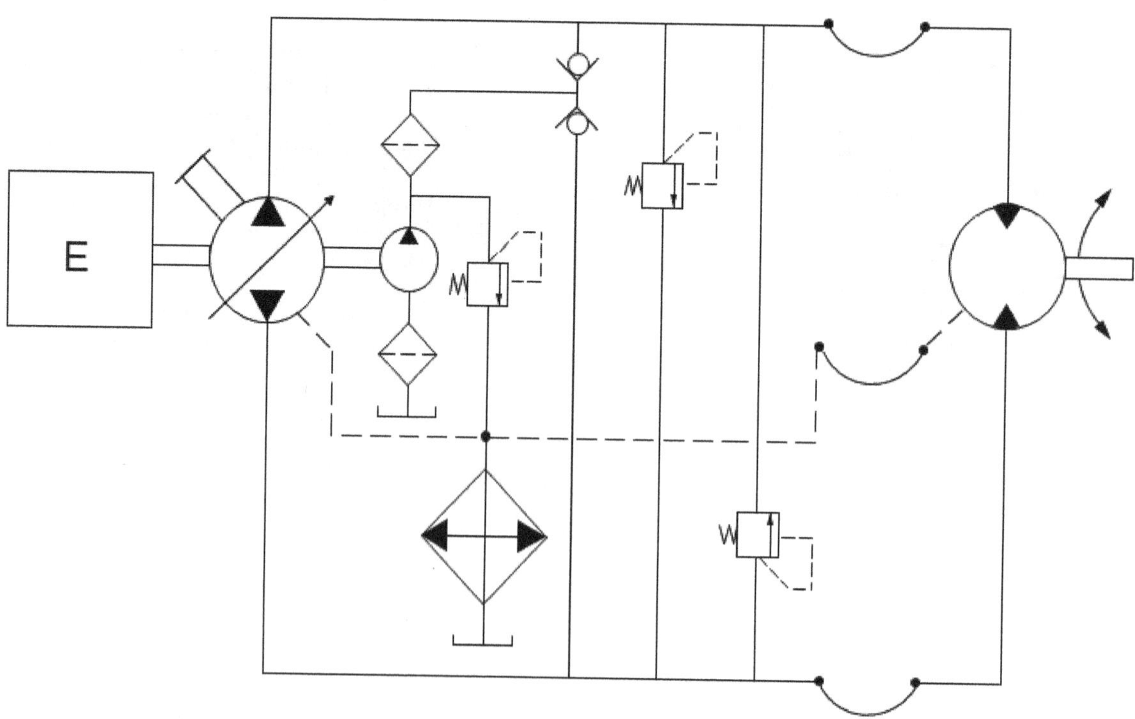

Chapter 8 Assignment

Student Name: -- Student ID: -------------------

Date: --- Score: ------------------------

Draw a complete hydraulic circuit to drive the cylinder shown below with all the required accessories and elements to make the system work properly and safely.

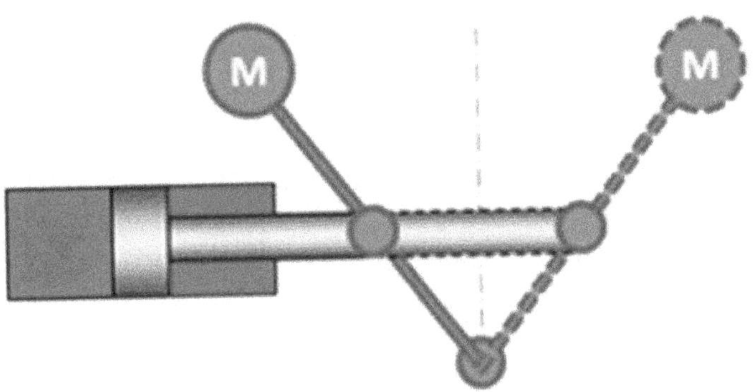

Answers to Chapter Reviews

Chapter 1:

1	2	3	4	5	6	7	8	9	10
D	B	C	B	C	D	C	D	C	C

Chapter 2:

1	2	3	4	5	6	7	8	9	10
A	C	B	D	D	D	B	C	D	D

Chapter 3:

1	2	3	4	5
D	A	C	B	B

Chapter 4:

1	2	3	4	5	6	7	8	9	10
C	B	A	B	B	C	C	D	D	C

11	12	13	14	15	16	17	18	19	20
B	D	D	A	A	A	D	D	A	C

21	22	23	24	25
A	C	A	C	D

Chapter 5:

1	2	3	4	5	6	7	8	9	10
A	B	C	B	A	D	C	D	D	B

11	12	13	14	15	16	17	18	19	20
D	A	D	B	C	D	B	D	B	A

21	22	23	24	25
A	D	C	C	A

Chapter 6:

1	2	3	4	5	6	7	8	9	10
C	C	C	C	A	B	C	D	C	B

Chapter 7:

1	2	3	4	5
A	C	C	A	B

Chapter 8:

1	2	3	4	5	6	7	8	9	10
D	A	C	A	B	D	B	B	D	D

11	12	13	14	15	16	17	18	19	20
B	B	D	A	B	C	A	B	C	D

CPSIA information can be obtained
at www.ICGtesting.com
Printed in the USA
FSHW020531190219
55770FS

9 780692 655443